Landscape-scale Conservation Planning

Stephen C. Trombulak · Robert F. Baldwin
Editors

Landscape-scale Conservation Planning

Springer

Editors

Stephen C. Trombulak
Department of Biology and Program
in Environmental Studies
Middlebury College
Middlebury, VT 05753
USA
trombulak@middlebury.edu

Robert F. Baldwin
Department of Forestry and Natural
Resources
Clemson University
Clemson, SC 29634-0317
USA
baldwi6@clemson.edu

ISBN 978-94-017-8463-4   ISBN 978-90-481-9575-6 (eBook)
DOI 10.1007/978-90-481-9575-6
Springer Dordrecht Heidelberg London New York

© Springer Science+Business Media B.V. 2010
Softcover re-print of the Hardcover 1st edition 2010
No part of this work may be reproduced, stored in a retrieval system, or transmitted in any form or by any means, electronic, mechanical, photocopying, microfilming, recording or otherwise, without written permission from the Publisher, with the exception of any material supplied specifically for the purpose of being entered and executed on a computer system, for exclusive use by the purchaser of the work.

Cover illustration: © 2010 JupiterImages Corporation
Photo text: Aerial view of landscape and lake

Printed on acid-free paper

Springer is part of Springer Science+Business Media (www.springer.com)

Foreword

Hugh P. Possingham

Landscape-scale conservation planning is coming of age. In the last couple of decades, conservation practitioners, working at all levels of governance and all spatial scales, have embraced the CARE principles of conservation planning – Comprehensiveness, Adequacy, Representativeness, and Efficiency. Hundreds of papers have been written on this theme, and several different kinds of software program have been developed and used around the world, making conservation planning based on these principles global in its reach and influence. Does this mean that all the science of conservation planning is over – that the discovery phase has been replaced by an engineering phase as we move from defining the rules to implementing them in the landscape? This book and the continuing growth in the literature suggest that the answer to this question is most definitely 'no.'

All of applied conservation can be wrapped up into a single sentence: what should be done (the action), in what place, at what time, using what mechanism, and for what outcome (the objective). It all seems pretty simple – what, where, when, how and why. However stating a problem does not mean it is easy to solve. Although conservation planners have enjoyed a number of spectacular successes in the last decade, like the re-zoning of the Great Barrier Reef (that achieved no-take areas over 33% of the region with at least 20% of every bioregion being conserved) and the influence of The Nature Conservancy's ecoregional plans, the theory and practice of making conservation planning decisions unfortunately continues to be filled with pitfalls and unanswered questions. Is most progress made by working from the top down or from the bottom up? Have planners focused too much on achieving efficiency and not enough on sufficiency? And how much of an ecosystem or population is sufficient for conservation? Is large-scale connectivity an objective in its own right, or an untested surrogate for adequacy? How do the merits of small-scale and large-scale planning trade off each other? Should natural scientists step aside in the belief that decades of research have revealed as much as needs to be known about the ecological bases for conservation and let the social scientists take over and deliver more effective strategies for implementation? Should conservation planners take refuge behind Dwight D. Eisenhower's insight that 'plans are *useless*, but *planning* is indispensable,' or simply decide that planning itself is just another form of intellectual displacement activity?

Ultimately, the complexities of social and economic contexts for conservation mean that the best way forward may be to walk away from generalities about single topics and run (and I do mean *run*) towards a better understanding of how the principles of conservation planning are applied together in more specific cases. This is largely what this book does. It brings together a wide variety of authors from disparate disciplines to tackle the issue of conservation planning in a large and well-defined landscape using the most up-to-date tools and ideas. It is only by trying to apply the theory and tools in a place with a specific ecology and socio-politico-economic setting that we can see how they fall short and need to evolve or be replaced. Hence, while this book appears to be about a specific place, it is about any place because only by learning lessons locally can we make the theory of conservation planning work anywhere.

Unlike most of my colleagues in biology departments – my scientific and technical inspiration comes from trying to solve real problems. In that sense this book provides much food for thought – rarely has so much intellectual and diverse thinking been focused on a specific region. Readers will be challenged and engaged, and thus hopefully the next phase of conservation planning will about evaluating the successes and failures of landscape-scale interventions.

Acknowledgments

This book could not have come to life without the participation and cooperation of numerous individuals and organizations. We are grateful to all of the colleagues we have worked with in and through Two Countries, One Forest over the past several years. These include Mark Anderson, Charlie Bettigole, Karen Beazley, Alice Chamberlain, Roberta Clowater, Patrick Doran, Kathleen Fitzgerald, Graham Forbes, Bill Ginn, Louise Gratton, Greg Kehm, John Nordgren, Jim Northup, Justina Ray, Conrad Reining, Ted Smith, James Sullivan, and Gillian Woolmer. We would also like to gratefully acknowledge financial support over the years for the science work of 2C1Forest from both the Henry P. Kendall Foundation and the Robert and Patricia Switzer Foundation. We would like to particularly thank Emily Bateson, whose early vision about the importance of a collective focus on the Northern Appalachian/Acadian ecoregion was pivotal in bringing this collaborative effort to life.

We would also like to thank all of the authors who contributed to this volume not only for the conservation work they do but for their boundless patience and continued engagement throughout the entire project. Readers should note that a large proportion of the authors work with NGOs or government agencies, all of which were affected dramatically during the global economic downturn that began in October 2008. All of these authors could have quite understandably backed out of their agreements to participate in order to focus exclusively on their organization's more immediate financial challenges. That they did not is a testament to their commitment to conservation and their desire to help pave the way for more effective conservation planning for the twenty-first century.

We are enormously thankful for the editorial and production staff at Springer, who were unfailingly patient, supportive, and engaged throughout the entire production of this book. We are especially grateful to Takeesha Moerland-Torpey for her guidance through so many aspects of the publication process.

SCT would like to thank several people and organizations for their support and encouragement. I gratefully acknowledge the Trustees and Administration of Middlebury College for the on-going support – both moral and financial – of my scholarly work. My work on this project was helped in particular by release time funded by the Andrew W. Mellon Foundation Cluster grant for Faculty Career Enhancement and by the resources made available to me by having the endowed chair in Environmental and Biosphere Studies. I would also like to thank the students in my Conservation Biology class (Spring 2010) for their invaluable assistance in checking all the citations used in the book, as well as my colleagues in both Biology and Environmental Studies, especially Bill Hegman, Matt Landis, Mark Lapin, Chris McGrory Klyza, and Bill McKibben, for their willingness to share their expertise and opinions whenever I needed help. I am grateful to my children, Sage and Ian, who have always been the most important reasons I think about the future. I will also be forever grateful to my partner, Josselyne Price, for her love, encouragement, and never-ending patience. I could never have finished this book without her support.

Finally, I would like to thank the pioneers of the field of landscape-scale conservation, who during the twentieth century saw clearly both the importance and challenge of thinking big about land and wildlife. Many had profound effects on my development as a professional, in particular David Brower, Dave Foreman, Larry Hamilton, Harvey Locke, Reed Noss, Hugh Possingham, Jamie Sayen, Michael Soulé, and John Terborgh.

RFB would like to thank the numerous institutions and people who have enriched my career. These include in chronological order, Jay Labov at Colby College, John Seidensticker at the National Zoological Park, Luther Brown at George Mason University (now Delta State University), Paul Durr in the Great Smoky Mountains, Seth Benz at the Audubon Expedition Institute, Aram Calhoun at the University of Maine, Phillip deMayandier at Maine Department of Inland Fisheries and Wildlife, SCT at Middlebury College, and faculty at Antioch University New England. Lissa Widoff and the Robert and Patricia Switzer foundation supported two phases of my research including much of the work reported here. Most currently I need to express sincere appreciation for my colleagues and mentors in Forestry and Natural Resources at Clemson University – anchoring the other end of the Appalachian Chain.

In all of my travels and studies there has been no greater pleasure than having and being with my children – Ella S. and Julian A. Baldwin. My partner in all of this is Elizabeth (Betty) Baldwin, whom I can't thank enough; also, her parents, Nancy and Bucky Dennis, and siblings. There would be no me without two adventurous souls – Mark M. and Margaret W. Baldwin – who introduced me to the Northern Forest at the age of 7 when the family packed up and moved from suburban Washington D.C. to Maine. And there would be much less of me without my brothers and their loved ones: Andrew, Ben, Charles, Martha, and Christopher.

Finally, I thank Maine itself. Without those forests and fields, seascapes, rocky headlands and mountains I would not have discovered nature. Now is the time to give back.

Contents

Foreword v
Hugh P. Possingham

Acknowledgments vii

**1 Introduction: Creating a Context for
Landscape-Scale Conservation Planning** 1
Stephen C. Trombulak and Robert F. Baldwin

2 Identifying Keystone Threats to Biological Diversity 17
Robert F. Baldwin

3 Why History Matters in Conservation Planning 33
Elizabeth Dennis Baldwin and Richard W. Judd

**4 Developing Institutions to Overcome Governance
Barriers to Ecoregional Conservation** 53
Robert B. Powell

**5 Changing Socio-economic Conditions for
Private Woodland Protection** 67
Robert J. Lilieholm, Lloyd C. Irland, and John M. Hagan

6 Aquatic Conservation Planning at a Landscape Scale 99
Keith H. Nislow, Christian O. Marks, and Kimberly A. Lutz

**7 From the Last of the Large to the Remnants of the Rare:
Bird Conservation at an Ecoregional Scale** 121
Jeffrey V. Wells

8 The Transboundary Nature of Seabird Ecology 139
Patrick G.R. Jodice and Robert M. Suryan

x Contents

9 Conservation Planning with Large Carnivores and Ungulates in Eastern North America: Learning from the Past to Plan for the Future 167
Justina C. Ray

10 Protecting Natural Resources on Private Lands: The Role of Collaboration in Land-Use Planning 205
Jessica Spelke Jansujwicz and Aram J.K. Calhoun

11 Integrating Expert Judgment into Systematic Ecoregional Conservation Planning 235
Karen F. Beazley, Elizabeth Dennis Baldwin, and Conrad Reining

12 The GIS Challenges of Ecoregional Conservation Planning 257
Gillian Woolmer

13 The Human Footprint as a Conservation Planning Tool 281
Stephen C. Trombulak, Robert F. Baldwin, and Gillian Woolmer

14 Assessing Irreplaceability for Systematic Conservation Planning 303
Stephen C. Trombulak

15 Conservation Planning in a Changing Climate: Assessing the Impacts of Potential Range Shifts on a Reserve Network 325
Joshua J. Lawler and Jeffrey Hepinstall-Cymerman

16 Modeling Ecoregional Connectivity 349
Robert F. Baldwin, Ryan M. Perkl, Stephen C. Trombulak, and Walter B. Burwell III

17 A General Model for Site-Based Conservation in Human-Dominated Landscapes: The Landscape Species Approach 369
Michale J. Glennon and Karl A. Didier

18 Integrating Ecoregional Planning at Greater Spatial Scales 393
Mark Anderson

Index 407

Contributors

Mark Anderson is Director of the Eastern Resource Office of The Nature Conservancy, based in Boston, Massachusetts (manderson@TNC.org).

Elizabeth Dennis Baldwin is Assistant Professor of Parks and Protected Areas Management in the Department of Parks, Recreation and Tourism Management at Clemson University in South Carolina (ebaldwn@clemson.edu).

Robert F. Baldwin is Assistant Professor of Conservation Biology/GIS in the Department of Forestry and Natural Resources at Clemson University in South Carolina (baldwi6@clemson.edu).

Karen F. Beazley is an Associate Professor in the School for Resource and Environmental Studies at Dalhousie University in Halifax, Nova Scotia (karen.beazley@dal.ca).

Walter B. Burwell III is a graduate of the Program in Environmental Studies at Middlebury College in Vermont (wburwell@middlebury.edu).

Aram J.K. Calhoun is an Associate Professor of Wetland Ecology in the Department of Wildlife Ecology at the University of Maine in Orono and a wetland scientist with Maine Audubon Society (calhoun@maine.edu).

Karl A. Didier is a spatial ecologist and conservation planner at the Wildlife Conservation Society, based in Gainesville, Florida (kdidier@wcs.org).

Michale J. Glennon is an Associate Conservation Scientist for the Wildlife Conservation Society's Adirondack Program in Saranac Lake, New York (mglennon@wcs.org).

John M. Hagan is President of Manomet Center for Conservation Sciences, based in Manomet, Massachusetts and Brunswick, Maine (jmhagan@manomet.org).

Jeffrey Hepinstall-Cymerman is an Assistant Professor in the Warnell School of Forestry and Natural Resources at the University of Georgia in Athens (jhepinstall@warnell.uga.edu).

Lloyd C. Irland is Lecturer and Senior Research Scientist, Yale School of Forestry and Environmental Studies in New Haven, Connecticut, and president of The Irland Group (lloyd.irland@yale.edu).

Jessica Spelke Jansujwicz is a doctoral student in the Ecology and Environmental Sciences Program at the University of Maine in Orono (jessica.jansujwicz@maine.edu).

Patrick G.R. Jodice is the Leader of the USGS South Carolina Cooperative Fish and Wildlife Research Unit and an Associate Professor in the Department of Forestry and Natural Resources at Clemson University in South Carolina (pjodice @clemson.edu).

Richard W. Judd is Adelaide & Alan Bird Professor of History in the History Department at the University of Maine in Orono (Richard.Judd@umit.maine.edu).

Joshua J. Lawler is an Assistant Professor in the School of Forest Resources at the University of Washington in Seattle (jlawler@u.washington.edu).

Robert J. Lilieholm is the E.L. Giddings Associate Professor of Forest Policy in the School of Forest Resources at the University of Maine in Orono (robert. lilieholm@maine.edu).

Kimberly A. Lutz is the Director of the Connecticut River Program at The Nature Conservancy, located in Northampton, Massachusetts (klutz@tnc.org).

Christian O. Marks is the floodplain forest ecologist for the Connecticut River Program at The Nature Conservancy, located in Northampton, Massachusetts (cmarks@tnc.org).

Keith H. Nislow is a Research Fisheries Biologist and leader of the Fish and Wildlife Habitat Relationships Team of the USDA Forest Service Northern Research Station in Amherst, Massachusetts (knislow@fs.fed.us).

Ryan M. Perkl is a doctoral candidate in Environmental Design and Planning in the Department of Planning and Landscape Architecture at Clemson University in South Carolina (rperkl@clemson.edu).

Hugh P. Possingham is a Professor in Mathematics and Biology, and Director of the Ecology Centre at the University of Queensland in Australia (h.possingham@ uq.edu.au).

Robert B. Powell is an Assistant Professor in the Department of Parks, Recreation, and Tourism Management at Clemson University in South Carolina (rbp@clemson.edu).

Justina C. Ray is the Executive Director of Wildlife Conservation Society Canada, headquartered in Toronto, Ontario (jray@wcs.org).

Conrad Reining is the Eastern Program Director for the Wildlands Network (conrad@ wildlandsnetwork.org).

Robert M. Suryan is an Assistant Research Professor in the Department of Fisheries and Wildlife at Oregon State University's Hatfield Marine Science Center in Newport, Oregon (rob.suryan@oregonstate.edu).

Stephen C. Trombulak is the Professor of Environmental and Biosphere Studies in the Department of Biology at Middlebury College in Vermont (trombulak@ middlebury.edu).

Jeffrey V. Wells is the Director of Science and Policy for the Boreal Songbird Initiative and a Visiting Fellow with the Cornell University Lab of Ornithology in Ithaca, New York (jeffwells@borealbirds.org).

Gillian Woolmer is the Assistant Director for the Wildlife Conservation Society Canada, headquartered in Toronto, Ontario (gwoolmer@wcs.org).

Chapter 1
Introduction: Creating a Context for Landscape-Scale Conservation Planning

Stephen C. Trombulak and Robert F. Baldwin

Abstract Over the last 130 years, conservation practitioners have increasingly enlarged their view of the important spatial scales on which to base the development and implementation of conservation plans. For example, even though national parks have been an essential tool in the global conservation toolbox since the late 1800s, it is now well understood that critical conservation goals can only be achieved if parks are viewed as being connected to each other ecologically and embedded within a larger landscape that includes a diverse mixture of ownerships, histories, and uses. The tools for planning at these greater spatial scales, from both the natural and social sciences, are only slowly being developed, tested, and refined. This book represents a step in that process, bringing together lessons on a variety of perspectives – including history, economics, wildlife biology, computer modeling, and climate change science – on how to achieve landscape-scale conservation planning. Although the authors represented in this book primarily describe their work on conservation planning in Eastern North America, these chapters serve as case studies on how conservation planning can be successfully approached in landscapes anywhere in the world.

Keywords Case study · Conservation · Ecoregion · Landscape · Planning

---

S.C. Trombulak (✉)
Department of Biology, Middlebury College, Middlebury, VT 05753, USA
e-mail: trombulak@middlebury.edu

R.F. Baldwin
261 Lehotsky Hall, Department of Forestry and Natural Resources, Clemson University, Clemson, SC 29634-0317, USA
e-mail: baldwi6@clemson.edu

S.C. Trombulak and R.F. Baldwin (eds.), *Landscape-scale Conservation Planning*, DOI 10.1007/978-90-481-9575-6_1, © Springer Science+Business Media B.V. 2010

1.1 Introduction

This book is about the future of conservation. Society now stands at the brink of many interrelated environmental challenges: widespread collapse of the diversity of life on Earth, from genes and species to ecosystems (MEA 2005); global warming and its associated transformation of seasons, storms, and sea level (IPCC 2007), or what Anne Raver (2002) dubbed 'global weirding'; and an increase in human population size that is not projected to plateau for at least another 50 years and until after the likely addition of another three billion people (Cohen 2003; Lutz et al. 2001), all with legitimate demands to food, water, energy, and shelter. None of these challenges will have easy or quick solutions and will dramatically shape both the cultural and natural landscapes through this new century and beyond (Clark and Dickson 2003; Holdren 2008).

What awaits society and what will remain of the natural world at the end of the twenty-first century is uncertain. What is clear is that it will depend largely on the choices society makes today about how to address the challenge of conservation, choices that will call upon society to use the best available sciences – both natural and social – to forecast how actions in the present will unfold in the future. Only through careful planning of how society confronts the challenge of conservation in the coming years will options for successfully meeting the needs of both people and nature remain open. Conservation planning, therefore, is an essential key to navigating the pathways into the future.

Understanding where conservation planning needs to go in the twenty-first century requires an awareness of how both the theory and practice of conservation developed during the past 140 years. Apart from isolated efforts to institute hunting regulations, the early history of conservation focused predominantly on local-scale initiatives aimed at protecting land itself. Beginning in the closing decades of the nineteenth century, the major conservation initiatives dealt with the establishment of parks – typically (although not exclusively) large tracts of government-owned lands – predominantly in locations that offered exceptional opportunities for public use and recreation. While a more complete narrative about land protection, in North America or anywhere else in the world, would require attention to more categories of protected areas than just parks – including national forests, Crown lands, wilderness areas, and wildlife refuges – the story of parks is perhaps the clearest lens through which to view the salient transitions in the relationships among conservation goals, protected areas, and the surrounding landscape.

In North America at the end of the nineteenth century, such national parks as Yellowstone (Wyoming, Montana, and Idaho; 1872), Banff (Alberta; 1885), Yoho (British Columbia; 1886), and Yosemite and Sequoia (California; 1890) were created, as well as state and provincial parks such as Adirondack and Catskill (1892) in New York and Algonquin (1893) in Ontario.

Other countries on other continents followed a similar pattern. Australia established the Royal National Park in 1879, and in New Zealand, Tongoriro National Park was established in 1887. In Africa, protected areas were created in South

1 Introduction: Creating a Context for Landscape-Scale Conservation Planning

Africa as early as 1895 (Hluhluwe-umfolozi Game Reserve). In Europe, the first national parks were established in Sweden in 1909 (a group of nine, most notably Sarek), and in South America in Chile in 1926 (Vincente Perez Rosales).

Yet these parks were founded and managed with similar underlying philosophies: (1) they could be managed effectively in isolation from the surrounding landscape, (2) ecological connectivity with locations outside of the park was unimportant – indeed, even unacknowledged – for achieving conservation goals, such as species (typically game) preservation, and (3) culturally imposed boundaries, such as park boundaries and international borders, had ecological meaning.

Despite its local, single-site focus, this early phase of conservation had some success in protecting wild species and landscapes as a result of several interrelated factors. First, protected areas created during the earliest period of parks were large. Algonquin Provincial Park encompasses over 7,600 km², Yellowstone National Park almost 9,000 km², Wood Buffalo National Park (Alberta and Northwest Territories) is 17,300 km², and Adirondack Park over 24,000 km² (although slightly less than half of this is publicly owned). At the time they were established, large expanses of relatively uncontested lands were generally available, especially in remote, sparsely populated regions, and therefore the creation of large protected areas was possible even in locations with great potential for timber and agriculture. Second, the human population was considerably smaller even 100 years ago than it is today. In 1900, the entire human population is estimated to have been just 1.6 billion (U.S. Census Bureau 2010a), less than one quarter of its present size. In the U.S., it was 76 million (U.S. Census Bureau 2010b) and in Canada, it was 5.3 million (Statistics Canada 2010), just 24% and 16%, respectively, of their sizes today. Thus, not only were fewer people contesting the establishment of protected areas, but more importantly, fewer people were living and working in the matrix of lands that surround and separate these areas. Third, technology, and in particular the automobile, was not in widespread use. Although first manufactured in the mid-1880s, vehicles powered by internal combustion engines were not mass-produced until 1914 (Georgano 2000). The associated network of roads soon followed, and even though the system of high-speed, high-volume roads did not begin to be developed in Canada until 1950 (Transport Canada 2009) and in the U.S. until 1956 (FHWA n.d.), it was noted as early as 1924 by Aldo Leopold that the automobile was a serious threat to wilderness in the Western US (Leopold 1924 [1991]).

Therefore, the need to view conservation lands within the context of the larger landscape was not especially critical for conservation up through the first half of the twentieth century for two equally important reasons. The first was that human population densities were lower and the polarity of landscape transformation and the magnitude of threats across boundaries were less. Even though many areas of North America and other developed lands were degraded by poor agricultural techniques and war, much of the planet was still only lightly transformed from a pre-industrial state. The degree of ecological connectivity both within and among protected areas was great even without expressed attention to these issues in the design and management of protected areas.

By the second half of the twentieth century, however, these conditions had changed dramatically. Human population increased fourfold between 1900 and 2000, and the magnitude of human transformation of the landscape – both in aggregate and per capita – had increased more than that. As examples, in the U.S., per capita energy consumption increased 2.8-fold from 1900 to 2000 (U.S. Energy Information Administration 2010), and globally per capita energy consumption increased 6.6% in just the 15 years between 1990 and 2005 (IEA 2007). The number of motor vehicles per capita increased 39% in the U.S. between 1960 and 2000 (RITA 2010).

The second reason that a landscape-scale perspective was not important for conservation up through the first half of the twentieth century is rooted in perceptions of what was important to conserve. The initial focus of conservation was largely on wildlife species, particularly those that were useful to society for food or products or were charismatic. Even within the National Park Service in the U.S., the scientific view of conservation was that it was necessary to kill 'bad' species, such as large predators, in order to promote 'good' species, such as deer and elk (Leopold 1949; Sellars 1997). Gradually, however, recognition grew among scientists, and ultimately society, of the importance of an ecological view that included all species, most notably predators, and all levels of biological organization, from genes to ecosystems. This gave rise to a concern for biological diversity in general, particularly to species at risk of extinction and ecosystems at risk of elimination (Wilson 1988), leading to the passage in the U.S. of the Endangered Species Act in 1973. This act was intended to protect not just species but also the ecosystems on which they depend, further elevating the importance of conservation planning for landscapes as a whole.

Ecologists and wildlife biologists began noting with increasing frequency that protected areas designed in the latter part of the nineteenth century and first half of the twentieth century, despite their large size, were increasingly unable to achieve their conservation goals. For example, in the 1970s, studies on the autecology of grizzly bears (*Ursus arctos*) in Yellowstone National Park noted that even with an area of 9,000 km^2, the park itself was too small to support sufficient numbers of these animals to allow for a viable population (Craighead 1979). This highlighted the critical importance of an approach to conservation planning for the grizzly bear that viewed the park as part of a larger landscape, perhaps as much as ten times again as large as the park itself, now called the Greater Yellowstone Ecosystem. Extending these analyses, Newmark (1985, 1987) noted that many different species of mammals found in national parks all throughout Western North America had area requirements that were greater than the sizes of the parks by themselves, and that extinction rates of mammalian species in these parks in fact exceeded the rates of colonization.

Furthermore, an appreciation of all dimensions of biological organization – not just composition but also structure and function – became manifested in such public policies in the U.S. as the Water Pollution Control Act Amendments of 1972, which called for the protection not just of biological diversity but of the biological integrity of the nation's waters (Frey 1975). As the goals of conservation expanded to

1 Introduction: Creating a Context for Landscape-Scale Conservation Planning 5

include more levels and more dimensions of biological organization, it became increasing clear that only a landscape-scale approach to conservation could provide the flexibility and scope to plan conservation at the appropriate scale to achieving these broad goals.

The historic model for conservation planning – draw a boundary around an important location, and manage it in isolation of all other locations – could no longer be successful. The world had changed, and conservation planning would need to change as well.

The limitations of the historic model were well appreciated within both academic and practitioner circles. In the mid-1970s, the Man and the Biosphere Program of UNESCO advanced a new model for protected areas (UNESCO 1974) that incorporated new design criteria based on two fundamental principles that would become central aspects of most subsequent conservation planning. First, hard administrative boundaries separating the theoretical 'inside' and 'outside' of a protected area were no longer ecologically tenable. Because people, animals, air, and water all readily move across culturally defined boundaries, land-use designations associated with protected areas needed to be crafted in such a way as to establish gradients of influence and purpose, ranging from wholly ecological within core areas to predominantly cultural within the surrounding landscape matrix, grading from one to the other across a buffer zone that allowed for complementary mixed uses, neither solely cultural nor ecological. The incorporation of buffer zones into the fundamental design of protected areas forced a re-imagining of what protected areas were, transforming them from areas with strict nature protection independent from all other lands to areas that were part of a broader conservation area, where strict nature protection and natural-resource stewardship were both valued in complementary locations and forms (Trombulak 2003).

Second, connectivity among conservation areas was increasingly acknowledged as important. The magnitude of human transformation of the landscape, a result of both human numbers and per capita human impact, had grown to the point where connectivity that allowed for movement in both the short-term (e.g., animal, especially predator, dispersal) and long-term (e.g., plant community response to climate change) needed to be explicitly considered in the design and management of conservation areas.

These two principles, and the design elements crafted in response to them, signaled an important shift in thinking about conservation. Planning solely on small spatial scales – sites and isolated locations – was no longer sufficient. Planning needed to incorporate active consideration of large spatial scales: how would conservation areas scattered across large geographic regions relate to and interact with each other ecologically, and how would they all collectively relate to and interact with the people who lived there, a perspective now generally referred to as ecosystem management (Agee and Johnson 1988; Grumbine 1994; Layzer 2008).

The development of this 'core-corridor-buffer' model, although still not fully realized successfully anywhere, defined a new paradigm in conservation planning, one that quickly spread and became the basis for numerous conservation initiatives. In the U.S., conservation planning for grizzly bears within the Greater Yellowstone

Ecosystem continued with vigor (Shaffer 1992). In the Pacific Northwest, conservation initiatives focused on the protection of the Spotted Owl (*Strix occidentalis*) increasingly began to consider the broader landscape within which the owls lived (e.g., distinctions between source and sink habitats, movement corridors, and area requirements for viable effective population sizes) as well as the health of the region's timber industry (Simberloff 1987).

Numerous conservation initiatives that encompassed large landscapes were launched beginning in the mid-1980s. The Wildlife Conservation Society and the Caribbean Conservation Corporation began development in 1990 of Paseo Pantera, a wildlife movement corridor, ostensibly designed for mountain lions (*Puma concolor*), along the length of the Central American isthmus (Marynowski 1992). Building off of earlier efforts to design protected area networks (e.g., Newman et al. 1992; Noss 1987; Noss and Harris 1986), a group of conservation biologists and advocates formed The Wildlands Project in 1991 to foster the development of an integrated system of conservation lands throughout North America (Foreman et al. 1992; Noss 1992). The Wildlands Project brought landscape-scale conservation planning into the mainstream of conservation work and thought, directly or indirectly stimulating the development of landscape-scale conservation plans in many regions in North America, including the Sky Islands region of the Southwestern U.S. and Northwestern Mexico (Foreman et al. 2000), the Northern Rocky Mountains from Yellowstone to Yukon (Y2Y; Y2Y Initiative 2010), the Adirondack Park to Algonquin Provincial Park corridor in Northeastern North America (A2A; Quinby et al. 2000), and the Northern Appalachian Mountains (Reining et al. 2006).

Realization of the importance of landscape-scale thinking for achieving conservation goals quickly spread to other organizations. In the mid-1990s, The Nature Conservancy, historically focused on management of discrete locations acquired for the protection of rare species and ecosystems, redesigned their approach to identifying priorities for conservation by developing ecoregional portfolios (Groves 2003; Groves et al. 2000). Such portfolios were envisioned to be sets of sites, identified and managed through the lens of their collective contribution to conservation throughout an ecoregion. Other organizations in North America, particularly the Sierra Club through their Critical Ecoregions Program (Elder 1994), followed suit. Moreover, similar initiatives developed on other continents, most notably in Australia on the Great Barrier Reef as early as 1975 (Great Barrier Reef Marine Park Authority n.d.).

That the future of conservation must focus on landscape-scale planning seems certain (Olson et al. 2001; Groves 2003). What is less clear is how to accomplish it, successfully making the transition from theory to practice. This book offers a variety of perspectives on conservation planning across large areas and transcending multiple boundaries, political and ecological – perspectives drawn from the authors' direct experiences with the application of conservation planning principles. These chapters serve as case studies to explore the details of key elements involved in conservation planning. Yet they are framed not just with an eye toward reporting what the authors did or understand with respect to the issues or locations that are

1 Introduction: Creating a Context for Landscape-Scale Conservation Planning

the focus of their specific chapters; rather, they explain broader concepts applicable across multiple regions, what particular challenges their systems or regions posed, why they made or advocated for certain decisions, what barriers they see for future applications, and what their results were – both good and bad – so that others can compare, adopt, and adapt.

1.2 What Is a Landscape?

We use the term 'landscape' liberally throughout this book as a generic label for a large expanse of land and water. While it has no precise definition in terms of size, composition, or defining characteristics, its usage implies an area that represents a heterogeneous mosaic of local land forms aggregated over greater and greater areas (Forman and Godron 1986; Noss and Cooperrider 1994; Urban et al. 1987). While some authors try to place the landscape context in a specific place within the hierarchy of land forms (Bailey 1996), the lexicon of landscape categories within both conservation and colloquial usage is quite varied – including ecoregion, biome, and ecozone – reflecting both the imprecise nature of landscape boundaries as well as their nested, hierarchical organization.

The landscape designation most commonly referenced by the authors here is that of 'ecoregion,' typically defined as a large area that contains a relatively distinct assemblage of plants and animals (Dinerstein et al. 1995). Practically speaking, the concept of ecoregion provides a way to classify locations based on biological communities, geography, and ecological processes; the flora and fauna that are present and the specifics of the ecological processes that go on at any two places within an ecoregion is typically more similar than to any location outside the ecoregion. From a conservation planning perspective, ecoregions provide a geographic framework for making decisions based on ecosystem processes rather than political boundaries.

Unsurprisingly, given the fluid composition of plant and animal assemblages and the wide range of criteria that could be used to classify a location's geography, ecoregions rarely have unambiguous boundaries. Numerous ecoregional classification schemes have been developed over the past 20 years (reviewed by Groves 2003), each based on different criteria and resulting in the identification of different ecoregions. Thus, different agencies and organizations, from the U.S. Forest Service and Environmental Protection Agency to The Nature Conservancy and World Wildlife Fund, have developed their own, often interrelated, ecoregional classification systems. With respect to conservation planning, it matters less which ecoregional classification system is used than that the system is appropriate for advancing the conservation goals.

A landscape, as we define it, is a collection of habitat patches sufficient enough in size to allow population processes to take place at a multigenerational time scale. What constitutes a landscape for one organism may simply be a portion of a landscape for another. Thus, some of the authors in this volume describe conservation issues and initiatives taking place within relatively local landscapes, such as Northern Maine (Chaps. 3, 5, 10), the Connecticut River watershed (Chap. 6), and the Adirondack Mountains (Chap. 17). Other authors discuss issues and initiatives that encompass landscapes for which these smaller regions are but a part, especially the Northern Appalachian/Acadian ecoregion (Sect. 1.5; Chap. 2, 4, 11–16). Still others consider larger landscapes including the entire Appalachian Mountain chain (Chap. 18), Eastern North America (Chap. 9), North and Central America (Chap. 7), and entire ocean basins (Chap. 8).

Thus, even such an amorphous concept as landscape can be used as an effective tool for conservation planning if the scale selected for planning is appropriate for the conservation goals to be achieved. The spatial expanse used for planning is easily adapted to best suit the requirements for achieving the conservation goals (even if political boundaries can make implementation of the plans difficult), and the tools and perspectives critical to successful planning – both social and ecological – are effective at multiple spatial scales.

1.3 What Is 'Planning'?

This book is about landscape-scale conservation planning, and as such also makes liberal use of the term 'planning,' for which we have isolated two distinct meanings. The first is the most widely held and refers to the official function of making land-use decisions in the context of laws, regulations, economics, aesthetics, and local environmental concerns. Most of these local land-use decisions are made by county, state, provincial, or municipal governments. Rarely are they made at the federal or multinational level (Theobald et al. 2000), although notable exceptions include spatial planning initiatives for transboundary projects within the European Union (Baldwin and Trombulak 2007) and when the Endangered Species Act is invoked in the U.S. to forcefully encourage the development of regional habitat conservation plans (Babbitt 2005). A more recent evolution of the land-use decision-making paradigm of planning is one closely related to the primary examples in this book. This is a highly relevant development for landscape-scale conservation and happens when a grassroots ecoregional conservation planning initiative acts as an umbrella for landowners, managers, and local planners from multiple jurisdictions to collaborate on making decisions to achieve regional conservation goals.

Theoretically, land-use planners weigh all available information and make an optimized decision for uses based on a number of considerations. However, this is not always the case. Most planning decisions are greatly biased by local concerns and as a consequence, negative local effects of bad land-use decisions accumulate to affect conditions over larger areas (Theobald et al. 1997). This fact has given rise to a new more inclusive view of planning, and the one relied upon throughout this book: conservation planning is a multilayered, systematic process that progresses in an orderly fashion from conservation vision to science, to communication of results and engagement of stakeholders, to design, and finally to implementation.

1 Introduction: Creating a Context for Landscape-Scale Conservation Planning

This second meaning for 'planning' assumes that the greater good is regional ecological integrity. This new field is summarized as 'systematic conservation planning' and includes decisions on scientific issues as model parameters and data inputs, conservation targets and goals, means of communicating to greater audiences, inclusion of specific stakeholders, boundary decisions in design phase, and strategic land-use decisions during implementation phase. The whole purpose of systematic conservation planning as presented in this book is to avoid the risk of biased, opportunistic decisions that are likely when all decisions are made locally and based on local information alone.

Taken as a whole, the chapters in this book illustrate that landscape-scale conservation planning is an interdisciplinary, collaborative process inclusive of more academic traditions than science alone. At its most fundamental level, landscape-scale conservation planning is a biological science concerned with ecological pattern and process. However, conservation planners recognize that biology alone won't actually save any species: it is the effective communication of scientific results and meaningful policy interpretations that will determine if the best available science is put into practice. Further, studies of landscape-scale conservation cannot ignore the human dimension of conservation. Landscape-scale conservation planning is closely related to the field of landscape ecology, an interdisciplinary field concerned with how human and natural processes interact to shape function of ecosystems in time and space and how landscapes should be designed to foster sustainability (Opdam et al. 2002). Reflecting the integration of the cultural, social, and biological, this book includes discussions of land-use history and cultural identity (Chap. 3), policy and governance (Chap. 4), land-use economics (Chap. 5), planning theory (Chap. 10), decision making processes (Chaps. 10, 17), integration of expert opinion into the planning process (Chap. 11), public education and outreach (Chap. 12), and forecasting of land-use transformation (Chap. 13).

We hope to contribute to the general movement to redefine landscape-scale conservation planning as a multi-scale, systematic, and repeatable process (Moilanen et al. 2009). Land-use decisions will always be primarily local, but our goal is to ensure that these are made with the full knowledge of the implications of local decisions for biological conditions at greater spatial scales. We also hope that the chapters in this book will contribute to ensuring that land-use decisions will be made in a temporal context, taking into account historical trajectories and projected future conditions.

In this way we hope our book will influence the field of planning as a whole. Most land-use decisions around the world are made by officials with little or no training in the biological sciences. They may appreciate the environment and be committed to protecting it by upholding local laws, but they normally do not understand the impacts of their land-use decisions on the greater ecosystems in which their communities reside. Consequently, the cumulative effects of their decisions can wreak ecosystem damage across extensive scales. Landscape-scale conservation planning offers a way out of this locally-controlled ecosystem degradation while ensuring local participation in the conservation planning process.

1.4 Case Studies for Conservation Planning

Each chapter serves as a case study in how specific issues have been evaluated and addressed for specific locations, ecosystems, or groups of organisms. In themselves, they each tell a story about the conservation needs of plants and animals in a landscape, the successes achieved and the failures faced in meeting those needs, and the people who have carried out the work. Each of these case studies is embedded in the broader body of knowledge of the principles and practices of conservation planning as specific stories that exemplify and amplify on these principles and practices.

Case studies, in this sense, become the vehicle through which the implementation of these principles can be evaluated, much in the same way that case studies are used in law, medicine, and business to understand how broader principles come to play in individual instances. Negotiating a contract, treating a patient, and evaluating a business plan are structurally the same as planning a conservation initiative: build an approach based on common principles, adapt to specific conditions, and learn from each case how to do it better. The cases presented here are explicit in each of these regards: (1) general principles that underlie the issue – whether they emerge from the natural sciences or social sciences – are presented, (2) the details of the issue are explored in specific landscapes, and (3) the lessons learned are detailed so that others can take advantage of the successes and avoid the mistakes as these issues are addressed in other landscapes.

However, the use of case studies to explore issues in conservation planning carries a risk. Because cases, by their very nature, are rooted in specific landscapes, readers might assume that the cases are only relevant to them if they actually work in that landscape. For example, a book such as this, nearly exclusively focused on examples of conservation planning in North America, and even then primarily on the eastern part of the continent, might be viewed as being only of regional interest. This would be a mistake. Despite the uneasy tension between 'globally significant' and 'regionally relevant' that always seems to surround discussion of specific conservation initiatives, we believe that these cases offer insight to anyone who is trying to implement a conservation planning process anywhere. Certainly the details associated with other landscapes and those found in those presented here will differ; what is of value is the exploration of how those details were considered and confronted.

1.5 The Book in Context

As noted in Sect. 1.2, most of the chapters in this book develop cases around the Northern Appalachian/Acadian ecoregion, which lies in the Northeastern U.S. and Southeastern Canada, encompassing over 330,000 km^2 and including all or a part of Northern New York, Vermont, New Hampshire, Maine, Southern Québec, New Brunswick, Nova Scotia, and Prince Edward Island (Fig. 1.1). This focus is

1 Introduction: Creating a Context for Landscape-Scale Conservation Planning 11

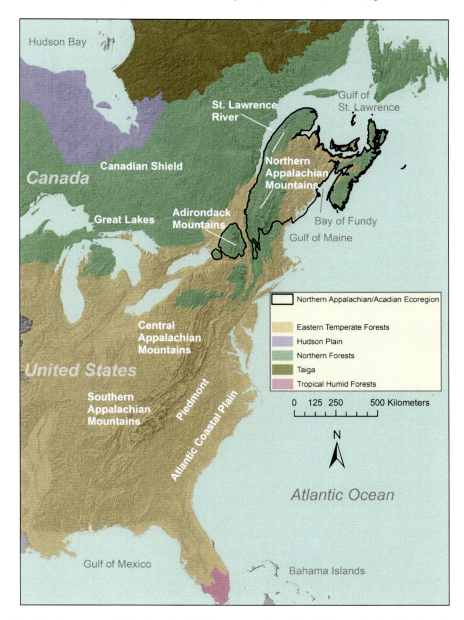

Fig. 1.1 Eastern North America, highlighting the location of the Northern Appalachian/Acadian ecoregion as delineated by The Nature Conservancy/Nature Conservancy Canada. Major geographical features are noted, and *colors* correspond to Level I ecoregions delineated by the Commission for Environmental Cooperation in partnership with the governments of Canada, Mexico, and the US

not by accident. The original impetus for this book emerged from a group of conservation practitioners – from a range of U.S. and Canadian academic institutions and environmental non-government organizations – who were working together to develop a set of planning tools and analyses to advance landscape-scale conservation planning throughout this ecoregion.

Our work was carried out under the auspices of Two Countries, One Forest (2C1Forest), a confederation of conservation organizations, foundations, and scientists active within the ecoregion. Founded in 2001, 2C1Forest's explicit mission was 'to protect the natural beauty, native species and ecosystems of the Northern Appalachian/Acadian ecoregion while maintaining economically healthy and culturally vibrant local communities' (Bateson 2005).

The ecological character of this region demanded that future conservation planning here take greater account of its shared threats and responsibilities. The region is ecologically diverse, dominated by spruce-fir and northern hardwood forests, extensive coastlines, inland mountain ranges, and glacially carved landscapes. It is an ecological transition zone between northern boreal and southern temperate forests, and it will increasingly come to serve as a north–south biological corridor for species as their ranges shift in response to climate change.

The ecological character of this region demanded that future conservation planning here take greater account of the environmental threats and conservation responsibilities collectively shared by the states and provinces that make up the ecoregion. Patterns of air and water pollution (Miller et al. 2005), species extirpations (Laliberte and Ripple 2004), and forest diseases and insect pests (Fraver et al. 2007) pay scant heed to the borders that separate the two countries or its states and provinces. Neither do the dominant geographic features, central to both the present distributions of species and their future responses to dynamic changes in climate and land use (e.g., mountain ranges, coastlines, major rivers and lakes), highlighting the importance of an ecoregional focus on planning for connectivity. Moreover, despite the pronounced cultural diversity within the ecoregion, which is home to two dominant language groups (French and English) and numerous First Nations, the entire social landscape – economic and cultural – is strongly dependent on the region's natural resources and will be dramatically pressed by human immigration in a warming world. Further, it has a shared vulnerability to numerous social forces that threaten biological diversity, including road (Baldwin et al. 2007) and amenities development (Baldwin et al. 2009).

Over the past 9 years, this group of conservation scientists has worked to realize the vision of the conservation collaborative by developing spatially explicit measures of human transformation of the landscape, forecast spatially explicit threats to the landscape over the coming decades, identify sites within the ecoregion that are irreplaceable with respect to their contribution to the ecoregion's biological and geographical diversity, and identify priorities for conservation action needed to address threats and protect irreplaceable sites both now and in the future, all of which are reviewed in this volume.

While the work of 2C1Forest broke much new ground with respect to landscape-scale conservation planning, in terms of tools, analyses, and engagement of a diverse

1 Introduction: Creating a Context for Landscape-Scale Conservation Planning

community of conservation practitioners with a shared vision, we came to realize that even the work of this one organization was only a part of the larger narrative of what needs to be considered, with respect to both natural and cultural features, to achieve conservation across this – or any other – landscape. Thus, additional authors were invited to share their experiences and expand the boundaries of what this volume represents.

This volume attempts to highlight a number of themes that are universal to landscape-scale conservation planning, themes that need to be addressed regardless of where the initiative is geographically located. It is not a manual for landscape-scale conservation planning. Rather, it is a lesson book, reminding and instructing through example the importance of selecting the proper temporal and spatial scale for the conservation goals chosen, considering both cultural and natural history, responding to present and emerging economic trends, engaging both stakeholders and experts, developing multivariate measures of threats and opportunity, and practicing patience, creativity, and collaboration. In one sense, it is about the future of conservation. More importantly, however, it is a collection of examples about how we all can engage in the work needed to move society toward the future that we want.

References

Agee, J. K., & Johnson, D. R. (1988). *Ecosystem management for parks and wilderness*. Seattle, WA: University of Washington Press.

Babbitt, B. (2005). *Cities in the wilderness: A new vision of land use in America*. Washington, DC: Island Press/Shearwater Books.

Bailey, R. G. (1996). *Ecosystem geography*. New York: Springer.

Baldwin, R. F., & Trombulak, S. C. (2007). Losing the dark: A case for a national policy on land conservation. *Conservation Biology, 21*, 1133–1134.

Baldwin, R. F., Trombulak, S. C., Anderson, M. G., & Woolmer, G. (2007). Projecting transition probabilities for regular public roads at the ecoregion scale: A Northern Appalachian/Acadian case study. *Landscape and Urban Planning, 80*, 404–411.

Baldwin, R. F., Trombulak, S. C., & Baldwin, E. D. (2009). Assessing risk of large-scale habitat conversion in lightly settled landscapes. *Landscape and Urban Planning, 91*, 219–225.

Bateson, E. M. (2005). Two Countries, One Forest – Deux Pays, Une Forêt: Launching a landscape-scale conservation collaborative in the Northern Appalachian region of the United States and Canada. *The George Wright Forum, 22*(1), 35–45.

U.S. Census Bureau (2010a). *Historical estimates of world population*. Retrieved February 28, 2010, from http://www.census.gov/ipc/www/worldhis.html

U.S. Census Bureau (2010b). *Historic national population estimates*. Retrieved February 28, 2010, from http://www.census.gov/popest/archives/1990s/popclockest.txt

Clark, W. C., & Dickson, N. M. (2003). Sustainability science: The emerging research program. *Proceedings of the National Academy of Sciences of the United States of America, 100*, 8059–8061.

Cohen, J. E. (2003). Human population: The next half century. *Science, 302*, 1172–1175.

Craighead, F. C., Jr. (1979). *Track of the grizzly*. San Francisco, CA: Sierra Club Books.

Dinerstein, E., Olson, D. M., Graham, D. J., Webster, A. L., Pimm, S. A., Bookbinder, M. P., et al. (1995). *A conservation assessment of the terrestrial ecoregions of Latin America and the Caribbean*. Washington, DC: The World Bank.

Elder, J. (1994). The big picture: Sierra club critical ecoregions program. *Sierra, 79*(2), 52–57.

IEA [International Energy Agency] (2007). *Energy balances of OECD countries (2008 edition) and energy balances of non-OECD countries (2007 edition)*. Paris: IEA. Retrieved February 28, 2010, from http://earthtrends.wri.org/text/energy-resources/variable-351.html

U.S. Energy Information Administration (2010). International total primary energy consumption and energy intensity. Retrieved February 28, 2010, from http://www.eia.doe.gov/emeu/international/energyconsumption.html

FHWA [Federal Highway Administration] (n.d.). Welcome to the Eisenhower interstate highway system web site. Retrieved February 28, 2010, from http://www.fhwa.dot.gov/interstate/homepage.cfm

Foreman, D., Davis, J., Johns, D., Noss, R., & Soulé, M. (1992). The Wildlands Project mission statement. *Wild Earth*, Special Issue, 3–4.

Foreman, D., Dugleby, B., Humphrey, J., Howard, B., & Holdsworth, A. (2000). The elements of a Wildlands Network Conservation Plan: An example from the Sky Islands. *Wild Earth, 10*, 17–30.

Forman, R. T. T., & Godron, M. (1986). *Landscape ecology*. New York: Wiley.

Fraver, S., Seymour, R. S., Speer, J. H., & White, A. S. (2007). Dendrochronological reconstruction of spruce budworm outbreaks in northern Maine, USA. *Canadian Journal of Forest Research, 37*, 523–529.

Frey, D. G. (1975). Biological integrity of water: An historical approach. In R. K. Ballentine & L. J. Guarraia (Eds.), *The integrity of water* (pp. 127–139). Washington, DC: U.S. Environmental Protection Agency.

Georgano, G. N. (2000). *Vintage cars 1886 to 1930*. Gothenburg, Sweden: AB Nordbok.

Great Barrier Reef Marine Park Authority (n.d.). *Legislation and regulations*. Retrieved February 20, 2010, from http://www.gbrmpa.gov.au/corp_site/about_us/legislation_regulations

Groves, C. R. (2003). *Drafting a conservation blueprint: A practitioner's guide to planning for biodiversity*. Washington, DC: Island Press.

Groves, C., Valutis, L., Vosick, D., Neely, B., Wheaton, K., Touval, J., et al. (2000). *Designing a geography of hope: A practitioner's handbook to ecoregional conservation planning*. Arlington, VA: The Nature Conservancy. Retrieved March 3, 2010, from http://marineplanning.org/resources/GOH2-v1.pdf

Grumbine, R. E. (1994). What is ecosystem management? *Conservation Biology, 8*, 27–38.

Holdren, J. P. (2008). Science and technology for sustainable well-being. *Science, 319*, 424–434.

RITA [Research and Innovative Technology Administration] (2010). *Passenger car and motorcycle fuel consumption and travel*. Retrieved February 28, 2010, from http://www.bts.gov/publications/national_transportation_statistics/html/table_04_11_m.html

IPCC [Intergovernmental Panel on Climate Change]. (2007). *Climate change 2007: Impacts, adaptation and vulnerability*. Cambridge: Cambridge University Press.

Laliberte, A. S., & Ripple, W. J. (2004). Range contractions of North American carnivores and ungulates. *BioScience, 54*, 123–138.

Layzer, J. A. (2008). *Natural experiments: Ecosystem-based management and the environment*. Cambridge, MA: MIT Press.

Leopold, A. (1924 [1991]). The river of the mother of God. In S. L. Flader & J. B. Callicott (Eds.), *The river of the mother of God and other essays by Aldo Leopold* (pp. 123–125). Madison, WI: University of Wisconsin Press.

Leopold, A. (1949). *A Sand County almanac, and sketches here and there*. New York: Oxford University Press.

Lutz, W., Sanderson, W., & Scherbov, S. (2001). The end of world population growth. *Nature, 412*, 543–545.

Marynowski, S. (1992). Paseo Pantera: The Great American biotic interchange. *Wild Earth*, Special Issue, 71–74.

MEA [Millennium Ecosystem Assessment]. (2005). *Ecosystems and human well-being: synthesis*. Washington, DC: Island Press.

Miller, E. K., Vanarsdale, A., Keeler, G. J., Chalmers, A., Poissant, L., Kamman, N. C., et al. (2005). Estimation and mapping of wet and dry mercury deposition across Northeastern North America. *Ecotoxicology, 14*, 53–70.

Moilanen, A., Wilson, K. A., & Possingham, H. P. (Eds.) (2009). *Spatial conservation prioritization: Quantitative methods and computational tools*. New York: Oxford University Press.

UNESCO [United Nations Educational, Scientific and Cultural Organization] (1974). *Task force on: criteria and guidelines for the choice and establishment of biosphere reserves (Man and the Biosphere Report No. 22)*. Paris: UNESCO.

Newman, B., Irwin, H., Lowe, K., Mostwill, A., Smith, S., Jones, J. (1992). Southern Appalachian wildlands proposal. *Wild Earth*, Special Issue, 46–60.

Newmark, W. D. (1985). Legal and biotic boundaries of western North American national parks: A problem of congruence. *Biological Conservation, 33*, 197–208.

Newmark, W. D. (1987). A land-bridge island perspective on mammalian extinctions in western North American parks. *Nature, 325*, 430–432.

Noss, R. F. (1987). Protecting natural areas in fragmented landscapes. *Natural Areas Journal, 7*, 2–13.

Noss, R.F. (1992). The Wildlands Project land conservation strategy. *Wild Earth*, Special Issue, 10–25.

Noss, R. F., & Cooperrider, A. Y. (1994). *Saving nature's legacy: Protecting and restoring biodiversity*. Washington, DC: Island Press.

Noss, R. F., & Harris, L. D. (1986). Nodes, networks, and MUMs: Preserving diversity at all scales. *Environmental Management, 10*, 299–309.

Olson, D. M., Dinerstein, E., Wikramanayake, E. D., Burgess, N. D., Powell, G. V. N., Underwood, E. C., et al. (2001). Terrestrial ecoregions of the world: A new map of life on Earth. *BioScience, 51*, 933–938.

Opdam, P., Foppen, R., & Vos, C. (2002). Bridging the gap between ecology and spatial planning in landscape ecology. *Landscape Ecology, 16*, 767–779.

Quinby, P., Trombulak, S., Lee, T., Long, R., MacKay, P., Lane, J., et al. (2000). Opportunities for wildlife habitat connectivity between Algonquin Provincial Park and the Adirondack Park. *Wild Earth, 10*, 75–80.

Raver, A. (2002, November 7). Bananas in the backyard. *New York Times*, Section F, p. 1.

Reining, C., Beazley, K., Doran, P., Bettigole, C. (2006). *From the Adirondacks to Acadia: A wildlands network design for the Greater Northern Appalachians (Wildlands Project Special Paper No. 7)*. Richmond, VT: Wildlands Project.

Sellars, R. W. (1997). *Preserving nature in the national parks: A history*. New Haven, CT: Yale University Press.

Shaffer, M. L. (1992). *Keeping the grizzly bear in the American West: A strategy for real recovery*. Washington, DC: Wilderness Society.

Simberloff, D. (1987). The Spotted Owl fracas: Mixing academic, applied, and political ecology. *Ecology, 68*, 766–772.

Statistics Canada (2010). Population and demography. Retrieved February 28, 2010, from http://www40.statcan.gc.ca/l01/ind01/l2_3867-eng.htm

Theobald, D. M., Miller, J. R., & Hobbs, N. T. (1997). Estimating the cumulative effects of development on wildlife habitat. *Landscape and Urban Planning, 39*, 25–36.

Theobald, D. M., Hobbs, N. T., Bearly, T., Zack, J. A., Shenk, T., & Riebsame, W. E. (2000). Incorporating biological information in local land-use decision-making: designing a system for conservation planning. *Landscape Ecology, 15*, 35–45.

Transport Canada (2009). *The trans-Canada highway*. Retrieved February 28, 2010, from http://www.tc.gc.ca/eng/mediaroom/backgrounders-b04-r007e-1669.htm

Trombulak, S. C. (2003). An integrative model for landscape-scale conservation in the twenty-first century. In B. A. Minteer & R. E. Manning (Eds.), *Reconstructing conservation: Finding common ground* (pp. 263–276). Washington, DC: Island Press.

Urban, D. L., O'Neill, R. V., & Shugart, H. H., Jr. (1987). Landscape ecology. *BioScience, 37*, 119–127.

Wilson, E. O. (Ed.). (1988). *Biodiversity*. Washington, DC: National Academy Press.

Y2Y [Yellowstone to Yukon] Initiative (2010). *Yellowstone to Yukon conservation initiative*. Retrieved March 1, 2010, from http://www.y2y.net/

Chapter 2
Identifying Keystone Threats to Biological Diversity

Robert F. Baldwin

Abstract Human beings have become the dominant force for environmental change and the task of conservation planning is to counter those changes most threatening biodiversity by identifying key areas providing resiliency and refuge. Landscape-scale conservation planners need to dissect those 'threats' (human activities that have driven ecological processes beyond the range of natural variability) to understand exactly what anthropogenic activities are influencing which aspects of ecosystem pattern and process. This chapter reviews two aspects of land use/land cover change (disaggregated and aggregated transitions), and introduces other anthropogenic activities that are treated in more depth in other chapters (i.e., pollution, disease, and climate change) before describing an ecoregional threat assessment project focused on identifying, mapping, and forecasting 'keystone threats.' Keystone threats are those strongly-interacting human activities – e.g., land use/land cover change – that if captured and modeled with some degree of accuracy can provide insights into where and when to protect habitats. The chapter suggests that in more wild or pristine areas, pollution or disease may be keystone threats while land use/land cover change will be the primary driver of biodiversity loss elsewhere. Given enough information any of these threats can be mapped and modeled to assist conservation planners in making decisions.

Keywords Human activity • Keystone threat • Land use/land cover • Modeling • Threats to biological diversity

---

R.F. Baldwin (✉)
261 Lehotsky Hall, Department of Forestry and Natural Resources,
Clemson University, Clemson, SC 29634-0317, USA
e-mail: baldwi6@clemson.edu

S.C. Trombulak and R.F. Baldwin (eds.), *Landscape-scale Conservation Planning*,
DOI 10.1007/978-90-481-9575-6_2, © Springer Science+Business Media B.V. 2010

2.1 Introduction

As an agent of ecological change, *Homo sapiens* has vastly increased its influence in recent centuries as our own populations and capabilities for resource extraction have grown exponentially. No longer are we driving species to extinction only because we overexploit them. Our global influence – in particular our ability to transform land cover – is causing widespread changes to biotic communities. Change is part of nature, but of special concern to conservation biologists is the degree to which human activities have caused increases in the severity and extent of disturbances, rate of species invasions, and alteration of global biogeochemical processes. Taken together, these changes suggest that humanity is a global ecological force equivalent to the great geological events that caused past mass extinctions (Vitousek et al. 1997; Wilson and Peter 1988). The task of conservation biology generally and of ecoregional conservation planning specifically is to identify and counter these threats.

This chapter is a general review of threats to biological diversity and how their processes and patterns can be assessed. I introduce the concept of 'keystone threats,' threats that are strongly interacting, which when identified can be used to make conservation planning more effective. Finally, I outline an example of this approach by applying it to the Northern Appalachian/Acadian ecoregion in eastern North America.

2.2 What Is a Threat, and What Is Threatened?

In many ways, the terms 'threat' and 'threatened' are value-laden and are viewed only in relation to what humans wish to protect. Conservation biologists are generally concerned with native biological diversity and generally view human activities to be potential threats to the integrity of natural systems. Yet, the field of conservation biology increasingly recognizes that some human activities – particularly the low-intensity practices of pre-industrial indigenous cultures – may enhance or maintain biological diversity (Nabhan et al. 2002). For the purpose of this chapter, a 'threat' is a human activity that has driven ecological processes beyond the range of natural variability to such a level and scale that the system is unlikely to recover its form and function within a timescale that is relevant to that system. In other words, a threat occurs at a location when human activities have become the dominant force shaping ecological and evolutionary processes there.

The impacts of human activities on biological diversity are scale-dependent and interwoven with characteristics of the life histories of individual species. Species with narrow geographical ranges and specific habitat requirements are much more vulnerable to extinction through isolated local extirpations than are species with extensive geographic ranges and broader habitat requirements. On the other hand, species that as individuals use a great deal of space tend to be longer lived, have lower reproductive rates, and are thus quite vulnerable to overexploitation, large-scale fragmentation, and other factors that might not influence a localized species to the same extent. If conservation planners understand the life histories of the species upon which their efforts are focused, and the spatial and temporal scales at which the population and community processes are carried out, planning will be more likely to succeed. For example, pool-breeding amphibians carry out their life histories on a local scale, such that the land-use decisions of a single farmer, developer, or town planner could have a meaningful impact on population viability (Chap. 10). Alternatively, wide-ranging mammals (Chap. 9) are vulnerable to changes in landscape connectivity, habitat availability, and patterns of exploitation at much greater spatial scales. In this case, the decisions of a single town planner make little difference unless coordinated with neighboring towns, whereas a state governor or provincial premier is in a position to make meaningful policy decisions that influence these populations.

2.3 What Are the Threats?

Most organisms that are at risk from human actions are threatened by multiple, interacting factors that couple broad-scale processes such as climate change, atmospheric deposition of toxic chemicals, and changes to the atmosphere (e.g., increased UV-B radiation due to decreased stratospheric ozone) with local processes such as overexploitation, physical habitat destruction, and disease (Alford and Richards 1999; Cardillo et al. 2006; Muths et al. 2003). Many of these threats, however, are linked to underlying drivers, which when identified can be targeted for conservation action. This builds upon the keystone species concept in ecology – the idea that one or a few species have disproportionate effects on natural systems and if removed, may result in the loss of numerous interactions and a dramatic restructuring of the system (Mills et al. 1993; Soulé et al. 2005). In the same manner, a keystone threat is one that is strongly interacting and if mitigated or removed could have cascading, positive effects for conservation. Threats can be grouped into major clusters, which when compared makes the case for change in land use/land cover as a keystone threat.

2.3.1 Change in Land Use/Land Cover

Physical habitat alteration as a result of the disturbance of the earth's surface and loss of natural cover is currently the single-most acute factor threatening biological diversity worldwide (Hunter and Gibbs 2007). This is for the simple reason that native diversity depends on structural and compositional diversity of habitat. This diversity is the result of eons of ecological and evolutionary change, processes that cannot produce new species as rapidly as humans can fragment and destroy natural land cover. The Global Human Footprint – a multivariate index of human influence on terrestrial systems (Chap. 13) – estimates that more than three-quarters of the

Earth's surface feels a human impact measureable at a 1-km resolution (Sanderson et al. 2002), a figure that is comparable to what is seen at even a much finer 90-m resolution (Woolmer et al. 2008). These indices of impact are actually conservative because they do not include consideration of major changes to aquatic, marine, or atmospheric systems. For example, atmospheric mercury deposition has been shown to influence a wide range of species that occupy terrestrial systems (Driscoll et al. 2007; Evers et al. 2007).

Changes in land use/land cover include such factors as direct conversion of forests and grasslands to agriculture, mining, proliferation of early succession habitats at the expense of late succession habitats, construction of transportation networks, expansion of dispersed human settlements, and growth of urban areas. Generally, the more human infrastructure is present in a landscape, the more that the ecological processes found there are under human control and the less habitat is available for native biological diversity (Theobald 2004).

The ability of people to access erstwhile remote areas drastically increases the likelihood that such areas will undergo changes in land use/land cover and associated anthropogenic ecological change (Laurance et al. 2001). Transportation networks (e.g., roads, railways, ATV trails, and, in many parts of the world, waterways) provide access for people and avenues for settlement and resource extraction. Hunter and Gibbs (2007) synthesize the work of Forman (1995) and others to posit access as a key component of the first of four stages of land use/land cover change – dissection, perforation, fragmentation, and attrition. Dissection facilitates access because it frequently occurs with road building. A road network is by design a means to access and move resources. As such, roads increase the overexploitation of natural resources. They also cause direct mortality and behavioral avoidance in wildlife populations, which results in population fragmentation (Hitchings and Beebee 1998; Kramer-Schadt et al. 2004; Reh and Seitz 1990).

It would be tempting to think that massive road building projects are a thing of the past. However, the Amazon Basin of Brazil is expected to receive over 6,000 km of new, paved roads in the coming decade, which will cause between 100,000 and 300,000 km^2 of additional deforestation (Carvalho et al. 2001). Even in long-settled landscapes – for example, the Northern Appalachians of the U.S. and Canada – road building is an on-going process. Over 2,000 new kilometers of road were built over a 17-year period in just one state – Maine – and it has been estimated that regular, public roads – especially those that provide access to subdivisions – in the ecoregion as a whole will double in the coming 2 decades (Baldwin et al. 2007).

To assess threats to biological diversity from changes in land use/land cover, it is helpful to examine multiple processes and patterns by which such changes arise, and in turn, how these conditions have varying effects on biological diversity. These processes and patterns can be grouped into two categories based on whether they are disaggregated (cumulative and incremental changes in land use/land cover on local scales) or aggregated (rapid land use/land cover change on regional scales) in nature.

Disaggregated Changes Local land-use decisions (e.g., roads and subdivisions) accumulate over time and space to have effects at regional scales (Theobald et al. 1997). Economists studying changes in land use/land cover refer to decisions that cause instantaneous changes to large swaths of land as aggregated (Bell and Irwin 2002). Such decisions are often made by individual landowners who manage large areas (i.e., tens of thousands of hectares), such as private companies with extensive land holdings or governments. By contrast, land-use decisions that are local (e.g., made by landowners operating at more local scales, such as a 100-ha family farm) are disaggregated yet can still accumulate over time and space to have region-wide effects. Disaggregated decisions operate on the hedonic principle: a given local landowner is said generally to operate in such a way as to maximize economic return on the land, and once current use returns diminish in relation to anticipated returns from future uses, a landowner sells to new uses (e.g., a farmer would sell land for a housing development if his or her neighbors have done so, driving up both taxes and land prices) (Bell and Irwin 2002; Irwin et al. 2003).

Most changes in rural land use/land cover over the last half-century can be attributed to disaggregated processes, accumulating over ever-increasing spatial scales. Approximately 22% of cropland in the U.S. has been lost since 1950; at the same time, exurban areas increased from 5% to 25% (Brown et al. 2005). Much of this conversion resulted from farms being sold, one by one, and transformed to housing developments (Bell and Irwin 2002). Eventually, entire regions become exurban-urban complexes, as envisioned more than 40 years ago by John McPhee, who predicted in *The Pine Barrens* (1968) that the Eastern U.S. from Richmond, VA, to Boston, MA, would become a single 'suburban-industrial corridor.'

Aggregated changes Many institutions own great swaths of natural habitat. Given that only about 13% of the land surface globally is covered by protected lands owned or managed by governments, the vast majority of the Earth's surface is in private hands. Thus, decisions made by large public and/or private institutions can have rapid effects on extensive swaths of land and water (Chap. 5). These aggregated decisions may include the development of new infrastructure, such as dams, industry, mines, industrial agriculture, electrical power lines, and roads. Changes in global markets for raw materials and land values put into place a series of economic changes that can result in the restructuring of many companies so that they can manage for both extractive resources (e.g., timber, cattle, minerals) and for development of high-value amenities (e.g., lakeshores, coastlines, ski areas). Traded publicly as Real Estate Investment Trusts (REIT's), these companies have developed plans for resorts and service communities in remote, lightly settled (i.e., 'wilderness') landscapes (Hagan et al. 2005). In so doing, REIT's place extensive new human infrastructure in remote areas, allowing for new, cumulative development around these new 'nodes' and more disaggregated change (see above). For example, Plum Creek Company in the Moosehead Lake region of Maine proposed a sweeping new development of timberlands that would transform use on approximately 160,000 ha (Austin 2005). Many such projects are viewed by conservation groups as strategic opportunities because they can negotiate a range of conservation prescriptions for the land, including short- and long-term easements, timber management agreements, and specific protections for biological diversity (Babbit 2005; Ginn 2005; Milder 2007).

Aggregated and disaggregated changes in land use/land cover are linked together in that an aggregated decision may produce a new growth node, which can lead to disaggregated growth in an otherwise lightly settled area. For example, a company may build a resort on private forestland and new growth patterns will emerge around that resort as the existing land uses (e.g., timber harvesting and hunting) become less profitable than the new ones (e.g., strip malls, restaurants, and other businesses that service the resort). Once a new land-use pattern is created for an area, the old uses rapidly diminish and are eventually replaced by new ones. Alternatively, disaggregated processes may set the stage for a nearby aggregated decision; for example, exurban growth will put pressure on adjacent landowners within a certain driving distance to shift to amenity development (Schnaiberg et al. 2002).

2.3.2 Pollutants

Probably no part of the planet remains free from human influence simply because byproducts of human industry reach every known place on Earth. Airborne pollutants, including mercury, acids, and radiation, travel thousands of kilometers away from their industrial sources (Anspaugh et al. 1988; Driscoll et al. 2001; Evers et al. 2007). Although ground-level (tropospheric) ozone (O_3) most severely affects the local area where it is produced, it can still cause damage hundreds of kilometers from its urban sources. Much mercury-laden and acidic precipitation in the Northern Appalachians originates in the coal-fired power plants of the Midwestern U.S. (Driscoll et al. 2001, 2007). Even relatively small predators, such as salamanders and brook trout, have shown levels of mercury accumulation sufficient to damage their nervous systems (Bank et al. 2005). Ground-level ozone originating from adjacent urban areas affects trees in protected areas (Karnosky et al. 2006). Radiation derived from the 1986 disaster at the Chernobyl Nuclear Power Plant at levels sufficient to influence human health was recorded throughout the Northern Hemisphere, including North America (Anspaugh et al. 1988). Organochlorines (e.g., PCB's and dioxin) and other chemical discharges accumulate up to hundreds of kilometers downstream from industrial sources and both bioaccumulate and biomagnify in organisms at higher tropic levels (Champoux 1996; Muir et al. 1996). Given these pervasive effects, one of the primary endeavors for conservation of biological diversity continues to be continued identification and mitigation of chemical risks.

In terms of assessing threats, pollutants are best viewed as one of many interacting factors. Compared to protecting physical habitat to support wildlife populations, pollutants are frequently a secondary concern for the conservation of biological diversity. This is not always the case, however, as pollutants can cause 'empty habitats.' Perhaps the most obvious North American example is the effects of the pesticide DDT on raptor populations in the second half of the twentieth century. Without the removal of DDT from the food web, populations of Bald Eagles (*Haliaeetus leucocephalus*), Osprey (*Pandion haliaetus*), and other predatory birds would likely have never recovered. In the 1970s, these populations

2 Identifying Keystone Threats to Biological Diversity 23

were seriously depleted and listed under the U.S. Endangered Species Act; by the 1990s many had recovered to such a level that they could be delisted, owing largely to aggressive conservation efforts and the banning of DDT (Bowerman et al. 1998). As an interacting factor, pollutants can weaken wildlife populations to the point where they are less resilient to environmental perturbations (Sparling et al. 2003). The best approach to addressing these threats may be to continue – through the policy process – to regulate toxins and implement cleaner technologies while continuing to focus conservation dollars on the protection and restoration of physical habitat and ecosystem processes.

2.3.3 Invasive Species

Species invasions are not new – range expansions and contractions are an integral part of the history of biological diversity. Invasions that have occurred in current times can, from one perspective, simply be considered anthropogenic additions to the suite of evolutionary forces that a species faces (Sax et al. 2002). However, this detached view breaks down when one is concerned primarily with conservation of native biological diversity, as are most conservation biologists. Many species, when given the opportunity to establish themselves in new environments as a result of the breakdown of historical biogeographic barriers, rapidly expand their populations and displace causing the extirpation of native species (Mooney and Hobbs 2000). These effects are particularly pronounced on islands and disproportionately influence island fauna (Sax and Gaines 2008). However, many non-native species have become established on continents with serious negative consequences. In North America, kudzu (*Pueraria* spp.) in southeastern forest edges (Webster et al. 2006), tamarisk (*Tamarix* spp.) in southwestern riparian areas (Busch and Smith 1995), and Japanese barberry (*Berberis thunbergii*) (Silander and Klepeis 2004) in the northeastern deciduous forests occupy habitats in great densities, which change the structure and function of local communities to the detriment of native species.

While invasion is natural, human actions have greatly aided the spread of some species by facilitating dispersal, providing favored – often disturbed – habitats, culturing hosts, and eliminating predators and competitors (Mack et al. 2000). Because many other threats enhance or facilitate invasions, threat assessments need to consider the interacting effects of changes in land use/land cover, transportation networks, wildlands recreation, and other human activities in facilitating dispersal and establishment of populations of potentially destructive plants and animals. For example, many land-development projects replace native palustrine, semipermanent wetlands with more open, emergent, and permanent wetlands, thereby facilitating invasions of bullfrogs (*Rana catesbeiana*), cattails (*Typha* spp.), purple loosestrife (*Lythrum salicaria*), and phragmites (*Phragmites australis*), and displacing local populations of plants and animals (e.g., pool-breeding amphibians) dependent on forested or shrubby, more ephemeral wetland habitats (Boone et al. 2004; Marks et al. 1994; Mitsch and Gosselink 2000; Tiner 1998). Considering the far-reaching effects of such structural changes to wetlands and forests prior to granting permission for developments would greatly enhance local conservation efforts.

2.3.4 Disease

Threats from wildlife and plant diseases are often intensified by the same set of circumstances favoring other invasive species: changes in land use/land cover, human transportation networks, and increased wildlife-human contact. It is difficult to predict when and where a new disease will become introduced into a system, and even more difficult to predict its effects. At the most extreme, entire ecosystems can be transformed by the outbreak of a disease that affects a critical species. For example, a handful of tree pathogens, including chestnut blight (*Cryphonectria parasitica*), Dutch elm disease (*Ophiostoma ulmi*), and gypsy moth (*Lymantria dispar*), have in the past 100 years greatly influenced the structure and function of Eastern North American forests, and others exotic species promise impacts of equal severity (e.g., hemlock wooly adelgid; *Adelges tsugae*) (Lovett et al. 2006). Such threats arise suddenly and often spread rapidly, which makes assessment and conservation action virtually impossible.

However, diseases can have strong interacting effects with land use that can be mitigated by ecological practices that better emulate the natural scale and intensity of disturbance. For example, the extent and severity of the spruce budworm (a native insect; *Choristoneura fumiferana*) outbreaks in stands of balsam fir (*Abies balsamea*) during the 1970s and 1980s in the Northern Appalachians were thought to have been intensified by the even-aged management practices of preceding decades (Seymour 1992). Alternatively, other diseases appear to influence wildlife populations in the absence of obvious interacting factors, at least at the local scale. For example, the chytrid fungus that affects frogs (i.e., chytridiomycosis caused by *Batrachochytrium dendrobatidis*) can cause catastrophic population losses even in relatively pristine areas. Eradications of populations of the midwife toad (*Alytes* spp.) from 86% of the ponds in a protected area in Spain were blamed on the fungus (Bosch et al. 2001) as were declines of the boreal toad (*Bufo boreas*) in Rocky Mountain National Park, Colorado (Muths et al. 2003).

2.3.5 Climate Change

The Earth's climate has a record of dramatic change. As such, it is difficult to categorize it as a threat to biological diversity except in the context of human-induced climate change, which is most likely going to be rapid and result in catastrophic effects in particular ecosystems. An analysis by Thomas et al. (2004) suggests that between 15–37% of species will be committed to extinction due to climate by 2050.

Assessing the threat to biological diversity of changing climate over near time scales (e.g., the next 50–100 years) involves coupling climate models, that are in turn linked to emissions scenarios and circulation models, with particular habitats (Chap. 15). The key obstacles in making strong predictions are scale – both temporal and spatial – and model uncertainty. Currently, such predictions are limited to broad-scale change and/or organisms whose climate-habitat links are particularly well understood (Carroll 2007; Lawler et al. 2009). As climate predictions improve and are increasingly scaled to assess local effects on plant and animal communities, conservation planners will be able to design reserve systems that represent not only current diversity, but accommodate future range shifts. At the same time, climate and land use are likely to interact strongly with each other (Dale 1997; Vitousek 1994); these interactions may be mitigated in the absence of better projections on the consequence of climate change on biological diversity by focusing on traditional land-conservation projects.

2.4 Threat Assessment: an Ecoregional Example

An ecoregion is a useful scale to assess threats for conservation planning purposes. This is because the relative ecological homogeneity that generally characterizes an ecoregion also typically gives rise to particular human activities that give rise to threats, including modes of resource extraction, transportation networks, and agricultural practices. Even so, as focus for conservation widens to encompass larger areas, the landscape becomes more heterogeneous, and threats must increasingly be assessed at multiple scales. For example, the Northern Appalachian/Acadian ecoregion encompasses mountains, forests, coastlines, and temperate to boreal climates, and has a complex social geography including multiple national, provincial, and state jurisdictions. Several primary environmental gradients exist, including temperate-boreal (latitudinal), temperate-alpine (altitudinal), and coastal-inland. Climate change will influence movement along these gradients of plant and animal distributions over extended timescales. For example, pollen records show that red spruce (*Picea rubens*), a cold-adapted species typifying much of the Northern Appalachians today, had a distribution that was much more restricted to coastal refugia about 5,000 years ago, from which it has since expanded inland (Schauffler and Jacobson 2002).

At the macro-scale, the continental position of the Northern Appalachian/Acadian ecoregion exposes it to prevailing air masses from Midwestern North America, which deposit toxin-laden precipitation, including heavy metals (e.g., mercury) and acids, while also changing nutrient dynamics in forest ecosystems (Driscoll et al. 2001, 2007). While these air masses blanket large portions of the region, the deposition does not have uniform impacts (Evers et al. 2007). Hotspots of deposition and/or impacts may be mapped and used in conjunction with other mapped threats to help guide landscape conservation efforts.

Aldo Leopold observed that a 'conservationist is one who is humbly aware that with each stroke [of his axe] he is writing his signature on the face of the land'

(Leopold 1949). The Northern Appalachian/Acadian ecoregion exists as a patchwork of land-use histories, whose effects may be read upon the land today. Forest communities throughout much of the region are artifacts of agricultural and silvicultural practices that have interacted with natural successional processes and changing climate (Foster 1992; Lorimer 1977; Russell et al. 1993). Foster (1999) traced the influence of intensive farming practices of the early nineteenth century as observed by Henry David Thoreau on ecosystems of that time and how these have influenced the ecosystems of today. Red maple (*Acer rubrum*) has become one of the most abundant and widespread trees in Eastern North America following centuries of logging that favored the adaptable tree (Abrams 1998). Populations of many species of wildlife have waxed and waned with varying land uses and hunting and trapping activities. Populations of species that preferentially inhabit open land (e.g., Bobolink [*Dolichonyx oryzivorus*], Eastern Meadowlark [*Sturnella magna*]) increased considerably from pre-settlement levels to the peak of agriculture in the mid-1800s and then have declined with reforestation; large mammals have shown opposite trends and many are now increasing (Foster et al. 2002; Chap. 9). Such land-use legacies and their influences on species distributions are well studied in some areas (Foster et al. 2002; Scott 2005) and therefore should be included in landscape conservation plans (Foster et al. 2003).

2.4.1 *Northern Appalachian/Acadian Ecoregion Threat Assessment*

In the Northern Appalachian/Acadian ecoregion, which straddles the Northeastern U.S. and Southeastern Canada, a collaborative conservation effort (Two Countries, One Forest) began during the 2000s to assess threats in order to prioritize landscapes for conservation action (Chap. 1). As discussed above, a key component of conservation planning is to identify threats present and future. My colleagues and I examined the literature to identify the major threats themselves and assess which were 'keystone threats' and how their expansion over time could plausibly be modeled with available data. Our region was dominated by forest that in many areas was becoming rapidly converted to new uses and specifically new settlement. The primary keystone threat we identified that was most important ecologically and could be modeled was change in land use/land cover. As this chapter suggests (Sect. 2.3.1), changes in land use/land cover are the drivers for considerable degradation of biological diversity and strongly interact with many other threats, including climate, toxins, and invasive species (Dale 1997; Foster et al. 2003; Vitousek 1994). Further, given the availability of remotely sensed data on land use/land cover, this threat can be modeled with some level of confidence.

Changes in land use/land cover are disturbances that influence biological diversity at every spatial scale, from local to regional (Theobald 2004; Theobald et al. 1997). Thus, we needed an assessment tool that could function well at both ends of this spectrum. In a conservation planning sense, we defined 'local' to mean conservation actions that occur at the town and municipal level and 'regional' to be those processes influencing wide swaths of land at the regional (e.g., multistate) level. We decided to use the Human Footprint methodology (Chap. 13) because it can be rescaled and it incorporates multiple sources of land-cover transformation (Sanderson et al. 2002). However, rather than use the Global Human Footprint developed by Sanderson et al. (2002), which has a 1-km resolution, we developed a region Human Footprint map at a finer resolution (90 m) that could be used to assess threats at the local level and then aggregated upward to the regional level (Woolmer et al. 2008).

Because we wished to project threats into the future, we had to transform a static, temporally specific assessment of land-cover transformation (The Human Footprint) into something that was dynamic. Thus, we developed a modeling approach that would allow us to project the Human Footprint into the future by incorporating land-cover transformation into a set of plausible future scenarios we termed the Future Human Footprint (Chap. 13). Several options for projecting changes in land use/land cover exist (Baldwin and deMaynadier 2009; Bell and Irwin 2002; Theobald 2003). However, in researching the drivers for these threats, we found that because of the heterogeneous land use/land cover characteristics of this ecoregion (e.g., intensely settled regions interspersed with lightly settled and 'wild' regions), we needed an threat assessment tool that integrated both aggregated and disaggregated processes (Sect. 2.3.1). We learned that human settlement and roads account for greater than 90% of the variation in Human Footprint scores and thus were the most influential features in terms of their ecological effects (Woolmer et al. 2008). This led us to develop mechanisms for projecting the development of these features into the future (Baldwin et al. 2007, 2009), integrating cumulative processes (road growth and human population) with rapid, large scale change. The results of both these analyses are discussed further in Chap. 13.

Ultimately we learned that conservation planning benefits from a system of projecting threats into the future that recognizes the complex dynamics by which threats themselves arise. Both disaggregated and aggregated processes must be integrated into forecasts of land-use change. Also we learned that to a great degree, the processes by which threats arise and thus may be modeled are ecoregion-specific. Specifically, land-use histories, current trajectories of change, and change that could plausibly occur in the future arise from the particular geography of an ecoregion and influence how current and future threat models should be constructed.

2.5 Lessons Learned

To achieve landscape-scale conservation planning, it is critical to identify the keystone threats, those threats that are so strongly interacting that if they were removed, other threats would be considerably weakened. The exact characteristics of keystone threats are dependent on the nature of the landscapes within which they act: change in land use/land cover is a driver of much environmental transformation but may not always be the most important in a given place. For example, in relatively pristine areas, atmospheric deposition of toxins may be the keystone threat (Driscoll et al. 2007).

Threat assessment involves understanding a range of biogeochemical, socioeconomic, and ecological processes. Thus, it is an interdisciplinary exercise, and understanding how threats influence specific landscapes often requires an understanding of multiple fields. A threat operates via mechanisms that need to be understood whether they are chemical (e.g., toxics), socioeconomic (e.g., land-use change), or ecological (e.g., species invasions).

Furthermore, it is important to understand threats on a systems level: how they interact with each other to influence the ecosystems of interest. Change in land use/land cover has strong interactions with nearly every other anthropogenic activity; it is important to understand the pathways by which this happens, and the degree to which threats are amplified or reduced by interactions.

Because different focal species that may be used in the process of conservation planning respond differently to a threat, a thorough understanding of the behavior, ecology, and evolution of the species believed to be threatened is needed. A threat to one species may well benefit, at least partially, another. Furthermore, species differ in the scale at which they respond to threats (Chap. 17).

Landscape conservation involves many interacting levels of stakeholders, including scientists, landowners, environmentalists, government officials, and resource managers. Scientists who work on such projects want to take care to develop models that are transparent and easily communicated yet do not sacrifice information content. Systematic, repeatable threat assessment depends upon mathematical models and software that require technical expertise. At the same time, models should be clear enough to demonstrate both spatial and temporal aspects of threats to a variety of participants and end-users. GIS currently offers the best way to illustrate complex spatial models to broad audiences (Theobald et al. 2000; Chap. 12).

A promise of ecoregional conservation planning is that stakeholder groups can come together, learn what can be done locally, and act with shared vision. In such open, collaborative processes it is important to acknowledge one's biases and assumptions, especially when communicating particular value-laden terms such as 'threat.' Some stakeholders may actually be viewed as 'threat vectors' by conservationists because those stakeholders work in development, forestry, mining, recreation, or other industries. Care needs to be taken in explaining scientifically what is threatened and what particular human activities are causing the threat. Only through careful use of language can conservation planning take advantage of the considerable value that threat assessments bring to identifying effective strategies for achieving conservation goals.

References

Abrams, M. D. (1998). The red maple paradox. *BioScience, 48*, 355–364.

Alford, R. A., & Richards, S. J. (1999). Global amphibian declines: A problem in applied ecology. *Annual Review of Ecology and Systematics, 30*, 133–165.

Anspaugh, L. R., Catlin, R. J., & Goldman, M. (1988). The global impact of the Chernobyl reactor accident. *Science, 242*, 1513–1519.

Austin, P. (2005). *Plum Creek's big plan*. Hallowell, ME: Maine Environmental Policy Institute.

Babbit, B. (2005). *Cities in the wilderness: A new vision of land use in America*. Washington, DC: Island Press/Shearwater Books.

Baldwin, R. F., & de Maynadier, P. G. (2009). Assessing threats to pool-breeding amphibian habitat in an urbanizing landscape. *Biological Conservation, 142*, 1628–1638.

Baldwin, R. F., Trombulak, S. C., Anderson, M. G., & Woolmer, G. (2007). Projecting transition probabilities for regular public roads at the ecoregion scale: A Northern Appalachian/Acadian case study. *Landscape and Urban Planning, 80*, 404–411.

Baldwin, R. F., Trombulak, S. C., & Baldwin, E. D. (2009). Assessing risk of large-scale habitat conversion in lightly settled landscapes. *Landscape and Urban Planning, 91*, 219–225.

Bank, M. S., Loftin, C. S., & Jung, R. E. (2005). Mercury bioaccumulation in Northern two-lined salamanders from streams in the Northeastern United States. *Ecotoxicology, 14*, 181–191.

Bell, K. P., & Irwin, E. G. (2002). Spatially explicit micro-level modeling of land use change at the rural-urban interface. *Agricultural Economics, 27*, 217–232.

Boone, M. D., Little, E. E., & Semlitsch, R. D. (2004). Overwintered bullfrog tadpoles negatively affect salamanders and anurans in native amphibian communities. *Copeia, 2004*, 683–690.

Bosch, J., Martinez-Solano, I., & Garcia-Paris, M. (2001). Evidence of a chytrid fungus infection involved in the decline of the common midwife toad (*Alytes obstetricans*) in protected areas of central Spain. *Biological Conservation, 97*, 331–337.

Bowerman, W. W., Best, D. A., Grubb, T. G., Zimmerman, G. M., & Giesy, J. P. (1998). Trends of contaminants and effects in bald eagles of the Great Lakes basin. *Environmental Monitoring and Assessment, 53*, 197–212.

Brown, D. G., Johnson, K. M., Loveland, T. R., & Theobald, D. M. (2005). Rural land use trends in the conterminous United States, 1950–2000. *Ecological Applications, 15*, 1851–1863.

Busch, D. E., & Smith, S. D. (1995). Mechanisms associated with decline of woody species in riparian ecosystems of the Southwestern U.S. *Ecological Monographs, 65*, 347–370.

Cardillo, M., Mace, G. M., Gittleman, J. L., & Purvis, A. (2006). Latent extinction risk and the future battlegrounds of mammal conservation. *Proceedings of the National Academy of Sciences of the United States of America, 103*, 4157–4161.

Carroll, C. (2007). Interacting effects of climate change, landscape conversion, and harvest on carnivore populations at the range margin: Marten and lynx in the Northern Appalachians. *Conservation Biology, 21*, 1092–1104.

Carvalho, G., Barros, A. C., Moutinho, P., & Nepstad, D. (2001). Sensitive development could protect Amazonia instead of destroying it. *Nature, 409*, 131.

Champoux, L. (1996). PCBs, dioxins and furans in Hooded Merganser (*Lophodytes cucullatus*), Common Merganser (*Mergus merganser*) and mink (*Mustela vison*) collected along the St. Maurice River near La Tuque, Quebec. *Environmental Pollution, 92*, 147–153.

Dale, V. H. (1997). The relationship between land-use change and climate change. *Ecological Applications, 7*, 753–769.

Driscoll, C. T., Lawrence, G. B., Bulger, A. J., Butler, T. J., Cronan, C. S., Eagar, C., et al. (2001). Acidic deposition in the Northeastern United States: Sources and inputs, ecosystem effects, and management strategies. *BioScience, 51*, 180–198.

Driscoll, C. T., Han, Y. -J., Chen, C. Y., Evers, D. C., Lambert, K. F., Holsen, T. M., et al. (2007). Mercury contamination in forest and freshwater ecosystems in the northeastern United States. *BioScience, 57*, 17–28.

Evers, D. C., Han, Y. -J., Driscoll, C. T., Kamman, N. C., Goodale, M. W., Lambert, K. F., et al. (2007). Biological mercury hotspots in the Northeastern United States and southeastern Canada. *BioScience, 57*, 29–43.

Foster, D. R. (1992). Land-use history (1730–1990) and vegetation dynamics in central New England, USA. *Journal of Ecology, 80*, 753–772.

Foster, D. R. (1999). *Thoreau's country: Journey through a transformed landscape*. Cambridge, MA: Harvard University Press.

Foster, D. R., Motzkin, G., Bernardos, D., & Cardoza, J. (2002). Wildlife dynamics in the changing New England landscape. *Journal of Biogeography, 29*, 1337–1357.

Foster, D. R., Swanson, F. J., Aber, J. D., Burke, I., Brokaw, N., Tilman, D., et al. (2003). The importance of land-use legacies to ecology and conservation. *BioScience, 53*, 77–88.

Ginn, W. (2005). *Investing in nature: Case studies of land conservation in collaboration with business*. Washington, DC: Island Press.

Hagan, J. M., Irland, L. C., & Whitman, A. A. (2005). *Changing timberland ownership in the Northern Forest and implications for biodiversity*. Brunswick, ME: Forest Conservation Program, Manomet Center for Conservation Sciences.

Hitchings, S. P., & Beebee, T. J. C. (1998). Loss of genetic diversity and fitness in common toad (*Bufo bufo*) populations isolated by inimical habitat. *Journal of Evolutionary Biology, 11*, 269–283.

Hunter, M. L., & Gibbs, J. P. (2007). *Fundamentals of conservation biology*. Oxford: Blackwell.

Irwin, E. G., Bell, K. P., & Geoghegan, J. (2003). Modeling and managing urban growth at the rural-urban fringe: a parcel-level model of residential land use change. *Agricultural and Resource Economics Review, 32*, 83–102.

Karnosky, D. F., Skelly, J. M., Percy, K. E., & Chappelka, A. H. (2006). Perspectives regarding 50 years of research on effects of tropospheric ozone air pollution on US forests. *Environmental Pollution, 147*, 489–506.

Kramer-Schadt, S., Revilla, E., Wiegand, T., & Breitenmoser, U. (2004). Fragmented landscapes, road mortality and patch connectivity: Modeling influences on the dispersal of Eurasian lynx. *Journal of Applied Ecology, 41*, 711–723.

Laurance, W. F., Cochrane, M. A., Bergen, S., Fearnside, P. M., Delamonica, P., Barber, C., et al. (2001). The future of the Brazilian Amazon. *Science, 291*, 438–439.

Lawler, J. J., Shafer, S. L., White, D., Kareiva, P., Maurer, E. P., Blaustein, A. R., et al. (2009). Projected climate-induced faunal change in the Western Hemisphere. *Ecology, 90*, 588–597.

Leopold, A. (1949). *A Sand County almanac, and sketches here and there*. New York: Oxford University Press.

Lorimer, C. (1977). The presettlement forest and natural disturbance cycle of northeastern Maine. *Ecology, 58*, 139–148.

Lovett, G. M., Canham, C. D., Arthur, M. A., Weathers, K. C., & Fitzhugh, R. D. (2006). Forest ecosystem responses to exotic pests and pathogens in Eastern North America. *BioScience, 56*, 395–405.

Mack, R. N., Simberloff, D., Lonsdale, W. M., Evans, H., Clout, M., & Bazzaz, F. A. (2000). Biotic invasions: Causes, epidemiology, global consequences, and control. *Ecological Applications, 10*, 689–710.

Marks, M., Lapin, B., & Randall, J. (1994). Phragmites australis (*P. communis*): Threats, management, and monitoring. *Natural Areas Journal, 14*, 285–294.

Milder, J. C. (2007). A framework for understanding conservation development and its ecological implications. *BioScience, 57*, 757–768.

Mills, L. S., Soulé, M. E., & Doak, D. F. (1993). The keystone-species concept in ecology and conservation. *BioScience, 43*, 219–224.

Mitsch, W. J., & Gosselink, J. G. (2000). *Wetlands*. New York: Wiley.

Mooney, H. A., & Hobbs, R. J. (Eds.). (2000). *Invasive species in a changing world*. Washington, DC: Island Press.

Muir, D. C. G., Ford, C. A., Rosenberg, B., Nordstrom, R. J., Simon, M., & Beland, P. (1996). Persistent organochlorines in beluga whales (*Delphinapterus leucas*) from the St. Lawrence River estuary. 1. Concentrations and patterns of specific PCBs, chlorinated pesticides and polychlorinated dibenzo-p-dioxins and dibenzofurans. *Environmental Pollution, 93*, 219–234.

Muths, E., Corn, P. S., Pessier, A. P., & Green, D. E. (2003). Evidence for disease-related amphibian decline in Colorado. *Biological Conservation, 110*, 357–365.

Nabhan, G. P., Pynes, P., & Joe, T. (2002). Safeguarding species, languages, and cultures in the time of diversity loss: From the Colorado Plateau to global hotspots. *Annals of the Missouri Botanical Garden, 89*, 164–175.

Reh, W., & Seitz, A. (1990). The influence of land use on the genetic structure of populations of the common frog, *Rana temporaria. Biological Conservation, 54*, 239–249.

Russell, E. W. B., Davis, R. B., Anderson, R. S., Rhodes, T. E., & Anderson, D. S. (1993). Recent centuries of vegetational change in the glaciated north-eastern United States. *Journal of Ecology, 81*, 647–664.

Sanderson, E. W., Jaiteh, M., Levy, M. A., Redford, K. H., Wannebo, A. V., & Woolmer, G. (2002). The human footprint and the last of the wild. *BioScience, 52*, 891–904.

Sax, D. F., & Gaines, S. D. (2008). Species invasions and extinction: the future of native biodiversity on islands. *Proceedings of the National Academy of Sciences, 105*, 11490–11497.

Sax, D. F., Gaines, S. D., & Brown, J. H. (2002). Species invasions exceed extinctions on islands worldwide: A comparative study of plants and birds. *The American Naturalist, 160*, 766–783.

Schauffler, M., & Jacobson, G. L. (2002). Persistence of coastal spruce refugia during the Holocene in northern New England, USA, detected by stand-scale pollen stratigraphies. *Journal of Ecology, 90*, 235–250.

Schnaiberg, J., Riera, J., Turner, M. G., & Voss, P. R. (2002). Explaining human settlement patterns in a recreational lake district: Vilas County, Wisconsin, USA. *Environmental Management, 30*, 24–34.

Scott, M. C. (2005). Winners and losers among stream fishes in relation to land use legacies and urban development in the southeastern US. *Biological Conservation, 127*, 301–309.

Seymour, R. S. (1992). The red spruce-balsam fir forest of Maine: Evolution of silvicultural practice in response to stand development patterns and disturbances. In M. J. Kelty (Ed.), *The ecology and silviculture of mixed-species forests* (pp. 217–244). New York: Kluwer.

Silander, J. A., & Klepeis, D. M. (2004). The invasion ecology of Japanese Barberry (*Berberis thunbergii*) in the New England landscape. *Biological Invasions, 1*, 189–201.

Soulé, M. E., Estes, J. A., Miller, B., & Honnold, D. L. (2005). Strongly interacting species: conservation policy, management, and ethics. *BioScience, 55*, 168–176.

Sparling, D. W., Krest, S. K., & Linder, G. (2003). Multiple stressors and declining amphibian populations: An integrated analysis of cause-effect to support adaptive resource management. In G. Linder, S. K. Krest, & D. W. Sparling (Eds.), *Amphibian decline: An integrated analysis of multiple stressor effects* (pp. 1–7). Pensacola, FL: Society for Environmental Toxicology and Chemistry.

Theobald, D. M. (2003). Targeting conservation action through assessment of protection and exurban threats. *Conservation Biology, 17*, 1624–1637.

Theobald, D. M. (2004). Placing exurban land-use change in a human modification framework. *Frontiers in Ecology and the Environment, 2*, 139–144.

Theobald, D. M., Miller, J. R., & Hobbs, N. T. (1997). Estimating the cumulative effects of development on wildlife habitat. *Landscape and Urban Planning, 39*, 25–36.

Theobald, D. M., Hobbs, N. T., Bearly, T., Zack, J. A., Shenk, T., & Riebsame, W. E. (2000). Incorporating biological information in local land-use decision-making: Designing a system for conservation planning. *Landscape Ecology, 15*, 35–45.

Thomas, C. D., Cameron, A., Green, R. E., Bakkenes, M., Beaumont, L. J., Collingham, Y. C., et al. (2004). Extinction risk from climate change. *Nature, 427*, 145–148.

Tiner, R. W. (1998). *In search of swampland: a wetland sourcebook and field guide*. New Brunswick, NJ: Rutgers University Press.

Vitousek, P. M. (1994). Beyond global warming: Ecology and global change. *Ecology, 75,* 1861–1876.

Vitousek, P. M., Mooney, H. A., Lubchenko, J., & Melillo, J. M. (1997). Human domination of earth's ecosystems. *Science, 277,* 494–499.

Webster, C. R., Jenkins, M. A., & Jose, S. (2006). Woody invaders and the challenges they pose to forest ecosystems in the eastern United States. *Journal of Forestry, 104,* 366–374.

Wilson, E. O., & Peter, F. M. (1988). *Biodiversity.* Washington, DC: National Academy Press.

Woolmer, G., Trombulak, S. C., Ray, J. C., Doran, P. J., Anderson, M. G., Baldwin, R. F., et al. (2008). Rescaling the human footprint: A tool for conservation planning at an ecoregional scale. *Landscape and Urban Planning, 87,* 42–53.

Chapter 3
Why History Matters in Conservation Planning

Elizabeth Dennis Baldwin and Richard W. Judd

Abstract Temporal scale analysis is important to fully understand a place and the multigenerational connections that form the basis of local resident's reaction to any conservation plan. Environmental history and conservation social science, specifically qualitative methods are useful to uncover and reveal important information regarding the history of land use and place attachment in a particular region. This study used both tools with an embedded case study designed to examine an intense conflict related to a conservation initiative in the heart of the Appalachian/Acadian ecoregion. Primary data for this study came from interviews with 21 opinion leaders in the region. The data were explored using a three part conceptual framework; cultural memory, essentialized images and vernacular conservation. The findings revealed clear fixed points in time, cultural memory, that define the local narrative of place. Not knowing these may have caused undue conflict from misunderstanding between conservation planners and local residents. Evidence of essentialized images escalation of the conflict was found, and clear examples were found, that may have helped form a conservation initiative rooted in the vernacular of the place. Understanding these elements can lead to a better process and ultimately one that preserves the dignity of local residents while creating a resilient conservation plan.

Keywords Cultural memory • Environmental history • Qualitative methods • Temporal scale • Vernacular conservation

---

E.D. Baldwin (✉)
Parks, Recreation, and Tourism Department, Clemson University, 271A Lehotsky Hall, Clemson, SC 29634
e-mail: ebaldwn@clemson.edu

R.W. Judd
History Department, University of Maine, 100 Stevens Hall, Orono, ME 04469
e-mail: RichardJudd@umit.maine.edu

S.C. Trombulak and R.F. Baldwin (eds.), *Landscape-scale Conservation Planning*,
DOI 10.1007/978-90-481-9575-6_3, © Springer Science+Business Media B.V. 2010

3.1 Introduction

Conservation is more than a matter of protecting ecosystems: it involves cultural associations that give the land a sense of mystery, adventure, peace, tranquility, and beauty – associations produced by multi-generational memories of work and recreation. Understanding the story of conservation in a particular landscape requires one to develop or use tools to uncover the often hidden meanings of place and the historical narrative of the people in a particular place (Kruger 2001; Wagner 2002). This chapter attempts to answer the question of why environmental history and conservation social science matter in conservation planning, and further, why such histories must consciously consider the relevance of spatial scale. Conservation planning has increased in scale due to a need for a global perspective and scientific collaboration to maintain biological diversity and plan for large-scale changes from natural and anthropogenic causes. This increase in scale can create a contest over the meanings of place that will influence acceptance of conservation plans (Cheng et al. 2003). Conservation is ultimately a social act, and its success depends on understanding the connections that people have to landscapes at multiple scales that may span generations (Black et al. 1998; Marcucci 2000; Runte 1997).

The landscape that people live and spend time in builds their identity. The emotional bond people have with a landscape is often through particular places; a single tree, a trail, or a point of land. However, when scientists target a region for conservation action, they often focus on much larger scales; the ecological importance of the entire region, a grouping of habitats, or the range of an important species. These different scales of perception and time make communication between conservation scientists and local residents difficult if not impossible (Black et al. 1998). Even worse, a dismissal of these local connections to place can be interpreted as a dismissal of the people who have knitted them through time across landscapes (Schenk et al. 2007). This in turn can lead to the loss of dignity of the people living in a region and thus promote fear that can lead to irrational or conflictual actions (Berkes 2004). An in-depth understanding of the conservation history of any area should reveal connections and values useful in communication and collaboration at small scales that will in turn lead to a more resilient large-scale conservation reality (Foster et al. 2003).

The use of multiple scales, including the temporal, is complex and has been used by conservation planners in a variety of ways. Black et al. (1998), for example, used data on land-use history to identify areas of conflict between conservation and development in order to steer the search for solutions on a less volatile path. Foster et al. (2003) called environmental history 'an integral part of ecological science and conservation planning' by helping us understand land-use legacies and how they may express themselves in the future, reveal previously unseen cultural connections to natural areas, and reduce 'missteps' in conservation planning (Foster et al. 2003). Participation by anthropologists in conservation planning has been called for to better understand local communities and their social definition of conservation, as well as to build local partnerships to strengthen large-scale conservation efforts with small-scale incentives (Brosius and Russell 2003). This all takes a considerable number of people and amount of time to incorporate this type of qualitative data, and although efforts to quantify such incorporation have been made, specifically related to place attachment and meanings (Williams and Vaske 2003), some researchers have found this will not 'uncover' or 'reveal' hidden meanings that may determine the ultimate success of conservation planning (Kruger 2001; Schenk et al. 2007).

In 1994, a proposal for a new national park of 1.3 million hectares in Northern Maine, the heart of the Northern Appalachian/Acadian ecoregion, fostered extreme reactions from residents in the region and surrounding landscape due to an apparent lack of understanding of the local perspective by those making the proposal. An examination of this case provides an excellent example of why understanding the environmental history and the social meaning of place is an important step in conservation planning. The case that follows traces the environmental history of the proposal for a new national park, as well as reactions of opinion leaders in the region. The primary data for the case are drawn from interviews conducted with 21 opinion leaders in Maine reflecting a pluralistic set of values regarding conservation planning at the landscape level in Northern Maine. These interviews were used to gain insight into the complexity of the land-use dilemma facing Maine. These data were supplemented with document analysis and informal meetings with state and non-profit groups between July 2003 and January 2006.

Today, close to 6% of Maine's forestland is publicly owned, and state ecological reserves are a only small fraction of that total (Lansky 2001). However, the legacy of the large industrial landowners in Northern Maine has been one of quasi-public land (Irland 1999); although privately held, public access to any part of Northern Maine was guaranteed unless posted. During the 1980s, much of the land in Northern Maine went up for sale (Chap. 5), a sign that anyone with the money might own a piece of the 'North Woods of Maine' or the 'Maine North Woods,' the traditional names for the northern 50% of the state. Although much of the land that changed ownership was simply transferred among different pulp and paper industries, some was also sold to private individuals, some for business investments, and others for conservation goals. Many of the new owners were not familiar with the long history of the traditional open access people enjoyed in the North Woods of Maine or did not care to accommodate it. For the first time in recent memory, Maine people began to feel restrictions on their access to the North Woods. This change, coupled with a depressed regional economy, created an opportunity for conservation advocates to participate in the debate about the future of the North Woods of Maine once again (Harper et al. 1990).

3.2 Methods

We used a qualitative case study approach for this research, which relied on both the environmental history of conservation in the region to discover the 'story,' as well as an examination of the motivations of different players and divergent meanings of the area. Tools from environmental history help one concentrate first on the collection of stories of place, as opposed to focusing on a problem to be solved (Cronon 1993). This starting point helps a researcher begin with an open mind regarding the actors. These stories, revealed through documents and interviews, can illuminate the context of a region at multiple scales, which may be critical in understanding a holistic narrative of place or the different ways people connect to and define themselves and their relationship to a particular place. National, regional, and local historical trends regarding these relationships can also be useful for developing this understanding and lead to more sensitivity on the part of the conservation planner and in turn lead to the building of trust from data sources, and thus increased validity of data.

The analysis of the case is organized in a conceptual framework of three themes: (1) cultural memory – fixed points in history of reference for people in a locale; (2) essentialized images – stereotypes built and supported for political power and gain; and (3) vernacular conservation – conservation design that includes the 'native' perceptions of place in its design. Integrating techniques to build a more holistic understanding of an area is an incredible challenge and may never be perfected, but working toward that end may lead to greater acceptance of conservation planning and, in turn, may help lead us out of the paradigm of seeing people in a region solely as a threat to conservation instead of as partners for achieving it (Brosius and Russell 2003; Marcucci 2000; Schenk et al. 2007).

3.3 Environmental History: A Modern-Day National Park Proposal in a Mostly Privately Owned Forest Landscape

In 1994, the newly formed environmental advocacy group RESTORE: the North Woods (RESTORE) proposed a 1.3 million-hectare national park in Northern Maine's mostly industrial forest of nearly 5 million hectares (Irland 1999). The proposed Maine Woods National Park and Preserve was based on an area proposed for protection in the late 1980s by the Wilderness Society (Watkins 1988). Increased clearcutting in Maine during the 1980s and large land sales created a fever of anxiety about the future of Maine's forests (Rolde 2001).

Land protection often generates conflict because it challenges the values and associations people have about the land. Popular associations are sometimes contradictory, involving assumptions about wood and wood fiber, hydropower, mass recreation, or wilderness, but they nevertheless are tangible cultural attachments that must be recognized if conservation efforts are to succeed. In an effort to mitigate conflict, models of compromise have been developed, particularly multiple-use management and large-scale conservation easements (Rondinini et al. 2005). Yet there have been criticisms of both of these models in their effort to be a win–win solution to the conflict of land protection and socio-economic uses of the land (Merenlender et al. 2004; Pidot 2003; Trombulak 2003).

3 Why History Matters in Conservation Planning   37

Maine is a rural state, dependant on a natural resource-based economy, and faces challenges of conservation in a mostly worked, humanized, and private landscape that can offer national lessons about new models of conservation (Judd 2003). What follows is a description of the setting of the park proposal, the opportunity perceived by park advocates, a description of the conflict that ensued, and the context for the conflict at multiple scales.

3.3.1 Setting

The state of Maine covers nearly 8 million hectares, and the North Woods of Maine is a little more than half that size. This northern half of Maine is also called the 'unorganized territories' and is managed by the state's Land Use Regulation Commission (LURC). It is mostly private land, which has historically been managed for timber and later pulp and paper production. It includes one large (80,000 ha) state park, Baxter, surrounding the highest point in Maine, Mt. Katahdin.

European settlers moved to Northern Maine in the mid-1800s (Barringer 1993). Most early timber harvesting was done in the southern half of the state and along river corridors of the north (Irland 1999). Harvesting of single trees was the trend in the early years, with harvesting conducted in the winter. It was not until the 1920s and 1930s that machines allowed harvesting to occur year-round in Maine (Rolde 2001). Chainsaws came into use in the 1940s, and the combination of the skidders and chainsaws led to early road building. The last log drive down the Penobscot River took place in 1975. It was during this period (the late 1970s and early 1980s) that extensive clearcutting, road building, and herbicide spraying occurred as a consequence of technological advances and an outbreak of the native spruce budworm (*Choristoneura* spp.). These practices began to raise an alarm with environmentalists (Lansky 1991) just as the state's new Department of Conservation took over the forestry division, and political power began to shift to southern Maine (Rolde 2001). With their attention now on the North Woods of Maine, the public got its first views of the results of heavy cutting from films like the *Paper Plantation* and presses like the *Maine Times* and the *Northern Forest Forum*. The first of several large land sales and layoffs in the paper industry started shortly after this in the mid-1980s, and this gave the conservation community the idea that if these lands were for sale then the time was ripe for a new plan for how they could be managed (Klyza and Trombulak 1994).

A unique aspect of Maine's people is that they know their land-use history. They may not know all the details, but the legacy of the landscape providing a livelihood and recreation are part of the psyche of Mainers (Judd 1997). Natural resource issues find themselves on the front page of the local newspapers daily. Mainers are adamant about local control of their natural resources and fear any loss of this to outside interests of any kind. The 'outside' lumber, timber, and pulp and paper companies caused concern just before the turn of the last century but are no longer considered by most as 'outsiders' (Bennett 2001; Irland 1999). Maine has a group of experienced professional outdoor guides licensed by the state, called Maine Guides. One such guide, who prefers to remain anonymous, said 'by a happy coincidence of history the industrial ownership has been good for Maine people for the past 100 years, but that is all ending, and people just don't want to see this.' Another said, 'They (Maine people) have a fear of the big system collapsing on them and yet they worship it; we come from a culture of victimhood, and you can't change that.'

3.3.2 Opportunity

In 1988, the Wilderness Society, after surveying the lower 48 states of the U.S., chose three places where they believed opportunities remained to protect or restore natural ecological coherence (Watkins 1988): the Greater Yellowstone Ecosystem, the Southern Appalachians, and the North Woods of Maine. The first two places were already predominately in public ownership, but Maine was mostly privately owned. After this call for protection, the Wilderness Society opened an office in Maine to study both the potential for a reserve and the array of conservation tools available to protect this area from perceived conservation threats.

In 1988, after an article in their magazine (Watkins 1988), the Wilderness Society began to investigate options for protecting the North Woods of Maine. In March 1989, they produced a report called 'A New Maine Woods Reserve: Options for Protecting Maine's Northern Wildlands' (Kellett 1989). This report identified over-harvesting and large land sales as key threats to the region. With the premise that the future of the North Woods of Maine was in jeopardy without a bold vision or a comprehensive conservation plan, the Wilderness Society called for immediate action, beginning with further research into the complexity of the issue and the best options to bring about a solution to protect their identified 1.1 million hectares.

Michael Kellett, the author of the 1989 report, left the Wilderness Society in 1991, and in 1992 he founded RESTORE: the North Woods. He began working immediately on a larger reserve or 'green line' area of 1.3 million hectares (similar to the 1.1 million hectare area identified in the Wilderness Society plan) that encompassed Baxter State Park. Kellett traveled to towns and schools with Jym St. Pierre, formerly of Maine's Land Use Regulatory Commission, to promote this concept. During these meetings people seemed to be confused by the 'green line' concept. Over the months that followed, RESTORE refined their proposal to its current form, the Maine Woods National Park and Preserve.

In 1994, the proposal outlined five proposed outcomes of a new national park and preserve in Maine (Kellett 2000):

1. Restore and protect the ecology of the Maine North Woods.
2. Guarantee access to a true Maine North Woods wilderness experience.
3. Interpret Maine's cultural heritage.
4. Anchor a healthy economy in Northern Maine.
5. Raise national awareness of the Maine North Woods.

Our research used the five proposed outcomes as a basis for interviews with decision leaders in the state in order to explore both the understanding of the goals as well as the perception of the process that RESTORE followed to achieve these goals.

3.3.3 Conflict

The 1994 proposal came in the midst of great controversy about the future of the North Woods of Maine. During this time, other groups launched their own visions for the area: the Northern Forest Alliance rolled out their list of important areas, The Nature Conservancy launched their plan of large-scale easements on industrial forestland, and the Forest Society of Maine began purchasing large easements in the Northern Forest and on the West Branch of the Penobscot River through the National Forest Legacy program (part of the U.S. Forest Service). Initiatives were launched by National and Maine Audubon, the Sierra Club, and the Natural Resource Council of Maine as well as other smaller groups with an interest or perceived stake in the North Woods of Maine.

The RESTORE proposal was described as having 'shaken the region' (Rolde 2001). For the first time, the environmental community could imagine some level of control of part of the North Woods of Maine. Proponents of the proposal believed that a national park was the best way to protect natural values, while providing a desperately needed economic surge and diversification. The idea of a national forest was discarded early in the planning because of the road building and harvesting that occurs in national forests. The North Woods of Maine is an area where road building has been on the rise since the 1970s, when heavier equipment made logging inland possible and roads replaced rivers as the method for timber transport (Irland 2000). However, it is still an area with relatively low road density compared to the rest of the Northern Forest landscape (Ritters and Wickham 2003).

Four excerpts from the *Bangor Daily News Letters* section of the paper reveal some of the issues and tensions expressed in public opinion.

National Park Potential, from the *Bangor Daily News*, June 6, 2003

Walter Plaunt Jr. (BDN letter, May 26) seems to think that because he is against a Maine Woods National Park everyone else in Maine thinks the same way. If Plaunt would leave Trescott Township long enough, he might discover that many citizens think a national park would be of great value to this state....This is particularly true of the area around Millinockett. The only thing keeping that area viable is Baxter State Park. A National Park encompassing northwestern Maine would be a shot in the arm for this region. It might even inspire the state and federal government to extend Interstate 95 to Fort Fairfield.-John Blaisdell, Bangor.

Legal Land Transaction, From the *Bangor Daily News*, July 7, 2003

Do the no-park protesters really believe that groups that want a national park to be established in Northern Maine will steal the required 3.2 million acres from their current owners? If these groups obtain the necessary acreage, it will likely be perfectly legal, through willing-seller and eager-buyer transactions. Since we can't tell the landowners to whom they can sell their lands, the worst thing that could happen is they would sell it to developers…and it will be posted…Those who cling to the dream that the current situation will forever remain the same could be disappointed. While the current landowners are generous in the area of public access, though for a fee, who says future landowners will feel obligated to follow this tradition?-Irvin Dube, Madawaska.

Who Needs Devastation? From the *Bangor Daily News*, July 22, 2003

A feasibility study is required before Congress can establish the proposed 3.2 million-acre North Woods National Park….Once a feasibility study is in hand, Congress could establish a park here as soon as the political climate is favorable….in spite of opposition from the public, the state government, and the state's congressional delegation….In actuality, such decisions are made [access, logging permits, snowmobile use and hunting] by federal park staff officials, Washington legislators and by the environmental organizations with the money and political clout to influence both groups….A North Woods National Park would be financially devastating for Maine.-William J. Peet, Harfords Point.

Many park opponents, from the *Bangor Daily News*, May 31, 2004

I read the editorial 'Conservation Conversation' (BDN, April 29), calling for an economic study of the north woods economy, including the possibility of a new national park. Contrary to the BDN's ending comments, the supporters and opponents of a park have not found 'common ground' around this issue…The Maine Woods Coalition was formed more than 3 years ago with the primary purpose of stopping the park. A thinly veiled study that would include the park possibility is of no interest to those of us who live in the area of its impact. A serious study that would look at the Northern Maine economy in a comprehensive manner and build on our existing strengths and opportunities should be further discussed.-Eugene J. Conlogue, Chairman Maine Woods Coalition Steering Committee, Millinocket.

3.3.4 Context

Opposition to national parks has a history as long as that of parks themselves. Gifford Pinchot wanted the first national parks to be open for timber harvesting, and he battled with John Muir to keep preservation values out of the public estate (Nash 2001). History usually presents park detractors as materialistic, and those in favor are usually characterized as forward-thinking (Hampton 1981). Hampton (1981) also noted that 'Both sides in the many specific controversies based their positions upon identifiable values that – despite changes in social and economic factors – have remained fairly constant over the last century. Both have relied upon polemics and propaganda, and both have appealed to arguments and values that are strikingly similar.' This dualism between utilitarian and preservation agendas related to land use is a two-century-old debate in the U.S.

An economic argument has often been made for the establishment of national parks on private lands (Pierce 2000). Pierce (2000) explains that the peak of the timber removal in the Southern Appalachians, and the following decline of the timber industry, led advocates for Great Smoky Mountains Park to extol the financial success of the western parks as a remedy for their rural region tied to its natural resources. The Great Smoky Mountains became a park in 1938 but not until after intense debates that lasted for 40 years and after the park service allowed for some lifetime leases of land inside the park boundaries. However, access to hunting and other extractive practices were now under the regulation of the federal government (Runte 1997).

Concerns about the RESTORE proposal have their roots in a long tradition of state sovereignty and anti-federalism that at times has become very strident. The perception of the federal government as a threat to local sovereignty has again complicated Maine preservationist policy. This is illustrated by events in Maine's land conservation history. The first example is the 1911 passage of the Weeks Act that set in motion the federal purchase of eastern forests. There was resistance to this in Northern Maine. Later, in 1931, 'when Congress proposed federal acquisition of tax-delinquent timberlands for a national forest in Maine, as was occurring throughout the eastern United States, Maine declined to be part of the plan. In fact, the proposal was so unpopular that no state legislator would sponsor an enabling bill' (Judd and Beach 2003).

A second example is the resistance to the number of attempts throughout Maine's history to create a national park in the heart of the North Woods of Maine. Although the Millinocket town council did support a plan for a Roosevelt National Park in the current proposal area, World War II derailed this proposal and it did not move forward (Rolde 2001). Probably the most well-supported initiative was the 1937 proposal for a Katahdin National Park in the area that is today Baxter State Park (National Park Service 1937). The federal government supported a feasibility study of the area, but it did not get congressional support, and many worried that inviting too many people to the North Woods of Maine could change its character forever (Irland 1999). Additionally, the authors of the report did not all agree on a national park designation. The Branch of Forestry representative, John F. Shanklin, supported instead a national monument, citing legislation that stated that a national park is land 'essentially in primeval condition,' and noting the evidence of human use on the landscape (National Park Service 1937). Percival Baxter, past governor of Maine, had his own plan for the region, which he began working on in 1931 (Rolde 2001). He eventually bought land and deeded it to the State of Maine for an 80,000-ha state park with a clear mandate and management structure.

A third example of a federal initiative, ultimately turned over to the state, is the Allagash Wilderness Waterway. The plan to build a dam and flood the Allagash Valley brought to a head the debate about the future of this wild river (Judd and Beach 2003). The ideas for protection included a national park and a river protection corridor managed by the state. Preservation groups and industry landowners joined forces in opposing federal designation, citing the increase in outside visitors that would bring about more development and increase the tax base for industry landowners. They and some state officials promoted the idea of a 'working wilderness' (Judd and Beach 2003; Rolde 2001). The waterway was established in 1966 by the Maine legislature, and in 1970 it became the first state-managed unit of the Wild and Scenic River System (Judd and Beach 2003; Maine Bureau of Parks and Lands 2005; Rolde 2001).

Beyond the anti-federalism, it is also useful to specifically examine the perception of the concept of 'wilderness' as presented in the original Wilderness Act for three distinct reasons. First, unlike classic western 'wilderness,' most of Maine's 'wilderness' is privately owned but with legal traditions that secure public access dating from the early colonial period and with the added understanding that landowners make these access concessions so that the state will not employ eminent domain to ensure public access rights. These traditions have complicated the preservation debate enormously in Maine.

Second, there is no pretense of 'purity' in Maine wilderness; these lands have become part of a traditional working rural landscape, and they have been shaped and reshaped by cultural and economic transformations like changing wood markets, agricultural decline, and a growing appreciation for the spiritual and recreational significance of wilderness landscapes. Wilderness is a viable tradition in Maine but under a much different guise than manifested in Western North America. Western wilderness involves vast natural ecosystems that are visually and culturally perceived as devoid of almost all human impact. Maine has no such 'pristine' environments; nor does ecological succession fit the Western wilderness ideal, where severe climate, altitude, and competition for soil moisture create open, park-like forests of relatively stable composition: forest succession in Maine is 'messy,' since the forest is so much more dynamic (Seymour et al. 2002). These considerations again complicate the debate over preservation.

Third, the North Woods of Maine is proximate to some of the most urbanized portions of North America, and this has enhanced its cultural significance and sharpened the political debate over its use and preservation. In contrast, Western wilderness is typically very remote from urban areas and abstract. The thinking about the North Woods of Maine has been shaped subtly by a century of urban wilderness fantasies – portrayed in volume after volume of travel-adventure books and tourist literature. Thus, the North Woods of Maine have been a cultural icon at least since the mid-nineteenth century Romantic era and the advent of tourism as an industry in the Northeastern U.S. For this reason, it is a natural feature with immense cultural significance not only for those who live nearby, but for the entire region. Here at the interface of two vastly different value systems – rural and urban – debate over forest use and preservation is a matter of wildly conflicting expectations.

The RESTORE approach – wholesale, blanket preservation – challenged a history of low-keyed conservation policy in Maine that began with the arrival of the paper industry and the portable sawmill in the 1880s. These developments touched off a long (and continuing) debate among Maine people about climate and watershed effects, stream flow, fish and game conservation, visual scars, the maintenance of small local woodworking mills, forest fires, and the fate of the tourist industry. In short, Maine harbored a tradition of subdued conservation consciousness that was predicated on state and private initiatives, small-scale conservation projects, pressure from women's clubs and fish and game associations, and subtle adjustments through year-by-year legislative acts, beginning with the 1909 Maine Forestry District. Most of this effort was premised on the idea that wildlands would be left at least to some degree in the hands of private owners. How much of this old conservation legacy remains is difficult to say, but it does need to be acknowledged in present-day policy debate: Maine people are not averse to conservation initiatives, but they are not very expansive in their thinking about it.

3.4 Results and Discussion

The environmental history explored in the previous section at both state and regional scales can now be used as the context for understanding the interview data. As described above, the interview results will be explored through the conceptual framework of three themes: cultural memory, essentialized images, and vernacular conservation. Qualitative data analysis built on an understanding of the stories of the region is helpful in interpreting motivations and will lead to greater understanding of the interwoven cultural and natural context for a more lasting and relevant conservation plan to be built upon. William Cronon emphasizes that all human history has a natural context, neither nature nor culture is static, and all environmental knowledge is culturally constructed and historically contingent (Cronon 1993). The findings are, for the most part, critiques of RESTORE. However, there has been recognition of their role as a catalyst for the discussion that is now on the table: the future of conservation in Northern Maine and the entire Northern Appalachian/Acadian ecoregion.

3.4.1 Cultural Memory

Cultural memory has to do with the memory people in a community have of events and fixed points in time that define ways of knowing. It has been defined as collective memory based on fixed events that define behaviors; it is repeated through generations and falls outside of everyday memory (Assmann and Czaplicka 1995). Taken as an a priori set of knowledge claims in a community or region, the cultural memory of any place or collection of places is important to understand in order to develop any type of conservation initiative. Exploring the history of any region in depth and the people living there can reveal these fixed points in time that define later reactions to policies and events. This can be done with both documents and interviews, looking for stories of important events that reveal the collective identity people share with each other in a region.

As described earlier, events in Maine have shaped a fear of the federal government, a more utilitarian view of conservation as well as suspicion of all outsiders. Reasons for this may be Maine's geographic isolation at a large scale; Maine is large enough to have had its own economy based on private lumber corporations, and the people chose to shape policy and sentiment to support their businesses. The interviewees strongly perceived that RESTORE did not fully understand Mainers' fear of the federal government and other historical factors that created the firestorm around the proposal. Another perception was that RESTORE knew the history and chose to ignore it in their urgency to ensure land protection. Many felt that if this was the case, RESTORE underestimated the resolve of Mainers to fight outside control of their region and lifestyles. The following quote illustrates this point.

> But you see people's perceptions about a lot of issues, at least the people here. They have some of these things in their minds, arguments about snowmobiles in parks, debates about whether there should be more motorized access. Even proposals that have been made to ban motors from Allagash Lake or to ban flying in and out of Allagash Lake because of the violation of the solitude of the people that work there, that's out there. That goes back a long time, it has been heard of, but the point is that people who then address this new issue of a park from this area have that stuff in the back of their mind. They remember. So they are going to have a view or they are going to have some distrust of other kinds, this brings in baggage for them, whereas...an eager, young kind of pro-national park activist who shows up out of who knows where, they don't even know about it...in a way they haven't interacted with the community, they haven't figured out what the culture contains, what experiences are out there that relate to this, and they have something to do with how people react to some of this stuff.

This person was referring to the controversy over a request to allow snowmobiles in Baxter State Park, which was eventually denied. Local people have wanted the restrictions in that park to ease. This was given as an example of the history and identity of the region the interviewee believed was ignored by RESTORE. There was also a sense of grief that came out in many of the interviews; grief for the loss of a life that is changing at record speed and what that means to local people in Northern Maine. The following quote refers to the ownership of large tracts of land by philanthropist Roxanne Quimby and her decision to make her land open only to non-motorized recreation, a source of great tension.

> ...But in that deeper rural Maine public consciousness, one could enjoy the fishing, the hunting, the recreation, the timber, the logging, the jobs – and all of that was embedded deeply in these interior counties at the community level and the family level, for that matter. And all of this in the last decade has introduced a picture that is perceived as relatively unstable compared to the long-standing prior history. And the national park proposal, RESTORE and Roxanne Quimby are lightning rods. And it gets particularly, I think, problematic, even for me, who is conservation-minded, I mean; when snow-mobile trails are closed off with new owners...I think Roxanne happens to be the lightning rod because she's out there and she's visible. So that's, you know, where things show up.

The following quote notes that decisions about land use are often based on a value-driven response that comes from the meaning of place and fear of the unknown. It is important because RESTORE relied heavily on an economic argument for the park.

> I don't think you can ever explain it [the RESTORE proposal] enough in a general public way to get people to sign onto it. I don't care what economic studies you come out with; they are not going to believe them.... People don't care; they wouldn't care if the governor said that a national park would put $5,000 extra into your pocket every year. People don't care. They don't believe it. They don't want to hear it and they don't care because what they care most about is their bias, their political perception, and the way it has been done, i.e., we don't like change.

Another interesting quote came from a long-time resident in Maine reflecting on some events that signaled a change in the North Woods of Maine that may have been early sources of opportunity for some and sources of great fear for others.

3 Why History Matters in Conservation Planning

You could make a case that when Great Northern announced in 1986, after the defeat of the Big A project that it was going to be downsizing. That was really the clanging bell, the first one that things are going to be different here in the North Woods and I well remember Bob Bartlett who was the president at the time, making that announcement that they were going to be reducing their work force severely over the years, and life was not going to be the same. As of that moment, 1986, this was before the Diamond Occidental fell, that was really big news, that was the biggest news in that decade in a way, because it said our history as we have known it for the last 100 years up this way is going to be changing and...so that began the circumstances and events that lead us up to today. Because, let me say this, because people could have thought it would be great to have a park, but if nobody was willing to sell they sure weren't going to get it from eminent domain, and so with the sales and the downsizing, first of all the mills, and then as more people got involved in looking at, well, do you really need to own all this land, that's when it became possible for a willing seller and a willing buyer to get together … until the Diamond sale, I don't think there had been any other major investing in land, but that was the first time that I think people might have let the hairs get raised on their back with excitement that maybe this was the start of something really big and maybe these lands would be up for sale for the first time since when.

3.4.2 Essentialized Images

Essentialized images is a simplistic characterization of a person, group, or community of people used as a means to build political power, and can allow conflicts to spin out of control into intractable situations that are ultimately destructive to conservation and to the local people in an area, creating 'brittle' arguments for conservation initiatives (Redford et al. 2006). The term 'essentialized images,' means the use of images in a way that objectify and dehumanize (Brosius 1999). This in turn may allow actors holding such images to ignore the contributions and different ways of knowing or creating meaning about a particular landscape. The only way to understand this multiplicity of place values is through discourse, either through research or as community meetings where real effort is made to understand, not stake one's claim to the landscape either through science or tradition.

The most controversial issue regarding RESTORE was their process and not the content of their proposal. The leaders interviewed felt that RESTORE was in a rush, and that they acted as if they were riding in to be the hero of the North Woods. This in and of itself worked against RESTORE's fifth goal for the park, which was to build pride. How can local people have pride in something that they did not participate in creating? They acted with a perception that the local residents needed to be saved from 'outside,' not as part of a conversation about how protection of shared values might be achieved. A specific example that created a focus on essentialized images in this case is a brochure RESTORE released to look exactly like an actual national park brochure (Fig. 3.1). Although this helped RESTORE communicate their vision, the brochure had the opposite effect on the local population as evidenced by the following interview responses:

> …it was clear to me early on that their aspirations are to establish a national park. And I'm a photographer on the side, right? So I appreciate the value of images. And I have to admit, I took them to task at one point, one-on-one, I said, 'What kind of b.s. is this, you know?'

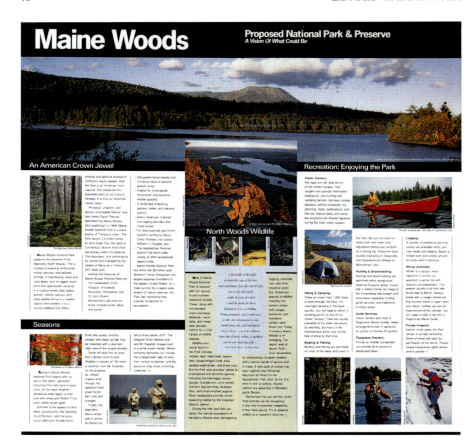

Fig. 3.1 Maine Woods National Park and Preserve brochure, created by RESTORE: the North Woods as part of their promotion campaign for the park

> They've got the north Maine woods, the national park in the same format, the same color-scheme. You can call it what you want, but I said, 'That is bogus. That is misleading.'

Another interviewee commented on the map created in the format of the National Park Service as having the result of making the local people look stupid. They were the ones explaining to visitors that there was no national park. This in turn created a deeper divide, leading locals to believe there was no room for compromise. The interviewees saw RESTORE as wedded to their proposal, not interested in adapting it.

> This isn't coming from me – I've had people who work in the visitor-tourism sector [who] said they have people that have showed up going, 'Where's the park?' And that does us a big disservice and I never quite got to the core of this until today, though. And you know people come with expectations. It misled people and it's like, they land in our dooryards and what? In consciousness, the realization light goes on, 'Boy, there's no national park.' And…who and where do they associate that with? They associate here. They don't associate that with RESTORE. We're the ones that wind up taking the heat.

3 Why History Matters in Conservation Planning

The following comment came from someone present at a meeting to debate the park that included what were at the time the opposite sides of the issue, represented by the Fin and Feather Club in Millinocket and RESTORE. This 1995 meeting was sponsored by the Maine Wildlife Society. This quote is evidence of the deep divide that seemed to create an impenetrable barrier to discourse.

> ...if you really listened to what they were all saying, it was incredible common ground; incredible common ground. Yet they're up there hating each other, ...they were caught on points of rhetoric and they weren't listening to each other, I felt. And I just, I was like, 'My gosh, they're arguing over here and over here, but if they really listened that there was so much mutual interest in seeing the future of the North Woods secured,' if they could ever just sit and get through that and talk, what great allies. ...after the meeting broke up...they were packing up and...I gave them my feedback. I said... 'I don't know if you heard that because you were caught up in it.' 'But it's incredible in terms of how much common thinking there is.' And I said, 'From what I could hear,' I said, 'it's just simply this question of it being a national park. You know, if it was somehow something other than a national park, all of the functions that you're talking about wanting to protect, are exactly all of the functions and values that the Fin and Feather Club want to protect. You know, some very minor little tweaking,' I said. I said, 'My – if you can see your way to do that, if you could somehow just shed the national park as the big handle – because that's what people...seem to be responding so negatively to – and really focus on what the values and functions are you're trying to protect,' I said, 'I think you probably have one of the strongest allies in the world right up there in these folks in Millinocket and you could make this all work. It could happen. Just don't make it a national park...get rid of the park as your goal and focus on the values you're trying to protect.' And it fell on deaf ears. It fell on deaf ears.

This interviewee is a prominent member of the conservation community in Maine and said at this point they realized that RESTORE had no intent of including locals at all. The overwhelming finding from the research interviews regarding essentialized images is that all interviewees generally agreed with the RESTORE goals but disagreed so much with the process that RESTORE followed that a proposal for a large protected area would never move forward if it was promoted by RESTORE. Their perceived lack of regard for the local people left locals suspicious and distrustful of their motives.

> Nobody would disagree with the values and goals on this list. However the fact that they are advanced by RESTORE and the way they have been advanced doesn't necessarily suggest that that's what's intended.

3.4.3 Vernacular Conservation

Vernacular conservation is a term to describe the use of the common or native (vernacular) meanings of place as a basis of conservation. Pimbert and Pretty (1995) define it as 'conservation based on site-specific traditions and economies; it refers to ways of life and resource utilization that have evolved in place and, like vernacular architecture, is a direct expression of the relationship between communities and their habitats.'

Just as the scale of conservation initiatives is increasing to more regional and continental approaches, so too is the recognition of historical and qualitative data that builds our knowledge of conservation at temporal and small spatial scales. Resources for conservation always exist in a place, one that will be imbued with many other meanings they will come to bear on any conservation initiative related to that place. Many of these meanings are tied to self-identity, a powerful force that is necessary to understand in any place or assemblage of places in order to create more resilient approaches to conservation (Cheng et al. 2003). This has also been called cross-scale conservation in recognition of the challenge of the social-ecological system within which decision making takes place (Berkes 2004). The following quote explores one such option for incorporating the cultural and social aspects of place more explicitly:

> I've been kicking the doors around here saying, 'For God sakes, the Maine Woods Forest Heritage. What the hell have we been about forever?' I mean, this is, to me, this is the opportunity. We need to get a limited study group of yea sayers and nay sayers, and put together a learning agenda, develop them into a learning community;...go visit some of these areas and look at what the tangible issues are that people have to deal with, and look at what the costs or benefits are and then come back and report on that. If it makes some sense, fine. If it doesn't make sense, fine. Or if it's a split report, fine. But we're interested in that, admittedly, because from a more selfish perspective, in the region, we think that they don't have the constraints that go with the national park. But what that brings us is maybe some additional resources, some visible recognition, and some financial resources to help us do our diversified economic development work here, at the same time protecting the rural life that we appreciate.

The quote above was one of many that explore alternative large-scale conservation options to a national park that may fit better within the region. There was consistent support from the many different viewpoints that a large-scale conservation vision was needed and that even the goals of the RESTORE proposal were a good guide, but that the fact that they left local wisdom out of the design was a direct insult to the local traditions and culture.

> If there's going to be a new entity here, the people need to be a major, major part of it. They need to say what's in their hearts and what their fears are and help to offer solutions.

And another interviewee echoed this sentiment in regards to large-scale easements:

> Easements are a direct response to the public interest in conservation of these lands, and they are moving us toward better use. However, they do not in any way say that we as Maine citizens are masters of our own destiny.

This last quote explores the pride and dignity that can come from a conservation plan that includes the local vision of place. This can create a sense of empowerment and can indeed be used to help foster long-term support of a conservation initiative even after planners are long gone from a region or on to the next initiative.

3.5 Lessons Learned

The conclusion among decision leaders in Maine today is that there is no political will in the state for the RESTORE proposal for a Maine Woods National Park. There is also a sense that RESTORE did not listen to the local people or pay enough attention to the cultural memory of the region that reacts vigorously against all things federal. This is not to say that locals did not notice the rise in unemployment, closing of schools, and increased regional tension and insecurity. There was and still is a palpable grief felt by the people in the North Woods of Maine as they continue to lose a life they thought was their birthright. The RESTORE proposal made many residents of Northern Maine feel like RESTORE was there to save the day and that this took away the last shred of their dignity, which arguably was central to the manner with which they confronted the sea change in social realities the region faces. As a result, the proponents of the RESTORE proposal were perceived as enemies independent of what the goals of the proposal actually were.

RESTORE's relentless pressure was based on a deep love for the North Woods of Maine, but it made local people fear conservation. The debate became one of Park vs. No Park, and participants somehow lost the ability to take a few steps back and define common goals and visions for the region and to look at the alternative options for large-scale conservation that could protect the myriad values and definitions of place. One interviewee summed this point up nicely:

> We'd be well served to get to the point where we started talking about how much and where instead of yes or no. That's the problem with the park debate – the park debate is yes or no, and never what's the good of the park proposal and what is the bad of the park proposal. What's the good of the way industry, tourism, and recreation use the forest and what is the bad of the way they use the forest?

This information is valuable for any advocacy group interested in conservation initiatives in a rural region. Without a deep understanding and respect for the local people, their lives as well as their values, insurmountable obstacles will remain in the path of conservation. Conservation cannot be done to people; it has to be done with them.

This research suggests that RESTORE, in its urgency, left out an important step in any planning process, which is to include the local players before you have a plan. However, many argue that the discussion about conservation only becomes real when we draw 'lines on a map' (Trombulak 2003), and so in RESTORE's defense they were bold enough to draw these lines. How can RESTORE's work and passion be used to help inspire a twenty-first century model for conservation in a forested landscape? The competing definitions of place and value systems in the North Woods of Maine are important to include in any forest management or conservation initiatives in the state.

On the issue of the North Woods of Maine, we never found anyone reluctant to speak with us. There was great interest in the 'telling of stories' about the landscape. Too often, people don't really listen to one another. Using tools and working with historians and conservation social scientists will help develop this understanding and social meanings of place. These ties to the land, which form the basis of identity of self, family, and culture, will ultimately be the stories that protect the landscape for the long term.

Our experience with this case study taught us some specific lessons:

First, it is important to be a student. Come to a place to learn from the 'natives,' as one would learn about an important member of their family. What is the story of place and how does one's conservation knowledge fit into the narrative of place? How can it be made relevant? This includes learning about local institutions and gatekeepers of information that will be useful in gaining both understanding and credibility. It may be too hard to talk to a large number people in an area, but if one talks to the *right* people, they usually can convey information that represents many perspectives in the locale.

Second, a holistic view of knowledge needs to be developed. People must often argue or present what they value in the context of others' values. This, to be done well (meaning that other value systems are respected), requires one to understand the story and context leading to those values in an open, transparent manner, using research methods that are free from emotional judgments but that can measure them. Environmental history and qualitative inquiry are two such tools and, if done well, will benefit all parties – scientists and planners as well as local residents. People will not support what they do not understand, and when the conservation planners are gone to a new place in need of their skills, it is the local residents who remain. Their partnership is essential. Therefore, earning their respect is, too.

Finally, flexibility is essential. A landscape-scale conservation initiative needs to be based on the context of the different cultures represented in the entire area. The fine line will always be how to incorporate the best possible science driving a conservation plan with local people and their intense love of place, however they display it. Understanding the cultural memory, the essentialized images, and interest or potential for including vernacular elements in the conservation plan can lead to a better process and ultimately one that preserves the dignity of local residents while creating a resilient conservation plan.

The greatest resistance to conservation in North Woods of Maine came out of fear of a loss of access to places important to people. Interestingly, it was the number one reason given by those who supported large-scale conservation in the region as well. Think for a moment of a place that is embedded deep in your soul, part of your identity, a place you will never see again, and is with you only as a memory. The fear of this loss is a major social driver that conservation planners engage with either unwittingly or in a fully cognizant way that builds compassion. It is this compassion that can drive the interest in a fully interdisciplinary approach to conservation planning that can be good for ecosystems and people.

References

Assmann, J., & Czaplicka, J. (1995). Collective memory and cultural identity. *New German Critique, 65(Cultural History/Cultural Studies), 65,* 125–133.

Barringer, R. (1993). Land use and ownership in Maine: A historical perspective. Unpublished Address to the Georges River Land Trust Annual Meeting, August 15, 1993. Edmund S. Muskie Institute of Public Affairs, University of Southern Maine.

Bennett, D. B. (2001). *The wilderness from chamberlain farm: A story of hope for the American wild.* Washington, DC: Island Press.

Berkes, F. (2004). Rethinking community-based conservation. *Conservation Biology, 18*, 621–630.

Black, A. E., Strand, E., Wright, R. G., Scott, J. M., Morgan, P., & Watson, C. (1998). Land use history at multiple scales: Implications for conservation planning. *Landscape and Urban Planning, 43*, 49–63.

Brosius, J. P. (1999). Analyses and interventions: Anthropological engagements with environmentalism. *Current Anthropology, 40*, 277–310.

Brosius, J. P., & Russell, D. (2003). Conservation from above: An anthropological perspective on transboundary protected areas and ecoregional planning. *Journal of Sustainable Forestry, 17*, 39–65.

Cheng, A. S., Kruger, L. E., & Daniels, S. E. (2003). "Place" as an integrating concept in natural resource politics: Propositions for a social science research agenda. *Society and Natural Resources, 16*, 87–104.

Cronon, W. (1993). The uses of environmental history. *Environmental History Review (Fall), 1993*, 1–22.

Foster, D., Swanson, F., Aber, J., Burke, I., Brokaw, N., Tilman, D., et al. (2003). The importance of land-use legacies to ecology and conservation. *BioScience, 53*, 77–88.

Hampton, H. D. (1981). Opposition to national parks. *Journal of Forest History, January*, 37–45.

Harper, S. C., Faulk, L. L., & Rankin, E. W. (1990). *The northern forest lands study of New England and New York*. Rutland, VT: USDA Forest Service.

Irland, L. (1999). *Northeast's changing forests*. Cambridge, MA: Harvard University Press.

Irland, L. (2000). Maine forests: A century of change 1900–2000 … and elements of policy change for a new century. *Maine Policy Review (Winter 2000)*, 66–77.

Judd, R. (1997). *Common lands, common people: The origins of conservation in Northern New England*. Cambridge, MA: Harvard University Press.

Judd, R. W. (2003). Writing environmental history from East to West. In B. A. Minteer & R. E. Manning (Eds.), *Reconstructing conservation: Finding common ground* (pp. 16–31). Washington, DC: Island Press.

Judd, R. W., & Beach, C. S. (2003). *Natural states: The environmental imagination in Maine, Oregon, and the nation*. Washington, DC: Resources for the Future.

Kellett, M. J. (1989). *A new Maine woods reserve: Options for protecting Maine's northern wildlands*. The Wilderness Society: Washington, DC.

Kellett, M. J. (2000). Maine Woods National Park: The best way to restore the wild. *Wild Earth, 10*(2), 60–64.

Klyza, C. M., & Trombulak, S. C. (Eds.). (1994). *The future of the Northern Forest*. Hanover, NH: University Press of New England.

Kruger, L. (2001). What is essential is invisible to the eye: Understanding the role of place and social learning in achieving sustainable landscapes. In S. R. J. Sheppard & H. W. Harshaw (Eds.), *Forests and landscapes: Linking ecology, sustainability and aesthetics* (pp. 173–187). Wallingford, UK: CABI.

Lansky, M. (1991). *Beyond the beauty strip: Saving what's left of our forests*. Gardiner, ME: Tilbury House.

Lansky, M. (2001). *An ecological reserve system for Maine: Are we really making progress?* Retrieved February 5, 2010, from http://www.meepi.org/files/ml051901.htm

Maine Bureau of Parks and Lands. (2005). *The Allagash wilderness waterway*. Unpublished brochure. Augusta, ME: Maine Department of Conservation.

Marcucci, D. J. (2000). Landscape history as a planning tool. *Landscape and Urban Planning, 49*, 67–81.

Merenlender, A. M., Huntsinger, l., Guthey, G., & Fairfax, S. K. (2004). Land trusts and conservation easements: Who is conserving what for whom? *Conservation Biology, 18*, 65–75.

Nash, R. (2001). *Wilderness and the American mind*. New Haven, CT: Yale University Press.

National Park Service. (1937). *Proposed Mount Katahdin National Park, Maine*. Washington, DC: USDI National Park Service.

Pidot, J. (2003). *Benevolent wizard or sorcerer's apprentice? A critical examination of the conservation easement phenomenon.* Washington, DC: Georgetown University Law Center Continuing Legal Education Environmental Law and Policy Institute.

Pierce, D. S. (2000). *The great smokies: From natural habitat to national park.* Knoxville, TN: University of Tennessee Press.

Pimbert, M., & Pretty, J. N. (1995). *Parks, people and professionals: Putting participation into protected area management (UNRISD Discussion Paper 57).* Geneva: United Nations Research Institute for Social Development.

Redford, K. H., Robinson, J. G., & Adams, W. M. (2006). Parks as shibboleths. *Conservation Biology, 29*, 1–2.

Ritters, K. H., & Wickham, J. D. (2003). How far to the nearest road? *Frontiers in Ecology and the Environment, 1*, 125–129.

Rolde, N. (2001). *The interrupted forest: A history of Maine's wildlands.* Gardiner, ME: Tilbury House.

Rondinini, C., Stuart, S., & Boitani, L. (2005). Habitat suitability models and the shortfall in conservation planning for African vertabrates. *Conservation Biology, 19*, 1488–1497.

Runte, A. (1997). *National parks: The American experience* (3rd ed.). Lincoln, NE: University of Nebraska Press.

Schenk, A., Hunziker, M., & Kienast, F. (2007). Factors influencing the acceptance of nature conservation measures – a qualitative study in Switzerland. *Environmental Management, 83*, 66–79.

Seymour, R. S., White, A. S., & deMaynadier, P. G. (2002). Natural disturbance regimes in northeastern North America-evaluating silvicultural systems using natural scales and frequencies. *Forest Ecology and Management, 155*, 357–367.

Trombulak, S. C. (2003). An integrative model for landscape-scale conservation in the twenty-first century. In B. A. Minteer & R. E. Manning (Eds.), *Reconstructing conservation: Finding common ground* (pp. 263–276). Washington, DC: Island Press.

Wagner, M. B. (2002). Space and place, land and legacy. In B. J. Howell (Ed.), *Culture, environment, and conservation in the Appalachian South* (pp. 121–132). Urbana, IL: University of Illinois Press.

Watkins, T. H. (1988). Wilderness America: A vision for the future of the nation's wildlands. *Wilderness, 52*, 13–17.

Williams, D. R., & Vaske, J. J. (2003). The measurement of place attachment: Validity and generalizability of a psychometric approach. *Forest Science, 49*, 830–840.

Chapter 4
Developing Institutions to Overcome Governance Barriers to Ecoregional Conservation

Robert B. Powell

Abstract Threats to biodiversity occur at local, regional, and landscape level scales. As a result responses to these threats increasingly use a systematic process to identify important habitat at large enough scales necessary to support biodiversity that is currently or potentially threatened by human activity. However despite the relative agreement regarding emerging best practices for identifying and ranking areas within an eco-region for conservation and the wide use of eco-regional planning and ecosystem management in both developed and developing nations, biodiversity and the habitats they rely on continue to degrade. In most cases, one of the major barriers to implementing these landscape scale conservation plans appears to be poor institutional coordination and cooperation (horizontal and vertical fragmentation) across eco-regional scales. This paper describes some of the common barriers to effective eco-regional governance which hamper the implementation of conservation planning efforts and proposes specific steps and conditions necessary for the development of eco-regional institutions, which are thought to overcome governance fragmentation. As complex and transboundary threats such as climate change, pollution, and land conversion increase, it is thought that without this transformation in governance, biodiversity will continue to decline.

Keywords Governance • Ecoregional planning • Networks • Public participation • Transboundary institutions

4.1 Introduction

Currently the world's biological diversity is at risk from a range of localized, regional, and transboundary threats (e.g., Rockström et al. 2009; Wilson 1992; Chap. 2). The conservation of biological diversity is no longer seen as being driven solely by localized

---

R.B. Powell (✉)
Department of Parks, Recreation, and Tourism Management, Clemson University, 128 McGinty Court, Clemson, SC 29634-0735
e-mail: rbp@clemson.edu

S.C. Trombulak and R.F. Baldwin (eds.), *Landscape-scale Conservation Planning*, 53
DOI 10.1007/978-90-481-9575-6_4, © Springer Science+Business Media B.V. 2010

threats that require singular action but instead is seen as being associated with complex interrelated problems that require adaptive and multi-scale approaches (e.g., Holling 1995; Meffe et al. 2002). In response to this crisis, international organizations such as the IUCN, non-governmental organizations (NGOs) such as The Nature Conservancy and World Wildlife Fund, and government agencies in the U.S. such as the U.S. Forest Service and the Environmental Protection Agency now promote ecoregional conservation planning and ecosystem management (e.g., Bailey 2002; Groves et al. 2002; Olson and Dinerstein 1998; Whittaker et al. 2005). However, despite the relative agreement regarding emerging best practices for identifying and ranking areas within an ecoregion for conservation and the wide use of ecoregional planning and ecosystem management in both developed and developing nations, species and the habitats they rely on continue to be degraded and lost (Adger et al. 2005; Rockström et al. 2009). In most cases, one of the most significant barriers to implementing these regional and ecoregional conservation plans appears to be their lack of integration with the governance milieu (e.g., Brosius and Russell 2003; Fall 2003), caused especially by poor intergovernmental coordination and cooperation across ecoregional scales (e.g., Karkkainen 2004; Powell et al. 2009; Yario 2009).

In this paper I describe some of the common barriers to effective ecoregional governance that hamper the implementation of conservation planning efforts, and propose specific steps and conditions necessary for the development of ecoregional institutions, which are thought to be a solution to governance fragmentation. As complex and transboundary threats such as climate change, pollution, and land conversion increase, it is thought that without this transformation in governance, biological diversity will continue to decline.

4.2 Ecoregional Conservation Planning

With the development of scientific inquiry and technological innovations, humankind's understanding of the natural environment and the causes of ecological degradation and extinction have expanded. It is now understood that ecosystems, natural processes, stressors, and threats are embedded in and part of a complex interrelated and interdependent socio-ecological system (Levin 1999; Machlis et al. 1997). In particular, human land use that disturbs or destroys vital habitat and erodes ecosystem function is one of the most pressing issues affecting biological diversity worldwide (Hunter and Gibbs 2007; Chap. 2). Despite the fact that the principles of ecosystem ecology are well accepted and acknowledge the interconnectedness of natural and human systems, the complexity, dynamism, and the scale of these interconnections are often overlooked (Folke et al. 2002; Gunderson and Holling 2002). In particular, conservation activities that seek to preserve the world's biological diversity have traditionally employed short-term, simplistic, and localized approaches, which have ignored their critical connections to broader, complex socio-ecological systems and, therefore, have minimized their long-term contributions (Olson and Dinerstein 1998). These localized conservation efforts often prove unsuccessful due to socio-ecological perturbations and trends that occur across expansive temporal and spatial scales and do not conform to traditional geopolitical boundaries (Gutzwiller 2002). With this advance in understanding of both scale and complexity, conservation in the twenty-first century has embraced more systematic, ecologically representative, and larger-scale planning and management endeavors that seek to protect biological diversity, while also providing a framework for adapting to changing conditions (Groom et al. 2006; Meffe et al. 2002). One such approach used to identify important habitat and to plan on a wide-scale is ecoregional conservation planning (Groves et al. 2002; Margules and Pressey 2000).

As I discuss it here, ecoregions refer to 'relatively large units of land or water that contain distinct assemblages of natural communities and share a large majority of species, dynamics, and environmental conditions' (Dinerstein et al. 2000). Ecoregional conservation planning, therefore, refers to a systematic process that is used to identify important habitat within an ecoregion necessary to support biological diversity that is currently or potentially threatened by human activity (Groves et al. 2002; Margules and Pressey 2000). Currently, scientists have many useful tools, data sources, and ecological criteria to construct and develop ecoregional conservation plans (Olson and Dinerstein 1998). The intent of these plans is to provide information for decision-makers to systematically select and prioritize important habitat for future conservation or preservation (Chap. 14) and, therefore, to influence future land-use management practices and optimize conservation of biological diversity (Margules and Pressey 2000; Olson and Dinerstein 1998). Although some researchers suggest that ecoregional planning emphasizes a technocratic and top-down approach to conservation that may marginalize local communities (Brosius and Russell 2003; Chap. 3), proponents argue that planning efforts that don't account for the larger scale and context will lead to uncoordinated action across ecoregions and unsustainable results.

While the development of ecoregional conservation plans is admittedly reliant on expert analysis and based on scientific assumptions, the plans are not designed to be finalized or implemented without public involvement. The plans, as they emerge from such exercises as irreplaceability analyses (Chap. 14), are simply tools to assist in raising awareness regarding the scale and interconnectedness of ecological systems, identifying important habitat, and prioritizing future resource allocation (Trombulak et al. 2008). Similarly, ecoregional conservation planning is not intended to inform how best to protect or manage these areas, it simply provides critical information to ecoregional stakeholders and land-use decisionmakers as an important part of ecosystem management (Groves et al. 2002).

4.3 Ecoregional Conservation and Ecosystem Management

Management and planning are different processes; management implies action, while planning refers to a process that provides information to inform and guide future action. In this context, ecoregional conservation planning is a process that collects, analyzes, prioritizes, and presents ecoregional data that can be used to develop coordinated management activities and policies across an ecoregion. This coordinated ecoregional or landscape scale management approach is called ecosystem management (EM) and it provides a framework for addressing the diverse needs and challenges identified during the ecoregional planning process. EM can be defined as a science-based and holistic approach to managing interrelated social and ecological resources to conserve biological diversity and promote resilient and sustainable ecosystems and economies (Cicin-Sain and Knecht 1998; Grumbine 1994; Hale and Olsen 2003; Margules and Pressey 2000). This definition acknowledges that human and ecological systems are actually one complex interrelated and interdependent system and that EM is designed as an alternative to traditional top-down, uncoordinated, and simplistic small-scale and sector organized management approaches (Meffe et al. 2002). The EM approach also recognizes that both natural and social change are inevitable and arise from acute natural and human-caused perturbations such as hurricanes or civil unrest, as well as chronic events such as climate change, predominant weather patterns, and market trends across wide geographic areas. These acute and chronic events often produce unexpected outcomes because complex socio-ecological systems do not necessarily respond in linear or socially or environmentally predictable ways and, therefore, necessitate an adaptive approach toward management (Holling 1995, 2001).

Attempting to incorporate ecoregional conservation planning into the management of complex socio-ecological systems is admittedly difficult. Although some relatively successful ecoregional conservation planning and EM projects exist, such as the Yellowstone toYukon conservation initiative (Chester 2003; Levesque 2001), many barriers and challenges confront the conservation of biological diversity at an ecoregional scale. These challenges exist despite the fact that conservation planners who use new technologies have the ability to identify and map areas of high conservation value that will enhance ecological connectivity across ecoregions (Chaps. 14 and 15). In fact, some suggest that these challenges to implementing conservation planning at an ecoregional scale actually stem from a reliance on scientific and ecological approaches while underestimating the importance and complexity of the socio-economic and governance context of ecoregions (Brosius and Russell 2003; Mascia et al. 2003; Pfueller 2008; Scott et al. 1999). Specifically, land-use decisions do not occur at the ecoregional scale but instead are made by individual, corporate, and governmental land owners/managers who are influenced by a wide range of social forces, economic trends, and governmental policies and regulations that operate at many levels. In other words, existing legal, geopolitical, and jurisdictional boundaries coupled with other social forces drive a high degree of both horizontal and vertical fragmentation in land-use management (Dietz et al. 2003).

4.4 Fragmentation

Government has traditionally been organized in a hierarchical fashion based on increasing scales of territorial and jurisdictional boundaries (Blatter 2004). Research suggests that developing an integrated and coordinated approach to governance at the ecoregional scale has been difficult to realize because of the complexity of the

4 Developing Institutions to Overcome Governance Barriers to Ecoregional Conservation   57

human-ecological system coupled with traditionally simplistic and sector approaches to governance (Berkes 2006; Cicin-Sain and Knecht 1998; Knight and Meffe 1997). Because these traditional governance approaches to resource management typically use 'one law, one agency, and one set of regulations' for a particular sector, spatial region, or resource (Ehler 2003), ecoregions often contain multiple overlapping and competing agencies and institutions with different mandates and laws, attempting to manage interdependent and interrelated activities and resources that affect the ecoregion's health, biological diversity, productivity, and resilience (Folke et al. 2003; Karkkainen 2004; Neuman 2007). This uncoordinated and inefficient approach to human-ecosystem management occurs when local, state or provincial, and national institutions (vertical fragmentation) and/or multiple institutions across one level of governance (horizontal fragmentation) attempt to manage an ecoregion without 'communication, coordination, or integration' (Cicin-Sain and Knecht 1998; Karkkainen 2004).

In addition, ecoregional approaches to planning and management assume that humans identify and informally organize at this same broad scale. Although ecoregions are useful for describing ecological boundaries, research suggests that a majority of people do not identify with this large scale but instead develop social and emotional connections to place at local scales based on social, historical, and cultural realities (e.g., Ardoin 2009; Fall 2003; Chap. 3). Founded on local perspectives, public reactions to ecoregional conservation planning and management can be misinformed and oppositional, leading to assumptions that these efforts are designed to marginalize human needs within the ecoregion in preference for biological diversity (Chester 2003). Because of the prevalence of governance fragmentation and the public's tendency to form strong identification at the local scale, ecoregional conservation planners and researchers agree that developing interactions and cooperation between the public, economic sectors, and governmental levels are necessary if we are to overcome complex social, economic, and ecological challenges and sustain long-term ecoregional health and resilience (e.g., Adger et al. 2005; Berkes 2004; Cicin-Sain and Knecht 1998; Meffe et al. 2002; Ostrom 1990; Powell et al. 2009; Westley 1995; Yario 2009). One mechanism for accomplishing this interaction is the development of institutions, both formal and informal, that operate across scales, sectors, and governmental levels.

4.5   The Emergence of Governance Through Ecoregional Institutions and Networks

Although humans traditionally organize, identify, and govern themselves in more localized and hierarchical structures, many emerging environmental and social issues (i.e., climate change, growing metropolises, and ecoregional conservation) transcend traditional boundaries and jurisdictions. Consequently, an emerging trend is a transition from formal hierarchical governmental structures to governance through the formation of networks to overcome fragmentation, facilitate improved flows of information, and fulfill functions necessary to deal with transboundary issues (Blatter 2004; Castells 2000). National, state or provincial, and local governmental institutions; non-governmental organizations, businesses, and the public are increasingly forming coordinated multi-scalar (operates across multiple jurisdictional and geographic scales), polycentric (contains many centers of authority), and multi-level (incorporates local to national level stakeholders and institutions) institutions and networks to address these complex and dynamic issues and to manage large transboundary socio-ecological systems and their resources (e.g., Karkkainen 2004; Baker et al. 2005).

In this context, governance refers to collaborative institutional structures that unite governmental, private, and non-governmental organizations, as well as individuals into both formal and informal arrangements (Benz 2001). The purpose of these multi-scalar, polycentric, and multi-level institutions and networks is not to replace current governmental institutions that perform day-to-day tasks and operations but instead to organize around a collective goal, issue, or resource that transcends usual socio-political boundaries (e.g., Benz 2001; Ostrom 1990). In relation to ecoregional conservation planning and EM, these institutions form to deal with natural resource use and often focus on management of the interconnected ecological and social processes that reduce ecological function and threaten biological diversity. The ecoregion and associated issues that affect biological diversity conservation in effect become the 'functional-action space' that helps to identify potential actions and provide loose boundaries for defining members and participants (Benz 2001; Karkkainen 2004). Because of the complexity, uncertainty, and dynamism of transboundary issues, these institutions need a high degree of flexibility and adaptive capacity to respond to emerging and changing conditions (Blatter 2003). Therefore, the scale, focus, and membership of the organization must evolve over time based on feedback from the monitoring and evaluation of existing conditions. Finally, because these types of institutions are generally informal and attract membership based on a particular goal, they are inherently cooperative in nature. Internal processes and procedures are typically egalitarian and collaborative and seek to overcome the fact that different participants will have different levels of power and resources (e.g., Benz 2001; Karkkainen 2004). In summary, these emerging governance institutions share several general characteristics: they are enduring polycentric networks of governmental, public, private, and non-governmental organizations and individuals organized around a complex issue/goal that operate cooperatively and in a flexible and adaptive way across multiple scales and sectors to accomplish specific tasks.

4.6 Development of a Functioning Ecoregional Institution

The development of governance structures that operate beyond traditional territorial and jurisdictional boundaries and account for the scale and complexity of socio-ecological systems is necessary to accomplish ecoregional conservation (Lebel et al. 2006). Research suggests that these intergovernmental institutions

4 Developing Institutions to Overcome Governance Barriers to Ecoregional Conservation 59

build social capital and networks that enhance efficiency, promote partnerships, foster learning, and promote adaptive governance to respond to complex and changing social and environmental conditions (Knight and Meffe 1997; Lubell 2004; Olsson et al. 2004; Walker et al. 2006). But what steps and conditions are necessary for the formation of an institution that operates across spatial and jurisdictional scales for ecoregional conservation?

A review of ecosystem planning and management case studies as well as literature from political ecology, regional planning, and organizational and management theory suggests several necessary conditions and specific steps. First and foremost, the relevance of the primary issue, in this case conservation of biological diversity at an ecoregional scale, must be clearly communicated and provide compelling motivation for stakeholders to participate (Blatter 2004; Ostrom 1990). In particular, a shared sense of interdependence across scales is necessary for instilling commitment in stakeholders (Griffin 2003; Powell et al. 2009). The premise is that if stakeholders recognize that biological diversity is jointly affected by the activities of others across multiple scales, they will also understand that uncoordinated activities will not support either the common or their own individual interests and that collective action is necessary. Ultimately, potential members of any ecoregional institution and their respective agency or organization must be willing and motivated to participate or they will not be viable and productive members of the institution (Gray 1985).

Second, initiators of ecoregional conservation planning and the development of an ecoregional governance institution are often non-governmental organizations that are, by nature, issue driven and not constrained by jurisdictional and geographic boundaries or mandates. However, potential members of any ecoregional institution and their respective organization must consider the initiator or convener of the initial meetings as legitimate and acceptable; this condition appears paramount for future success (Gray 1989). This is particularly important in the context of planning in developing countries, where many ecoregional and ecosystem management projects have been externally driven and funded (e.g., Christie 2005).

Third, ecoregional institutions, if they are to be effective in fostering multi-scalar, polycentric, and multi-level governance, must attract membership that is representative of the breadth of scales, sectors, and levels of governance in the 'functional-action space' (Benz 2001). This includes the general public, especially residents of locations where conservation activities will be promoted. Although outside the scope of this essay, involving the public requires providing information and education; exchanging information through hearings, comments, surveys, and focus groups; and actively engaging the public in the decision-making process through collaboration, negotiation, mediation, and co-management (e.g., Force and Forester 2002; Leach 2006; Chap. 10).

Fourth, during the development of any collaborative organization, such as an ecoregional institution, regular opportunities for formal and, more importantly, informal face-to-face communication, need to occur. Researchers consider this social process a critical element for organizational success (Elsbach and Glynn 1996; Fritz et al. 1999; Roberts 2001). These opportunities for members to interact are thought to build social capital, trust, and organizational commitment (Lebel et al. 2006; Riketta 2002; Schoemaker and Jonker 2005), as well as facilitate organizational learning and enhance problem solving and adaptive management (Quinn 1985; Westley 1995). In particular, the development of social capital, defined as personal, social, and organizational networks of relationships (e.g., Bordieu 1986; Halpern 2005), enhances the development of social norms and reciprocity and is thought to facilitate individual and collective action (Coleman 1990). However, these benefits from open communication between members of an ecoregional institution are contingent upon a participatory and collaborative process, which is often dictated by the group's leader (Edmondson 2003; Elsbach 2004).

Fifth, the quality of leadership strongly influences inter-group processes and communication and, ultimately, the success of an ecoregional institution (e.g., Ostrom 1990). Therefore, a good leader for a collaborative organization such as an ecoregional institution needs to possess numerous qualities and skills related to decision making, group diplomacy, availability, problem solving, personnel management, planning, organization, and communication (Kreske 1996). Preskill and Torres (1999) identified a leader as a person who facilitates teamwork and is '(a)...skilled in the areas of group process, collaborative problem solving, team development, active listening and conflict management; (b) [facilitates] learning as a process; and (c) models dispositional ideals,' including being non-dominating, friendly, empathetic, open to input, and inclusive (Willemyns et al. 2003). In particular, when dealing with an adaptive ecoregional institution focusing on complex issues related to ecoregional conservation, a leader should relinquish his or her role and allow people with special areas of expertise an opportunity to facilitate and lead when appropriate (Edmondson 2003; Preskill and Torres 1999; Willemyns et al. 2003).

Sixth, because of the cooperative and informal nature of ecoregional institutions, consensus regarding the rules and normative standards for accountability need to be reached (Benz 2001; Ostrom 1990). Shared authority established through collaborative and consensus driven processes appear important for developing successful long-term commitment and successful outcomes (Karkkainen 2004; Ostrom 1990).

Seventh, due to their informal nature, ecoregional institutions must clearly define the problems they seek to address and articulate their specific purposes (Cicin-Sain and Knecht 1998; Gray 1989; Olsen 2003; Olsen et al. 1997). This is generally thought of as an iterative and facilitated planning process (e.g., Ehler and Douvere 2009; Healey 1997; Randolph 2004) where emergent and evolving problems are identified. Often this process facilitates the identification of new potential partners and members that may be advantageous and necessary for addressing particular emergent issues (Westley 1995). Also through the problem identification process, consensus regarding the major problems facing an area is developed, which facilitates future collaboration (Heikkila and Gerlak 2005; Lebel et al. 2006).

Eighth, as an institution identifies the major problems and generates specific goals, individual participants, who are representatives from multiple organizations, will need to learn to bridge differing perspectives and legal mandates to establish new strategies for achieving their goals (Bidwell and Ryan 2006; Grumbine 1994). However, each individual member may be constrained by their affiliated

4 Developing Institutions to Overcome Governance Barriers to Ecoregional Conservation   61

organization's capacity, resources, culture, and bureaucracy (Bidwell and Ryan 2006; Kanter 1989).

Next, once problems and goals are identified, prioritization of activities is vital to ensure efficient use of resources and social capital (Bidwell and Ryan 2006; Olsen 2003; Olsen et al. 1997). Once prioritized, tasks are assigned and the ecoregional institution must give responsible agencies the latitude to independently manage and perform their activities (Huda 2004; Kanter 1989). Because many of these identified projects will require specific resources and expertise, multiple independent and informal partnerships within the broader ecoregional institution will emerge based on members' organizational affiliation, scale of operation, resource availability, and expertise (Benz 2001; Cicin-Sain and Knecht 1998; Heikkila and Gerlak 2005). Research suggests that in these informal partnerships, tasks are typically managed and administered by one lead organization, and the other partnering organizations facilitate the accomplishment of the task by removing bureaucratic barriers or providing additional resources (Powell et al. 2009). Additionally, the most fruitful partnerships for accomplishing on-the-ground conservation appear to involve vertical partnerships (local with state, regional, or national institutions) (e.g., Powell et al. 2009), reiterating the importance of local involvement (e.g., Christie et al. 2005; Fall 2003; Western and Wright 1994). It should be noted that due to the complexity of ecoregional conservation, members will often need to operate without complete information and with a high level of uncertainty (Evans and Klinger 2008), which will require continual monitoring and adaptation of activities.

Finally, although members are employed by their respective organization, the intergovernmental institution must develop mechanisms for evaluation, reassessment, and adjustment of activities if adaptive integrated management is to occur over the long term (e.g., Holling 1995; Meffe et al. 2002; Olsen et al. 1998; Salafsky et al. 2001). Because of the complexity of most ecoregional conservation issues, many efforts will fail in the short term; without a long-term commitment to adaptive management, these efforts will never have the opportunity to succeed.

4.7 Lessons Learned

Protecting biological diversity at the ecoregional scale should be thought of as a continual process. Long-term commitment to ecoregional governance is critical if conservation is to be successful. In particular, the formation of multi-level, polycentric ecoregional institutions and networks can facilitate coordinated management action and overcome fragmentation. This paper sought to draw insight from case studies as well as organizational, planning, political ecology, and governance research to develop a list of theoretical conditions and steps necessary for the formation of a multi-level, polycentric ecoregional institution for conservation:

1. Communicate importance of goal or overarching issue.
2. Develop a sense of interdependence.

3. Ensure acceptance of convener/facilitator.
4. Attract interorganizational membership that is representative of 'functional-action space.'
5. Provide regular informal and formal meetings.
6. Identify and develop effective leadership.
7. Institute collaborative and consensus driven internal processes.
8. Develop mechanisms for evaluation and accountability.
9. Add new members to ensure representativeness and to perform specific tasks.
10. Undertake planning processes to identify (a) the goals and objectives of the group, (b) the major problems and issues, and (c) the actions necessary for meeting these goals and objectives.
11. Assign responsibility for actions to individual members or groups of members.
12. Form informal partnerships to accomplish action items based on resources, expertise, jurisdictions, and geography.
13. Identify and monitor indicators of success.
14. Adapt and learn by following adaptive management cycle.
15. Commit to long-term involvement.

Several conditions appear to be important antecedents to their formation. Multi-scalar, polycentric, and multi-level ecoregional institutions and networks tend to focus on one central and overarching issue. This overarching issue must be clearly articulated and important enough to motivate a range of stakeholders to participate in the formation and subsequent activities of an ecoregional institution. In addition, the source and communicator of this information must be trusted; similarly, the convener of initial meetings must also be deemed legitimate.

With these precursors met, successful formation and functioning of the institution appears dependent on attracting and involving participants that represent the 'functional-action space'; recruiting effective leadership; meeting regularly in both formal and informal settings; and establishing collaborative and consensus driven procedures and rules including mechanisms for evaluation and accountability. Once functioning, the institution will need to undertake iterative planning processes to identify the goals and objectives of the group, the major issues facing the ecoregion, and the actions necessary to attain these goals and objectives. As specific action items are identified, a member or group of members will take responsibility for implementation. These informal partnerships appear necessary to efficiently allocate and focus resources to accomplish these specific tasks. Finally, because of the complexity of the issues threatening biological diversity at the ecoregional scale, monitoring and evaluation of these activities need to occur so that the ecoregional institution may adapt to inevitable set-backs and emergent changes in the ecoregional context.

Admittedly, effective ecoregional governance for conservation faces many challenges and barriers that are not specifically addressed here. Wavering commitment by member organizations may arise due to institutional amnesia caused by staff turnover, which requires concerted training of new individuals (Powell et al. 2009). Stakeholder fatigue, due to inherent failures and set-backs, and the long time horizon for achieving successful completion of actions may undermine

4 Developing Institutions to Overcome Governance Barriers to Ecoregional Conservation   63

the sustainability of ecoregional efforts. Tensions between local development interests and ecoregional management will undoubtedly develop. Without providing meaningful public involvement as part of local on-the-ground efforts and linking the results of conservation actions to human well being, the public may develop mistrust and actively oppose conservation efforts. However, despite the many challenges, the world is increasingly complex and interconnected, which requires the development of ecoregional institutions to overcome governmental fragmentation in order to reverse the current threats to the world's biological diversity.

References

Adger, W. N., Hughes, T. P., Folke, C., Carpenter, S. R., & Rockstrom, J. (2005). Social-ecological resilience to coastal disasters [viewpoint]. *Science, 309*, 1036–1039.

Ardoin, N. (2009). *Sense of place and environmental behavior at an ecoregional scale*. Dissertation, Yale University.

Bailey, R. G. (2002). *Ecoregion-based design for sustainability*. New York: Springer.

Baker, A., Hudson, D., & Woodward, R. (2005). Introduction: Financial globalization and multi-level governance. In A. Baker, D. Hudson, & R. Woodward (Eds.), *Governing financial globalization: International political economy and multi-level governance* (pp. 3–23). New York: Routledge.

Benz, A. (2001). From associations of local governments to "regional governance" in urban regions. *Deutsche Zeitschrift für Kommunalwissenschaften, 40*(2), Article 4. Retrieved February 1, 2010, from http://www.difu.de/index.shtml?/publikationen/dfk/en/01_2

Berkes, F. (2004). Rethinking community-based conservation. *Conservation Biology, 18*, 621–630.

Berkes, F. (2006). From community-based resource management to complex systems: The scale issue and marine commons. *Ecology and Society, 11*(1), Article 45. Retrieved February 1, 2010, from http://www.ecologyandsociety.org/vol11/iss1/art45/

Bidwell, R. D., & Ryan, C. M. (2006). Collaborative partnership design: The implications of organizational affiliation for watershed partnerships. *Society and Natural Resources, 19*, 827–843.

Blatter, J. (2003). Beyond hierarchies and networks: Institutional logics and changes in transboundary spaces. *Governance: An International Journal of Policy, Administration, and Institutions, 16*, 503–526.

Blatter, J. (2004). From 'spaces of place' to 'spaces of flows'? Territorial and functional governance in cross-border regions in Europe and North America. *International Journal of Urban and Regional Research, 28*, 530–548.

Bordieu, P. (1986). The forms of capital. In J. Richardson (Ed.), *Handbook of theory and research for the sociology of education* (pp. 241–258). New York: Greenwood Press.

Brosius, J. P., & Russell, D. (2003). Conservation from above: An anthropological perspective on transboundary protected areas and ecoregional planning. *Journal of Sustainable Forestry, 17*, 39–66.

Castells, M. (2000). *The rise of the network society*. Malden, MA: Blackwell.

Chester, C. C. (2003). Responding to the idea of transboundary conservation: An overview of public reaction to the Yellowstone to Yukon (Y2Y) conservation initiative. *Journal of Sustainable Forestry, 17*, 103–125.

Christie, P. (2005). Is integrated coastal management sustainable? *Ocean and Coastal Management, 48*, 208–232.

Christie, P., Lowry, K., White, A., Oracion, E., Sievanen, L., Pomeroy, R. S., et al. (2005). Key findings from a multidisciplinary examination of integrated coastal management process sustainability. *Ocean and Coastal Management, 48*, 468–483.

Cicin-Sain, B., & Knecht, R. W. (1998). *Integrated coastal and ocean management: Concepts and practices*. Washington, DC: Island Press.

Coleman, J. S. (1990). *Foundations of social theory*. Cambridge, MA: The Belknap Press.

Dietz, T., Ostrom, E., & Stern, P. C. (2003). The struggle to govern the commons. *Science, 302*, 1907–1912.

Dinerstein, E., Powell, G., Olson, D. M., Wikramanayake, E., Abell, R., Loucks, C., et al. (2000). *A workbook for conducting biological assessments and developing biodiversity visions for ecoregion-based conservation: Part I, terrestrial ecoregions*. Washington, DC: World Wildlife Fund.

Edmondson, A. C. (2003). Speaking up in the operating room: How team leaders promote learning in interdisciplinary action teams. *Journal of Management Studies, 40*, 1419–1452.

Ehler, C. N. (2003). Indicators to measure governance performance in integrated coastal management. *Ocean and Coastal Management, 46*, 335–345.

Ehler, C. N., & Douvere, F. (2009). *Marine spatial planning: A step-by-step approach toward ecosystem-based management (IOC Manual and Guides No. 53 ICAM Dossier No. 6)*. Paris: Intergovernmental Oceanographic Commission and Man and the Biosphere Programme, UNESCO.

Elsbach, K. D. (2004). Managing images of trustworthiness in organizations. In R. M. Kramer & K. S. Cook (Eds.), *Trust and distrust in organizations: Dilemmas and approaches* (pp. 275–292). New York: Russell Sage.

Elsbach, K. D., & Glynn, M. A. (1996). Believing your own PR: Embedding identification in strategic reputation. *Advances in Strategic Management, 13*, 65–90.

Evans, K. E., & Klinger, T. (2008). Obstacles to bottom-up implementation of marine ecosystem management. *Conservation Biology, 22*, 1135–1143.

Fall, J. J. (2003). Planning protected areas across boundaries: New paradigms and old ghosts. *Journal of Sustainable Forestry, 17*, 81–102.

Folke, C., Carpenter, S. R., Elmqvist, T., Gunderson, L. H., Holling, C. S., & Walker, B. (2002). Resilience and sustainable development: Building adaptive capacity in a world of transformations. *Ambio, 31*, 437–440.

Folke, C., Colding, J., & Berkes, F. (2003). Synthesis: Building resilience and adaptive capacity in social-ecological systems. In F. Berkes, J. Colding, & C. Folke (Eds.), *Navigating social-ecological systems: Building resilience for complexity and change* (pp. 352–387). Cambridge: Cambridge University Press.

Force, J. E., & Forester, D. J. (2002). Public involvement in National Park Service land management issues. *National Park Service Social Science Research Review, 3*, 1–28.

Fritz, J. M. H., Arnett, R. C., & Conkel, M. (1999). Organizational ethical standards and organizational commitment. *Journal of Business Ethics, 20*, 289–299.

Gray, B. (1985). Conditions facilitating interorganizational collaboration. *Human Relations, 38*, 911–936.

Gray, B. (1989). *Collaborating: Finding common ground for multiparty problems*. San Francisco, CA: Jossey-Bass.

Griffin, J. G., Jr. (2003). Transboundary natural resources management: Conservationists' dreams or solid means to achieve real benefits for real people. *Journal of Sustainable Forestry, 17*, 229–230.

Groom, M. J., Meffe, G. K., & Carroll, C. R. (2006). *Principles of conservation biology* (3rd ed.). Sunderland, MA: Sinauer.

Groves, C. R., Jensen, D. B., Valutis, L. L., Redford, K., Shaffer, M. L., Scott, J. M., et al. (2002). Planning for biodiversity conservation: Putting conservation science into practice. *BioScience, 52*, 499–512.

Grumbine, R. E. (1994). What is ecosystem management? *Conservation Biology, 8*, 27–38.

Gunderson, L. H., & Holling, C. S. (Eds.). (2002). *Panarchy: Understanding transformations in human and natural systems*. Washington, DC: Island Press.

Gutzwiller, K. (Ed.). (2002). *Applying landscape ecology in biological conservation*. New York: Springer.

Hale, L. Z., & Olsen, S. B. (2003). Coastal governance in donor-assisted countries. In S. B. Olsen (Ed.), *Crafting coastal governance in a changing world* (pp. 37–60). Narragansett, RI: The Coastal Resources Management Program.

Halpern, D. L. (2005). *Social capital*. Cambridge: Polity Press.

Healey, P. (1997). *Collaborative planning: Shaping places in fragmented societies*. Vancouver, BC: University of British Columbia Press.

Heikkila, T., & Gerlak, A. K. (2005). The formation of large-scale collaborative resource management institutions: Clarifying the roles of stakeholders, science, and institutions. *The Policy Studies Journal, 33*, 583–612.

Holling, C. S. (1995). What barriers? What bridges? In L. H. Gunderson, C. S. Holling, & S. S. Light (Eds.), *Barriers and bridges to the renewal of ecosystems and institutions* (pp. 3–36). New York: Columbia University Press.

Holling, C. S. (2001). Understanding the complexity of economic, ecological, and social systems. *Ecosystems, 4*, 390–405.

Huda, A. T. M. S. (2004). Interagency collaboration for integrated coastal zone management: A Bangladesh case study. *Coastal Management, 32*, 89–94.

Hunter, M. L., Jr., & Gibbs, J. P. (2007). *Fundamentals of conservation biology* (3rd ed.). Malden, MA: Blackwell.

Kanter, R. M. (1989). Becoming PALs: Pooling, allying and linking across companies. *Academy of Management Executive, 3*, 183–193.

Karkkainen, B. C. (2004). Post-sovereign environmental governance. *Global Environmental Politics, 4*, 72–96.

Knight, R. L., & Meffe, G. K. (1997). Ecosystem management: Agency liberation from command and control. *Wildlife Society Bulletin, 25*, 676–678.

Kreske, D. L. (1996). *Environmental impact statements: A practical guide for agencies, citizens, and consultants*. New York: Wiley.

Leach, W. D. (2006). Public involvement in USDA Forest Service policymaking: A literature review. *Journal of Forestry, 104*, 43–49.

Lebel, L., Anderies, J. M., Campbell, B., Folke, C., Hatfield-Dodds, S., Hughes, T. P., et al. (2006). Governance and the capacity to manage resilience in regional social-ecological systems. *Ecology and Society, 11*(1), Article 19. Retrieved February 1, 2010, from www.ecologyandsociety.org/vol11/iss1/art19/ES-2005-1606.pdf

Levesque, S. L. (2001). The Yellowstone to Yukon conservation initiative: Reconstructing boundaries, biodiversity, and beliefs. In J. Blatter & H. M. Ingram (Eds.), *Reflections on water: New approaches to transboundary conflicts and cooperation* (pp. 123–162). Boston: MIT Press.

Levin, S. A. (1999). *Fragile dominion: Complexity and the commons*. New York: Perseus.

Lubell, M. (2004). Collaborative environmental institutions: All talk and no action? *Journal of Policy Analysis Management, 23*, 549–574.

Machlis, G. E., Force, J. E., & Burch, W. R., Jr. (1997). The Human Ecosystem, Part I: The Human Ecosystem as an organizing concept in ecosystem management. *Society and Natural Resources, 10*, 347–367.

Margules, C. R., & Pressey, R. L. (2000). Systematic conservation planning. *Nature, 405*, 243–253.

Mascia, M. B., Brosius, J. P., Dobson, T. A., Forbes, B. C., Horowitz, L., McKean, M. A., et al. (2003). Conservation and the social sciences. *Conservation Biology, 17*, 649–650.

Meffe, G. K., Nielsen, L. A., Knight, R. L., & Schenborn, D. A. (2002). *Ecosystem management: Adaptive, community-based conservation*. Washington, DC: Island Press.

Neuman, M. (2007). Multi-scalar large institutional networks in regional planning. *Planning Theory and Practice, 8*, 319–344.

Olsen, S. B. (2003). Frameworks and indicators for assessing progress in integrated coastal management initiatives. *Ocean and Coastal Management, 46*, 347–361.

Olsen, S. B., Tobey, J., & Kerr, M. (1997). A common framework for learning from ICM experience. *Ocean and Coastal Management, 37*, 155–174.

Olsen, S. B., Tobey, J., & Hale, L. Z. (1998). A learning-based approach to coastal management. *Ambio, 27*, 611–619.

Olson, D. M., & Dinerstein, E. (1998). The global 200: A representation approach to conserving the Earth's most biologically valuable ecoregions. *Conservation Biology, 12*, 502–515.

Olsson, P., Folke, C., & Berkes, F. (2004). Adaptive co-management for building resilience in social-ecological systems. *Environmental Management, 34*, 75–90.

Ostrom, E. (1990). *Governing the commons: The evolution of institutions for collective action.* Cambridge: Cambridge University Press.

Pfueller, S. L. (2008). Role of bioregionalism in Bookmark Biosphere Reserve, Australia. *Environmental Conservation, 35*, 173–186.

Powell, R. B., Cuschnir, A., & Peiris, P. (2009). Overcoming governance and institutional barriers to integrated coastal zone, marine protected area, and tourism management in Sri Lanka. *Coastal Management, 37*, 633–655.

Preskill, H. S., & Torres, R. T. (1999). *Evaluative inquiry for learning in organizations.* Thousand Oaks, CA: Sage.

Quinn, J. B. (1985). Managing innovation: Controlled chaos. *Harvard Business Review, 63*(May/June), 73–84.

Randolph, J. (2004). *Environmental land use planning and management.* Washington, DC: Island Press.

Riketta, M. (2002). Attitudinal organizational commitment and job performance: A meta-analysis. *Journal of Organizational Behavior, 23*, 257–266.

Roberts, J. (2001). Trust and control in Anglo-American system of corporate governance: The individualizing and socializing effects of processes of accountability. *Human Relations, 54*, 1547–1572.

Rockström, J., Steffen, W., Noone, K., Persson, A., Chapin, F. S., III, Lambin, E. F., et al. (2009). A safe operating space for humanity. *Nature, 461*, 472–475.

Salafsky, N., Margoluis, R., & Redford, K. (2001). *Adaptive management: A tool for conservation practitioners.* Washington, DC: Biodiversity Support Program, World Wildlife Fund.

Schoemaker, M., & Jonker, J. (2005). Managing intangible assets: An essay on organizing contemporary organizations based upon identity, competencies and networks. *Journal of Management Development, 24*, 506–518.

Scott, J. M., Norse, E. A., Arita, H., Dobson, A., Estes, J. A., Forster, M., et al. (1999). The issue of scale in selecting and designing biological reserves. In M. E. Soulé & J. Terborgh (Eds.), *Continental conservation: Scientific foundations of regional reserve networks* (pp. 19–38). Washington, DC: Island Press.

Trombulak, S. C., Anderson, M. G., Baldwin, R. F., Beazley, K., Ray, J. C., Reining, C., et al. (2008). *The Northern Appalachian/Acadian ecoregion: Priority locations for conservation action (Special Report 1).* Warner, NH: Two Countries, One Forest.

Walker, B., Gunderson, L. H., Kinzig, A., Folke, C., Carpenter, S. R., & Schultz, L. (2006). A handful of heuristics and some propositions for understanding resilience in social ecological systems. *Ecology and Society, 11*, Article 13. Retrieved February 1, 2010, from www.ecologyandsociety.org/vol11/iss1/art13/ES-2005-1530.pdf

Western, D., & Wright, R. M. (Eds.). (1994). *Natural connections: Perspectives in community-based conservation.* Washington, DC: Island Press.

Westley, F. (1995). Governing design: The management of social systems and ecosystems management. In L. H. Gunderson, C. S. Holling, & S. S. Light (Eds.), *Barriers and bridges to the renewal of ecosystems and institutions* (pp. 391–427). New York: Columbia University Press.

Whittaker, R. J., Araújo, M. B., Jepson, P., Ladle, R. J., Watson, J. E., & Willis, K. J. (2005). Conservation biogeography: Assessment and prospect. *Diversity and Distributions, 11*, 3–23.

Willemyns, M., Gallois, C., & Callans, V. J. (2003). Trust me, I'm your boss: Trust and power in supervisor-supervisee communication. *International Journal of Human Resource Management, 14*, 117–127.

Wilson, E. O. (1992). *The diversity of life.* Cambridge, MA: Harvard Press.

Yario, P. (2009). Marine protected areas, spatial scales, and governance: Implications for the conservation of breeding seabirds. *Conservation Letters, 2*, 171–178.

Chapter 5
Changing Socio-economic Conditions for Private Woodland Protection

Robert J. Lilieholm, Lloyd C. Irland, and John M. Hagan

Abstract Protecting conservation values on privately owned lands is a significant issue in many parts of the world. Early conservation strategies, which focused on setting-aside public lands from largely unpopulated 'frontier' regions, are becoming an increasingly limited option as populations grow, settlements spread, ownership patterns solidify, and land values rise. Yet as these human-defined boundaries proliferate across the globe and divide lands once wild into privately-owned parcels, the lessons of landscape ecology beckon us toward another view – a view where sharp lines and divisions in ownership are blurred to protect the ecological processes that ultimately sustain us all. These processes have been shaping the human face of New England's landscape for well over 200 years. Here, we recount these changes and pay particular attention to some recent innovations in protecting conservation values on private lands. As we demonstrate, the region's long conservation tradition has spawned some uncommon approaches for sustaining human and natural systems across a landscape that is largely under private ownership. The approaches taken and lessons learned have much to offer other regions of the world seeking ways to creatively bridge the divide between private property and public values.

Keywords Conservation easements • Ecosystem services • Forestry • Landscape change • Land trusts • Land-use planning • New England

---

R.J. Lilieholm (✉)
School of Forest Resources, The University of Maine, 5755 Nutting Hall
Orono, ME 04469-5755
e-mail: robert.lilieholm@maine.edu

L.C. Irland
The Irland Group and Yale University, 174 Lord Road, Wayne, ME 04284
e-mail: lcirland@gmail.com

J.M. Hagan
Manomet Center for Conservation Sciences, 14 Maine St., Suite 305, Brunswick, ME 04011
e-mail: jmhagan@manomet.org

S.C. Trombulak and R.F. Baldwin (eds.), *Landscape-scale Conservation Planning,*
DOI 10.1007/978-90-481-9575-6_5, © Springer Science+Business Media B.V. 2010

5.1 Introduction

In 1783, the aftermath of the American Revolution had left Massachusetts broke and in debt. One avenue through which to rebuild the State's coffers was to sell-off large, remote, and unsettled northern sections of Massachusetts – present day Maine – to wealthy individuals (Coolidge 1963). Massachusetts had wanted to encourage settlement in this remote corner of the state, but it was a tough place to make a living. Glaciers from 12,000 years ago had left the ground full of rocks and unfriendly to tillage. Border disputes with Canada created uncertainties over governance, and shifting alliances among the region's Native Wabanaki peoples led to periodic bouts of conflict and insecurity (Baker and Reid 2004). As a result, by the mid-1800s, the region was still largely unsettled. Then, to compound matters, the opening of the Erie Canal in 1825, followed by railroads and the 1862 Homestead Act, further lured would-be settlers to the fertile soils of the Midwest. In short, it was easier to walk to Ohio than clear trees and grub boulders out of Maine's forests.

These events conspired to leave several largely uninhabited regions across the Northeastern U.S. and Canada, ranging from New York's Adirondack Mountains in the west, across Northern Vermont, New Hampshire and Maine, to large portions of the Maritime Provinces to the east. The limited extent of human settlement in this heart of the Northern Appalachian/Acadian ecoregion (The Nature Conservancy 2007) means that, relative to most inhabited lands elsewhere, the forest and wildlife are remarkably intact. A few species such as wolf (*Canis lupus* or *lycaon*) and woodland caribou (*Rangifer tarandus*) have been extirpated, but most continue to thrive in a vast landscape that has proven surprisingly resilient to nearly two centuries of logging – the primary human activity in the region.

Yet while ecosystems and species have largely persisted, changing land-ownership patterns and global competitive forces are fueling trends that could undermine this resilience. Forest management for timber income continues to be the dominant land use. But land ownership – especially in Maine, which remains almost entirely in private ownership in sharp contrast to nearby New Hampshire and New Brunswick – has shifted away from vertically integrated forest products companies, with a decades-long commitment to sustained-yield timber management, to financial investors with much shorter planning horizons (Irland and Lutz 2007). Further, the production of pulp and paper – the economic bread-and-butter of the region for nearly a century – is gradually re-locating to plantation-based operations in the Southern Hemisphere as North American demand for printing and writing grades of paper decline. Meanwhile, increasingly volatile energy costs are a reminder that wood can still produce heat as it did when humans first harnessed fire thousands of years ago. In fact, heat from wood is becoming just as important to some in the twenty-first century as it was when the Pilgrims settled at Plymouth in 1620.

5 Changing Socio-economic Conditions for Private Woodland Protection

So, what will become of this vast forest given these new and emerging trends? In this chapter, we recount the social and economic forces that have created the landscapes seen today and explore the emerging trends likely to shape its future. We then describe the conservation risks and opportunities these trends pose for one of the most remarkable landscapes in North America.

Our discussion largely focuses on Maine's North Woods, Thoreau's 'damp and intricate wilderness' that remains the largest contiguous block of land without paved road access east of the Mississippi River (Chap. 3). This 4-million-hectare void – visible by satellite at night as a noticeable absence of lights and still recognizable on present-day public road maps – is roughly 95% privately owned, fully exposing it to the market pressures that drive changes in land ownership and use (Birch 1996; Lansky 1992). To place our discussion in context, we include some coverage of the broader 10-million-hectare Northern Forest region that extends from Maine's North Woods across northern New Hampshire and Vermont, to Tug Hill and the Adirondack Mountains of Upstate New York (Dobbs and Ober 1995). We also visit the periphery of this forested region, where growing human pressures for development and recreation foreshadow future drivers of landscape change. Finally, we include limited discussion of the adjacent Canadian portion of the Northern Appalachian/Acadian ecoregion. There, however, despite a similar forested landscape and similar industrial structure, human influences differ, with nearly half of the forest under publicly-held 'Crown' ownership that is increasingly managed for nature protection and the provision of ecosystem services (Floyd and Chaini 2008).

5.2 Changing Human Impacts in the Northern Forest

Like all forested ecosystems, change is a recurring theme in the Northern Forest – especially with respect to human influences. Rumors that the region's large forest industries were ready to sell and move elsewhere were commonplace in the 1970s and 1980s, fueled by concerns that old-time timber families would succumb to the ravages of spruce budworm defoliation, estate taxes, and the steady fragmentation of ownership over successive generations (Irland 1999a). At the same time, fears were widespread that public access to the North Woods could be curtailed through exclusive leasing arrangements with private groups.

While these fears were, in retrospect, overblown, the pace and scale of change since then has accelerated, piloting the region into an uncertain future. Here, we describe the major socio-economic forces affecting the region and assess their likely impact on ecosystem health and protection efforts. In most cases, it is still too early to determine whether these forces will enhance or diminish conservation opportunities. Indeed, making such predictions would be folly in such an unsettled time when financial markets and institutions are mired in recession around the world. Nonetheless, the pace and scale of change warrant a critical inspection of

5.2.1 Changing Patterns of Forest Ownership

Since Colonial times, a succession of forestland owners have come and gone in the Northern Forest, each leaving to varying degrees their mark upon the land. In large measure, the evolution of the region's current forest landscape can be traced back to the late 1890s, when paper companies began amassing vast holdings to supply their growing mills (Irland 2009a; Judd 1997; Wilkins 1978). These land holdings were often dispersed across different portions of a state or even across several states. At the time, 'blocking-up' lands to form large contiguous units was not particularly important for these companies because they often had mills in multiple locations. In fact, until river-driving logs to market was banned by the Maine Legislature in 1974, access to water for transport was more important than overland proximity to a mill.

Many of the large forest-industry holdings that persisted into the 1990s had their historical beginnings in this period, but in most instances, these ownerships did not reach their maximum extent until the 1960s. The largest holdings of Northern Maine were typically in squares called townships (9,324 ha, or 36 mi^2) – and oftentimes included clusters of townships. Even today, it is not uncommon to see 40,000–200,000-ha contiguous blocks in single ownerships.

While the forest industry dominated the region for nearly a century, ownership was by no means static. For example, a number of large land transactions occurred during the 1960s and 1970s – some the result of mergers, others stemming from acquisitions within the industry intended to increase or consolidate existing holdings. But importantly, these sales did not include significant marketing of development tracts, sales outside the forest industry, or even the division of previous ownerships into smaller ones. For all practical purposes, while new owners came and went, the pattern of ownership remained unchanged (Harper et al. 1990).

The calm was illusory, however, as what would become a seismic change in ownership began in 1982 (Anonymous 2005; Northern Forest Lands Council 1994). Indeed, just when it seemed that the large paper companies would reign forever, Sir James Goldsmith, a British financier, purchased Diamond International Corporation, its mills, and 400,000 ha of timberland stretching from Maine to New York (Fallon 1991). Then, in a radical departure from earlier practices, Goldsmith in 1988 began selling-off parcels to a host of new buyers, including both real estate developers and conservationists. The shift was not completely unexpected, however. In fact, Hagenstein (1989) had noted that

> Changes in ownership and use of the large forest holdings are already occurring, and more changes are likely.... The increasing spread between the value of this land for timber growing and for recreation and development puts pressure on current owners.

Goldsmith's actions – and his widely reported profits – signaled the opening of what was to become a massive shift in forest ownership from vertically integrated forest products companies to a host of new owners with mostly or entirely financial interests (Whitney 1989). If forestry and wood products were the highest and best use for a given piece of land, then that was fine for these investors, but if other uses could provide a better return on investment, then that was fine too – at least for many. Moreover, although some new owners entered into decades-long wood supply contracts with paper companies to secure a future income stream, they were also free to consider other options. For example, they might harvest particularly hard on one portion of their holdings to meet contractual obligations, while selling-off another portion for real estate development. For the conservation community and many recreationists, it was easy to imagine the worst.

To observers outside the forest industry, the prospect of a mill selling its 'wood-basket' to financial investors seemed short-sighted, but with the poor economy of the late 1980s and early 1990s and declines in newspaper and magazine advertising, paper companies were struggling and had few options. A convenient solution was to convert hard assets like land into cash. Because Wall Street tended to undervalue timberlands, companies could increase their market value by converting land to cash while securing future wood needs through long-term supply contracts with the new forest owners (Block and Sample 2001; Irland 1999b).

Changes in the federal tax code helped motivate the transfers in ownership away from these publicly-traded corporations. For example, the 1974 Federal Employee Retirement Income Security Act encouraged institutional investors holding pension plans to diversify from traditional fixed-income investments to other assets such as stocks, real estate, and timberlands. Then, the Tax Reform Act of 1986 nearly doubled the effective tax rate for corporate timberland by eliminating preferential treatment of capital gains. Preferential treatment was later restored to individual taxpayers but not to corporations. As a result, individual taxpayers who received pass-through income from investment partnerships, pension plans, and Real Estate Investment Trusts (REITs; Chap. 2) enjoyed significantly lower tax rates (Block and Sample 2001). By 2005, the results were striking, with vast areas of Northern Maine experiencing a sweeping shift in ownership from the forest products industry to these new financial investors (Fig. 5.1) (Hagan et al. 2005).

The transition in ownership raised a host of questions over the future of forest management (Binkley and Hagenstein 1989). For example, some hypothesized that these new financial investors, with timberland decoupled from the cyclic and erratic appetites of pulp and paper mills, might herald the onset of a new era in stewardship (Binkley et al. 1996). The result would be sound silvicultural investments focused on growing the best trees for the highest-paying markets, with non-industrial timberland owners focused on maximizing long-term return on investment. Overlooked was the prospect that investors might instead favor short-term returns over those derived from longer planning horizons. As it turned out, many of these new owners did in fact have comparatively short 10- to 15-year investment horizons. For example, Hancock Timber Resources Group, which purchased 275,000 ha of timberland in the mid-1990s, had completely divested its lands by 2005. In the end, many of these new owners sought a balance between generating cash flow from

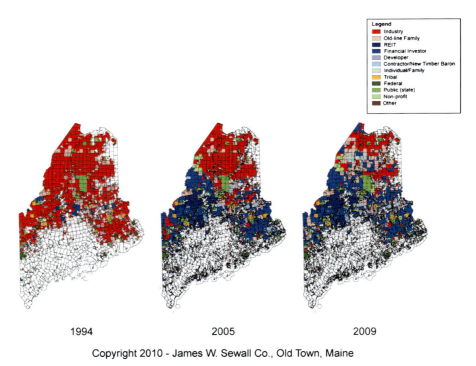

Fig. 5.1 Map of Maine timberland ownership by owner type in 1994 and 2005 (reprinted with permission from Hagan et al. 2005, 2009). Land ownership data provided by James D. Sewall Co. of Old Town, Maine

timber sales and maintaining asset value through some level of forest management, albeit at levels below prevailing industry standards.

5.2.2 Rising Development Pressures

As the once-stable ownership patterns of the Northern Forest unraveled to include a host of new owners and interests, growing pressures for real estate development further clouded the region's future. Here, remoteness and limited access failed to provide shelter from prevailing national trends. Indeed, over the previous 30 years, the U.S. had witnessed an explosion in suburban and second-home 'recreational' development, driven by growing populations, rising incomes, and favorable tax policies (Alig et al. 2004; Stein et al. 2005; White and Mazza 2008; White et al. 2009; Chap. 2). And by the mid-1980s, four-lane highways had opened-up even the farthest reaches of the region to fast and convenient auto travel. Ready access to credit and a widespread belief that real estate values would rise forever further fueled land speculation, parcelization, and home construction. In the Northern

5 Changing Socio-economic Conditions for Private Woodland Protection

Forest and its fringes, much if not all of these changes were for leisure and recreational uses.

These trends, evident across the region, were not entirely new (Payne et al. 1975). In rural and coastal Maine, for example, amenity-based second-home and recreational development had long been a driver of rural land-use change and local economic growth, with Maine ranking first in the nation in the percent of its housing stock in seasonal use (16% – or five times the national average) (Bell 2007; Ednie et al. 2010). But the scale and pace of development were new, and conservationists viewed these changes with alarm. Indeed, prime amenity development targeting the region's most remote lakes, rivers, and shorelines can diminish recreational opportunities, erode the health of forest and aquatic systems, and introduce into previously natural areas a host of human influences, including noise and light pollution, access roads and traffic, and a growing list of invasive species (Small and Lewis 2009; Wiersma 2009).

As available lakefront properties were developed, builders moved upslope to 'view lots' – locations where evening lights can undermine the feeling of remoteness for great distances. Furthermore, as construction increased, so too did prices. At one time, even citizens of modest means could and did enjoy having a rustic lakeside 'camp.' By 2005, however, escalating prices made such aspirations within reach of only the wealthy, and these newer 'camps' were increasingly full-featured high-end homes, oftentimes described as 'executive retreats.' During this land boom, many tracts of 16.4 ha were sold to avoid subdivision regulations – raising the prospect of further subdivision in the future (Chap. 13).

Unlike other regions of the U.S., rising population has been a minor driver of subdivision and new construction. For example, Maine lost an average of 440 residents per year during the 1990s (Brookings Institution 2006), and since 2000 added just 4,300 new residents each year – a rate less than half the national average. Canada's Maritimes also grew slowly – increasing just 5.2% from 1981 to 2006, compared to a 30% gain for all of Canada (Statistics Canada 2008). Indeed, more important to the Northern Forest than population increase was the dispersed or 'sprawling' nature of new development, especially along its outer fringes where forests met more settled regions.

For example, according to the Brookings Institution (2006), 77% of Maine's population growth between 2000 and 2005 occurred outside of established municipal centers or 'regional hubs.' In Northern Maine, virtually all new growth took place outside of these hubs, with most hubs actually losing population. In addition to being scattered beyond established town centers, much of Maine's new residential growth was land intensive, averaging 4 ha per dwelling unit. Compared with national data, between 1980 and 2000 Maine ranked second in the percent of rural land converted to development, and third in the amount of land developed per housing unit (Brookings Institution 2006). A recent U.S. Forest Service study warns of more to come, with several Maine watersheds ranked among the nation's top 15 in the amount of forested area projected to experience increased housing densities by 2030 (Stein et al. 2005; White and Mazza 2008; White et al. 2009).

5.3 The Rise of New England's Land Conservation Movement

Efforts to conserve important tracts or characteristics of the New England landscape have waxed and waned over the past 150 years, driven by changing people, policies, and public consciousness. Here, we recount several important eras in conservation and highlight some of the changing roles played by individuals, government, and, more recently, conservation-based non-governmental organizations (NGOs). We illustrate these trends with a series of case studies and conclude the section with an overview of accomplishments in Maine over the last two decades.

5.3.1 Early Land Conservation Efforts

The late 1800s witnessed the birth of New England's land conservation movement – a movement that would have profound impacts across the U.S. and around the world (Foster 2009). The creation of Adirondack State Park in 1892 proved seminal (Porter et al. 2009) and was followed by a series of Progressive Era laws that led to the creation of the National Forest System in 1897, the U.S. Forest Service in 1905, and the National Park Service in 1916. The Weeks Act of 1911 was particularly significant for the Eastern U.S., clearing the way for a series of new National Forests in the East that would soon include the White Mountains in New Hampshire and Southwestern Maine (1918) and the Green Mountains in Vermont (1932). Also important was the 1906 Antiquities Act, which, through some creative thinking allowed President Theodore Roosevelt to establish the first National Monuments and Wildlife Refuges (Dana and Fairfax 1980).

Ironically, these early federal-level conservation efforts, often fueled by New England intellectuals, politicians, and philanthropists, largely bypassed the State of Maine. Instead, many trace Maine's land conservation tradition to 1919, when John D. Rockefeller, Jr., and other wealthy benefactors provided early support for the creation of Acadia National Park along Maine's rugged 'Downeast' coast (Cronan et al. 2010). Later, in 1931, Governor Percival Baxter donated to the people of Maine what would eventually become today's 83,000-ha Baxter State Park (Fig. 5.2). These early actions – pioneered by private citizens – helped forge the foundation of the publicly protected lands in this region today.

As the Progressive Era waned, the nation entered tumultuous times – the Great War, the Roaring Twenties and subsequent collapse of the stock market, the Great Depression, and World War II. Throughout this period, public and private land conservation efforts across the Northern Forest entered a period of hibernation, only to be revived by the environmental movement of the 1960s and 1970s. This new era saw the creation of the Allagash Wilderness Waterway and its designation as a National Wild and Scenic River, the creation of land-use institutions for the 'unorganized' regions of Northern Maine that lacked local government, and the recovery of 'public reserved lands' into public management and their consolidation

Fig. 5.2 Maine's Mount Katahdin, shown here from Sandy Stream Pond, draws some 55,000 recreationists to Baxter State Park each year (Photo by Spencer R. Meyer)

into premier conservation tracts such as Grafton Notch State Park, the Bigelow Range, Round Pond on the Allagash River, and several important lakes along Maine's Downeast coastal region.

Toward the end of that period, perhaps unknowingly, Maine pioneered what was to become the future of land conservation – negotiating Great Northern Paper Company's historic 1980 donation to the State of an 8,000-ha conservation easement along a heavily used recreational corridor of the Penobscot River. This was one of the State's first large-scale conservation easements, whose significance was unfortunately overshadowed by controversy over Great Northern's ill-fated 'Big A' dam proposal (Palmer et al. 1992).

5.3.2 The Public Sector's Role in Land Conservation

As ownership patterns unraveled across New England's forests in the 1980s, state and federal governments were the first to respond (Lilieholm and Romm 1992). The Northern Forest Lands Study and the Governors' Task Force on Northern Forest Lands, both created in 1989, were charged with assessing the future outlook for the entire 10-million-hectare Northern Forest region (Lilieholm 1990). These efforts culminated with the final report by the Northern Forest Lands Council (1994), which generated widespread awareness of the region and the threats to it, and served a valuable role in educating policy makers and, perhaps more importantly, the press. While the report led to only minor policy changes (Kingsley et al. 2004), the Council's actions were important in garnering significant land protection resources across the Northern Forest.

In 1987, Maine's Legislature created the Land for Maine's Future (LMF) Program to protect important open spaces across the state. Since then, four public bonds totaling $117 million have passed with overwhelming support, allowing LMF to work with a wide array of partners to assist in the voluntary protection of over 200,000 ha, including 100,000 ha under conservation easements (Barringer et al. 2004; Cronan et al. 2010; Irland 1998).

In recent years, the federal role in conservation has been largely that of cash provider to state and local governments. Important funding sources include the 1964 Land and Water Conservation Fund and the Forest Legacy Program (FLP) – a voluntary program created under the 1990 Farm Bill and administered by the U.S. Forest Service that works with states, land trusts, and others to protect ecologically significant private forests from development. The availability of funding through FLP was pivotal as it enabled a succession of important earmarks. The institutionalizing of this process was evidenced by glossy, full-color acquisition proposals complete with maps and tables prepared by environmental coalitions itemizing their wish lists for each Congressional season. Since its first appropriations in FY 1992, FLP funding has helped protect over 273,000 ha of working forests in Maine (440,000 ha regionally), mostly through cost-sharing arrangements in the acquisition of conservation easements. In contrast, direct federal land acquisition for conservation purposes was much more limited, including, for example, the 13,000-ha Appalachian National Scenic Trail corridor in Western Maine between 1986 and 2002, and the 4,000-ha Sunkhaze Meadows National Wildlife Refuge in 1988.

5.3.3 The Rise of Conservation-Based Non-governmental Organizations

Beginning in the 1990s, national-level conservation NGOs began to emerge as major policy actors in the New England states. An early landmark was The Wilderness Society's 1992 conference on the Northern Forest held at the University of Vermont (Lepisto 1992). Serving as advocates, mobilizers, and dealmakers with largely private capital, NGOs like The Nature Conservancy, the New England Forestry Foundation, the Forest Society of Maine, Appalachian Mountain Club, and others came to be land- and easement-holders across large swathes of the Northern Forest, protecting vast acreages with dollar amounts far exceeding what state and federal governments could muster. These NGOs offered industrial forest owners 'bankable' deals, buying large tracts all at once. In some instances, it is likely that had a conservation buyer not appeared, no other buyer could have assembled the resources needed to acquire these parcels in their entirety (Ginn 2005).

Approaches to land conservation evolved over time and gained in sophistication, moving from fee simple acquisition to include the purchase of conservation easements

5 Changing Socio-economic Conditions for Private Woodland Protection

that sever development rights from land, working forest easements that limit land use to third-party certified sustainable forestry, and long-term timber supply contracts between landowners and wood processors. The transition was rapid, with business executives, private conservation groups, and public officials implementing novel concepts, almost, it seems, improvising as they went. Through all this, the bulk of the land remained as working forest, owned, in a novel way, by private investors.

Three case studies, described below, help to illustrate just a small range of actors and methods used to conserve landscapes in Maine.

The Pingree Easement David Pingree was a successful Salem shipping magnate who first purchased timberland in Maine in the 1840s. Totaling nearly 400,000 ha in Northern and Western Maine, these lands had remained relatively intact, passed down through successive generations. By the 1990s, many of Pingree's heirs were concerned that as their number grew through time, increasingly diverse interests might lead to the eventual break-up of the ownership. In 1996, the New England Forestry Foundation (NEFF) – a non-profit forest conservation and management organization based in Massachusetts – approached Pingree Associates, which represents the approximately 100 landowning heirs, about purchasing a conservation easement on the Pingree lands.

NEFF's interest in conserving working forestland in New England coincided with the growing interest of many Pingree heirs to keep their ownership intact. NEFF's philosophy is that private forest ownership creates strong connections between people and the land, and that easements ensure that these connections are maintained through time (K. Ross, pers. comm.). The easement mechanism was a perfect fit. Negotiations were completed in March 2001 when $28 million was paid to Pingree Associates for a permanent conservation easement on 308,448 ha of working forestland – a cost of roughly $91/ha. The remaining 102,000 ha of ownership were excluded, thus giving heirs some options for future development.

The agreement was an historic event, representing the largest conservation easement in U.S. history. The land would remain as forest in perpetuity, managed under sustainable forest management practices. As noted in the easement's language:

> It is the purpose of this Conservation Easement to maintain the Property forever in its present and historic primarily undeveloped condition as a working forest, and to conserve and/or enhance forest and wildlife habitats, shoreline protection, and historic public recreation opportunities of the Property for present and future generations.

The easement's restrictions were simple compared to the language in later agreements that would guide future management on other parcels. Indeed, unlike many subsequent easements involving the new forest owners who were primarily financial investors, the Pingree Easement was driven by a deep-seated family land ethic combined with a goal of preserving forever the legacy of David Pingree. No particular certification system was required, although the lands were and remain dual-certified under the standards of the Forest Stewardship Council and the Sustainable Forestry Initiative. In this respect, the simplicity of the restrictions was central to the deal's success. The Pingree Easement took place independent of the many land ownership changes occurring across New England between 1995 and 2005.

A few other conservation easements with long-held family forestland occurred, but most of the large-scale conservation opportunities came about as a result of industry sales and new investor owners seeking to monetize the conservation assets of their newly-acquired holdings over and above timber values.

The Downeast Lakes Forestry Partnership In 1999, Typhoon LLC, a relatively new timber investment management organization (TIMO), purchased 180,000 ha of timberland from Georgia-Pacific in Eastern Maine. Nearby residents – fearing large-scale parcelization, lakefront development, and loss of access for both recreation and forest products – responded by forming the Downeast Lakes Land Trust (DLLT) – a community-based partnership of lodge owners, foresters, and Registered Maine Guides whose livelihoods depended upon continued access to the woods. Inspired by the recent Pingree Easement, DLLT approached NEFF to explore ways in which the former Georgia-Pacific lands could be protected from future development while continuing to function as working forests.

The result – the Downeast Lakes Forestry Partnership (DLFP) – was completed in 2008 and protected roughly 138,000 ha through three separate transactions. The first involved the sale of an 80-km conservation corridor (1,222 ha) along Spednic Lake and the Upper St. Croix River, terminating at the U.S.-Canada border. The second transaction involved the purchase by DLLT of nearly 11,000 ha near the town of Grand Lake Stream to be managed as a community forest to support sustainable local ecotourism. Finally, NEFF purchased a 126,000-ha conservation easement – the second largest in the U.S. history – at a cost of roughly $100/ha.

The DLFP is noteworthy on several counts. First, the Partnership's origin was truly grassroots, yet it grew to include a diverse set of interests, including WalMart, The Conservation Fund, The Nature Conservancy, U.S. Fish and Wildlife Service, the Woodie Wheaton Land Trust, Passamaquoddy Tribe of Indian Township, the National Wildlife Federation, the State of Maine, Wagner Forest Management, the National Fish and Wildlife Foundation, and others. Moreover, the majority of funding came from private sources, with protection achieved at modest cost. Finally, while not as large as the Pingree Easement, the DLFP lands cover roughly one-quarter of Washington County and protect 2,400 km of stream and river shoreline, 60 lakes and ponds with 700 km of shoreline, and 22,000 ha of wetlands. Moreover, these lands remain open to public use, with two million-dollar endowments in place to fund future monitoring and management needs.

Plum Creek's Concept Plan for the Moosehead Lake Region Moosehead Lake is considered by many to be the crown jewel of Maine's North Woods. Covering 31,000 ha and with 640 km of mostly undeveloped shoreline, the lake's crystal-clear waters have drawn visitors from near and far for over a century. Like elsewhere in Northern Maine, large industrial ownerships controlled much of the Moosehead Lake region, providing stability to the landscape for residents and recreationists alike.

The region's future as a source of wood and backcountry recreation was called into question, however, in 1998, when Seattle-based Plum Creek Timber Company acquired 365,000 ha of land from Scott Paper Company – including large areas around Moosehead Lake and 100 km of shoreline. Plum Creek, the largest private landowner in the U.S., had earlier evolved from a timber company to a REIT in order to take advantage of tax laws and opportunities to develop parcels on its massive holdings. As a REIT, Plum Creek viewed its newly acquired lands as more than just timberland and began identifying areas for 'highest and best use' development.

Plum Creek's Moosehead properties are located within the jurisdiction of Maine's Land Use Regulation Commission (LURC), created in 1971 to manage growth and protect environmental quality on the 4.2 million hectares of unincorporated lands spread across Northern and 'Downeast' Maine (Bley 2007). In 2005, Plum Creek submitted to LURC a 30-year *'Concept Plan for the Moosehead Lake Region'* (Plum Creek 2007). The plan sought to rezone approximately 6,400 ha from forestry to development. Included were plans for two large resorts, 975 house lots, and scattered commercial development. Approximately 4,330-ha – just 4% – would be permanently protected.

Many residents in the Moosehead Lake region supported Plum Creek's plan, welcoming the prospects of new jobs and economic development in the beleaguered region. Indeed, unknown to many, the Moosehead Lake region had reached its peak as a tourist destination in the late 1800s and early 1900s, when four rail lines and several steamships ferried visitors to the region's inns, resorts, camps, and lodges. Surrounding farms – now forests – provided fresh meat and produce to the thousands of people that lived in and visited the region. But as tourism declined in the 1930s and 1940s, the great lodges and hotels fell into disrepair, leaving logging as the region's main source of income.

At the state level and beyond, however, Plum Creek's proposal drew intense criticism as diverse groups expressed outrage over the company's plan for 'sprawl in the wilderness,' describing it as Maine's 'largest-ever development proposal.' After years of public comment, hearings, and study, LURC approved a modified version of the plan in October 2008 (Land Use Regulation Commission 2009). Under the new plan, Plum Creek would permanently protect about 160,000 surrounding hectares through the donation and sale of lands and easements to be held by The Nature Conservancy, the Forest Society of Maine, and the Appalachian Mountain Club. Plum Creek agreed to the modifications, and LURC unanimously accepted their rezone petition in September of 2009, although in October 2009, the Natural Resources Council of Maine filed a lawsuit in Maine Superior Court to appeal LURC's Plum Creek decision.

The Moosehead controversy is noteworthy for several features, including the sharply contrasting views expressed toward the plan. For example, many critics lament the planned development of 6,400 ha in the heart of this relatively undeveloped region, viewing it as a dangerous foothold and harbinger of things to come. Others focus on the permanent protection of nearly 160,000 ha – more than 96% of the total project area – and view it as a conservation milestone and benchmark for other development proposals. For example, the Moosehead Forest Conservation Project – a regional vision created by The Nature Conservancy, the Forest Society of Maine, and the Appalachian Mountain Club – places the Plum Creek Concept Plan within a regional setting of interconnected conservation lands that exceed 810,000 ha and span from Mount Katahdin to the Canadian border (Fig. 5.3). Going

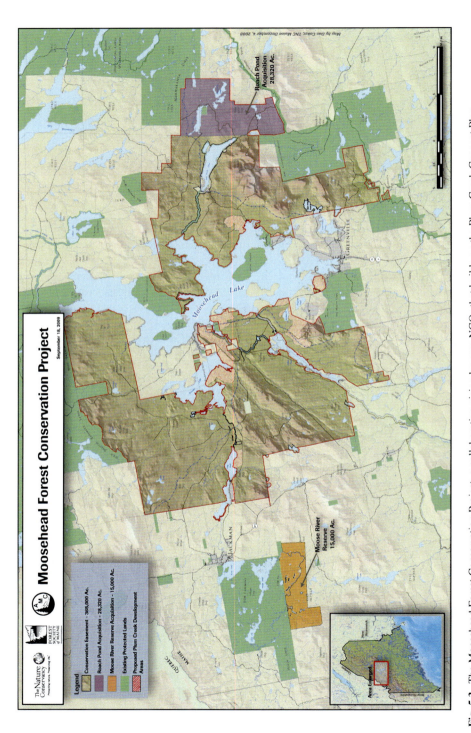

Fig. 5.3 The Moosehead Forest Conservation Project, a collaborative vision by area NGOs that builds on the Plum Creek Concept Plan

forward, the Moosehead controversy suggests that future large-scale development proposals will also include large-scale conservation opportunities – a 'carrot and stick' approach that may serve to bridge both public and private interests in conservation and development (Lilieholm 2007).

As demonstrated by these three case studies, over time, as ownerships changed and the goals and capacities within the conservation community grew, forestland – even when recently cut-over – had significant value to the conservation community and others as either fee simple ownership or under no-development easements. Moreover, easements were becoming commonplace, and tens of millions of dollars were raised to fund conservation deals across the region.

After the success of the Pingree Easement, a major 'land rush' for both fee purchases and easements began, and many new financial investors were anxious to enter the market. During these years, opportunities arose at a rapid pace, oftentimes constrained by the supply of willing sellers. Ironically, it was industry sales and these new landowning investors that made possible much of the conservation gains in this region over the last 20 years. Yet at the same time, conservationists were becoming better versed in their demands and expectations for how these lands were to be managed under easements.

Throughout this period, more than 100 Maine-based land trusts supported to varying degrees the dozen or so large conservation NGOs operating across the state (Cronan et al. 2010). These trusts, of differing size and capacity, were mostly based in Southern and Coastal Maine, but often had conservation interests that spanned the state. And unlike many other regions of the U.S., their missions typically surpassed ecosystem protection to include rural economic development goals through the protection of working forests (Fairfax et al. 2005).

Looking back, while state and federal governments initiated early protection efforts in the 1980s, few could have envisioned the dramatic role that would soon be played by private, nonprofit land conservation groups (Cronan et al. 2010; Ginn 2005). These efforts, championed by a new set of actors pursuing novel ideas and drawing on new sources of funding, would preside over the conversion of hundreds of thousands of hectares once held by private timber companies into quasi-public assets sheltered from speculative land markets and protected in perpetuity as forests for ecosystem health and recreational use. In many ways, New England's Post-Progressive Era conservation legacy is largely their story. It is also important to note, however, that individuals did not entirely leave the scene. Indeed, some contemporary actors are reminiscent of early public-minded benefactors, like Burt's Bees co-founder Roxanne Quimby, who acquired nearly 36,000 ha of Northern Maine for ecosystem protection – 75% of which lies adjacent to Baxter State Park (Chap. 3).

Through these combined resources, Maine's conservation lands (defined here as areas where development, or other land uses deemed incompatible with conservation objectives, are prohibited or are very strongly limited through ownership control or deed restrictions, including both public and private lands [Cronan et al. 2010]) increased from 445,000 ha in 1993, to 1.5 million hectares at the end of 2006 – an increase from 5% to 17% of the state's 8.6-million-hectare land base (Fig. 5.4)

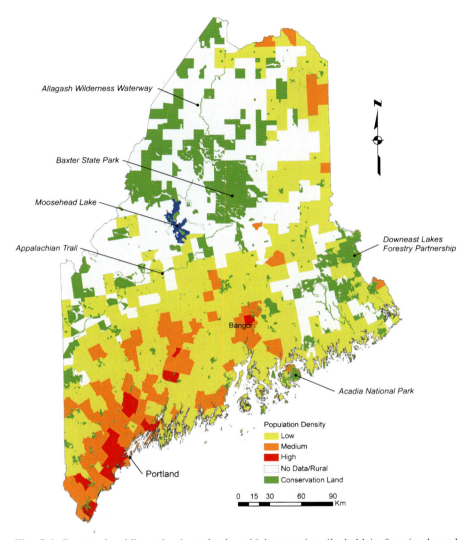

Fig. 5.4 Protected public and private lands, which are primarily held in fee simple and conservation easements, along with a small amount of leased land. Also shown is population by Census Place as of 2000 (From Lilieholm 2007) (Map by Jill Tremblay)

(Cronan et al. 2010). During this period, federal lands increased by 11%, state conservation lands increased by 52%, municipal holdings went up by a factor of 5.5, and land trust holdings increased 58-fold (Cronan et al. 2010). As of 2007, 55% of these lands were held in easements (Fig. 5.5). While the state is the largest fee owner of conservation lands (more than 400,000 ha), land trusts hold easements on nearly 700,000 ha and account for almost 850,000 ha of total lands conserved. In fact, the largest share of conservation lands protected through both ownership and easements is held by non-profit land trusts and conservation organizations (57%),

5 Changing Socio-economic Conditions for Private Woodland Protection

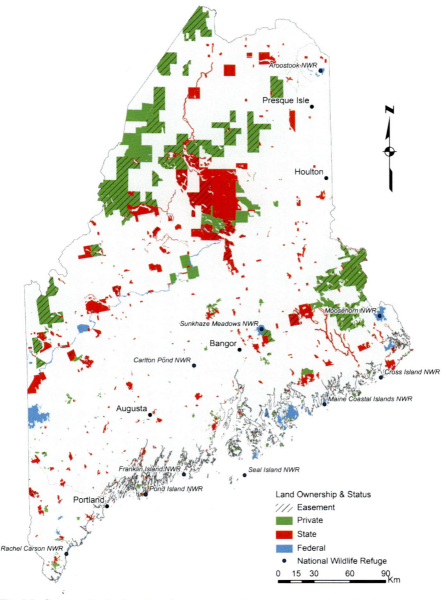

Fig. 5.5 Conservation lands under private, state and federal ownership, including lands protected through conservation easements (Sources: Land for Maine's Future and The Nature Conservancy; map by Jill Tremblay)

followed by the state (35%), the federal government (5%), and municipalities (3%) (Cronan et al. 2010).

Across the border in Canada, land conservation took a much different course. For example, unlike Maine, which is 5% publicly owned, Crown lands comprise half of New Brunswick's forests, and 31% of Nova Scotia's (Floyd and Chaini 2008). Of the remaining private forest lands, the forest industry owns one-third in Nova Scotia and one-half in New Brunswick (Floyd and Chaini 2008). As a result, private conservation efforts have developed with less urgency, largely due to the region's greater presence of public lands and established conservation policies.

For example, while over 100 local land trusts operate in Maine, as of 2010, just two national-level trusts operated in the two provinces – Nature Conservancy Canada (9,734 ha protected almost exclusively through fee ownership), and Ducks Unlimited (3,971 ha held in fee, plus an additional 8,118 ha where Ducks Unlimited holds a long-term restrictive covenant). In New Brunswick, three other land trusts operated as of early 2010. The largest, the Nature Trust of New Brunswick, was established in 1987 and is most active, having created 29 nature preserves covering roughly 2,000 ha. In Nova Scotia, the Nova Scotia Nature Trust was created in 1994 and protects 1,500 ha. While lagging behind U.S. efforts in area protected, some important polices are moving forward to encourage conservation on private lands in Canada, such as a 2008 law that exempts land trusts from property taxes, as well as a program that provides income tax benefits to landowners who donate land or undertake conservation easements under the Ecological Gifts Program administered by Environment Canada.

5.4 Emerging Issues

The changes playing out across the Northern Forest are many and complex, and present enormous challenges to sustaining the long-term ecological health of the region. Understanding these changes and anticipating future trends and impacts are critical first-steps in devising methods to sustain the region's human and natural systems. Below, we outline some of these changes and assess how protection efforts can continue to move forward in a period of fluid and dynamic change.

5.4.1 Increasing Ownership Fragmentation and Landscape Complexity

The changes occurring across the Northern Forest can be viewed from many perspectives. Fragmentation is one common feature that cuts across the landscape and, in the Northern Forest, manifests itself in a number of new and distinct forms. Fragmentation resulting from the conversion of forestland to development is typically the most visible form – and the form that often generates the greatest degree of public concern. But underlying and oftentimes fueling this conversion process are less-visible changes, such as the fragmentation of large landholdings into smaller parcels in an effort to maximize financial returns or raise quick capital. Another is ownership fragmentation – the transition from a landscape dominated by forest industry control, to one held by a myriad of new owners, most notably financial interests like TIMOs and REITs. These new owners hold diverse values and hence pursue a diverse and complex range of objectives. These in turn affect future land-use decisions ranging from the intensity of forest management to the degree to which parcels for subdivision and development are identified.

It is useful to note that the ownership fragmentation of large parcels in remote areas is not necessarily harmful in itself. Indeed, oftentimes the remaining fragments continue to constitute extensive tracts that are managed and function as before. Little noticed, however, is that all around the fringe of the Northern Forest is a landscape of former farms and small tracts. These parcels will continue to be broken into smaller bits by inheritance, 'liquidators,' and subdivision. The result is that the avenues by which visitors enter these wildlands will sprout an increasing number of 'No Trespassing' signs, even along remote roads and trails. In addition, major roadways to gateway communities will see increased linear sprawl, diminishing many natural and public values.

In addition to fragmenting forests, parcels, and ownership classes, the Northern Forest has experienced a relatively new form of fragmentation over the last 15 years – the identification, valuation, and sale of various aspects of value inherent in land (Lilieholm 2007). For example, Maine's forestlands were initially valued for their ability to supply timber and were thus held in fee simple ownership by the forest industry as a secure source of timber for its mills. By the 1990s, however, the diversification potential of holding forestland as part of a larger portfolio of assets was recognized, and vast areas were acquired by financial interests to diversify portfolios and reduce risk (Zinkhan et al. 1992). Next, 'highest and best use' values were identified and sold-off for development, followed by growing markets for conservation easements and, more recently, new markets for ecosystem services like carbon sequestration (Pagiola et al. 2002; Small and Lewis 2009). All of these market-based innovations expanded the perception of the values inherent in land, which in turn fueled rising land prices as new players entered the market to capture one or more of these values (Levert et al. 2007).

While conservationists typically shun fragmentation – especially on the landscape level – some of these changes presented extraordinary opportunities for ecosystem protection. For example, the growing acceptance of conservation easements, which offer for sale a parcel's development potential, has allowed for the protection of millions of hectares of Northern Forest at far less cost than traditional protection through fee simple ownership – for example, less than \$100/ha in the Pingree and Downeast Lakes case studies described above. For many landowners, including the forest industry, this innovation in the land market has allowed for the sale of development rights while retaining 'working forests' as part of a productive landbase, sometimes under long-term timber supply agreements. As they develop, emerging markets for carbon offsets and other ecosystem services have the potential to benefit an even broader array of forest landowners.

These changes, however, are not without risk. Indeed, the increased 'monetization' of land has fueled highest and best-use strategies that have raised investor expectations for financial returns and fueled development pressures on some of the region's most valued lakes, waterways, and mountain tops. Moreover, as the number of conservation lands increase across the landscape, these sites are beginning to act as magnets for new development, a process that undermines past conservation investments while limiting the ability to expand protected areas and create connecting corridors (Radeloff et al. 2010). Furthermore, on a more basic level, the rise of markets for ecosystem services could undermine long-standing public expectations regarding landowner duties to protect environmental quality. For example, to what extent do emerging markets for ecosystem services in effect transfer the ownership of long-established public goods like clean water and wildlife habitat to the private realm, with private interests demanding payment to provide what was previously required through regulation. This is an important point. In fact, under long-standing common law, many of these ecosystem services cannot be privately owned. They have always belonged to the state, so could never have been sold to the current holders of fee title. Aside from this legal note, it appears that the number of groups wishing to collect payments for ecosystem services is likely to far exceed those willing to write the checks. Despite this, interest in creating markets for ecosystem services has generated a vast literature (cf. Benson et al. 2009).

From a landscape-level conservation and management perspective, fragmentation in all its many forms increases the complexity of the landscape and thus presents challenges to identifying and realizing a comprehensive vision for the region's future. This fragmentation and complexity extends to the regulatory framework as well, with federal, state, and increasingly, local controls. In an interesting twist given recent trends in de-regulation, some entirely new quasi-private regulatory frameworks have been willingly created by buyers and sellers as part of the easement process, such as third-party oversight obligations to monitor development and certify forest management practices.

This public-private regulatory framework is complex. At the federal level it includes wetlands protections, air and water quality regulations, as well as species habitat protections under the 1973 Endangered Species Act. Here, decisions over critical habitat designation for the listed Canada lynx (*Lynx canadensis*) have the potential to affect 2.7 million hectares of Northern Maine (Vashon et al. 2008). At the state level, a host of regulations enforced by the state and/or municipalities affect forest practices and development, including the Mandatory Shoreland Zoning Act (1971), the Forest Practices Act (1989), provisions under the 2005 Timber Harvesting Standards to Substantially Eliminate Liquidation Harvesting, and newly-enacted rules to protect significant vernal pools under the Natural Resources Protection Act. The extent to which these regulations can effectively be implemented by fractured jurisdictions across a fragmented landscape remains to be seen (Chap. 4).

Looking forward, this complexity – embodied in more players holding diverse interests and a wide range of legal rights to common parcels of land – suggests a vast potential for conflict. Indeed, the emergence of conservation easements as the dominant form of landscape protection in Maine places much faith – in this region and elsewhere – in a relatively new and largely untested conservation approach (McLaughlin 2005). Already, first-generation easements are creating management challenges due to vague wording and new and unanticipated issues (Pidot 2005). These lessons are being incorporated into the language of newer easements, but the future is always uncertain, suggesting that challenges will remain (Anonymous 2008; deGooyer and Capen 2004; McLaughlin 2006).

Moreover, conflict over emerging issues will likely involve an ever-growing number of stakeholders with a legal interest in the condition and products of these lands. Some will favor development, some conservation. Some will push for recreational access, while others might seek to exclude all use. Still others will desire greater timber harvests in support of local communities. These issues will be contentious but will probably pale in comparison to the social, economic, and ecological stressors likely to arise from global climate change (Jacobson et al. 2009).

All of these factors highlight the need for accurate, cost-effective, and timely assessments of the many social, economic, and ecological values embodied in land. Some standards and protocols are already in-place, like timber appraisal methodologies for financial reporting and monitoring protocols required under third-party environmental certification. Other methods are being developed, such as the use of remotely sensed data to reduce the need for expensive on-the-ground monitoring (Williams et al. 2006). Maine's new law requiring that easements be recorded with the state adds another layer of accountability and should help protect the public's legitimate interest in ensuring easement compliance given the tax advantages that landowners often gain from such transactions.

5.4.2 The Need for a Landscape-Level Perspective in Conservation

The rapid increase in protected lands across the Northern Forest has raised concerns over the extent to which these isolated parcels represent important ecological zones and are able to sustain landscape-level processes that can support species migration and gene flow. The concern is not new. Maine's early public reserves were in fact widely-scattered and small patches, and in cases like Acadia National Park, included significant privately-owned 'in-holdings' as well. However, our improved understanding of landscape ecology calls for a more strategic and comprehensive approach to ecosystem protection (Baldwin et al. 2007a; Cronan et al. 2010; Chaps. 14 and 16).

An illustrative example is Baxter State Park – a green rectangle of over 80,000 ha that stands out on any map of the State of Maine. Basic concepts of landscape ecology were largely unknown when Governor Percival Baxter made his initial land purchases in the 1930s, and he simply acquired what he could at the time. As a result, the park is bounded by straight lines, which have little bearing on natural systems present there, and public and private efforts now seek to 'round out' the Park's linear borders to include environmentally important areas surrounding the core massif.

As ownership transition continues across the Northern Forest, remaining large contiguous private parcels become particularly important for the protection of biological diversity. Indeed, these large parcels represent a dwindling opportunity to realize any comprehensive vision across large, contiguous watersheds and landforms (New England Governors Conference 2009). Once these parcels are fragmented among different owners – especially those with differing goals and objectives – opportunities are lost, perhaps forever (Wiersma 2009).

Yet in the Northern Forest, an important distinction must be made between large parcels and large sales. For example, one recent offering of timberland totaling more than 80,000 ha across the Northern Forest included lands in three states, with the largest parcel being about 28,000 ha. Another example was the approximately 315,000 ha sold by Boise Cascade Mead and MeadWestvaco in the mid-1980s that was spread across three states – a significant portion of which consisted of numerous scattered blocks of a few hundred to a few thousand hectares. In short, it is easy to let a number like 315,000 ha become reified into an imaginary 315,000-ha contiguous block that, in fact, never existed. Fortunately, a number of efforts are gaining ground across the region to consolidate protection across landscapes (Foster et al. 2010). These efforts represent a nascent foundation for comprehensive approaches to protect the Northern Forest.

5.4.3 A Struggling Forest Products Sector

Forest management for timber and income has been a dominant activity across the Northern Forest since settlement, and the fate of the region and its communities has always been closely aligned with that of the forest products sector (Fig. 5.6). In the

Fig. 5.6 Forest products processing is Maine's largest manufacturing sector, contributing over $5.3 billion to the Maine economy each year, and providing nearly one-third of the state's manufacturing jobs (From North East State Foresters Association 2007) (Photo by Spencer R. Meyer)

5 Changing Socio-economic Conditions for Private Woodland Protection   89

1840s, the City of Bangor – the principal gateway to Maine's vast North Woods – was recognized as the Lumber Capital of the World (Lilieholm 2007). Later, the region's forests and abundant hydropower – at the time, the largest privately owned hydroelectric network in the world – supported a major paper industry that dominated North American markets for much of a century (Irland 1999b, 2004a, b, Irland 2009b).

Today, the social and economic importance of the forest sector is being questioned due to mill closures, recession, and steady declines in overall employment (Levert and Lawton 2009). Indeed, many of the region's historical advantages are fading or simply becoming irrelevant. For example, in an era of rising energy costs and regional power grids, why use hydropower – regardless of its low cost – to make low-value newsprint? Moreover, lumber made from small-diameter trees is a side-product of the paper industry, which requires woodchips. More to the point, the region's hardwoods are slow-growing and of low quality, and neither newsprint nor framing lumber from Maine's small trees are competitive in today's global markets – hardly a base for a prosperous future (Irland 2001, 2005).

In years past, many viewed the economic viability of the forest industry as an important restraint on real-estate development, but the ever-widening gap between forestland's value for timber versus development has diminished this role. Indeed, under industry control, investments in stand productivity through planting, herbicides, and pre-commercial thinning enhanced growth as much as 100% over that of unmanaged natural stands. But the shift away from industry control has diminished investment levels, and in future years, many intensively managed stands will not be maintained as such.

Yet despite these trends, overall harvest and production levels remain high by historic measures, and the industry continues to provide nearly one-third of Maine's manufacturing jobs, income, and value-added (North East State Foresters Association 2007). Moreover, the forest products sector is recognized as an important industrial cluster in Maine's economic development strategy, and despite limited state resources, receives widespread support (Bilodeau et al. 2009; Colgan and Baker 2003). Finally, emerging technologies for engineered wood products and forest-based 'bioproducts' and biofuels increasingly offer hope to a struggling industrial sector (Benjamin et al. 2009; Damery et al. 2009; Lilieholm 2007).

5.4.4 A Growing Call for Wood-Based Energy

Across New England, rural areas never entirely abandoned wood heating, and firewood piles have always been a common sight on farms and in rural communities. Beginning with the twin energy crises of the 1970s, every oil price spike has brought with it a spate of investment in woodstoves, as well as a steady procession of would-be developers promoting the latest proposals for biomass-fired power plants and blueprints for making diesel fuel from 'wood waste.'

Even given this long history, however, rising energy costs in 2008 set new records. As the price of home heating oil reached nearly $4.50/gallon, a new crisis emerged, and former Maine Governor Angus King warned that parts of the state could be 'uninhabitable this winter' because many residents literally could not afford to heat their homes. Six months later, crude oil prices collapsed as the global economy ground to a near stand-still, but the summer's peak prices nonetheless spurred an array of wood-to-energy initiatives, from new wood pellet plants and biomass-fired electrical generating facilities, to aggressive research and development on wood-based liquid biofuels like cellulosic ethanol (Benjamin et al. 2009; Governor's Wood-to-Energy Initiative 2008).

The public's new-found interest in wood-to-energy sparked an important insight for many in the Northeastern U.S. Indeed, though largely overlooked, wood was already playing a critical role in the region's energy mix, supplying 20% of Maine's electricity and 25% of its overall energy needs (North East State Foresters Association 2007). However, a host of factors conspired to limit wood's ability to expand its role in the region's energy balance, especially with respect to home heating and electrical power generation.

For example, Maine's forest resource is already balanced on a knife-edge, with aggregate harvest levels roughly equal to overall forest growth (Laustsen 2009; McWilliams et al. 2005). This is generally true for most of the Maritime Provinces in Canada as well. In some portions of the region, growth exceeds harvest, but the wood is scattered across small tracts, and many owners do not wish to see it harvested (Butler 2008). Moreover, manufacturing by-products like sawdust and planer shavings are already being used for on-site energy production and a host of products ranging from animal bedding to wood stove pellets. In short, the supply of low-cost wood available for energy is limited (Kingsley 2008), and in a rare moment of alignment, both industry and environmentalists generally agree that little room exists to sustainably increase wood harvests for energy.

In addition to resource availability, the price of fossil fuels is a key determinant in the future of wood-to-energy initiatives (Governor's Wood-to-Energy Initiative 2008). For example, wood is generally thought to be competitive at prices for fossil fuels higher than what generally prevailed up until about 2006. As such, crude oil's precipitous decline from nearly $150/barrel in the summer of 2008 to less than $40 just 6 months later undermines support for new wood-energy projects. Declining oil prices do more than decrease the price of wood energy's chief substitute, however – they also undermine the political will to sustain public policies needed to nurture the industry. And even if higher oil prices are sustained, the associated higher costs of logging and transport-to-mill limit the prospects for increased returns to landowners.

From a broader perspective, the wood-to-energy debate reveals a host of complex tradeoffs between ecosystem health and widely supported efforts to reduce carbon emissions, reliance on imported fossil fuels, and the transition to a more local-based and sustainable energy future. Indeed, substantially increasing timber harvests for bioenergy raises numerous concerns about forest health ranging from the retention of coarse woody debris and long-term site quality, to wildlife habitat and aesthetics (Benjamin et al. 2010; Marciano et al. 2009). Already, biomass harvesting guidelines are being prepared in Maine and New Brunswick to reduce ecological impacts in anticipation of increased harvest pressures (Benjamin 2010). While uncertain at this time, the growing importance of wood as an energy source has the potential to radically alter the future of the Northern Forest.

5.4.5 Growing Economic Uncertainty

As we write in early 2010, it is hard to gauge the future course of land development pressures and rising land values on the region's private forests. The historic collapse of the U.S. housing market, combined with a global financial crisis, economic recession, and near-record declines in stock markets around the world, have severely reduced investment capital and real estate demand. Added to this were, until recently, rising energy costs that had motorists altering their driving habits and analysts speculating that future development would be redirected towards established metropolitan centers – especially first-tier suburbs near urban areas that had lost their luster in recent decades. Already, some high-end, amenity-based developments are under severe financial pressure (Effinger and Lin 2008). Such failures, if sustained, could profoundly alter the pace, scale, and nature of leisure lot development in remote regions – at least in the medium-term.

How these events ultimately affect private woodlands is uncertain. At one end of the spectrum, the economic slowdown may represent a mere speed-bump on the road to continued fragmentation and development. At the other end, however, a prolonged economic recession could undermine demands for housing, lumber, and rural property, creating a downward spiral of collapsing rural land values just as U.S. home values have declined roughly 30% in recent years. This scenario – similar to events of the Great Depression of the 1930s – could lead to any number of outcomes. For example, during the Great Depression, millions of hectares across the U.S. reverted to state and local ownership through forfeiture from unpaid taxes. Barring this extreme, falling land values could pave the way for renewed control by the forest industry or an expansion of conservation lands through outright acquisition and purchase of easements, perhaps ultimately leading to new state or national forests and parks. Regardless of how these events unfold, the increasingly fluid nature of land markets in the region suggests continued opportunities for mixed-methods approaches to land protection, hopefully guided by a comprehensive regional framework.

5.5 Lessons Learned

We began our discussion examining current trends and what these changes portend for future conservation efforts. This is a complicated task, with multiple plausible yet oftentimes contradictory answers. As one ponders the future, it is first important to recognize that the conservation gains of the last 20 years have been truly impressive (Cronan et al. 2010). This result stems from many factors, but ironically, these lands would not be protected today had they not first been placed at risk. It can never be known with certainty what would have happened had sweeping changes in ownership, state and federal initiatives for protection, and private efforts to acquire lands and easements not occurred. Similarly, it can never be known to what extent improved regulatory policies constrained subdivision and fragmentation from occurring in the first place.

Despite these gains, however, looking forward many would agree that the current patchwork of protected lands reflects more reactions to unexpected opportunities than active, forward-looking strategy. As a result, the current map of conservation lands and easements lacks any kind of coherence when viewed through the lens of conservation biology. For example, there is inadequate protection for mature, late-successional and old-growth forests, many rare and unique habitats, key terrestrial and aquatic processes, and wildlife corridors – especially between protected parcels. Remedying these shortfalls makes the protection of existing large blocks all the more important. Yet each year brings fewer chances to hit conservation 'home runs' on the scale of the Pingree Easement. Such opportunities, as they arise, should be aggressively pursued.

A comprehensive approach to landscape-level protection will require looking beyond borders (Baldwin et al. 2007b). It will require surmounting data limitations that often fail to transcend municipal, state, and national boundaries. It will require working with new landowners to attract public and philanthropic support through well-considered projects backed by credible science. It will also require addressing the needs of local communities by fostering civic engagement, employment opportunities, and efforts to build and maintain social capital (Lilieholm 2007). Identifying compelling opportunities of this kind is a prerequisite in overcoming 'donor fatigue' and attracting federal funding.

Fragmentation of parcels and of interest in parcels means that many more people will have a say in the future of the Northern Forest (Raymond and Fairfax 2002). Conflict will continue. Successfully navigating such a complex socio-political landscape will require strong, resilient and, at times, shifting coalitions. Meeting this challenge will require the recognition that public support for conservation is changing (Richardson 2008). Indeed, recreational visits to Northern Maine are declining, along with the number of people engaged in traditional outdoor pursuits like backpacking, hunting, and fishing. In their place are suburban-based interests with other agendas (Fausold and Lilieholm 1999; Governor's Council on Maine's Quality of Place 2007).

Similarly, the near-departure of the forest industry as a major category of landowner has created a power-void in state and local representation, exacerbated by the fact that many new financial interests reside out-of-state (Beardsley 2003). In the past, industry's interest in limiting development helped retain unsettled spaces, but this interest is now gone. Given rising development pressures, the notion that profitable timber operations can protect against conversions in land use is increasingly obsolete. These changes will stress existing laws and institutions that once served to support private forest ownership. For example, what is the future role for Maine's Tree Growth current-use tax program in an environment where, even if local government waived all taxes on forestland, development would still prevail?

5 Changing Socio-economic Conditions for Private Woodland Protection

Given this background, it is striking that in the midst of the greatest economic recession since the Great Depression, with state and federal budgets under extreme pressure, the Governors of New England understand these challenges and in late 2009 adopted a resolution identifying land conservation as a priority for the region (New England Governors Conference 2009). Even more to the point, Maine Governor John Baldacci, in his January 2010 State of the State address, took the time to reinforce the importance of land conservation efforts and announced a new initiative – aptly named the Great Maine Forest Initiative (Baldacci 2010).

Going forward, we believe that private land conservation approaches will continue to dominate conservation efforts in the Northern Forest region, especially as government budgets remain under pressure. Indeed, considering the heightened public interest and the huge amounts of land that changed ownership since the 1980s, the area that actually ended-up in public ownership was quite small, with much of the public sector's role being confined to assisting the private sector in the securing of easements. Through all the storm and controversy, the region's preference for private action and private ownership prevailed.

Acknowledgments We thank Keith Ross of LandVest for helpful information on the aggregations program in Massachusetts and general observations on conservation easements. This research was supported by the Maine Sustainability Solutions Initiative (National Science Foundation Grant No. EPS-0904155), the University of Maine's Center for Research on Sustainable Forests, and the Maine Agricultural and Forest Experiment Station.

References

Alig, R. J., Kline, J. D., & Lichtenstein, M. (2004). Urbanization on the U.S. landscape: Looking ahead in the 21st century. *Landscape and Urban Planning, 69*, 219–234.

Anonymous. (2005). *Northern Forest Lands Council: 10th Anniversary Forum. Final report, April 25, 2005: Recommendations for the conservation of the Northern Forest.* Concord, NH: North East State Foresters Association. Retrieved February 2, 2010, from http://www.nefainfo.org/publications/nflc10thforumfinal.pdf

Anonymous. (2008). *Final report of the Governor's Task Force Regarding the Management of Public Lands and Publicly-held Easements in Maine.* Retrieved February 2, 2010, from State of Maine, Department of Conservation Web site: http://www.maine.gov/doc/parks/

Baker, E. W., & Reid, J. G. (2004). Amerindian power in the early modern Northeast: A reappraisal. *The William and Mary Quarterly, 61*(1), 77–106.

Baldacci, H. J. E. (2010). *State of the state address to the Maine legislature.* Retrieved February 2, 2010, from http://www.maine.gov/tools/whatsnew/index.php?topic=Gov+News&id=89381 &v=Article-2006

Baldwin, R., Trombulak, S., Beazley, K., Reining, C., Woolmer, G., Nordgren, J., et al. (2007a). The importance of Maine for ecoregional conservation planning. *Maine Policy Review, 16*(2), 66–77.

Baldwin, E. D., Kenefic, L. S., & LaPage, W. F. (2007b). Alternative large-scale conservation visions for northern Maine: Interviews with decision leaders in Maine. *Maine Policy Review, 16*(2), 78–91.

Barringer, R., Coxe, H., Kartez, J., Reilly, C., & Rubin, J. (2004). *Land for Maine's Future Program: Increasing the return on a sound public investment.* Orono, ME: Margaret Chase Smith Center for Public Policy, University of Maine. Retrieved February 2, 2010, from http://www.maine.gov/spo/lmf/docs/finalreport_forweb.pdf

Beardsley, W. H. (2003). *The future of forestland ownership in Maine. A Delphi study of Maine forestland ownership in 2020 AD (Occ. Papers No. 101)*. Bangor, ME: Husson College.

Bell, K. P. (2007). Houses in the woods: Lessons from the Plum Creek Concept Plan. *Maine Policy Review, 16*(2), 44–55.

Benjamin, J. G., & Ed. (2010). *Considerations and recommendations for retaining woody biomass on timber harvest sites in Maine*. Orono, ME: School of Forest Resources, University of Maine.

Benjamin, J. G., Lilieholm, R. J., & Damery, D. (2009). Challenges and opportunities for the Northeastern forest bioindustry. *Journal of Forestry, 107*(3), 125–131.

Benjamin, J. G., Lilieholm, R. J., & Coup, C. (2010). Forest biomass harvests: A "special needs" operation? *Northern Journal of Applied Forestry, 27*(2), 45–49.

Benson, C., & Rebelo, C. (Eds.). (2009). Financing for forest conservation: Payments for ecosystem services in the tropics. *Journal of Sustainable Forestry, 28*, 279–596.

Bilodeau, M., Lilieholm, R. J., Shaler, S., & Van Walsum, P. (2009). The meaning of a changing environment: Sector issues and opportunities – forest products. In G. L. Jacobson, I. J. Fernandez, P. A. Mayewski, & C. V. Schmitt (Eds.), *Maine's climate future: An initial assessment* (pp. 45–49). Retrieved February 3, 2010, from http://climatechange.umaine.edu/files/Maines_Climate_Future.pdf

Binkley, C. S., & Hagenstein, P. R. (1989). *Conserving the North Woods: Issues in public and private ownership of forested lands in Northern New England and New York (Bulletin 96)*. New Haven, CT: Yale School of Forestry & Environmental Studies.

Binkley, C. S., Raper, C. F., & Washburn, C. L. (1996). Institutional ownership of US timberland. *Journal of Forestry, 94*(9), 21–28.

Birch, T. W. (1996). *Private forest-land owners of the northern United States, 1994 (Research Bulletin NE-136)*. Radnor, PA: USDA Forest Service, Northeastern Forest Experiment Station.

Bley, J. (2007). LURC's challenge: Managing growth in Maine's unorganized territories. *Maine Policy Review, 16*(2), 92–100.

Block, N. E., & Sample, V. A. (2001). *Industrial timberland divestitures and investments: Opportunities and challenges in forestland conservation*. Washington, DC: Pinchot Institute for Conservation.

Brookings Institution. (2006). *Charting Maine's future: An action plan for promoting sustainable prosperity and quality places*. Washington, DC: Brookings Institution. Retrieved February 2, 2010, from http://www.brookings.edu/reports/2006/10cities.aspx

Butler, B. J. (2008). *Family forest owners of the United States, 2006 (General Technical Report NRS-27)*. Newtown Square, PA: USDA Forest Service, Northern Research Station.

Colgan, C. S., & Baker, C. (2003). A framework for assessing cluster development. *Economic Development Quarterly, 17*, 352–366.

Coolidge, P. T. (1963). *History of the Maine woods*. Bangor, ME: Furbish-Roberts.

Cronan, C. S., Lilieholm, R. J., Tremblay, J., & Glidden, T. (2010). A retrospective assessment of land conservation patterns in Maine based on spatial analysis of ecological and socioeconomic indicators. *Environmental Management, 45*(5), 1076–1095.

Damery, D., Kelty, M., Benjamin, J. G., & Lilieholm, R. J. (2009). Developing a sustainable forest biomass industry: Case of the U.S. Northeast. *Ecology and the Environment, 122*, 141–152.

Dana, S. T., & Fairfax, S. K. (1980). *Forest and range policy: Its development in the United States* (2nd ed.). New York: McGraw-Hill.

deGooyer, K., & Capen, D. E. (2004). *An analysis of conservation easements and forest management in New York, Vermont, New Hampshire, and Maine*. Prepared for the North East State Foresters Association. Retrieved on February 2, 2010, from http://www.nefainfo.org/publications/nefa_final_report_7.2004.pdf

Dobbs, D., & Ober, R. (1995). *The Northern Forest*. White River Junction, VT: Chelsea Green Publishing.

Ednie, A., Daigle, J., & Leahy, J. (2010). The development of recreation place attachment on the Maine Coast: User characteristics and reasons for visiting. *Journal of Park and Recreation Administration, 28*(1), 36–51.

5 Changing Socio-economic Conditions for Private Woodland Protection

Effinger, A., & Lin, A. (2008, December 19). Billionaire's ski club in Montana stiffs florists, blacksmiths. *Bloomberg News*. Retrieved February 3, 2010, from http://www.democraticunderground.com/discuss/duboard.php?az=view_all&address=103x410316

Fairfax, S. K., Gwin, L., King, M. A., Raymond, L., & Watt, L. A. (2005). *Buying nature: The limits of land acquisition as a conservation strategy, 1780–2004*. Cambridge, MA: MIT Press.

Fallon, I. (1991). *Billionaire: The life and times of Sir James Goldsmith*. London: Hutchinson.

Fausold, C. F., & Lilieholm, R. J. (1999). The economic value of open space: A review and synthesis. *Environmental Management, 23*, 307–320.

Floyd, D. W., & Chaini, R. (2008). *Atlantic Canada's forest industry: Current status, future opportunities*. Fredericton, NB: University of New Brunswick, Canadian Institute for Forest Policy and Communication.

Foster, C. H. W. (2009). *Twentieth-century New England land conservation: A heritage of civic engagement*. Cambridge, MA: Harvard University Press.

Foster, D. R., Donahue, B., Kittredge, D., Lambert, K. F., Hunter, M., Hall, B., et al. (2010). Wildlands and woodlands: A vision for the New England landscape. Cambridge, MA: Harvard University Press.

Ginn, W. J. (2005). *Investing in nature: Case studies of land conservation in collaboration with business*. Washington, DC: Island Press.

Governor's Council on Maine's Quality of Place. (2007). *People, place, and prosperity: 1st report of the Governor's Council on Maine's Quality of Place*. Augusta, ME: Maine State Planning Office.

Governor's Wood-to-Energy Initiative. (2008). *The Governor's wood-to-energy task force report*. Retrieved February 2, 2010, from http://www.maine.gov/doc/initiatives/wood_to_energy/task_force.html

Hagan, J. M., Irland, L. C., & Whitman, A. A. (2005). *Changing timberland ownership in the Northern Forest and implications for biodiversity (Forest Conservation Program, Report # MCCS-FCP-2005-1)*. Brunswick, ME: Manomet Center for Conservation Sciences. Retrieved February 2, 2010, from http://www.manomet.org/pdf/ForestOwnerChangeReport.pdf

Hagenstein, P. R. (1989). *Ownership and management of Maine forest land*. Falmouth, ME: Maine Audubon Society.

Harper, S. C., Faulk, L. L., & Rankin, E. W. (1990). *The northern forest lands study of New England and New York*. Rutland, VT: USDA Forest Service and the Governors' Task Force on Northern Forest Lands.

Irland, L. C. (1998). Policies for Maine's public lands: A long-term view. In *Maine Choices 1999 – A preview of state budget issues*. Augusta, ME: Maine Center for Economic Policy. Retrieved February 3, 2010, from www.mecep.org/av.asp?na=166

Irland, L. C. (1999a). *The Northeast's changing forest*. Cambridge, MA: Harvard University Press.

Irland, L. C. (1999b). *Forest industry and landownership in the Northern Forest: Economic forces and outlook. Report for the Open Space Institute*. Wayne, ME: The Irland Group.

Irland, L. C. (2001). *Northeastern paper mill towns: Economic trends and economic development responses (Miscellaneous Publication 750)*. Orono, ME: Maine Agriculture & Forestry Experiment Station, University of Maine.

Irland, L. C. (2004a). This evergreen empire: Maine's forest resources in a new century. Conference Report. In *Blaine House conference on Maine's Natural Resource Based Industries: Charting a New Course* (pp. 97–113). Augusta, ME: Maine State Planning Office. Retrieved February 2, 2010, from http://www.state.me.us/spo/specialprojects/docs/nrbi_chartingnewcourse/nrbiconf_appenh.pdf

Irland, L. C. (2004b). Maine's forest industry: From one era to another. In R. E. Barringer (Ed.), *Changing Maine* (pp. 362–387). Gardiner, ME: Tilbury House.

Irland, L. C. (2005). U.S. forest ownership: Historic and global perspective. *Maine Policy Review, 14*(1), 16–22.

Irland, L. C., & Lutz, J. (2007). Whither vertical integration? *Atlantic Forestry Review, 13*(3), 57–59.

Irland, L.C. (2009a). New England forests: Two centuries of a changing landscape. In R. Judd and B. Harrison, (Eds), New England: A Landscape History. MIT Press Cambridge, MA.

Irland, L.C. (2009b). Papermaking in Maine: Economic trends, 1894–2000. Maine History 45(1), 53–74.

Jacobson, G. L., Fernandez, I. J., Mayewski, P. A. & Schmidt, C. V. (Eds.). (2009). *Maine's climate future: An initial assessment*. Orono, ME: University of Maine. Retrieved February 2, 2010, from http://www.climatechange.umaine.edu/about/reports/climate-future/

Judd, R. W. (1997). *Common lands, common people: The origins of conservation in Northern New England*. Cambridge, MA: Harvard University Press.

Kingsley, E. (2008). The myth of free wood. *Northern Woodlands, 56*, 9.

Kingsley, E., Levesque, C. A., & Petersen, C. (2004). *The Northern Forest of Maine, New Hampshire, Vermont, and New York: A look at the land, economies, and communities, 1994–2004 (Draft)*. Concord, NH: Northeast State Foresters Association.

Land Use Regulation Commission. (2009). *Concept Plan for the Moosehead Lake region, Zoning Petition ZP 707*. Retrieved February 11, 2010, from LURC Web site: http://www.maine.gov/doc/lurc/reference/resourceplans/moosehead.html

Lansky, M. (1992). *Beyond the beauty strip: Saving what's left of our forests*. Gardiner, ME: Tilbury House.

Laustsen, K. M. (2009). *2006 mid-cycle report on inventory and growth of Maine's forests*. Augusta, ME: Maine Forest Service. Retrieved February 2, 2010, from http://www.maine.gov/doc/mfs/pubs/midcycle_inventory_rpt.html

Lepisto, T. A. (Ed.). (1992). *Conference report: Sustaining ecosystems, economies, and a way of life in the Northern Forest*. Washington, DC: The Wilderness Society.

Levert, M., & Lawton, C. (2009, December 14). *Paper goods: Is the glass half empty, half full or has it fallen completely off the table for Maine's paper industry? MaineBiz*. Retrieved February 3, 2010, from http://www.maine.gov/spo/economics/docs/publications/LeVertLawton_2-PaperGoods_Mainebiz.pdf

Levert, M., Colgan, C., & Lawton, C. (2007). Are the economics of a sustainable Maine forest sustainable? *Maine Policy Review, 16*(2), 26–36.

Lilieholm, R. J. (1990). Alternatives in regional land use planning: Forestry need not suffer. *Journal of Forestry, 88*(4), 10–12.

Lilieholm, R. J. (2007). Forging a common vision for Maine's North Woods. *Maine Policy Review, 16*(2), 12–25.

Lilieholm, R. J., & Romm, J. M. (1992). The Pinelands National Reserve: An intergovernmental approach to nature preservation. *Environmental Management, 16*, 335–343.

Marciano, J., Lilieholm, R. J., Leahy, J., & Porter, T. L. (2009). *Preliminary findings of the Maine forest and forest products survey (University of Maine Forest Bioproducts Research Initiative, Technical Report)*. Orono, ME: University of Maine. Retrieved February 3, 2010, from http://www.forest.umaine.edu/files/2009/05/Maine-Forest-Bioproducts-Survey-Report-7-17-09.pdf

McLaughlin, N. A. (2005). Rethinking the perpetual nature of conservation easements. *The Harvard Environmental Law Review, 29*, 421–521.

McLaughlin, N. A. (2006). Amending perpetual conservation easements: A case study of the Myrtle Grove controversy. *University of Richmond Law Review, 40*, 1031–1097.

McWilliams, W. H., Butler, B. J., Caldwell, L. E., Griffith, D. M., Hoppus, M. L., Laustsen, K. M., et al. (2005). *The forests of Maine: 2003 (Resource Bulletin NE-164)*. Newtown, PA: USDA Forest Service, Northeastern Research Station. Retrieved February 3, 2010, from http://www.fs.fed.us/ne/newtown_square/publications/resource_bulletins/pdfs/2005/ne_rb164.pdf

New England Governors Conference. (2009). *Report of the Blue Ribbon Commission on Land Conservation*. Boston, MA: New England Governors Conference. Retrieved February 3, 2010, from http://efc.muskie.usm.maine.edu/docs/NEGCLandConservationReport.pdf

North East State Foresters Association. (2007). *The economic importance and wood flows from Maine's forests*. Concord, NH: North East State Foresters Association.

Northern Forest Lands Council. (1994). *Finding common ground: Conserving the Northern Forest: The recommendations of the Northern Forest Lands Council*. Concord, NH: Northern Forest Lands Council.

5 Changing Socio-economic Conditions for Private Woodland Protection

Pagiola, S., Bishop, J., & Landell-Mills, N. (2002). *Selling forest environmental services: Market-based mechanisms for conservation and development*. London: Earthscan.

Palmer, K. T., Taylor, G. T., & LiBrizzi, M. A. (1992). *Maine politics and government*. Lincoln, NE: University of Nebraska Press.

Payne, B. R., Gannon, R. C., & Irland, L. C. (1975). *The Second-home recreation market in the Northeast*. Washington, DC: USDI Bureau of Outdoor Recreation.

Pidot, J. (2005). *Reinventing conservation easements: A critical examination and ideas for reform (Policy Focus Report Code PF013)*. Cambridge, MA: Lincoln Institute of Land Policy.

Plum Creek. (2007). *Plum Creek Concept Plan for the Moosehead Lake Region: Executive Summary*. Retrieved February 3, 2010, from http://www.plumcreekplanmaine.com/resource/d/46948/ExecSummary.pdf.

Porter, W. F., Erickson, J. D., & Whaley, R. S. (2009). *The great experiment in conservation: Voices from the Adirondack Park*. Syracuse, NY: Syracuse University Press.

Radeloff, V. C., Stewart, S. I., Hawbaker, T. J., Gimmi, U., Pidgeon, A. M., Flather, C. H., et al. (2010). Housing growth in and near United States protected areas limits their conservation value. *Proceedings of the National Academy of Sciences of the United States of America, 107*, 940–945.

Raymond, L., & Fairfax, S. K. (2002). The "Shift to Privatization" in land conservation: A cautionary essay. *Natural Resources Journal, 42*, 599–639.

Richardson, B. (2008). *Regional landscape conservation in Maine status report and interim summary: Best practices for enhancing quality of place*. Augusta, ME: Maine State Planning Office. Retrieved February 2, 2010, from http://www.maine.gov/spo/specialprojects/qualityof-place/documents/regionallandscapeconservation.pdf

Small, R. A., & Lewis, D. J. (2009). *Forest land conversion, ecosystem services, and economic issues for policy: A review (General Technical Report PNW-GTR-797)*. Portland, OR: USDA Forest Service, Pacific Northwest Research Station. Retrieved February 3, 2010, from http://www.fs.fed.us/openspace/fote/pnw-gtr797.pdf

Statistics Canada. (2008). *Canada's ecozones and population change, 1981–2006*. Retrieved February 11, 2010, from Statistics Canada Web site: http://www.statcan.gc.ca/pub/16-002-x/2008002/article/10625-eng.htm

Stein, S., McRoberts, R. E., Alig, R. J., Nelson, M. D., Theobald, D. M., Eley, M., et al. (2005). *Forests on the edge: Housing development on America's private forests (General Technical Report, PNW-GTR-636)*. Portland, OR: USDA Forest Service, Pacific Northwest Research Station. Retrieved February 3, 2010, from http://www.fs.fed.us/projects/fote/reports/fote-6-9-05.pdf

The Nature Conservancy. (2007). *The Northern Appalachian-Acadian ecoregion*. Retrieved February 3, 2010, from http://conserveonline.org/workspaces/ecs/napaj/nap_extras/web/ecoreg_intro.

Vashon, J. H., Meehan, A. L., Jakubas, W. J., Organ, J. F., Vashon, A. D., McLaughlin, C. R., et al. (2008). Spatial ecology of a Canada lynx population in Northern Maine. *Journal of Wildlife Management, 72*, 1479–1487.

White, E. M., & Mazza, R. (2008). *A closer look at forests on the edge: Future development on private forests in three states (General Technical Report PNW-GTR-758)*. Portland, OR: USDA Forest Service, Pacific Northwest Research Station. Retrieved February 3, 2010, from http://www.fs.fed.us/pnw/pubs/pnw_gtr758.pdf

White, E. M., Alig, R. J., Stein, S. M., Mahal, L. G., & Theobald, D. M. (2009). *A sensitivity analysis of "Forests on the Edge: Housing development on America's private forests" (General Technical Report PNW-GTR-792)*. Portland, OR: USDA Forest Service, Pacific Northwest Research Station. Retrieved February 3, 2010, from http://www.fs.fed.us/pnw/pubs/pnw_gtr792.pdf

Whitney, R. M. (1989). Forces for change in forest land ownership and use: The large landowners' situation. In C. S. Binkley & P. R. Hagenstein (Eds.), *Conserving the North Woods (Bulletin 96)* (pp. 72–96). New Haven, CT: Yale School of Forestry & Environmental Studies.

Wiersma, G. B. (2009). *Keeping Maine's forests: A study of the future of Maine's forests*. Orono, ME: Center for Research on Sustainable Forests, University of Maine. Retrieved February 2, 2010, from http://www.crsf.umaine.edu/pdf/KeepingMainesForests_2009.pdf

Wilkins, A. H. (1978). *Ten million acres of timber: The remarkable story of forest protection in the Maine forestry district (1909–1972)*. Woolwich, ME: TBW Books.

Williams, K., Sader, S. A., Pryor, C., & Reed, F. (2006). Application of geospatial technology to monitor forest legacy conservation easements. *Journal of Forestry, 104*(2), 89–93.

Zinkhan, F. C., Sizemore, W. R., Mason, G. H., & Ebner, T. J. (1992). *Timberland investments: A portfolio perspective*. Portland, OR: Timber Press.

Chapter 6
Aquatic Conservation Planning at a Landscape Scale

Keith H. Nislow, Christian O. Marks, and Kimberly A. Lutz

Abstract Inland surface waters provide vital ecosystem services and support a diverse and important biota. An overriding feature of freshwater ecosystems is connectedness, which has been compromised by a wide range of human actions. Strong connections between terrestrial watersheds and receiving waters, and upstream and downstream linkages within river systems, make a large-scale perspective essential in conservation planning. In this chapter, we present the essential elements of large-scale aquatic conservation planning, with emphasis on stream and river ecosystems of the Northern Appalachian/Acadian Ecoregion. We review relevant aspects of the structure and function of freshwater ecosystems, discuss different approaches to aquatic conservation, and provide a case study of large-scale conservation planning and implementation in the Connecticut River basin.

Keywords Anadromous fish • Aquatic conservation • Connecticut River • Floodplain forests • Freshwater ecosystems

---

K.H. Nislow (✉)
Research Fisheries Biologist and Team Leader, USDA Forest Service – Northern Research Station, Massachusetts, MA 01003, USA
e-mail: knislow@fs.fed.us

C.O. Marks
The Nature Conservancy, 25 Main Street, Suite 220, Northampton, MA 01060, USA
e-mail: cmarks@tnc.org

K.A. Lutz
Director, Connecticut River Program, The Nature Conservancy, 25 Main Street, Suite 220, Northampton, MA 01060, USA
e-mail: klutz@tnc.org

S.C. Trombulak and R.F. Baldwin (eds.), *Landscape-scale Conservation Planning*,
DOI 10.1007/978-90-481-9575-6_6, © Springer Science+Business Media B.V. 2010

6.1 Introduction

Although inland surface waters cover a small fraction of the Earth's surface, they represent critically important environments for landscape-scale conservation. Aquatic habitats vary in many important attributes and range in size from tiny forest pools and headwater streams to great rivers and large lakes. These habitats support a diverse and important biota, provide vital ecosystem services, and possess powerful esthetic, economic, recreational, and spiritual values. At the same time, increasing demands by an expanding human population have put immense pressure on aquatic habitats and resources and emphasize the need to support aquatic conservation and management (Dynesius and Nilsson 1994).

Aquatic resources have long been at the forefront of conservation efforts. A major impetus behind the inception of the U.S. National Forest System was the protection of water resources that had been threatened by destructive forestry practices (Glasser 2005). Initial efforts were largely focused on water quality and quantity related only to drinking water, and an extensive body of legislative and regulatory protections, ranging from the landmark Federal Clean Water Act of 1972 through a wide array of state and municipal regulations and statutes, has formed to protect this essential resource.

In addition to water quality, the extensive loss of freshwater wetlands, along with a belated recognition of their ecological importance, has resulted in significant regulatory protection for these habitats. Most recently, emphasis has increased on more inclusive aspects of aquatic habitats, including loss of aquatic biodiversity, which is both a global (Dudgeon et al. 2005) and regional problem (Saunders et al. 2006). In addition to the protections for freshwater species listed under the Endangered Species Act, specific legal protections in the U.S. for aquatic biota include the Anadromous Fish Restoration Act of 1965, which mandates conservation and management to conserve and protect fish species that migrate between freshwater and marine habitats.

The New England region of the U.S. (including the states of Connecticut, Maine, Massachusetts, New Hampshire, Rhode Island, and Vermont), embedded in large part within the Northern Appalachian/Acadian and the Lower New England/Northern Piedmont ecoregions, provides a prime example of these issues, both in terms of their impacts and efforts to address them on a landscape scale. This is a well-watered area, whose abundance of freshwater habitats has contributed greatly to the health and welfare of the human population resident there. Following European settlement, large-scale land conversion, along with major projects to engineer river flow that fueled early industrialization, seriously compromised the ecological integrity of aquatic ecosystems throughout this region. Since then, major shifts away from heavy industry and agriculture and an increasing understanding of the value of water resources have led to large-scale recovery of forestlands and major improvements in water quality. In addition, a public that increasingly appreciates the ecological values of aquatic habitats provides a strong public base of support for conservation.

6 Aquatic Conservation Planning at a Landscape Scale

However, the legacy of land use (Nislow 2005, 2010), atmospheric pollution (Driscoll et al. 2001), and hydrologic change (Magilligan and Nislow 2001; Nislow et al. 2002), combined with emerging threats from climate change (Sharma et al. 2007; Chap. 15), invasive species (Les and Mehrhoff 1999), urbanization, and residential development (McMahon and Cuffney 2000) remain significant challenges (Chap. 2). As is the case for terrestrial conservation, perhaps the biggest institutional challenge to large-scale aquatic conservation planning is the pattern of land ownership. In contrast to other ecoregions, where large blocks of land are managed under single jurisdiction, the Northeastern U.S. is made up almost entirely of small landholdings, which can greatly complicate landscape-scale planning. While distinct, these regional characteristics and threats are not unique relative to other landscapes. Lessons learned in this region about aquatic conservation planning should be broadly relevant to conservation practitioners elsewhere.

In this chapter, we review the opportunities and challenges of aquatic conservation in the Northeastern U.S., particularly in New England and the Adirondack Mountains in order to provide an ecoregion-appropriate perspective on aquatic conservation planning. We focus this chapter on running water ecosystems (e.g., streams, rivers, and their associated floodplain and riparian corridors), but many of the principles we consider apply to ponds and lakes as well. As an illustration of these concepts, we outline and discuss the approach to aquatic conservation currently being implemented by the Connecticut River Program of The Nature Conservancy (TNC).

6.2 Attributes of Rivers and Streams in the Northeastern U.S.: Implications for Conservation

A number of excellent reviews of the structure and function of stream and river ecosystems is available for a wide range of levels of expertise and background (cf., Allan and Castillo 2007; Karr and Chu 1999). In this section, we review some of the aspects of river and stream ecosystems that are particularly relevant to conservation planning in the Northern Appalachian/Acadian ecoregion. While many of the examples are specific to this ecoregion, the general patterns and processes identified are relevant to aquatic ecosystems everywhere, and thus need to be taken into account in any aquatic conservation program.

6.2.1 Terrestrial-Aquatic Linkages

A major consideration in aquatic conservation planning is the intimate relationship between the stream and its valley (Hynes 1970). The strong influence of the terrestrial environment – the watershed – determines the physiochemical conditions of surface waters (Golley 1996). The transformation of chemical constituents as they move through terrestrial ecosystems determines the chemical and nutrient composition of the surface water. The timing, magnitude, and seasonality of runoff, influenced by the type of parent material and land cover, acts to erode and deposit sediments and other materials and form the physical structure of the stream channel. Terrestrial ecosystems also influence the flow of solar energy into the aquatic ecosystem via interception by forest canopies.

For the most part, the interaction between aquatic ecosystems and their watersheds go in one direction – downhill – as flows of water, sediment, and nutrients follow the direction of gravity. These large, unidirectional influences have an important consequence for aquatic conservation planning, as conservation measures for terrestrial ecosystems can contribute to and, in some cases, accomplish important aquatic conservation goals. Thus, aquatic conservation essentially requires a watershed-based perspective on the landscape, focusing on both aquatic and terrestrial upland habitats within the watershed.

However, while processes and conditions at any place in the watershed can influence aquatic habitats, areas directly adjacent to streams and rivers – riparian areas – have a disproportionate influence. Direct interception of sunlight by riparian trees has a large influence on water temperature (Moore et al. 2005), which in turn determines the types of aquatic organisms a waterbody can support. Trees in the riparian zone also contribute the majority of coarse organic material, in the form of leaves and downed wood. Fallen leaves frequently are the base of the food webs of small streams (Vannote et al. 1980), while large woody debris (LWD) has a major influence on stream ecosystem structure and function (Dolloff and Warren 2003).

While the direction of influence generally flows from terrestrial uplands to aquatic ecosystems, there are some important exceptions. In large rivers flowing through broad lowland valleys, the 'balance of power' between terrestrial and aquatic habitats may shift as the flood and sediment regimes of large rivers create distinct soils, landforms, and disturbance regimes that provide habitat for distinct floral and faunal assemblages (Naiman and Decamps 1997).

One fundamental consideration of the importance of terrestrial-aquatic linkages in conservation planning is that conserving terrestrial habitats (such as intact forest blocks) can go a long way toward conserving aquatic ecosystems. In the Northeastern U.S., large-scale reforestation (Foster et al. 2002) and reduction of point-source terrestrially-derived pollution has made a substantial contribution to aquatic conservation via major increases in water quality (Mullaney 2004). Because such conservation goals are likely to be promoted for other reasons, a fundamental decision for prioritization in any landscape-scale aquatic conservation program might well be to target aquatic conservation goals that will not be achieved as a corollary to terrestrial conservation.

In spite of the recovery of terrestrial ecosystems in many locations following the nadir of their ecological condition, current and expected future threats to aquatic habitats in the context of aquatic-terrestrial interactions remain. First, even a century past the historical peak of deforestation in the Northeastern U.S. (Foster 1992), the legacy of these large-scale changes in land-use remains on the landscape because

6 Aquatic Conservation Planning at a Landscape Scale

some ecological processes critical to structure and function in aquatic ecosystems may take centuries to recover. In landscapes that have been subject to extensive timber harvest and land-use conversion, recovery of LWD to pre-disturbance levels lags behind forest recovery on the order of centuries (Bragg 2000). As a result, river systems in the Northeastern U.S. have some of the lowest levels of LWD recorded in North America (Magilligan et al. 2008). Given current trajectories for forest recovery, these levels are likely to increase substantially over the next 50 years (Nislow 2010). As another example, in spite of major legislation mandating pollution emission reductions, decades of base cation loss associated with acid rain will continue to make streams in the Northern Appalachian/Acadian ecoregion vulnerable to episodic acidification for decades to come (Driscoll et al. 2001). Finally, hydrologic alteration associated with the large number of dams and impoundments in this ecoregion will continue to affect river morphology and connectivity between rivers and adjacent riparian areas and floodplains (Magilligan and Nislow 2001).

6.2.2 Upstream–Downstream Linkages

Just as water flows from hill slopes to the stream channel, streams continue to flow downstream. In the process, they form predictable networks of channels as small streams meet and form larger streams, which in turn meet and form larger rivers. This characteristic network structure of stream and river ecosystems has important consequences for aquatic conservation planning. Due to the predictable longitudinal changes in physical habitat conditions, aquatic habitats at different points in the network support distinct natural communities (Vannote et al. 1980). Headwaters and large rivers have distinct fish communities, with overall fish species diversity tending to consistently increase in a downstream direction. At the same time, some species use the entire river network at different points in their life cycle. For example, a number of fish species spawn in small streams, putting their vulnerable eggs and fry in habitats with few predators, then move to more productive downstream areas that provide better conditions for growth. This pattern is most evident in anadromous fishes such as the Atlantic salmon (*Salmo salar*), which spawn in streams and rivers and then migrate to productive marine or lake environments. Fish may also use different parts of a river system as refugia from disturbances such as extreme temperatures, floods, or droughts.

The longitudinal connectivity of river systems has been seriously compromised by human activities in the Northern Appalachian/Acadian ecoregion, as it has in most ecoregions throughout the continent south of the boreal forest. Water power was the backbone of early industrialization in most of North America. In the Northeastern U.S., many of the small mill dams of that era still dot the landscape, along with major dams on all of the region's large rivers. These structures, combined with a more recent bout of flood control dams in the early-to-mid-twentieth century, have resulted in the Northeastern U.S. having the highest number of dams per square kilometer of any region in the U.S. (Graf 1999). While the effects of dams on river ecosystems are a national and global issue, the way that the impacts of dams are manifest in aquatic ecosystems in the Northeastern U.S. has important implications for conservation. In spite of the high density of dams, dams in the Northeastern U.S. impound a lower portion of the total annual runoff than in any other ecoregion (Graf 1999). This is due to a combination of the high annual runoff characteristic of this mesic region, combined with a large number of small dams, which are frequently either run-of-the-river or have only limited storage capacity. At the ecoregional scale, therefore, dams may impact rivers in the Northeastern U.S. more through effects on connectivity than through changes in hydrologic or sediment regimes (Graf 2006).

Further, in addition to dams, agricultural, residential, and urban development have resulted in very high road densities (Riiters and Wickham 2003), which often run along valley floors and cross streams at numerous points. Many of these road crossings are barriers to the passage of fish and other aquatic organisms (Warren and Pardew 1998). The combination of numerous small dams and high road densities underscores the importance of longitudinal connectivity as a conservation issue in this region.

6.2.3 Invasions, Extirpations, and Restorations in Aquatic Ecosystems

While the physico-chemical regime is an important target for aquatic conservation, major changes in aquatic community structure itself can have feedback effects at the species, community, and ecosystem level. These changes include invasions (the purposeful or accidental introduction and establishment of non-native species), range extensions (natural changes in species abundance and distribution), extirpations (elimination of a native species), and restorations (re-establishment of native species that have been extirpated).

As a function of its long post-European settlement history and comparatively early development, all of these factors have had major influences on aquatic and riparian ecosystems in the Northeastern U.S. In a sense, even the 'native' flora and fauna are composed of relatively recent colonists following the recession of the most recent glacial ice sheet beginning approximately 19,000 years ago (Curry 2007; Schmidt 1986). As a consequence, native aquatic assemblages in the region are naturally depauperate, with a low number of widely distributed species, in strong contrast to unglaciated rivers such as the Colorado River in the Southwestern U.S., which has a unique and specialized fauna that has evolved over millions of years (Stanford and Ward 1986). This low species diversity in the Northeastern U.S. may in itself contribute to vulnerability to invasion, as some evidence indicates that invasive species are more likely to become established in species-poor communities, particularly in highly human-modified watersheds (Gido and Brown 1999). Apart from obligate aquatic species such as fishes, many exotic plant species have invaded riparian areas, where open canopies and frequent disturbance provide ideal conditions for colonization (Zedler and Kercher 2004).

Invasive species present a special challenge for aquatic habitat conservation. Frequently, eliminating these species is a costly undertaking with high uncertainty that these efforts will work. In some cases, not dealing directly with invasive species may make other conservation efforts (such as habitat conservation) moot (Simberloff et al. 1999). At the same time, in the case of well-established and valuable sport fishes, such as introduced salmonids and black basses, these species have strong constituencies among sport and commercial fishers, and efforts for removal and control frequently meet with public resistance. Further, it is important to distinguish between range extensions and invasions, particularly in the context of species responses to global climate change (Chap. 15). All of these difficulties, however, emphasize the conservation value of sites that are relatively free from invaders and suggest the vital importance of efforts to prevent the establishment of invasive species in these areas whenever possible.

In addition to invasions, European settlement brought with it a wave of extinctions and extirpations of aquatic species. Two driving factors for this stand out in importance. First, barriers to migratory fish resulted in widespread extirpations at the regional and watershed levels (Saunders et al. 2006). For example, in the Connecticut River basin Atlantic salmon were completely extirpated, Atlantic and shortnose sturgeon (*Acipenser oxyrinchus* and *A. brevirostrum*) nearly extirpated, and the abundance of American eel (*Anguilla rostrata*), American shad (*Alosa sapidissima*), and blueback herring (*Alosa aestivalis*) reduced by an order of magnitude (Gephard and McMenemy 2004). Given that anadromous fishes make up as much as 30% of native fish faunas in some coastal rivers, this constitutes a substantial change in aquatic community structure. Second, intensive trapping caused major declines in and widespread extirpation of North American beaver (*Castor canadensis*). Beaver are a keystone species in aquatic and riparian ecosystems throughout the northern hemisphere, profoundly altering aquatic habitats by constructing dams and influencing riparian vegetation by their use of trees for forage and materials (Collen and Gibson 2000).

However, in the last century, coincident with a decline in water-powered industry, an increase in forested land cover, and major changes in public sentiment toward conservation, several extirpated aquatic and riparian species have been re-established in the Northeastern U.S. Beaver have been re-established via initial management reintroductions along with natural recolonization from local refugia following the regulation of trapping and now have reached high population densities in many areas (Foster et al. 2002). Also, for the last 30 years, migratory fish species such as the Atlantic salmon have been the subject of active restoration efforts throughout this region, involving substantial investments at the federal, state, and private levels. In contrast to the natural recovery of beaver, efforts to re-establish native anadromous fishes have met with only mixed results, and the majority of native anadromous fishes are still absent or at substantially reduced population sizes compared to historical levels (Saunders et al. 2006).

6.3 Aquatic Conservation Strategies in the Northeastern U.S.

6.3.1 Species-based Approaches to Aquatic Conservation

Many conservation efforts are explicitly tied to the population status of particular species or groups of species. Others are concerned with the conservation of overall species diversity. Both of these approaches require an understanding of the habitat requirements that support either particular species of concern or the habitat features associated with a high level of species diversity.

A species-based approach confers some important advantages (Chap. 17). Species that are economically important and have large public constituencies provide considerable support to conservation efforts. Conserving habitat and protecting environments for so-called 'umbrella' species can help to conserve other non-target species, as well as to protect key ecosystem services such as erosion control and maintenance of water quality. Species-based approaches also can provide specific, measurable targets (e.g., species persistence, increased abundance and distribution) to evaluate the success of the conservation action. Finally, powerful legislation (such as the Endangered Species Act) can provide significant support for species-based conservation programs. Anadromous fishes in the Northeastern U.S. are a major focus of species-based conservation efforts (Gephard and McMenemy 2004). These efforts are backed up by two major pieces of federal legislation. The Anadromous Fish Conservation Act of 1965 applies to all native anadromous species, while two native species, the Atlantic salmon and the shortnose sturgeon are also listed under the Endangered Species Act.

In addition to their strong constituencies and legal support, anadromous fish such as Atlantic salmon, alewife (*Alosa pseudoharengus*), and native populations of sea lamprey (*Petromyzon marinus*), which require a wide range of habitats to complete their complex life cycles, may serve as useful umbrella species for all those species that require only a subset of these habitats. For example, the listing of the last remaining wild Atlantic salmon stocks in Maine under the Endangered Species Act (National Research Council 2004) has resulted in the purchase of conservation easements along hundreds of kilometers of riparian forests (Haberstock et al. 2000; National Research Council 2004) as well as the removal of dams and other barriers in many watersheds (Gephard and McMenemy 2004).

The ecological realities associated with the landscape-scale context of the New England and the Adirondack Mountains, as well the rest of the formerly glaciated portions of North America, present some major challenges to species-based conservation, as well as provide examples of some of the intrinsic limitations of this approach. Compared to other regions on the continent, where high species diversity and high rates of endemism make resident freshwater fishes important conservation targets, the Northern Appalachian/Acadian ecoregion has a generally depauperate stream and river fauna, made up of common, widely-distributed habitat-generalist species. However, even for species whose habitat requirements have been extensively studied, species-based approaches have some important pitfalls. Habitat factors may be limiting to species abundance and persistence under some environmental contexts but not others, and the effect of habitat conservation or restoration may be therefore quite uncertain. For example, re-establishment of riparian forests may increase the population abundance of stream salmonids in warmer streams where riparian shade prevents temperatures from increasing beyond tolerable levels, but may reduce abundance in colder streams (Nislow 2005, 2009). Perhaps most problematic, in many situations, species abundance and persistence may have a strong stochastic component or may be largely determined by factors external to conservation efforts. For example, in spite of major habitat conservation efforts in rivers and streams throughout the Northeastern U.S., anadromous fish populations continue to decline precipitously throughout the region (Saunders et al. 2006). While this may be in large part due to factors influencing marine survival, and despite the fact that improvements in freshwater habitat may have wide-ranging positive effects, conservation efforts undertaken on behalf of species that continue to decline run the risk of being judged as failures.

6.3.2 Process- and Services-based Approaches to Aquatic Conservation

As an alternative to species-based conservation planning, process- and services-based approaches focus on conserving or restoring critical processes and habitat conditions that have been altered by human activity. In this approach, the explicit goal is the process (e.g., sediment balance, flow regime, longitudinal connectivity) with the implicit assumption that these processes, if restored to their natural state, will help conserve species of concern and biological diversity at multiple spatial scales.

This approach acknowledges the large indeterminacy in species response to habitat management and change. In addition, the process-restoration approach may help to avoid the conflicts that can emerge when managing separately for multiple species of interest. Also, because the target of a process-based approach is the process or condition itself, targets may be easier to set, monitor, and achieve. Finally, protecting key processes protects key ecosystem services derived from freshwaters, including protection of water supply and mitigation of catastrophic floods, along with recreational and associated economic opportunities.

Process-based restoration is increasingly used in river management (Beechie and Bolton 1999; Rheinhardt et al. 1999). In particular, the restoration of natural flow regimes has become an important goal in river conservation and restoration, with the expectation that restoring this key process will result in across-the-board improvements in habitat conditions for a wide range of riverine and riparian species (Poff et al. 1997). The process-based approach has also been widely incorporated into floodplain and river channel restoration efforts (Beechie et al. 1996; Berg et al. 2003). More recently, it has been expanded to include an emphasis on restoration of a natural range of variability (Richter et al. 1997) as opposed to targeting specific conditions.

In spite of important advantages compared to species-based approaches, process-based conservation approaches pose significant challenges, particularly in the Northeastern U.S. For example, large tracts of wilderness in the Northwestern U.S. and Canada and eastern Siberia provide useful reference conditions that help guide process-based restoration in those regions (Naiman et al. 2002). In contrast, the majority of rivers in the Northeastern U.S. have a long history of anthropogenic modification. Apart from making it difficult to determine appropriate reference conditions, environmental change may dramatically alter both the magnitude and direction of process restoration impacts. For example, restoration of historical flood regimes with the expectation of restoring native floodplain vegetation may have the opposite effect in the presence of exotic invasive vegetation, which can often take advantage of flood-disturbed soils (Zedler and Kercher 2004). In addition, large-scale impacts, particularly global climate change, may dramatically alter the way in which processes affect target species and communities. Finally, it is unclear whether the corollary effects of either species-based approaches (with process and services conservation as a byproduct) or process-based approaches (with species conservation as a byproduct) are most effective at achieving the goals of aquatic conservation. In an explicit comparison, Chan et al. (2006) found that targeting biological diversity achieved a high percentage of service-based goals, whereas targeting services failed to protect a large percentage of species.

Given these considerations, it seems that incorporating both species-based and process-and services-based approaches would have a number of benefits for achieving aquatic conservation on a landscape scale. To further explore this point, in the following section we discuss in detail an example of a major aquatic conservation program in the Northeastern U.S. that uses both these approaches.

6.4 Case Study: The Nature Conservancy's Connecticut River Program

6.4.1 The Geographical and Cultural Context for Conservation of the Connecticut River

The 660 km-long Connecticut River is New England's longest river. Its headwaters are in the Fourth Connecticut Lake at the Canadian border in Québec, and it empties into Long Island Sound at Old Saybrook, Connecticut. The watershed encompasses an area of over 28,000 km^2 and has 44 major tributaries each with drainage areas greater than 75 km^2. All told, there are over 32,000 km of streams in the watershed. The Connecticut River drops 730 m from its source to the sea, and has a daily average flow of nearly 450 m^3/s. The flow has ranged as high as 8,000 m^3/s and as low as 27 m^3/s. The lower 100 km of the river are tidal, with the boundary between salt and freshwater about 27 km from its mouth under normal conditions. It has a major influence on the coastal marine ecosystems near its mouth, as its waters represent 70% of the freshwater inflow to Long Island Sound.

6 Aquatic Conservation Planning at a Landscape Scale 109

Because it runs predominantly north-south, the Connecticut River valley encompasses almost the entire range of environmental and socio-economic conditions in the Northeastern U.S. and provides an example of nearly all of the threats and issues found in the region (Chap. 2). During the colonial period up through the early part of the twentieth century, most of the basin was heavily deforested with associated erosion of soils, followed by a dramatic recovery of forest cover starting in the mid-1800s in Southern New England (Foster et al. 2005). The watershed is now 80% forested, 12% agricultural, 3% developed, and 5% wetlands and water.

The southern half of the Connecticut River valley was among the first parts of the country to industrialize, which brought with it increasing levels of water pollution. The mills from that era left a legacy of altered fluvial geomorphology (Walter and Merritts 2008) and numerous dams that continue to obstruct fish passage (Fig. 6.1) and alter the hydrologic regime of the river and its tributaries (Magilligan and Nislow 2001). Since the widespread initiation of wastewater treatment following the 1972 Federal Clean Water Act, along with the loss of heavy industry throughout the watershed, water quality has improved in the Connecticut River, with downward trends in total phosphorus, total nitrogen, and indicator bacteria, and upward trends in pH and dissolved oxygen (Mullaney 2004). At the same time, major efforts to restore key anadromous fishes that had been extirpated or greatly reduced in abundance are active in the basin (Gephard and McMenemy 2004), including extensive efforts by the U.S. Fish and Wildlife Service to restore Atlantic salmon to the watershed and improve passage for other species.

The conservation opportunity in this recovery of natural forest cover and water quality in the Connecticut River is challenged by most of the major threats to ecosystems in the region. Urbanization and residential development, particularly in the southern part of the Connecticut River watershed, is a significant emerging threat. The watershed has 390 towns, villages, and cities, which are home to 2.3 million people. Urbanization and residential development are challenges for river conservation in many ways including polluted runoff, increased sediment, and greater runoff from impervious surfaces that increase flash flooding. Specifically, much of the urbanization and residential development, both historic and contemporary, is located in the river's floodplains. Consequently, protecting urban centers from flood damage has led to the construction of a system of 14 flood control dams on major Connecticut River tributaries.

The Connecticut River floodplains contain some of the richest farmland in the Northeastern U.S. Its deep, well-drained soils are a product of glacial Lake Hitchcock, which flooded much of the valley as the last ice sheet receded northward at the end of the most recent period of glaciation, in combination with more recent river floods (Klyza and Trombulak 1999). In a region with generally steep topography, these large flat sites are also, quite understandably, coveted for development. The construction of roads and other infrastructure associated with urbanization and residential development increasingly fragment both streams and riparian habitats (Fig. 6.1). This proliferation of edges and disturbance and cultivation of fertile soils often create ideal conditions for the spread of invasive plant species that further degrade remnant natural riparian forests.

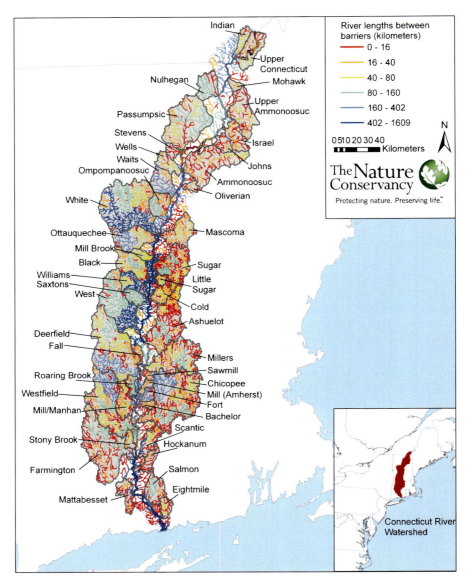

Fig. 6.1 Map of the Connecticut River watershed categorizing streams by the length without stream barriers (Datasources: Watersheds from TNC's size 2 and 3 watersheds, 2001; Connected lengths from USGS NHD-plus, 2006) (Copyright: The Nature Conservancy, Connecticut River Program)

The relatively recent origin of New England forests after agricultural abandonment and periodic logging have resulted in a forest structure with on average much smaller trees than in the pre-settlement forest. Consequently, fewer logs of sufficient size fall into streams to form important pool habitat (Magilligan et al. 2008; Nislow 2010). Streams flowing through agricultural fields also often lack the requisite riparian buffers with large trees. While much of the Connecticut River basin has higher buffering capacity than other major basins in the region due to a preponderance of calcium-rich bedrock, headwater streams, particularly in the upper watershed north of the Massachusetts border are in areas with acidic forest soils overlaying granitic parent materials, and are consequently vulnerable to acidification. Both trees (Driscoll et al. 2001) and fish have been severely affected (McCormick et al. 2009).

These current and emerging threats require regulation to keep pace, a considerable challenge give that the Connecticut River basin includes parts of four states (Connecticut, Massachusetts, New Hampshire, Vermont), and numerous municipal jurisdictions, which in New England play a strong role in manifesting specific conservation practices (Chap. 4). For example, the combination of increased precipitation and runoff due to projected climate change (Marshall and Randhir 2008) and development spreading further onto floodplains could have dire consequences for both people and nature. Regulatory foresight by municipalities and proper management of floodplains could reduce these potential impacts.

Non-governmental organizations and land trusts can also do much to meet these conservation challenges as they have done in the past. The Nature Conservancy has been working in the Connecticut River landscape for more than 40 years. The Conservancy's first land acquisition in the watershed was 18.6 ha at Burnham Brook in East Haddam, Connecticut in 1960. Acquisition of ecologically significant properties accelerated during the late 1990s and early 2000s as land-use patterns began to change and large forested tracts in the northern portion of the basin became available for purchase. These largest tracts include the protection of the 8,900 ha in the Nulhegan River watershed of Vermont, protecting a complex of northern hardwood forests, ponds, and lowland spruce-fir forests, and the acquisition with several key partners of three large tracts (totaling over 73,000 hectares) in New Hampshire that conserve mountain peaks, ponds, wetlands, and lowland forests and swamps in the New Hampshire headwaters region. In addition to significant land protection in the northern portion of the basin, several thousand hectares of tidal wetlands were purchased, following the Ramsar Designation, recognizing international significance of the wetlands of the Connecticut River. In total, The Nature Conservancy and its partners have protected over 100,000 ha in the watershed.

6.4.2 Project History

In the late 1990s, as regional (in contrast to site-based) planning efforts began in earnest throughout The Nature Conservancy, the Connecticut River emerged as an area of regional significance. The Nature Conservancy chapters located in the four states through which the Connecticut River flows initiated a coordinated Conservation Action Planning (CAP) effort designed to identify the most important sites within the basin. It was during this first CAP process that the vision of a watershed-scale project, as opposed to separate site-scale projects, was adopted.

From April through November 2004, the Connecticut River Program hosted three more basin-wide CAP workshops, but with a fundamentally different goal than the previous workshops. The goal of the second round was to explicitly address freshwater conservation at the basin scale. Close to 50 attendees from all four basin states, including federal and state natural resource staff, academics, non-profit organizations, and staff from The Nature Conservancy, participated in one or more workshops (Chap. 11). In addition, expertise in a wide variety of disciplines was represented: fisheries biologists, mussel experts, floodplain specialists, hydrologists, geomorphologists, botanists, and ecologists. During the CAP planning process, attendees focused on three tasks:

1. Identify the biological diversity of greatest interest, referred to in this process as conservation targets, and its current and desired status.
2. Identify the most critical threats currently or likely to degrade this biological diversity.
3. Develop strategies to abate the threats and maintain or restore biological diversity given existing constraints and opportunities.

The outcomes of these three tasks were as follows:

Connecticut River Conservation Targets The biological diversity of the Connecticut River basin is comprised of numerous species and communities, making it impractical to evaluate each for conservation planning. Conservation targets, therefore, represent a subset of species, communities, and ecological systems, which were selected to comprehensively represent the biological diversity of the basin. The CAP participants identified six conservation targets for the Connecticut River basin:

1. The Connecticut River's main stem
2. Its tributary ecosystems, which include 38 major tributaries encompassing over 38,000 km of river
3. Its tidal wetlands and estuaries, which include an extensive system of high-quality freshwater and brackish tidal marshes
4. Its floodplain ecosystems and riparian zones
5. Migratory fish, which include ten diadromous fish known to inhabit the river
6. Mussel assemblages, including 12 species tracked by state heritage programs, the rarest of which are the dwarf wedgemussel (*Alasmidonta heterodon*), brook floater (*A. varicosa*), and yellow lampmussel (*Lampsilis cariosa*)

Key Conservation Strategies During the workshop, more than 45 strategies were identified, many of which were already being implemented in whole or in part by the numerous organizations and agencies that have a stake in the health of the Connecticut River. Therefore, TNC decided to critically examine where its skills and expertise could best be used to take a leadership role in advancing a strategy that had yet to be fully implemented, or to be a catalyst for a strategy that had been implemented but hadn't gathered sufficient momentum to achieve its desired conservation outcomes.

The five strategies selected were as follows:

1. Restore the natural flow, form, and other dynamics of the river to improve aquatic diversity along the waterway.

6 Aquatic Conservation Planning at a Landscape Scale

2. Promote river connectivity – unbroken access to the river throughout the length of the river and its floodplain – which is essential for healthy floodplain forests and the movement of fish and other species.
3. Reduce the spread of invasive plant and animal species, which displace native species and their habitats, and safeguard uninvaded areas.
4. Restore floodplain forests along floodplain rivers.
5. Protect and preserve lands critical to the river's health.

This plan was adopted by the four state chapters of The Nature Conservancy, and the Connecticut River team was assembled. The original team consisted of Conservancy staff in all four basin states as well as a regional freshwater team leader. While the core team has remained much the same, numerous working groups have developed since 2004, and these working groups include key agency partners such as U.S. Army Corp of Engineers, U.S. Fish and Wildlife Service, U.S. Forest Service and U.S. Geological Survey. Although the objectives and action steps presented in November 2004 at the end of the project's planning phase have been refined over time, the same themes continue to form the basis of the work today. We believe that involvement of a wide diversity of stakeholders in the planning process was critical in creating a robust conservation plan. First, the participants' expertise in a variety of disciplines and in geographies allowed for a robust discussion of the important elements of biological diversity and of ecosystem process across the entire basin. The variety of perspectives on the development of strategies also required the diversity of perspectives from watershed-based NGO's to large federal agencies. Finally, the process itself was designed to bring groups to consensus decisions, which in turn empowered TNC to implement strategies that were selected by the group.

6.4.3 Current Program and Future Challenges

The vision for the Connecticut River Program resulting from the CAP is to improve the health of New England's largest river system by restoring both natural flow patterns and connectivity. Specifically, the program envisions the restoration of flow patterns that (1) display natural variations in magnitude, frequency, duration, timing, and rate of change, (2) transport appropriate loads of sediment and nutrients, and (3) maintain productive and diverse habitats supporting numerous species. Further, the program seeks to restore unfragmented, connected river and stream networks that permit the natural movement of nutrients, materials, and individual organisms and sustain populations and ecosystems.

The Connecticut River Program is further envisioned as a center of scientific excellence, actively exporting knowledge in environmental flow management, solutions to stream fragmentation, and floodplain restoration. Five years after the initial planning, substantial progress has been made on all fronts. Progress on each of these goals is described below.

Current Research The relatively large involvement of scientific research from the beginning of this project was needed in part because the CAP identified critical gaps in knowledge about how this aquatic system functions. Specifically, streams in the watershed have far more barriers on them than can be dealt with individually (Fig. 6.1). To prioritize stream barriers for removal, it is important to know the minimum distance of connected stream length that can support a viable fish population. Even for well-studied, widespread species such as Eastern brook trout (*Salvelinus fontinalis*), clear answers for this question are not known. To address this knowledge gap, a partnership was established with the U.S. Geological Survey and the U.S. Forest Service. By capitalizing on a long-term study site, TNC has a unique opportunity to learn from its partners to test the effects of increasing stream connectivity via culvert replacement on a wild brook trout population. Long-term monitoring of survival, growth, and movement will enable determination of the effects of increasing connectivity on a stream system that is highly representative of impacted streams throughout the Northeastern U.S. on a species that is a widespread sentinel of ecosystem change and is of strong management and public interest.

Similarly, floodplain forest ecology has been little studied in the Northeastern U.S., unlike in the southern and western regions of the country. The physical processes of a river not only determine its geomorphology but also act as environmental filters that determine potential species distribution and floodplain forest composition. Threshold values in physical processes that result in changes in species composition can be quantified and combined in a predictive model. For example, species differ in the maximum flood duration that they can survive. The purpose of the floodplain forest research is to quantify these ecological thresholds for floodplain forests and incorporate them into a model that makes spatially explicit predictions about the past, present, and future of Connecticut River habitats as a function of key drivers of environmental change like climate and dam operation. Determining how much flooding different types of floodplain forests in the New England require will be vital to guiding hydrologic restoration prescriptions.

In addition to these knowledge gaps, working at the scale of a whole watershed requires new scientific tools. With 70 major dams on the tributaries and 13 on the mainstem, most sub-watersheds have suffered considerable hydrologic alteration (Fig. 6.2). Major dams are defined as those with a storage capacity exceeding 10% of annual runoff for that sub-watershed. Modifying the operation at all these dams for ecological benefit will require knowing how changes to them interact with changes in other parts of the watershed, because it is crucial to maintain flood protection for downstream sites. To simulate different dam operation and climate scenarios, a model is needed that includes all 44 major tributaries and 70 major dams. Such a model is being developed in collaboration with the U.S. Army Corps of Engineers and the University of Massachusetts. Since watersheds in the Northeastern U.S. tend to have multiple dams rather than the simpler case of a single dominant dam, as is more typical in the Western and the Southern U.S., this modeling approach opens new possibilities for hydrologic restoration throughout much of the region.

With each of the goals above, every attempt has been made to disseminate lessons learned throughout the scientific community. This has been done this through

6 Aquatic Conservation Planning at a Landscape Scale

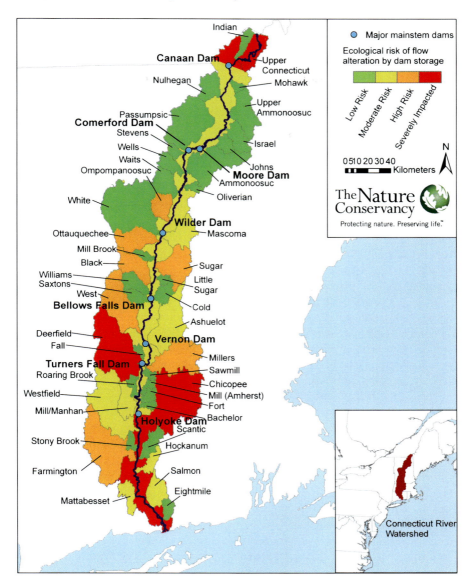

Fig. 6.2 Flow ratings in Connecticut River sub-watersheds, based on the combined storage capacity of large dams within each sub-watershed relative to its annual runoff (see also Zimmerman and Lester 2006) (Datasources: Watersheds from TNC's size 2 and 3 watersheds, 2001; Streams are NHD-plus from USEPA and USGS 2008; Flow rating data from Julie Zimmerman) (Copyright: The Nature Conservancy, Connecticut River Program)

peer-reviewed articles (Letcher et al. 2007; Zimmerman et al. 2010), presentations at national and international meetings (Ecological Society of America, Conservation Biology, Instream Flow Council), and extensive internal communications.

6.5 Lessons Learned

From our experience with the Connecticut River Program, we believe that planning for aquatic conservation at the regional or large-watershed scale brings with it considerable benefits. First, it yields a better understanding of the spatial distribution of threats and the system-wide consequences of management actions. For example, when the project was initiated, the focus was on dam modification at two priority tributaries in the central portion of the basin. However, it was quickly realized that flow alteration was pervasive (a large portion of the basin was altered) and connected (flow modifications at one point could have important impacts well downstream, and dams in many cases are operated in concert). Therefore, the analysis was expanded to all 70 large dams, which will allow the program to deal with altered flow regimes by working at different sites and with different dam operators to achieve basin-wide as well as tributary-specific conservation objectives. In addition, it will help to ensure that flow prescriptions accomplished at one site will not cause deleterious effects downstream.

Second, it allows incorporation of both target-based and process-based approaches. Planning at a larger scale naturally leads to thinking about critical river processes such as flow, sediment transport, and channel migration that are difficult to consider at a limited geographic scale. At the same time, it permits consideration of the relationships between these processes and targets (both species and communities) at scales that are relevant to the maintenance of viable populations.

Third, it gives more opportunities to engage with research early and often. Engagement with the research community is an essential part of the Connecticut River Program. We feel that working at large scales makes it more likely that the interests and expertise of researchers and conservation practitioners will overlap. Further, working at a large scale allows for the level of replication that is essential for generating results that both researchers and conservation practitioners can use.

Finally, it prevents an exclusive emphasis on 'showcase' sites. Yet another benefit of regional-scale planning is that it encourages conservation planning to move away from an exclusive emphasis on the 'best' sites. Given the uncertainty in determining which sites add most conservation value, particularly in the context of climate and other sources of large-scale environmental change, implementing a range of strategies from protection of the best habitats to restoration of degraded habitats will undoubtedly lead to a healthier watershed.

While offering many advantages, large-scale planning presents challenges. An example is the involvement of partners who, for reasons of organization limitations, cannot work outside of a specific geography. This has been a challenge for The Nature Conservancy, but one that as a multi-state organization, it can manage by deploying chapters to work closely with state agency staff on state-specific aquatic policies or with local watershed groups on tributary-specific issues. While planning should be done at a large scale, implementation usually comes down to site-specific actions. Demonstrating concrete accomplishments at the local scale is, therefore, somewhat paradoxically an essential element in successful landscape-scale aquatic conservation.

6 Aquatic Conservation Planning at a Landscape Scale

Acknowledgements The authors would like to thank A. Jospe, A. Lester, and J. Zimmerman for data and assistance in the preparation of the figures. We would also like to acknowledge the support of the Bingam Trust and The USDA Forest Service Northern Research Station.

References

Allan, J. D., & Castillo, M. M. (2007). *Stream ecology: structure and function of running waters* (2nd ed.). Berlin: Springer.

Beechie, T., & Bolton, S. (1999). An approach to restoring salmonid habitat-forming process in Pacific Northwest watersheds. *Fisheries, 24*, 6–15.

Beechie, T., Beamer, E., Collins, B., & Benda, L. (1996). Restoration of habitat-forming processes in Pacific Northwest watersheds: A locally adaptable approach to salmonid habitat restoration. In D. L. Peterson & C. V. Klimas (Eds.), *The role of restoration in ecosystem management* (pp. 48–67). Madison, WI: Society for Ecological Restoration.

Berg, D. R., McKee, A., & Maki, J. J. (2003). Restoring floodplain forests. In D. R. Montgomery, S. Bolton, D. B. Booth, & L. Wall (Eds.), *Restoration of Puget Sound rivers* (pp. 248–291). Seattle, WA: University of Washington Press.

Bragg, D. C. (2000). Simulating catastrophic and individualistic large woody debris recruitment for a small riparian system. *Ecology, 81*, 1383–1394.

Chan, K. M. A., Shaw, M. R., Cameron, D. R., Underwood, E. C., & Daily, G. C. (2006). Conservation planning for ecosystem services. *PLoS Biology, 4*, e379. doi:10.1371/journal.pbio.0040379.

Collen, P., & Gibson, R. J. (2000). The general ecology of beavers (*Castor* spp.) as related to their influence on stream ecosystems and riparian habitats, and the subsequent effects on fish – a review. *Reviews in Fish Biology and Fisheries, 10*, 439–461.

Curry, R. A. (2007). Late glacial impacts on dispersal and colonization of Atlantic Canada and Maine by freshwater fishes. *Quaternary Research, 67*, 225–233.

Dolloff, C. A., & Warren, M. L., Jr. (2003). Fish relationships with wood in large rivers. In S. V. Gregory, K. L. Boyer, & A. M. Gurnell (Eds.), *The ecology and management of wood in world rivers (Symposium 37)* (pp. 179–194). Bethesda, MD: American Fisheries Society.

Driscoll, C. T., Lawrence, G. B., Bulger, A. J., Butler, T. J., Cronan, C. S., Eagar, C., et al. (2001). Acidic deposition in the Northeastern United States: Sources and inputs, ecosystem effects, and management strategies. *BioScience, 51*, 180–198.

Dudgeon, D., Arthington, A. H., Gessner, M. O., Kawabata, Z., Knowler, D. J., Lévêque, et al. (2005). Freshwater biodiversity: Importance, threats, status and conservation challenges. *Biological Reviews, 81*, 163–182.

Dynesius, M., & Nilsson, C. (1994). Fragmentation and flow regulation of river systems in the northern third of the world. *Science, 266*, 753–762.

Foster, D. R. (1992). Land use history (1730–1990) and vegetation dynamics in central New England, USA. *Journal of Ecology, 80*, 753–772.

Foster, D. R., Motzkin, G., Bernardos, D., & Cardoza, J. (2002). Wildlife dynamics in a changing New England landscape. *Journal of Biogeography, 29*, 1337–1357.

Foster, D., Kittredge, D., Donahue, B., Motzkin, G., Orwig, D., Ellison, A., et al. (2005). *Wildlands and woodlands: A vision for the forests of Massachusetts*. Retrieved February 2, 2010. Harvard Forest Web site: http://harvardforest.fas.harvard.edu/publications/pdfs/HF_WandW.pdf

Gephard, S., & McMenemy, J. (2004). An overview of the program to restore Atlantic salmon and other anadromous fishes to the Connecticut River with notes on the current status of these species in the river. In P. M. Jacobson, D. A. Dixon, W. C. Leggett, B. C. Marcy Jr., & R. R. Massengill (Eds.), *The Connecticut River ecological study (1965–1973) revisited: Ecology of the lower Connecticut River, 1973–2003 (Monograph 9)* (pp. 287–317). Bethesda, MD: American Fisheries Society.

Gido, K. B., & Brown, J. H. (1999). Invasion of North American drainages by alien fish species. *Freshwater Biology, 42*, 387–399.

Glasser, S. P. (2005). A history of watershed management in the US Forest Service: 1897–2005. *Journal of Forestry, 103*, 255–258.

Golley, F. B. (1996). *A history of the ecosystem concept in ecology: More than the sum of the parts.* New Haven, CT: Yale University Press.

Graf, W. L. (1999). Dam nation: A geographic census of American dams and their large-scale hydrologic impacts. *Water Resources Research, 35*, 1305–1311.

Graf, W. L. (2006). Downstream hydrologic and geomorphic effects of large dams on American rivers. *Geomorphology, 79*, 336–360.

Haberstock, A. E., Nichols, H. G., DesMeules, M. P., Wright, J., Christensen, J. M., & Hudnut, D. H. (2000). Method to identify effective riparian buffer widths for Atlantic salmon habitat protection. *Journal of the American Water Resources Association, 36*, 1271–1286.

Hynes, H. N. B. (1970). *The ecology of running waters.* Liverpool, UK: Liverpool University Press.

Karr, J. R., & Chu, E. W. (1999). *Restoring life in running waters: Better biological monitoring.* Washington, DC: Island Press.

Klyza, C. M., & Trombulak, S. C. (1999). *The story of Vermont: A natural and cultural history.* Hanover, NH: University Press of New England.

Les, D. H., & Mehrhoff, L. J. (1999). Introduction of non-indigenous aquatic vascular plants in southern New England: A historical perspective. *Biological Invasions, 1*, 281–300.

Letcher, B. H., Nislow, K. H., Coombs, J. A., O'Donnell, M. J., & Dubreuil, T. L. (2007). Population response to habitat fragmentation in a stream-dwelling brook trout population. *PLoS-ONE, 2*, e1139. doi:10.1371/journal.pone.0001139

Magilligan, F. J., & Nislow, K. H. (2001). Hydrologic alteration in a changing landscape: Effects of impoundment in the Upper Connecticut River Basin, USA. *Journal of the American Water Resources Association, 37*, 1551–1569.

Magilligan, F. J., Nislow, K. H., Fisher, G. B., Wright, J., Mackey, G., & Laser, M. (2008). The geomorphic function and characteristics of large woody debris in low gradient rivers, coastal Maine, USA. *Geomorphology, 97*, 467–482.

Marshall, E., & Randhir, T. (2008). Effect of climate change on watershed system: A regional analysis. *Climatic Change, 89*, 263–280.

McCormick, S. D., Keyes, A., Nislow, K. H., & Monette, M. (2009). Impacts of episodic acidification on in-stream survival and physiological impairment of Atlantic salmon (*Salmo salar*) smolts. *Canadian Journal of Fisheries and Aquatic Sciences, 66*, 394–403.

McMahon, G., & Cuffney, T. F. (2000). Quantifying urban intensity in drainage basins for assessing stream ecological conditions. *Journal of the American Water Resources Association, 36*, 1247–1261.

Moore, R. D., Spittlehouse, D. L., & Story, A. (2005). Riparian microclimate and stream temperature response to forest harvesting: A review. *Journal of the American Water Resources Association, 41*, 813–834.

Mullaney, J. R. (2004). Summary of water quality trends in the Connecticut River, 1968–1998. *American Fisheries Society Monograph, 9*, 273–286.

Naiman, R. J., & Decamps, H. (1997). The ecology of interfaces: Riparian zones. *Annual Review of Ecology and Systematics, 28*, 621–658.

Naiman, R. J., Bilby, R. E., Schindler, D. E., & Helfield, J. M. (2002). Pacific salmon, nutrients, and the dynamics of freshwater and riparian ecosystems. *Ecosystems, 5*, 399–417.

National Research Council, Committee on Atlantic Salmon in Maine. (2004). *Atlantic salmon in Maine.* Washington, DC: National Academies Press.

Nislow, K. H. (2005). Forest change and stream fish habitat: lessons from 'Olde' and New England. *Journal of Fish Biology, 67*, 186–204.

Nislow, K. H. (2010). Riparian management: Alternative paradigms and implications for wild Atlantic salmon. In P. Kemp & D. Roberts (Eds.), *Salmonid fisheries: Freshwater habitat management* (in press). Hoboken, NJ: Wiley-Blackwell.

Nislow, K. H., Magilligan, F. J., Fassnacht, H., Bechtel, D., & Ruesink, A. (2002). Effects of dam impoundment on the flood regime of natural floodplain communities in the Upper Connecticut River. *Journal of the American Water Resources Association, 38*, 1533–1548.

Poff, N. L., Allan, J. D., Bain, M. B., Karr, J. R., Prestegaard, K. L., Richter, B. D., et al. (1997). The natural flow regime. *BioScience, 47*, 769–784.

Rheinhardt, R. D., Rheinhardt, M. C., Brinson, M. M., & Faser, K. E., Jr. (1999). Application of reference data for assessing and restoring headwater ecosystems. *Restoration Ecology, 7*, 241–251.

Richter, B. D., Baumgartner, J. V., Wigington, R., & Braun, D. P. (1997). How much water does a river need? *Freshwater Biology, 37*, 231–249.

Riiters, K. H., & Wickham, J. D. (2003). How far to the nearest road? *Frontiers in Evolution and Ecology, 1*, 125–129.

Saunders, R., Hachey, M. A., & Fay, C. W. (2006). Maine's diadromous fish community: Past, present and implications for Atlantic salmon recovery. *Fisheries, 31*, 537–547.

Schmidt, R. E. (1986). Zoogeography of the northern Appalachians. In C. H. Hocutt & E. O. Wiley (Eds.), *Zoogeography of North American freshwater fishes* (pp. 137–159). New York: Wiley.

Sharma, S., Jackson, D. A., Minns, C. K., & Shuter, B. J. (2007). Will northern fish populations be in hot water because of climate change? *Global Change Biology, 13*, 2052–2064.

Simberloff, D., Doak, D., Groom, M., Trombulak, S., Dobson, A., Gatewood, S., et al. (1999). Regional and continental restoration. In M. E. Soulé & J. Terborgh (Eds.), *Continental conservation: Scientific foundations of regional reserve networks* (pp. 65–98). Washington, DC: Island Press.

Stanford, J. A., & Ward, J. M. (1986). Fishes of the Colorado River basin. In B. R. Davies & K. F. Walker (Eds.), *The ecology of river systems* (pp. 385–402). Dordrecht, The Netherlands: W. Junk Publishers.

Vannote, R. L., Minshall, G. W., Cummins, K. W., Sedell, J. R., & Cushing, C. E. (1980). The river continuum concept. *Canadian Journal of Fisheries and Aquatic Sciences, 37*, 130–137.

Walter, R. C., & Merritts, D. J. (2008). Natural streams and the legacy of water-powered mills. *Science, 319*, 299–304.

Warren, M. L., Jr., & Pardew, M. G. (1998). Road crossings as barriers to small-stream fish movements. *Transactions of the American Fisheries Society, 127*, 637–644.

Zedler, J. B., & Kercher, S. (2004). Causes and consequences of invasive plants in wetlands: Opportunities, opportunists, and outcomes. *Critical Reviews in Plant Sciences, 23*, 431–452.

Zimmerman, J., & Lester, A. (2006). *Spatial distribution of hydrologic alteration and fragmentation among tributaries of the Connecticut River*. Northampton, MA: The Nature Conservancy, Connecticut River Program.

Zimmerman, J. K. H., Letcher, B. H., Nislow, K. H., Lutz, K., Magilligan, F. J. (2010). Determining the effects of dams on subdaily variation in river flows at the sub-basin scale. *Rivers Research and Application*. http://dx.doi.org/10.1002/rra.1324.

Chapter 7
From the Last of the Large to the Remnants of the Rare: Bird Conservation at an Ecoregional Scale

Jeffrey V. Wells

Abstract Because of the vast intercontinental distances that birds can travel in the course of a year between winter and summer grounds, bird conservation requires planning across large landscapes, even sometimes spanning the globe. I review a number of efforts to institute ecoregional and trans-ecoregional conservation planning efforts focused on birds, including Partners in Flight, U.S. Shorebird Conservation Plan, Waterbird Conservation for the Americas, the North American Bird Conservation Initiative and Joint Ventures, all of which seeks to overcome the parochial limitations of local-scale planning. These initiatives highlight the importance of (1) applying conservation values beyond that of simple rarity, (2) integrating conservation plans across political boundaries and even continents, (3) making conservation plans that are both spatially explicit and policy specific, and (4) emphasizing the conservation needs of birds over the research needs of science.

Keywords Bird conservation • Conservation planning • Ecoregional planning • Joint ventures • Neotropical migrants

7.1 Introduction

Bird conservation has come a long way since Martha, the last female Passenger Pigeon (*Ectopistes migratorius*), died in the Cincinnati Zoo in 1914. In North America, the loss of the Passenger Pigeon, Bachman's Warbler (*Vermivora bachmanii*), Carolina Parakeet (*Conuropsis carolinensis*), Eskimo Curlew (*Numenius borealis*), Imperial Woodpecker (*Campephilus imperialis*), and perhaps Ivory-billed Woodpecker (*Campephilus principalis*) showed us that species have demographic, ecological, and geographic limits across life-history stages and geographically disparate locations.

J.V. Wells (✉)
261 Water St., Suite 1, Gardiner, ME 04345, USA
e-mail: jeffwells@borealbirds.org

S.C. Trombulak and R.F. Baldwin (eds.), *Landscape-scale Conservation Planning*,
DOI 10.1007/978-90-481-9575-6_7, © Springer Science+Business Media B.V. 2010

In hindsight, we can see that even though the Eskimo Curlew's Arctic breeding range was remote and secure, it was driven to extinction by high mortality rates caused by hunting on its migration routes to and from its wintering grounds in Southern South America. Loss of the Eskimo Curlew taught us that bird conservation requires planning that can span vast areas, not only within single ecoregions but across many, sometimes even spanning the globe. The examples are many and varied. Tiny songbirds like the Cerulean Warbler (*Dendroica cerulea*) nest in the Eastern U.S. and Canada and winter in Northern South America, a wintering geography that they share with birds that breed in the boreal forest, like the Olive-sided Flycatcher (*Contopus cooperi*) and the Canada Warbler (*Wilsonia canadensis*). The rapidly declining Rusty Blackbird (*Euphagus carolinus*) nests across the North American boreal forest and winters in the Southeastern U.S. Conservation of even the rarest and most range-restricted birds like the Whooping Crane (*Grus americana*) or the Kirtland's Warbler (*Dendroica kirtlandii*) requires consideration of nesting and wintering areas thousands of miles apart as well as the issues they face on their long migratory routes.

Early bird conservation efforts were by necessity focused on finding solutions to imminent specific threats. Beginning especially in the late 1800s with the formation of the first Audubon Societies and the American Ornithologists' Union, bird conservation was focused on policy changes to remove, reduce, or mitigate the major threats to bird populations, starting with uncontrolled market hunting, then later wetland losses and the effects of pollution and toxins, but all without landscape-scale context (Belanger 1988; Ossa 1973; Wells 2007). Although not exclusively bird-focused or part of true ecoregional planning, land-based conservation efforts to set aside land for non-consumptive uses began at the federal and state level in the late 1800s with federal national parks established at Yellowstone, Yosemite, and Sequoia, and at the state level with New York State's Adirondack and Catskill Parks (Krech et al. 2004). In the early 1900s, protected areas began to be established specifically for birds through what would become the National Wildlife Refuge (NWR) system (Fischman 2003).

Bird-oriented land protection efforts generally focused on remnant habitat for endangered and declining species and for waterfowl (Belanger 1988). For example, Aransas NWR was established in 1937 to protect the remaining wintering habitat for Whooping Cranes, Red Rock Lake NWR for Trumpeter Swans (*Cygnus buccinator*) in the 1930s (Wells 2007), and numerous other national wildlife refuges for declining waterfowl starting in the 1920s (Fischman 2003). Further complicating matters, acquisition of land for national wildlife refuges was often driven by political considerations and opportunities, so much so that the resulting system of protected lands has been described as a 'hodgepodge' (Fischman 2003; Chap. 14). As a result, the NWR system currently protects fewer endangered and threatened species than other types of federal public lands (Stein et al. 2000). Protection efforts on private land, which began in greater earnest in the 1960s and 1970s with national, regional, and local land trusts, were similarly focused on species, habitats, or places that were thought to be rare or to have some significant feature or use that was threatened (Brewer 2003).

Weaknesses of this approach, however, soon came to light. In the 1970s and 1980s, more and more bird species were categorized as declining and of conservation concern despite the enactment of various regulatory tools and continuing land purchase and management initiatives (Groves 2003; Noss et al. 1995; Woodruff and Ginsberg 1998).

At the same time, the resources required for development and implementation of single-species management plans were seen as too great and requiring wasteful and sometime counterproductive redundancy (Kohm 1991; LaRoe 1993; National Research Council 1995). Public and private conservation agencies and organizations began to develop conservation plans for multiple species and for site-based conservation that took into account a broader set of influential factors from within the landscape (Grumbine 1994; Kessler et al. 1992; Salwasser 1992).

Eventually modern tools for conservation planning, such as use of geographic information systems (Chap. 12), population viability analysis, and site selection algorithms (Chap. 14), allowed planners to model the impacts that different management options and landscape changes would have on a given geographic area. One of the most striking results from the application of these new tools was to show that for many species, existing protected areas would be unlikely to ensure their long-term persistence and viability (Deguise and Kerr 2006; Kautz and Cox 2001; Rodrigues et al. 2004; Scott et al. 2001; Stein et al. 2000). This is because conservation activities had been carried out without understanding of the broadest landscape and ecological context (Grumbine 1994; Newmark 1987, 1995; Woodruff and Ginsberg 1998).

7.2 Ecoregional Conservation Planning for Birds

Although not specifically focused on birds, The Nature Conservancy (TNC) was one of the first organizations to implement a systematic ecoregional planning approach (Groves 2003). The organization began a process during the 1990s to develop conservation plans for 81 ecoregions across the U.S. These ecoregional plans focused on sites of occurrence of key plant and animal species, representative and rare ecosystems, as well as what are termed 'matrix' habitat, communities that are typically habitats that make up a large proportion of the land cover of the ecoregion. The conservation targets derived from this planning process provide a meaningful set of actions and recommendations for maintaining some priority bird populations, although they are often targeted to a very limited set of rare or peripheral bird species. For example, TNC's Northern Appalachian/Acadian ecoregion plan lists goals for peripheral species – Razorbill (*Alca torda*), Golden Eagle (*Aquila chrysaetos*), and Sedge Wren (*Cistothorus platensis*) – while not explicitly including Partners in Flight (PIF) priority species Black-throated Blue Warbler (*Dendroica caerulescens*), Canada Warbler, or Rusty Blackbird. Many of the TNC ecoregional plans eventually integrated bird species priorities identified through coalitions like PIF, the U.S. and Canadian Shorebird Conservation Plans, Waterbird Conservation for the Americas and others, at least as secondary conservation targets or within targets for ecological community representation.

The most significant effort to institute ecoregional conservation planning focused on birds came from the efforts of the Partners in Flight coalition, which began developing new ways to assess conservation values for birds at state, national, and ecoregional scales in the 1990s. One very important result of the evolution of the conservation assessment was the recognition of biological diversity values beyond a focus on species diversity, rarity, and endangerment to include the concepts of latent risk of extinction (Cardillo et al. 2006) and maintenance of abundance (Carter et al. 2000; Dunn et al. 1999; Rosenberg and Wells 2000; Wells and Rosenberg 1999). This broadening of consideration of conservation values provided a lens through which a region of virtually any size could evaluate its current and potential contributions to global conservation. For example, a small park, municipal or county government, or even a small state like Rhode Island, would be unlikely to support a great many globally rare or endangered species. Such small regions would often resort to developing species priorities based on peripheral species that were rare in the state but globally abundant and secure. In contrast, if that region knew that it harbored significant populations of a species of national or regional conservation priority or a species under latent extinction risk, then conservation priorities could be tailored to help the species most in need of conservation attention at larger, including global, spatial scales (Wells et al. 2010).

Research on this process in the Northeastern U.S. showed the striking contrast in conservation priorities that were obtained depending on whether short-term endangerment and risk values were applied compared to long-term stewardship responsibility (Rosenberg and Wells 2000). When endangerment values for bird species were used to rank ecoregions across the Northeastern U.S., the Appalachian Mountains of Virginia, West Virginia, and Pennsylvania were highlighted as 'hotspots' (Fig. 7.1). In contrast, a ranking that used a measure of each ecoregion's importance to sustaining bird species highlighted Northern New England and the Appalachian Mountains of West Virginia and Virginia (Fig. 7.2).

Other coalitions, including the U.S. Shorebird Conservation Plan, Waterbird Conservation for the Americas, and others have now also completed ecoregional conservation plans at varying levels of detail and focus. In 1999, through the efforts of the Commission for Environmental Cooperation, a new umbrella group called the North American Bird Conservation Initiative (NABCI) was formed by the governments of Canada, Mexico, and the U.S. to coordinate among the growing number of bird conservation initiatives, such as:

1. Canadian Shorebird Conservation Plan: http://www.cws-scf.ec.gc.ca/mbc-com/default.asp?lang=en&n=d1610ab7
2. North American Bird Conservation Initiative: http://www.nabci.net/International/English/about_nabci.html
3. North American Grouse Partnership: http://www.grousepartners.org/
4. North American Waterfowl Management Plan and Joint Ventures: http://www.fws.gov/birdhabitat/NAWMP/index.shtm
5. Northern Bobwhite Conservation Strategy: http://www.qu.org/seqsg/nbci/nbci.cfm

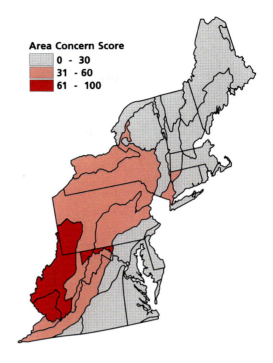

Fig. 7.1 Map of ecoregions in Northeastern U.S. ranked by their relative concentrations of 'high conservation risk' Neotropical migrant landbirds. *Dark red* and *pink areas* represent ecoregions with highest densities of at-risk species (Adapted from Rosenberg and Wells 2000)

6. Partners in Flight – U.S.: http://www.partnersinflight.org/
7. Partners in Flight – Canada: http://www.cws-scf.ec.gc.ca/mbc-com/default.asp?lang=en&n=7AEDFD2C
8. U.S. Shorebird Conservation Plan: http://www.fws.gov/shorebirdplan/
9. Waterbird Conservation for the Americas: http://www.waterbirdconservation.org/
10. Wings Over Water: Canada's Waterbird Conservation Plan: http://www.cws-scf.ec.gc.ca/mbc-com/default.asp?lang=en&n=B65F9B7E

To help standardize ecoregional planning units, NABCI and the Commission for Environmental Cooperation developed a consensus ecoregional map of Canada, Mexico, and the U.S. with mapped boundaries of Bird Conservation Regions (BCR's) for North America (Fig. 7.3; Table 7.1). These boundaries have been adopted as one form of implementation vehicle by many conservation groups, but the original vision of a common set of ecoregional planning units has not been achieved and each coalition has developed its own plans using different scales and amalgamations of ecoregions. For example, Partners in Flight has ten ecoregional plans in the Northeastern U.S. and adjacent Canada within an area encompassed by

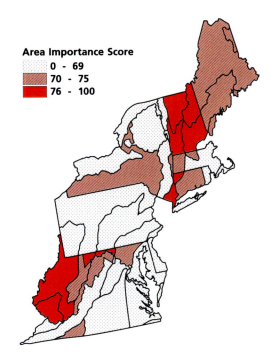

Fig. 7.2 Map of ecoregions in northeastern U.S. ranked by their relative concentrations of Neotropical migrant landbird species with high proportions of populations within the Northeastern U.S. *Dark red* and *pink areas* represent ecoregions with highest densities of these so-called 'high responsibility' species (Adapted from Rosenberg and Wells 2000)

five NABCI BCR's. In contrast, the Waterbird Conservation for the Americas coalition has a single plan for the same region that encompasses the five NABCI BCR's. The lack of a common planning unit among the initiatives has added an unfortunate complexity for those tasked with implementing bird conservation activities within states and regions.

The development of these bird-focused ecoregional plans continues to evolve as to how the planning process is carried out, which factors are considered, the detail in scale (for example, whether specific lands are identified as priorities), and the identification and prioritization of threats, issues, opportunities, and recommendations. However, the basic process of developing such plans typically contains these common elements:

1. Identify priority species based on endangerment and responsibility.
2. Group priority species by habitat types.
3. When possible, map significant occurrences of habitat/species groups within the target region.
4. Identify and, if possible, quantify the risks, threats, and management issues relevant to each habitat/species group.
5. Identify conservation opportunities.
6. Identify needs and develop recommendations to meet those needs.

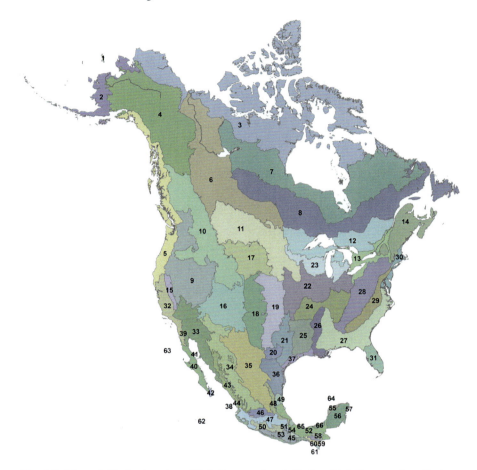

Fig. 7.3 Map of North American Bird Conservation Initiative Bird Conservation Regions. Numbers correspond to names in Table 7.1

7.3 The Benefits of Ecoregional Planning for Birds

Most bird conservation plans, even those specific to a state or other political entity, now explicitly recognize the spatial, demographic, and sometimes genetic connections of individuals, populations, species, and communities with other elements of the abiotic and biotic environment within the ecoregions, many of which overlap and extend beyond the planning region. The boundaries of many ecoregions overlap with more than one state or province, sometimes even with more than one country. For example, there are nine BCR's that overlap both the U.S. and Canada and six that overlap both the U.S. and Mexico. Because the majority of birds in North American are migratory, virtually all ecoregions of North America are demographically linked with ecoregions of Central and South America and the Caribbean. Thus state, provincial, and federal governments and other organizations have the opportunity

Table 7.1 Names of the North American Bird Conservation Regions. Numbers correspond to labels in Fig. 7.3

BCR number	BCR name
1	Aleutian/Bering Sea Islands
2	Western Alaska
3	Arctic Plains and Mountains
4	Northwestern Interior Forest
5	Northern Pacific Rainforest
6	Boreal Taiga Plains
7	Taiga Shield and Hudson Plains
8	Boreal Softwood Shield
9	Great Basin
10	Northern Rockies
11	Prairie Potholes
12	Boreal Hardwood Transition
13	Lower Great Lakes/St. Lawrence Plain
14	Atlantic Northern Forest
15	Sierra Nevada
16	Southern Rockies/Colorado Plateau
17	Badlands and Prairies
18	Shortgrass Prairie
19	Central Mixed Grass Prairie
20	Edwards Plateau
21	Oaks and Prairies
22	Eastern Tallgrass Prairie
23	Prairie Hardwood Transition
24	Central Hardwoods
25	West Gulf Coastal Plain/Ouachitas
26	Mississippi Alluvial Valley
27	Southeastern Coastal Plain
28	Appalachian Mountains
29	Piedmont
30	New England/Mid-Atlantic Coast
31	Peninsular Florida
32	Coastal California
33	Sonoran and Mojave Deserts
34	Sierra Madre Occidental
35	Chihuahuan Desert
36	Tamaulipan Brushlands
37	Gulf Coastal Prairie
38	Islas Marias
39	Sierras de Baja California
40	Desierto de Baja California
41	Islas del Golfo de California
42	Sierras y Planicies del Cabo
43	Planicie Costera, Lomerios y Canones de Occidental
44	Marismas Nacionales
45	Planice Costera y Lomerios del Pacifico Sure

(continued)

7 From the Last of the Large to the Remnants of the Rare

Table 7.1 (continued)

BCR number	BCR name
46	Sur del Altiplano Mexicano
47	Eje Neovolcanico Transversal
48	Sierra Madre Oriental
49	Planicie Costera y Lomerios Secos del Golfo de Mexico
50	Cuenca del Rio Balsas
51	Valle Tehuacan-Cuicatlan
52	Planice Costera y Lomerios Humedos del Golfo de Mexico
53	Sierra Madre del Sur
54	Sierra Norte de Puebla-Oaxaca
55	Planicie Noroccidental de Yucatan
56	Planicie de la Peninsula de Yucatan
57	Isla Cozumel
58	Altos de Chiapas
59	Depresiones Intermontanas
60	Sierra Madre de Chiapas
61	Planicie Costera del Soconusco
62	Archipielago de Revillagigedo
63	Isla Guadalupe
64	Arrecife Alacranes
65	Los Tuxtlas
66	Pantanos de Centla-Laguna de Terminos

to develop a shared vision for maintaining natural resources. This vision can include a clear accounting of the conservation values within each political unit's portion of the ecoregion and how collaboration with organizations in other states, provinces, or countries could benefit the species under consideration. Done well, this could be a major advantage in efficient use of financial and human resources to focus on conservation outcomes.

Another advantage of ecoregional planning is that it can help to clarify when the parochial focus on conservation priorities of a federal, state, or local government agency could cause it to unknowingly threaten the future of species that it values. For example, analysis of state government generated lists of bird species of conservation concern across all U.S. states show that not only is there great variability among states in the local versus global focus of their species priorities but also that many states highlight locally rare but globally common species, arguably at the expense of globally rare but locally abundant species (Wells et al. 2010).

Already the shift in conceptual thinking toward a broader ecoregional planning context over the last 20 years has resulted in major changes in bird conservation efforts. At the state and provincial levels, virtually all U.S. states and Canadian provinces now consider non-game species in planning, resource allocation, and for inclusion on endangered/threatened species lists (George et al. 1998; Wells et al. 2010). State, federal, aboriginal, and non-governmental organizations now regularly participate in a variety of coalitions to develop and implement joint bird conservation plans (Wells 2007). In 1998, U.S. state agencies collectively spent more than $130 million on non-game wildlife management activities (Richie and Holmes 1998). Further, funding decisions are increasingly tied to priorities developed and described in bird conservation ecoregional plans (International Association of Fish and Wildlife Agencies 2005).

7.4 Challenges

One of the greatest challenges that has only begun to be considered in conservation planning for birds is that of adaptation and resilience to climate change. It has already been well documented that bird distributions in North America and Europe have shifted northward, and that insect and plant ranges have shifted upward in elevation (Hitch and Leberg 2007; Mathews et al. 2004; Root et al. 2003, 2005; Rosenzweig et al. 2007; Chap. 15) . Few ecoregional conservation plans and fewer still of those focused on bird conservation have considered how to plan for the inevitable changes in and disruptions of natural communities that will occur with climate change (McLaughlin et al. 2002; Pimm 2009; Root and Schneider 2002) nor do they address the need for large reservoirs of habitat that will be necessary to buffer against those impacts.

Maintaining or restoring habitat connectivity across landscapes has been considered in conservation planning for large, wide-ranging mammals such as grizzly bears (*Ursus arctos*), timber wolves (*Canis lupus* and/or *lycaon*), and caribou (*Rangifer tarandus*) (Chap. 9). But because of the ability of most birds to travel freely across habitat barriers, connectivity has often been ignored in bird conservation planning except in regard to altitudinal migrants and greatly isolated and fragmented populations of rare species (Dobson et al. 1999). Climate change forces consideration of the issue in bird conservation planning since the plants that comprise the habitats that they rely on will survive only if they are able to move across landscapes to track the ecological envelope to which they are adapted. Highly fragmented landscapes with many barriers and low connectivity to similar habitats will likely prevent migration of some plant species and habitats, which will result in the decline in area of habitat suitable for many bird species (Collingham and Huntley 2000; Opdam and Wascher 2004; Wilson et al. 2004, 2005).

Higher elevation regions like the Appalachian Mountains in Eastern North America and the Rocky Mountains and Sierra Nevada Mountains in Western North America are likely to become increasingly important as regional refugia for species forced upward in elevation by climate change. At the continental scale, the boreal forest ecoregions of Canada and Alaska will likely become an increasingly important global refugium as the distributions of birds, as well as thousands of other species, move north at unprecedented rates (Carlson et al. 2009; Kharouba et al. 2009). The ability of these higher elevations and northern latitudes to serve as refugia will be enhanced if conservation planning takes into account this role so that very large habitat blocks with high levels of connectivity are maintained or restored on the landscape (Dobson et al. 1999). Perhaps the most obvious recommendation that follows from this observation is that the proportion of the land area designated for some form of protected status or managed with explicit consideration of the needs of wildlife must be greatly increased (Carlson et al. 2009; Innes et al. 2009; Mawdsley et al. 2009).

Many bird conservation plans for forested regions recommend an increase in the amount of forestlands that are maintained in older age classes because of a low proportion of such habitat in many areas as a result of faster timber harvest cycles in recent decades (Cyr et al. 2009; Rosenberg and Wells 2005). Mature forest habitats hold large amounts of carbon in storage, continue to sequester carbon at rates comparable to younger-age forests (Luyssaert et al. 2008), are more resilient to environmental change, and provide refugia for birds and other wildlife dependent on mature forest (Carlson et al. 2009; Secretariat of the Convention on Biological Diversity 2009). The combined benefits of increasing forest protection for birds, wildlife, and carbon storage could make for useful conservation synergies (Bradshaw et al. 2009; Secretariat of the Convention on Biological Diversity 2009; Thompson et al. 2009).

Many bird-focused ecoregional plans lack explicit goals and recommendations to guide (1) purchase of conservation land or easements, (2) identification of unprotected lands or changes in regulations, (3) legislation, or (4) land-use planning initiatives, or to measure progress in achieving meaningful bird conservation. Although some exceptions exist, including the listing of identified Important Bird Areas (Wells et al. 2005) and the spatially explicit identification of lands (Bird Conservation Areas) required for survival of key species in some Midwestern U.S. plans (e.g., Twedt et al. 1999), the general lack of spatially explicit goals and recommendations in most bird conservation ecoregional plans will hamper the ability of bird conservation priorities to be implemented in meaningful ways. Most U.S. states have some form of land protection initiative that typically has guidelines for prioritizing land purchases or set-asides of public land. These land protection programs could be excellent implementation tools for achieving bird conservation goals, but if bird conservation plans do not identify sites and their significance for priority bird species, it becomes more difficult to give such sites priority within state land protection programs.

A few plans have mapped specific areas that require increased protection or changes in management and have set out explicit habitat goals. For example, the Partners in Flight Mississippi Alluvial Valley plan identifies 87 Bird Conservation Areas (BCA's) spread across 9.7 million ha in seven states extending from Louisiana and Mississippi and north to Illinois (Twedt et al. 1999). Within the 87 BCA's, the plan recommends 101 sites to maintain or restore approximately 1.2 million ha of mature, wetland forests to sustain populations of Swallow-tailed Kite (*Elanoides forficatus*), Cerulean Warbler, and numerous other bird species. A useful next step for detailed plans like this would be to analyze the extent of overlap of these priority locations with other schemes for the prioritization of land protection, such as TNC's site portfolios within their ecoregional plans; municipal, county, state, and federal land protection initiatives; and land trust focus areas. The projected costs of implementing land protection and restoration actions to achieve the goals of such plans would be helpful in efforts to acquire adequate funding. A further step forward would be to analyze how well the identified portfolio sites would capture other conservation values, including the maintenance of other wildlife and plant species, representative ecosystems, ecological processes, and resilience to environmental change.

The integration of specific bird conservation goals into existing ecoregional and landscape-level land-use plans is currently fragmented and uneven from region to region and plan to plan. Even within existing bird conservation ecoregional plans, integration and coordination of goals and recommendations from different initiatives (e.g., landbirds, waterbirds, shorebirds, waterfowl) is needed.

Many land-use planning and land-protection programs at multiple scales do not include consideration of the needs of priority bird species. This is often true at local scales where land trusts and municipal governments may rely on volunteers and staff with little understanding of biology and no awareness of the existence of ecoregional bird conservation plans. Leaders of bird conservation initiatives would be well served to find ways to educate such conservation practitioners about the existence of bird conservation ecoregional plans and recommendations. It would be especially helpful to develop some simplified goals across the multiple bird conservation goals scaled to local applications to make it more explicit how local land trusts, for example, could focus their efforts at helping achieve ecoregional bird conservation goals.

7.5   Case Study: Joint Ventures

Waterfowl conservationists were among the first to consider both the broad geographic scales over which waterfowl breed, migrate, and winter as well as the diversity of governmental and non-governmental partners that needed to work together to achieve conservation success. They formed a series of interrelated initiatives that considered habitat protection and hunting policy. In order to achieve the level of integration among agencies and groups necessary to reach the habitat protection and enhancement goals for waterfowl, innovative integration partnerships called Joint Ventures were formed across North America beginning in the late 1980s. The 25 existing Joint Ventures bring together partners for land protection and management that can span from local municipal governments, land trusts, and chapters of Ducks Unlimited and the Audubon Society, to the Canadian Wildlife Service, U.S. Fish and Wildlife Service, Land Trust Alliance, and The Nature Conservancy, to name only a few of the thousands of partners that have been engaged over the years. The Atlantic Coast Joint Venture, for example, which covers an area of 115 million ha stretching from Maine to Florida has, with its many partners, protected over 2 million ha of wetland habitats and leveraged nearly $800 million of conservation funding since 1988. Collectively, Joint Ventures have now protected, restored, or enhanced over 6 million ha of habitat. Many have been at the forefront of finding ways to integrate waterfowl priorities with those of Partners in Flight, the U.S. and Canadian Shorebird Plans, Waterbird Conservation for the Americas, and priorities of partners in Mexico and Central and South America and the Caribbean. This is a model that all bird conservation work should strive for and that all ecological or land-use planning initiatives should consider.

7.6 Lessons Learned

First, conservation planners need to think bigger and more broadly. Long-term solutions that will increase or maintain bird populations require consideration of land-use practices on a much larger land base than most conservation practitioners have previously acknowledged. Recent reviews suggest that as much as 40–70% of a region's land base should be in some form of protected area in order to maintain a full suite of conservation values (Noss et al. 1995; Svancara et al. 2005). Conservation plans should also provide a more explicit accounting of the conservation values expected to be lost at lower proportional levels of land protection to allow for a full cost-benefit accounting for decision-makers and the public.

Bird conservation efforts at all levels also need to continue to apply a broader set of conservation values beyond those only of rarity, endangerment, and raw species diversity to include concepts like maintaining abundance, supporting a diverse range of ecological conditions and types of bird and wildlife communities, and considering co-occurring conservation values like carbon storage. This can only be accomplished by considering species priorities at multiple geographic scales and by asking the question, 'What bird conservation values in this region are globally significant?' The answer may include not only species that are globally rare but also species that are still relatively abundant but for which the region in question supports a significant proportion of the total population. A good place to start is to reference the bird species priorities and plans of Partners in Flight, the U.S. and Canadian Shorebird Plans, Waterbird Conservation for the Americas Plan, and the North American Waterfowl Management Plan.

Second, it is important to integrate rather than segregate conservation plans. Because of the complexity of the biological and political realms within which bird conservation is practiced, an infinite number of conservation initiatives, plans, and organizations sometimes seem to be involved. Each has their own set of species, geographic area, or issues on which they focus. The North American Bird Conservation Initiative and many Joint Ventures have produced web directories that list and link to the various bird conservation initiative plans and state plans that overlap each BCR to assist in the integration of conservation plans and priorities.

Third, it is important to be spatially explicit and policy specific. The less specific the recommendations in any conservation plan, the less likely that the plan will provide the necessary guidance to maintain or increase priority bird populations. Currently, very few ecoregional plans for bird conservation provide specific recommendations for areas that should be prioritized for land protection or management efforts or acreage goals. But spatially explicit recommendations have been derived for some federally endangered or threatened bird species within TNC ecoregional plans, in a few bird conservation initiative plans (e.g., MANEM Waterbird Working Group 2006; RHJV 2004; Twedt et al. 1999), and within a variety of regional and local initiatives (e.g., Florida Fish and Wildlife Conservation Commission 2005; Pronatura Noreste et al. 2004). These provide models for how site priorities and area goals can be included in ecoregional plans.

The Important Bird Areas Program uses carefully applied criteria for identifying sites of importance for birds across the world including in Canada, U.S., and Mexico (Devenish et al. 2009; Wells 1998, 2007; Wells et al. 2005) and the sites identified through this program should be more broadly included within ecoregional conservation plans. Over recent decades, researchers have pushed to better survey and document sites that are important for priority bird species (e.g., Atwood et al. 1996; Rosenberg et al. 2000; Shriver et al. 2005) so that more relevant information on important sites is now available for inclusion in ecoregional conservation plans.

Similarly, bird conservation ecoregional plans need to become more sophisticated in their analysis of policy challenges and opportunities and should not shy away from making recommendations that relate to policy changes that could help achieve plan goals.

Finally, research needs should not be emphasized at the expense of conservation needs. Many bird conservation ecoregional planning initiatives have been led by researchers who enumerate the research questions that, if answered, would provide a more precise understanding of the conservation actions likely to be most effective in increasing or sustaining bird populations. Because of this, many bird conservation ecoregional plans contain extensive and detailed lists of research questions and needs while conservation recommendations are so general and vague that they provide little useful guidance to conservation practitioners. Bird conservation ecoregional planning initiatives should strive to include clear goals and guidance that can be implemented for land conservation acquisition, public and private land management, and governmental conservation policy.

References

Atwood, J. L., Rimmer, C. C., McFarland, K. P., Tsai, S. H., & Nagy, L. R. (1996). Distribution of Bicknell's Thrush in New England and New York. *Wilson Bulletin, 108*, 650–661.

Belanger, D. O. (1988). *Managing American wildlife: A history of the International Association of Fish and Wildlife Agencies.* Amherst, MA: University of Massachusetts Press.

Secretariat of the Convention on Biological Diversity (2009). *Connecting biodiversity and climate change mitigation and adaptation: Report of the Second Ad Hoc Technical Expert Group on Biodiversity and Climate Change, CBD Technical Series No. 41.* Montreal, QC: Secretariat of the Convention on Biological Diversity.

Bradshaw, C. J. A., Warkentin, I. G., & Sodhi, N. S. (2009). Urgent preservation of boreal carbon stocks and biodiversity. *Trends in Ecology and Evolution, 24*, 541–548.

Brewer, R. (2003). *Conservancy: The land trust movement in America.* Lebanon, NH: University Press of New England.

Cardillo, M., Mace, G. M., Gittleman, J. L., & Purvis, A. (2006). Latent extinction risk and the future battlegrounds of mammal conservation. *Proceedings of the National Academy of Sciences, 103*, 4157–4161.

Carlson, M., Wells, J., Roberts, D. (2009). *The carbon the world forgot: Conserving the capacity of Canada's boreal forest region to mitigate and adapt to climate change*. Seattle, WA, and Ottawa, ON: Boreal Songbird Initiative and Canadian Boreal Initiative.

Carter, M. F., Hunter, W. C., Pashley, D. N., & Rosenberg, K. V. (2000). Setting conservation priorities for landbirds in the United States: the Partners in Flight approach. *Auk, 117*, 541–548.

Collingham, Y. C., & Huntley, B. (2000). Impacts of habitat fragmentation and patch size upon migration rates. *Ecological Applications, 10*, 131–144.

Cyr, D., Gauthier, S., Bergeron, Y., & Carcaillet, C. (2009). Forest management is driving the eastern North American boreal forest outside its natural range of variability. *Frontiers in Ecology and the Environment, 7*, 519–524. doi:10.1890/080088.

Deguise, I. E., & Kerr, J. T. (2006). Protected areas and prospects for endangered species conservation in Canada. *Conservation Biology, 20*, 48–55.

Devenish, C., Díaz Fernández, D. F., Clay, R. P., Davidson, I., & Yépez Zabala, Í. (Eds.). (2009). *Important bird areas in the Americas – priority sites for biodiversity conservation (BirdLife Conservation Series No. 16)*. Quito, Ecuador: BirdLife International.

Dobson, A., Ralls, K., Foster, M., Soulé, M. E., Simberloff, D., Doak, D., et al. (1999). Corridors: Reconnecting fragmented landscapes. In M. E. Soulé & J. Terborgh (Eds.), *Continental conservation: Scientific foundations of regional reserve networks* (pp. 129–170). Washington, DC: Island Press.

Dunn, E. H., Hussell, D. J. T., & Welsh, D. A. (1999). Priority-setting tool applied to Canada's landbirds based on concern and responsibility for species. *Conservation Biology, 13*, 1404–1415.

Fischman, R. L. (2003). *The National Wildlife Refuges: Coordinating a conservation system through law*. Washington, DC: Island Press.

George, S., Sharpe, W. J., III, & Senatore, M. (1998). *State endangered species acts: Past, present, and future*. Washington, DC: Defenders of Wildlife.

Groves, C. R. (2003). *Drafting a conservation blueprint: A practitioner's guide to planning for biodiversity*. Washington, DC: Island Press.

Grumbine, R. E. (1994). What is ecosystem management? *Conservation Biology, 8*, 27–38.

Hitch, A. T., & Leberg, P. L. (2007). Breeding distributions of North American bird species moving north as a result of climate change. *Conservation Biology, 21*, 534–539.

Innes, J., Joyce, L. A., Kellomaki, S., Louman, B., Ogden, A., Parrotta, J., et al. (2009). Management for adaptation. In R. Seppala, A. Buck, P. Katila (Eds.), *Adaptation of forests and people to climate change. A Global Assessment Report (IUFRO World Series Volume 22)*. Helsinki, Finland: International Union of Forest Research Organizations.

Kautz, R. S., & Cox, J. A. (2001). Strategic habitats for biodiversity conservation in Florida. *Conservation Biology, 15*, 55–77.

Kessler, W. B., Salwasser, H., Cartwright, C. W., Jr., & Caplan, J. A. (1992). New perspectives for sustainable natural resources management. *Ecological Applications, 2*, 221–225.

Kharouba, H. M., Algar, A. C., & Kerr, J. T. (2009). Historically calibrated predictions of butterfly species' range shift using global change as a pseudo-experiment. *Ecology, 90*, 2213–2222.

Kohm, K. A. (Ed.). (1991). *Balancing on the brink of extinction: The Endangered Species Act and lessons for the future*. Washington, DC: Island Press.

Krech, S., III, McNeill, J. R., Merchant, C. (2004). *Encyclopedia of World Environmental History (Vol. 2)*. New York: Routledge.

LaRoe, E. T. (1993). Implementation of an ecosystem approach to endangered species conservation. *Endangered Species Update, 10*(3&4), 3–6.

Luyssaert, S., Detlef Schulze, E., Börner, A., Knohl, A., Hessenmöller, D., Law, B. E., et al. (2008). Old-growth forests as global carbon sinks. *Nature, 455*, 213–215.

Mathews, S. N., O'Connor, R. J., Iverson, L. R., & Prasad, A. M. (2004). *Atlas of climate change effects in 150 bird species of the Eastern United States (General Technical Report NE-318)*. Newtown Square, PA: USDA Forest Service, Northeastern Research Station.

Mawdsley, J. R., O'Malley, R., & Ojima, D. S. (2009). A review of climate-change adaptation strategies for wildlife management and biodiversity conservation. *Conservation Biology, 23*, 1080–1089.

McLaughlin, J. F., Hellman, J. J., Boggs, C. L., & Ehrlich, P. R. (2002). Climate change hastens population extinctions. *Proceedings of the National Academy of Sciences, 99*, 6070–6074.

National Research Council. (1995). *Science and the Endangered Species Act*. Washington, DC: National Academies Press.

Newmark, W. D. (1987). A land-bridge perspective on mammalian extinctions in western North American parks. *Nature, 325*, 430–432.

Newmark, W. D. (1995). Extinction of mammal populations in western North American national parks. *Conservation Biology, 9*, 512–526.

Noss, R. F., LaRoe, E. T., III, & Scott, J. M. (1995). *Endangered ecosystems of the United States: A preliminary assessment of loss and degradation (Biological Report 28)*. Washington, DC: USDI National Biological Service.

Opdam, P., & Wascher, D. (2004). Climate change meets habitat fragmentation: Linking landscape and biogeographical scale levels in research and conservation. *Biological Conservation, 117*, 285–297.

Ossa, H. (1973). *They saved our birds*. New York: Hippocrene Books.

Pimm, S. L. (2009). Climate disruption and biodiversity. *Current Biology, 19*, R595–R601.

Pronatura Noreste, The Nature Conservancy, and World Wildlife Fund. (2004). *Ecoregional conservation assessment of the Chihuahuan Desert (2nd ed.)*. Retrieved February 5, 2010, from http://www.parksinperil.org/files/chihuahuan_desert_report_cover_and_toc.pdf

RHJV (Riparian Habitat Joint Venture). (2004). *The riparian bird conservation plan: A strategy for reversing the decline of riparian associated birds in California (Version 2.0)*. Stinson Beach, CA: Point Reyes Bird Observatory. Retrieved February 5, 2010, from http://www.prbo.org/calpif/pdfs/riparian.v-2.pdf

Richie, D., & Holmes, J. (1998). *State wildlife diversity program funding: A 1998 survey*. Washington, DC: International Association of Fish and Wildlife Agencies.

Rodrigues, A. S. L., Andelman, S. J., Bakarr, M. I., Boitani, L., Brooks, T. M., Cowling, R. M., et al. (2004). Effectiveness of the global protected area network in representing species diversity. *Nature, 428*, 640–643.

Root, T. R., & Schneider, S. H. (2002). Climate change: Overview and implications for wildlife. In T. L. Root & S. H. Schneider (Eds.), *Wildlife responses to climate change: North American case studies* (pp. 1–56). Washington, DC: Island Press.

Root, T. L., Price, J. T., Hall, K. R., Schneider, S. H., Rosenzweig, C., & Pounds, J. A. (2003). Fingerprints of global warming on wildlife animals and plants. *Nature, 421*, 57–60.

Root, T. L., MacMynowski, D. P., Mastrandrea, M. D., & Schneider, S. H. (2005). Human-modified temperatures induce species changes: Joint attribution. *Proceedings of the National Academy of Sciences, 102*, 7465–7469.

Rosenberg, K. V., & Wells, J. V. (2000). Global perspectives on Neotropical migratory bird conservation in the Northeast: Long-term responsibility versus immediate concern. In R. Bonney, D. N. Pashley, R. J. Cooper, & L. Niles (Eds.), *Strategies for bird conservation: The partners in flight planning process* (Proceedings of the 3rd Partners in Flight Workshop; 1995 October 1–5; Cape May, NJ (Proceedings RMRS-P-16), pp. 32–43). Ogden, UT: Rocky Mountain Research Station, Forest Service, U.S. Department of Agriculture.

Rosenberg, K. V., & Wells, J. V. (2005). Conservation priorities for terrestrial birds in the Northeastern United States. In C. J. Ralph & T. D. Rich (Eds.), *Bird conservation implementation and integration in the Americas* (Proceedings of the Third International Partners in Flight Conference, March 20–24, 2002, Asilomar, CA, Volume 1 (Gen. Tech. Rep. PSW-GTR-191), pp. 236–253). Albany, CA: USDA Forest Service, Pacific Southwest Research Station.

Rosenberg, K. V., Barker, S. E., Rohrbaugh, R. W. (2000). *An atlas of Cerulean Warbler populations. Final Report to U.S. Fish and Wildlife Service: 1997–2000 breeding seasons*. Ithaca, NY: Cornell Laboratory of Ornithology. Retrieved February 5, 2010, from http://www.birds.cornell.edu/cewap/cwapresultsdec18.pdf

Rosenzweig, C., Casassa, G., Karoly, D. J., Imeson, A., Liu, C., Menzel, A., et al. (2007). Assessment of observed changes and responses in natural and managed systems. In M. L. Parry, O. F. Canziani, J. P. Palutikof, P. J. vander Linden, & C. E. Hanson (Eds.), *Climate change 2007: Impacts, adaptation, and vulnerability. Contribution of Working Group II to the Fourth Assessment Report of the Intergovernmental Panel on Climate Change* (pp. 79–131). Cambridge, UK: Cambridge University Press.

Salwasser, H. (1992). From new perspectives to ecosystem management: Response to Frissell et al. and Lawrence and Murphy. *Conservation Biology, 6,* 469–472.

Scott, J. M., Davis, F. W., McGhie, R. G., Wright, R. G., Groves, C., & Estes, J. (2001). Nature reserves: Do they capture the full range of biological diversity? *Ecological Applications, 11,* 999–1007.

Shriver, W. G., Jones, A. L., Vickery, P. D., Weik, A., & Wells, J. V. (2005). The distribution and abundance of obligate grassland birds breeding in New England and New York. In C. J. Ralph & T. D. Rich (Eds.), *Bird conservation implementation and integration in the Americas* (Proceedings of the Third International Partners in Flight Conference, March 20–24, 2002, Asilomar, CA, Volume 1 (General Technical Report PSW-GTR-191), pp. 511–518). Albany, NY: USDA Forest Service, Pacific Southwest Research Station.

Stein, B. A., Kutner, L. S., & Adams, J. S. (Eds.). (2000). *Precious heritage: The status of biodiversity in the United States.* New York: Oxford University Press.

Svancara, L. K., Brannon, R., Scott, J. M., Groves, C. R., Noss, R. F., & Pressey, R. L. (2005). Policy-driven versus evidence-based conservation: A review of political targets and biological needs. *BioScience, 55,* 989–995.

Thompson, I., Mackey, B., McNulty, S., Mosseler, A. (2009). *Forest resilience, biodiversity, and climate change. A synthesis of the biodiversity/resilience/stability relationship in forest ecosystems (Technical Series no. 43).* Montreal, QC: Secretariat of the Convention on Biological Diversity, Montreal.

Twedt, D., Pashley, D., Hunter, C., Mueller, A., Brown, C., Ford, B. (1999). *Partners in flight bird conservation plan for the Mississippi Alluvial Valley.* Retrieved February 5, 2010, from Partners in Flight Web site: http://www.partnersinflight.org/bcps/plan/MAV_plan.html

MANEM Waterbird Working Group (2006). *Waterbird conservation plan for the Mid-Atlantic/New England/Maritimes Region: 2006–2010.* Retrieved February 5, 2010, from Waterbird Conservation for the Americas Web site: http://www.waterbirdconservation.org

Wells, J. V. (1998). *Important bird areas in New York State.* Albany, NY: National Audubon Society.

Wells, J. V. (2007). *Birder's conservation handbook: 100 North American birds at risk.* Princeton, NJ: Princeton University Press.

Wells, J. V., & Rosenberg, K. V. (1999). Grassland bird conservation in Northeastern North America. *Studies in Avian Biology, 19,* 72–80.

Wells, J. V., Niven, D. K., & Cecil, J. (2005). The Important Bird Areas Program in the United States: building a network of sites for conservation, state by state. In C. J. Ralph & T. D. Rich (Eds.), *Bird conservation implementation and integration in the Americas* (Proceedings of the Third International Partners in Flight Conference, March 20–24, 2002, Asilomar, CA, Volume 1 (General Technical Report PSW-GTR-191), pp. 1265–1269). Albany, CA: USDA Forest Service, Pacific Southwest Research Station.

Wells, J. V., Robertson, B., Rosenberg, K. V., Mehlman, D. W. (2010). Global versus local conservation focus of U.S. state agency endangered bird species lists. *PloS One, 5,* e8608. doi: 10.1371/journal.pone.0008608.

International Association of Fish and Wildlife Agencies (2005). *State wildlife action plans: Defining a vision for conservation success.* Retrieved February 12, 2010, from http://www.dnr.state.mi.us/publications/pdfs/HuntingWildlifeHabitat/WCS/AFWA_Overview.pdf

Florida Fish and Wildlife Conservation Commission (2005). *Florida's Wildlife Legacy Initiative: Florida's comprehensive wildlife conservation strategy.* Tallahassee, FL: Florida Fish and Wildlife Conservation Commission. Retrieved February 5, 2010, from http://www.myfwc.com/WILDLIFEHABITATS/Legacy_index.htm

Wilson, R. J., Thomas, C. D., Fox, R., Roy, D. B., & Kunin, W. E. (2004). Spatial patterns in species distributions reveal biodiversity change. *Nature, 432,* 393–396.

Wilson, R. J., Gutierrez, D., Gutierrez, J., Martinez, D., Agudo, R., & Monserrat, V. J. (2005). Changes in the elevational limits and extent of species ranges associated with climate change. *Ecology Letters, 8,* 1138–1146.

Woodruff, R., & Ginsberg, J. R. (1998). Edge effects and the extinction of populations inside protected areas. *Science, 280,* 2126–2128.

Chapter 8
The Transboundary Nature of Seabird Ecology

Patrick G.R. Jodice and Robert M. Suryan

Abstract The term 'seabird' is generally applied to avian species that forage in the marine environment over open water. Seabirds typically nest in colonies and are long-lived species with low annual reproductive rates. Seabird breeding sites typically occur on islands or along coasts and as such are often at the boundaries of ecological or political zones. During the breeding season, seabirds cross a very distinct terrestrial/marine ecological boundary on a regular basis to forage. Even relatively 'local' species cross multiple jurisdictions within a day (e.g., state lands and waters, and federal waters) while pelagic species may transit through international waters on a daily, weekly, or monthly time-frame. Seabird life-histories expose individuals and populations to environmental conditions affecting both terrestrial and marine habitats. The wide-ranging and transboundary nature of seabird ecology also exposes these species to various environmental and anthropogenic forces such as contamination, commercial fisheries and climate forcing that also are transboundary in nature. Therefore, wherever conservation of seabirds or the management of their populations is the goal, consideration must be given to ecosystem dynamics on land and at sea. Because the jurisdiction of agencies does not cross the land-sea boundary in the same manner as the seabirds they are managing, these efforts are facilitated by multi-agency communication and collaboration. By their very nature and by the nature of the systems that they must function within, seabirds embody the complexity of wildlife ecology and conservation in the twenty-first century.

Keywords Contaminants • Environmental forcing • Fisheries bycatch • Foraging • Marine conservation

---

P.G.R. Jodice (✉)
South Carolina Cooperative Fish & Wildlife Research Unit, Clemson University, G27 Lehotsky Hall, Clemson, SC 29643, USA
e-mail: pjodice@clemson.edu

R.M. Suryan
Oregon State University, Hatfield Marine Science Center, 2030 S.E. Marine Science Dr., Newport, Oregon 97365, USA
e-mail: rob.suryan@oregonstate.edu

S.C. Trombulak and R.F. Baldwin (eds.), *Landscape-scale Conservation Planning*, 139
DOI 10.1007/978-90-481-9575-6_8, © Springer Science+Business Media B.V. 2010

8.1 Introduction

In this chapter, we discuss the transboundary nature and multi-scale properties of seabird ecology and life history, considering examples from local to global spatial scales and at daily to decadal temporal scales. The examples we provide will demonstrate that seabirds use multiple spatial scales within relatively brief time frames, cross political boundaries on a regular basis, and thus exemplify transboundary and multi-scale concepts as they relate to wildlife ecology and conservation (Wolf et al. 2006).

Many wildlife species travel substantial distances and cross multiple political and ecological boundaries during migration periods. For example, many songbirds that breed in the Northern Appalachians migrate during the winter to the Southeastern U.S., the Caribbean, or Central and South America; therefore, management and conservation efforts for these species typically consider their winter, summer, and stop-over regions (Chap. 7). The crossing of ecological and political boundaries by wildlife in marine ecosystems also occurs readily, and although the transitions might appear subtle, they are equally striking. For example, gray whales (*Eschrichtius robustus*) and Atlantic bluefin tuna (*Thunnus thynnus*) may traverse entire ocean basins during their annual cycles.

Many seabirds undergo similar large-scale movements during migration. Northern Gannets (*Morus bassanus*) breeding in Atlantic Canada winter as far south as the Atlantic and Gulf coasts of the U.S. (Mowbray 2002). Cory's Shearwaters (*Calonectris diomedea*) breeding in the Mediterranean and on the Azores and Canary Islands winter throughout the South Atlantic, the Eastern Tropical Atlantic, and the Western Indian Oceans (González-Solís et al. 2007). Sooty Shearwaters (*Puffinus griseus*), upon completion of their breeding cycle in New Zealand, traverse the Pacific Ocean from the Southern to the Northern Hemisphere, crossing from the eastern to the western boundary in a figure-eight pattern (Shaffer et al. 2006) (Fig. 8.1). Most recently, the 30,000-km, round-trip migration route of the Arctic Tern (*Sterna paradisaea*) has been mapped using global location sensing units (i.e., geolocators) (Egevang et al. 2010). Seabirds also cross ecological and political boundaries on much shorter and more frequent time scales compared to those observed during migration. During the breeding season, seabirds cross a very distinct terrestrial/marine ecological boundary on a regular basis to forage. Even relatively 'local' species cross multiple jurisdictions within a day (e.g., state lands and waters, and federal waters) while more pelagic species may transit through international waters.

The environmental dynamics of the ecosystems inhabited by seabirds also incorporate large and variable spatial and temporal scales. For example, locally and short-term severe weather may interfere with chick-feeding, decrease chick growth, or increase chick mortality (Konarzewski and Taylor 1989; Velando et al. 1999). In contrast, the onset of an El Niño event may affect food availability and subsequently seabird productivity at the scale of months, while a shift in climate regimes such as the Pacific Decadal Oscillation or the North Atlantic Oscillation may alter foraging and breeding conditions of seabirds for years and affect entire

8 The Transboundary Nature of Seabird Ecology

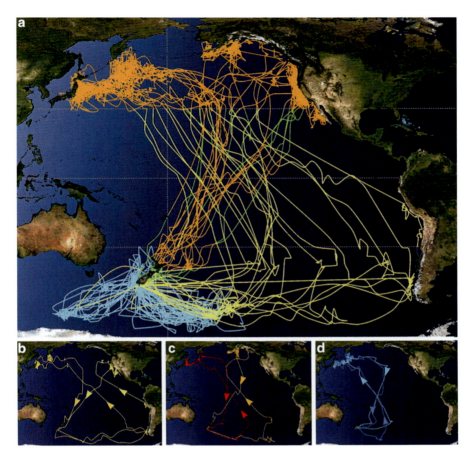

Fig. 8.1 Shearwater migrations from breeding colonies in New Zealand. (**a**) Nineteen sooty shearwaters tracked via miniature geolocation (light sensing) tags during breeding (*light blue lines*), post-breeding migration into the Northern Hemisphere (*yellow lines*), and wintering grounds (Northern Hemisphere summer) and southward return migration to the breeding colony (*orange and green lines*). (**b, c, d**) The three *lower panels* show migration paths of breeding pairs, demonstrating that some go to different wintering areas and meet back at the colony the following breeding season, while others go to the same areas – all exhibiting a figure eight migration pattern (From Shaffer et al. 2006)

ocean basins (Chavez et al. 2003; Velarde et al. 2004). Anthropogenic threats to seabirds, such as habitat disturbance at colonies, oil spills, or climate change, also may operate from local and short-term to global and long-term scales. The transboundary nature of seabirds thus differs from that of songbirds (Chap. 7) or even other large marine vertebrates because of their propensity to cross multiple ecological and political boundaries on short and frequent time scales and because they are similarly affected by large-scale anthropogenic events and ecosystem dynamics.

In the following sections, we provide examples of seabird behavior, ecology, and conservation that exemplify the concepts of both landscape-scale – in this case, referring to large spatial scales on both land and sea – and transboundary patterns and processes. Along with a review of seabird biology and life history, we also review the transboundary and landscape-scale nature of seabird foraging ranges and breeding habitats, and the effects of contaminants, environmental forcing, and fisheries bycatch on seabirds.

8.2 Seabirds: Taxa, Life-History Traits, and Foraging Ecology

In the following section, we provide a brief review of key life-history traits that exemplify the landscape-scale nature of seabird ecology. Seabird biology and natural history are also thoroughly reviewed by Furness and Monaghan (1987), Gaston (2004), and Schreiber and Burger (2001).

The term 'seabird' is generally applied to species that forage in the marine environment over open water. Typically included are all species from the orders Sphenisciformes (penguins) and Procellariiformes (albatrosses, petrels, storm-petrels, fulmars, and shearwaters), most species from the order Pelecaniformes (pelicans, boobies, frigate-birds, gannets, and cormorants), and some species from the order Charadriiformes (alcids, gulls, terns, skuas, and skimmers; Schreiber and Burger 2001). There are 65 seabird genera and approximately 222 wholly marine and 72 partially marine species (Gaston 2004). Seabirds include some of the most abundant birds on Earth, such as the Wilson's Storm-petrel (*Oceanites oceanicus*), which may number greater than 10 million individuals (Warham 1990); some of the rarest birds on earth, such as the Chatham Island Petrel (*Pterodroma magentae*) and the Chinese Crested Tern (*Sterna bernsteini*), each of which likely has only 10–20 breeding pairs (BirdLife International 2008); and numerous highly endemic birds such as the Bermuda Petrel (*Pterodroma cahow*) and Black-capped Petrel (*Pterodroma hasitata*), which now breed in only one or a few sites in the West Indies, the Fiji Petrel (*Pterodroma macgillivrayi*), found only near the island of Gau in the South Pacific, and the Christmas Island Frigatebird (*Fregata andrewsi*), which breeds only in that island group.

Seabirds can also be categorized by the marine zones in which they tend to forage. For example, albatrosses (Diomedeidae) are considered classic pelagic seabirds because they typically forage away from the coastal zone and over open ocean during both the breeding and non-breeding seasons. In contrast, most gulls (Laridae) and terns (Sternidae) are regarded as nearshore because they tend to forage in coastal waters and winter in coastal zones where they may often be found loafing on beaches. Some seabirds use both nearshore and pelagic zones. For example, many alcids and penguins forage in the nearshore and pelagic zones during both the breeding and non-breeding seasons and only rarely use terrestrial habitat outside of the breeding season. Although these categories present some ambiguities and are not strictly defined, they do provide an immediate and clear transboundary reference in terms of spatial scale.

Approximately 96% of seabird species nest in colonies (Wittenberger and Hunt 1985). Colony size can vary from tens of pairs to over 1 million, and the abundance of nesting birds varies based on attributes such as availability of nesting habitat, proximity of food, or size and proximity of nearby colonies. Seabirds use a wide variety of substrates for nesting habitat (Gaston 2004). The most common nest occurs on open ground. Ledges of cliff faces are also used where they are available. Shearwaters, storm-petrels, diving-petrels, puffins, and tropicbirds commonly use ground burrows or crevices in cliffs. Trees and shrubs are commonly used in tropical areas by Pelecaniformes, gulls, and terns, although one species of alcid, the Marbled Murrelet (*Brachyramphus marmoratus*), specializes in nesting on limbs of old-growth trees in the Pacific Northwest. Nests also can be found on human-made structures. Two examples include Least Terns (*Sternula antillarum*), which commonly nest on rooftops throughout the Southeastern U.S. (Gore and Kinnison 1991; Krogh and Schweitzer 1999) and Black-legged Kittiwakes (*Rissa tridactyla*) which nest on abandoned structures in the U.K. (Coulson 1968) and Alaska (Gill et al. 2002).

Seabirds tend to be long-lived, relatively slowly reproducing species especially at the 'pelagic' end of the spectrum. For example, while some nearshore species such as terns, skimmers, and gulls may breed at 2–4 years of age, pelagic species such as albatrosses and petrels may delay breeding until 10 years or more. Unlike waterfowl and songbirds, seabird clutches tend to be small (≤5 eggs). Nearshore species typically have larger clutches compared to pelagic species, most of which lay only one egg. Several species only breed every other year (e.g., albatrosses, frigatebirds; Warham 1990, Nelson 2005) and many seabirds will abandon current nesting attempts, especially when feeding conditions are poor, as a means to increase the probability of surviving to reproduce the following year (Golet et al. 1998). The incubation period of seabirds ranges from a fairly typical 28–30 days in many nearshore species to about 80 days in large seabirds such as Wandering Albatrosses (*Diomedea exulans*) and Northern and Southern Royal Albatrosses (*D. sanfordi* and *D. empomophora*) (Tickell 2000).

Nestling or chick-rearing periods are variable among seabirds and can be extensive. Gulls and terns may fledge in 30 days or less, Brown Pelicans (*Pelecanus occidentalis*) require approximately 75 days, Magnificent Frigatebirds (*Fregata magnificens*) 150–185 days, and Wandering and Royal albatrosses and King Penguins (*Aptenodytes patagonicus*) 240 or more days. In contrast, some seabird chicks depart the nest prior to developing the ability to fly. Many gulls and terns will depart the nest within a few days of hatching, some forming large crèches in intertidal zones. For these species, management during the breeding season thus requires secure nesting areas and secure chick-rearing areas. A unique trait among some alcids (e.g., Common Murres [*Uria aalge*], Ancient Murrelets [*Synthliboramphus antiquus*]) is for chicks to depart the nest prior to gaining flight and to complete the majority of pre-fledge chick-growth at sea, including chicks from the *Synthliboramphus* murrelets which depart the nest within days, as well as other alcids, which depart the nest beginning within 2 weeks after hatching.

Seabirds employ a variety of foraging techniques, forage in a variety of locations, and forage upon a variety of items. The dominant diet item among seabirds is fish, and the type and size taken depends in part on the foraging technique, geographic distribution, size of the bird, and marine habitat. In many northern and mid-latitude areas, fish such as herring, sardines, anchovies, and menhaden (Clupeiformes), sand eels (*Ammodytes* spp.), and smelts (Osmeridae) are common in diets, while in tropical latitudes flying fish (Exocoetidae) may be more common. Invertebrates such as cephalapods (e.g., squid) and zooplankton (e.g., krill) are also important food items, the latter particularly so in high latitude or highly productive regions. Seabirds also use anthropogenic food sources such as offal and discarded bycatch from commercial fisheries, and the availability and distribution of these food sources may alter seabird diets, distributions, and population dynamics (Furness 2003; Garthe et al. 1996).

Seabirds forage primarily by surface feeding (e.g., gulls, terns, albatrosses), plunge diving into the top few meters of the water column (e.g., pelicans), pursuit diving (e.g., alcids, penguins, shearwaters, diving-petrels, and cormorants, some of which can access waters as deep as 100–500 m during their pursuit dives), and kleptoparasitism (skuas, jaegers, and frigatebirds). Seabirds may forage individually, in small single- or multi-species flocks, or occasionally in large flocks numbering over 1 million. Surface-feeding seabirds may forage in association with sub-surface foragers such as alcids, penguins, tuna, dolphins, or whales that effectively drive prey toward the surface (Hebshi et al. 2008), and this habit can be common in nutrient poor, oligotrophic waters (Ballance et al. 1997). Seabirds tend to locate prey visually, although some procellariids use olfaction (Nevitt et al. 2008) and some specialized species such as skimmers (*Rynchops* spp.) use tactile senses.

The location of foraging depends to a certain extent on the foraging technique and the accessibility of prey. Seabirds often frequent locations that are character-ized by nutrient-rich surface waters such as upwelling zones, fronts and eddies, seamounts, or along the edge of the continental shelf. Ultimately foraging locations are dictated by a combination of habitat features that affect prey availability, including attributes such as ocean and wind circulation patterns, the extent of upwelling and productivity, turbidity, and distance from the breeding site. The spatial scale at which these features operate varies from local to global, and their temporal scale also varies from relatively predictable (e.g., upwelling generated via water currents and associated with a landmass or seamount) to highly ephemeral (e.g., local wind-generated aggregation of surface prey items).

Most seabirds are central-place foragers during the breeding season, returning to land on a regular basis to incubate or feed nestlings. The distance between the foraging area and the breeding site varies over four orders of magnitude across all seabirds. The frequency of food delivered to chicks also varies widely among species and is one of the primary factors that contribute to the transboundary habits of seabirds (i.e., regularly crossing from terrestrial to marine systems). Feeding fre-quency can vary within and among species based on factors such as distance to the food source, the extent and type of parental care required by the chick, weather, and chick age. For some species, feeding frequency ('feeds') is best measured on a per hour basis. For example, studies of chick feeding by Brown Pelicans in Mexico and South Carolina revealed that chicks received 1–4 feeds per hour, although the number of feeds decreased with age (Pinson and Drummond 1993; Sachs and Jodice 2009). In other species, feeds are best measured on a per day basis. Jodice et al. (2006) found that at six colonies during 5 years of study chicks of Black-legged Kittiwakes received on average 2–5 feeds per day with adults foraging primarily in nearshore waters (Suryan et al. 2002). Trivelpiece et al. (1987) measured feeding rates in three species of penguins raising chicks at King George Island. Chicks of the more nearshore Adelie Penguin (*Pygoscelis adeliae*) were fed about once per day while those of the more offshore and deep-diving Chinstrap and Gentoo penguins (*P. antarctica* and *P. papua*) received 1.5–2.0 feeds per day. Feeding also occurs less than daily in many pelagic species. For example, many albatrosses and petrels regularly feed chicks at 1–5 day intervals although the gap between feeds extends with chick age (Warham 1990). Very infrequent feedings occur in the King Penguin, which during the winter starvation period may deliver food to chicks only once per 30–90 days (Cherel et al. 1987). The variability associated with these provisioning rates is based in part on life-history traits but also can vary with environmental conditions. This fact becomes important when discussing the concept of ecoregions within the marine environment and the extent to which seabirds traverse both ecoregional and political boundaries.

8.3 Seabirds, Boundaries, and Scales

Large-scale conservation planning and the mapping of biological diversity for conservation purposes are more common in terrestrial compared to marine systems (Spalding et al. 2007 and included references). For example, despite the prevalence of marine environments across the globe, these habitats are underrepresented in global reserve networks, comprising less than 0.5% of the earth's surface (Chape et al. 2005). Only within the past 10 years have global classification systems been developed for the marine environment. Longhurst (2007) proposed biogeographical provinces for pelagic waters (approximately ten for each ocean basin). Within this scheme, boundaries are not fixed in space or time but instead can shift based on the temporal changes in physical forcing that regulate phytoplankton distribution. Spalding et al. (2007) developed a biogeographic system for coastal and shelf areas (Fig. 8.2). This hierarchy of 12 realms, 62 provinces, and 232 ecoregions provides a comprehensive and readily available framework for marine conservation planning within the area in which most marine diversity and most threats occur (Spalding et al. 2007; UNEP 2006).

Here we present several aspects of seabird ecology and management that highlight the landscape-scale properties of seabirds.

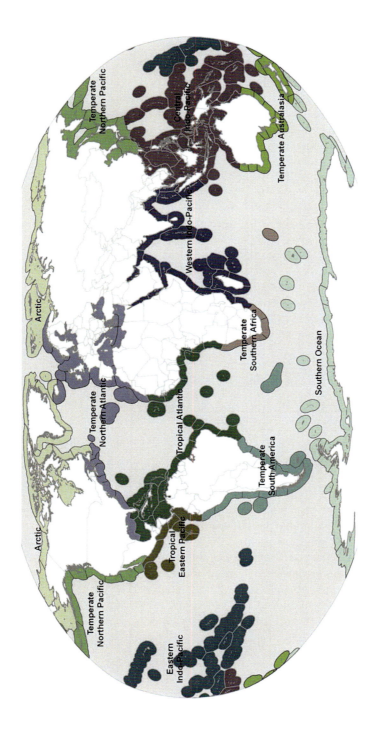

Fig. 8.2 Final biogeographic framework for coastal and inshore areas: Realms and provinces. Biogeographic realms with ecoregion boundaries outlined; definitions of bioregions in Spalding et al. (2007)

8.3.1 Breeding Habitats, Political Boundaries, and Ecological Boundaries

Seabird breeding sites typically occur on islands or along coasts and as such are often at the boundaries of ecological or political zones and hence influenced by the dynamics of both marine and terrestrial systems. Across the range of seabird species, the consistent use of a site as a nesting location varies from strongly philopatric to highly plastic. In addition, some species have a limited number of nesting sites while others occupy numerous sites. Seabirds that are philopatric and that nest in only one or a few locations can present a substantial conservation challenge. For example, the Short-tailed Albatross (*Phoebastria albatrus*) currently numbers about 2,500 individuals with breeding colonies on only two islands off the coast of Japan. Key threats to this species include the instability of soil, the threat of mortality and habitat loss from an active volcano, and vulnerability to other natural disasters such as typhoons at its main breeding site. Interestingly, the second remote breeding island for this species is currently disputed territory among three Asian nations (BirdLife International 2008), thus adding a different twist to the concept of 'transboundary.' Nonetheless, the Short-tailed Albatross demonstrates an 'all eggs in one basket' situation. In species that rely on a single location for a colony, a goal of conservation planning may be to reduce the risk to a species, perhaps from a stochastic event such as a storm or predator invasion, by developing an alternate nesting site (Miskelly et al. 2009).

Some seabirds, such as the Red-legged Kittiwake (*Rissa brevirostris*), have only a few nesting locations that are widely spaced. Major colonies are located on the Pribilof Islands in the Eastern Bering Sea, Bogoslof Island in the Aleutian chain which lies approximately 350 km south of the Pribilof Islands, Buldir Island which lies 1,000 km west of Bogoslof Island, and the Commander Islands which lie another 700 km west of Buldir Island and are within Russian waters. These few colonies occur in two realms (Arctic and Temperate Northern Pacific), two provinces (Arctic and Cold Temperate Northwest Pacific), and three ecoregions (Eastern Bering Sea, Aleutian Islands, and Kamchatka Shelf) as delineated by Spalding et al. (2007).

Unlike the previous examples, some seabirds are loosely philopatric and tend to move readily among multiple sites from 1 year to the next. This is very common in some beach-nesting terns, where the quality and size of breeding beaches are subject to a high degree of interannual variability due to winter storms and sediment transport. Management of these species, therefore, requires a network of readily available sites that can accommodate thousands of birds from 1 year to the next. For example, along the coast of South Carolina, Royal and Sandwich terns (*Sterna maxima* and *S. sandvicensis*) have nested on nearly a dozen sites over the past 3 decades (Jodice et al. 2007). These sites occur over about 175 km of coastline, and colonies of thousands of birds frequently move among sites in consecutive years. For example, between 1990 and 1991, the nest counts at one colony in South Carolina decreased from 8,200 to 200 while nest counts at another colony increased from 900 to 11,000. Additionally, between 1986 and 2005, six different sites were used only one to four times each and during that period nest counts ranged from several to nearly 4,000. While the reasons underlying such large-scale and natural relocations are varied and may include both natural and anthropogenic factors (e.g., beach erosion, human disturbance or development), the management message is that a single site cannot support a species having a low degree of colony philopatry.

Seabird breeding ranges also may cross multiple political boundaries. While many species of landbirds breed among multiple nations, seabirds may nest in multiple nations as well as cross these boundaries on a daily or weekly basis as they forage. For example, the West Indian Breeding Seabird Atlas (www.wicbirds.net) catalogs breeding locations and population estimates for 25 seabirds on nearly 800 islands spread across 39 countries from Bermuda to the islands off of Northeastern South America. Two wide-ranging species in the region are the Audubon's Shearwater (*Puffinus lherminieri*) and White-tailed Tropicbird (*Phaethon lepturus*). Each nests in over 20 countries throughout the West Indies (Lee 2000a; McGehee 2000) and in 5–6 ecoregions based on Spalding et al. (2007) including the Bermuda, Bahamian, Eastern Caribbean, Greater Antilles, and Southern Caribbean ecoregions. Conservation regulations, enforcement, education, and funding for wildlife management and conservation vary considerably across the region making management efforts spatially inconsistent and temporally variable. Although the need for transboundary conservation efforts in this region has been recognized for over a decade (Gochfeld et al. 1994), such efforts have yet to be fully realized.

Along with variability in the number of nesting sites used by a species and the consistency with which sites are used among years, seabirds also display variability in the types of habitats used for nesting. While most seabirds typically nest on cliffs or plateaus, or in burrows immediately adjacent to their marine foraging habitat, others do not. Seabirds also nest in forests and alpine areas, which are quite distinct from the marine zone. Here we provide four examples of seabirds that nest 'inland' and face management challenges associated specifically with their inland nesting habitat.

Inland nesting is not uncommon among the petrels and shearwaters, which often nest in burrows or cavities. The Hutton's Shearwater (*Puffinus huttoni*) is endemic to New Zealand and is considered threatened. The species currently nests at only two alpine sites in the Seaward and Inland Kaikoura Mountains at elevations of 1,200–1,800 m (Cuthbert et al. 2001; Cuthbert and Davis 2002). Nesting habitat of Hutton's Shearwater is considered to be endangered and has been lost to introduced nest predators and browsers, the latter of which are responsible for erosion in the alpine nesting areas (BirdLife International 2009a; Cuthbert et al. 2001). Like Hutton's Shearwater, Newell's Shearwater (*Puffinus newels*) is also considered to be endangered. The species is now confined to steeply sloped, forested sites at 160–1,200 m elevation and as far as 14 km inland on Kaua'i, Molokai, and Hawaii in the Hawaiian Islands (Ainley et al. 2001; Day and Cooper 1995). While their bones can be found in caves throughout the island chain, populations of Newell's Shearwaters persist in areas least affected by introduced predators and urbanization (adults collide with power lines while commuting inland) (BirdLife International 2009b). Another highly endangered, inland nesting seabird is the Black-capped Petrel, which nests on forested slopes and cliffs at elevations of 1,500–2,300 m at a limited number of sites in Haiti and the Dominican Republic, although it nested much more broadly throughout the Caribbean before humans arrived in the region (Lee 2000b). Deforestation for charcoal and small-scale agriculture is the primary factor underlying loss of nesting habitat. Another inland forest-nesting species, the Marbled Murrelet, nests in old-growth forests along the Pacific Northwest coast of North America up to 65 km inland. Nesting habitat has declined due to timber harvesting and fragmentation in coastal forests (Gaston and Jones 1998). Management actions, research, and planning for each of these four species have focused not only on the marine environment but also on issues related to forest management, urbanization, or grazing in the nesting environment and thus have incorporated transboundary and landscape-scale thinking.

8.3.2 Ranges of Seabirds: from Bays to Oceans

Like many birds, seabirds often cross ecoregional and political boundaries during post-breeding dispersal and migration. For example, many nearshore species common to the southeastern U.S., such as Royal Terns and Brown Pelicans, migrate across multiple state boundaries during the non-breeding season, although they typically remain within the region. In contrast, other seabirds engage in extensive post-breeding dispersal. The Great and Magnificent Frigatebirds (*Fregata minor* and *F. magnificens*), for example, travel 1,400–4,400 km from their breeding sites and continue to make foraging trips of many hundreds of kilometers once they relocate (Weimerskirch et al. 2006). Short-tailed Albatrosses breeding on Torishima Island off the coast of Japan disperse over 10,000 km to the Bering Sea off Alaska and Russia, with some crossing to the opposite side of the Pacific Ocean (Fig. 8.3; Suryan et al. 2006). Likewise, other species of albatrosses in the Southern Hemisphere are well known for their global circumnavigations in a region where ocean crossings are unimpeded by land masses (Croxall et al. 2005). In the Western Atlantic, the Great Shearwater (*Puffinus gravis*) breeds in the South Atlantic but disperses to the Bay of Fundy (http://www.tristandc.com/wildgreatshearwater.php).

Within the breeding season, both pelagic and nearshore species of seabirds frequently range over extensive areas and cross multiple habitats and political jurisdictions. In fact, many species do not commonly forage close to their colonies due to what is referred to as 'Ashmole's halo' (Birt et al. 1987), a zone around the colony that tends to be depleted of prey due to its proximity to the colony (Ashmole 1963, 1971). Typically the size of the halo shows a direct relationship with colony size although recent modeling efforts suggest that the halo effect may be undetectable for small colonies or for colonies of far-ranging pelagic species (Gaston et al. 2007). Nonetheless, this general pattern means that natural resource managers responsible for seabird colonies should consider not only ecological and management-related issues on and near the colony, but depending on the size and location of the colony, managers also may need to consider vast areas of marine habitat in which seabirds may forage even while rearing chicks. These areas are often in international

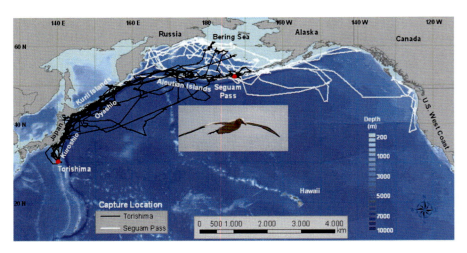

Fig. 8.3 Post-breeding migration paths of 14 satellite tracked short-tailed albatrosses. Albatrosses were tagged at their breeding colony on Torishima, Japan, and at-sea near Seguam Pass, Alaska. These results demonstrated that juvenile albatrosses (<1 year old; white lines) were ranging much farther than adults, which was later confirmed by additional tracking studies (From Suryan et al. 2006)

waters or in waters controlled by other governments or other wildlife or fisheries management agencies. Several examples of foraging ranges of breeding seabirds across four orders of magnitude (less than 10 km to more than 1,000 km) serve to demonstrate the need to address multiple spatial scales when considering conservation and management actions for this suite of species.

The Little Tern (*Sternula albifrons*) is a small (less than 60 g) seabird that breeds along coasts and inland waterways of temperate and tropical Europe and Asia. This species is declining in Europe, particularly in the U.K. where populations have declined by about 30% since the mid-1980s. Despite its small size, this species may cover 10–27 linear km during 1–2 h of foraging and regularly travels 2–3 km offshore during the breeding season, covering areas of 6–50 km^2 (Perrow et al. 2006). In the Southeastern U.S., the similarly-sized Least Tern, also considered to be a nearshore species, may be found up to 10 km offshore during the breeding season. Hence, even relatively small, inshore species may use offshore areas.

Many breeding seabirds have foraging ranges on the order of 20–100 km and hence forage not just locally but throughout a region. Adams et al. (2004) found that, on average, breeding Cassin's Auklets (*Ptychoramphus aleuticus*) at two colonies in California foraged within 30 km of their colonies and that colony-based foraging areas covered 500–1,200 km^2. Hatch et al. (2000) used satellite tags to track movement patterns of Common and Thick-billed Murres (*Uria lomvia*) from two colonies in the Gulf of Alaska and two colonies in the Chukchi Sea. They found that both species, when attending chicks, regularly foraged 50–80 km from colony sites and that foraging ranges of the two species at the same colonies overlapped considerably. However, the foraging ranges of both murre species differed considerably when examined at two colonies separated by about 50 km. Garthe et al. (2007) used GPS loggers to examine foraging ranges of Northern Gannets breeding on Funk Island, approximately 60 km northeast of Newfoundland. They found that gannets raising chicks regularly flew at speeds of 30–40 km/h to forage 32–70 km from the colony. However, other studies of Northern Gannets have revealed broader foraging ranges. Hamer et al. (2001) used satellite tags to track gannets rearing chicks at one colony in Southeast Scotland and another in Southeast Ireland. Individuals from the colony in Ireland foraged 14–238 km from the colony and covered an area of 45,000 km^2. In comparison, birds from the colony in Scotland foraged 39–540 km from the colony covering an area greater than 200,000 km^2. These last two studies demonstrate that foraging ranges of the same species may differ among colonies, and therefore management and conservation efforts also may require data from multiple locations.

At a larger scale are the pelagic species that may depart the nest for multiple days at a time and forage throughout or travel across ocean basins. This is especially common in the order Procellariiformes. For example, one of us (PGRJ) along with several colleagues documented a single Audubon's Shearwater (200 g) that was rearing a chick in the Northern Bahamas to have traveled over 1,000 km during a 1-week foraging trip, covering waters from the Charleston Bump to Cuba (Fig. 8.4).

Fig. 8.4 Locations of an Audubon's Shearwater determined via global location sensor during 1 week in June 2008. The path distance for this bird was ca. 3,000 km and the linear distance from the northernmost to southernmost point was ca. 1,200 km. This individual was tagged at the Long Cay colony, Bahamas, and was rearing a chick (unpublished data collected by P. Jodice, W. Mackin. R. Phillips, and J. Arnold)

The larger Black-capped Petrel breeds primarily in Haiti and the Dominican Republic but is commonly observed near the shelf break off of Cape Hatteras, North Carolina, during the breeding season, a distance of about 2,000 km from Haiti (Lee 2000b). One of the more extreme cases of long-distance foraging during chick-rearing occurs in the Wandering Albatross. During a single foraging trip, this species may travel 900 km per day and up to 15,000 km during the entire foraging trip, and may range from colony sites on South Georgia Island over 2,000 km north to waters off the coast of Brazil (Jouventin and Weimerskirch 1990; Prince et al. 1992).

These examples demonstrate that even nearshore seabirds can range over a substantial area on a daily basis while pelagic species may cover thousands of kilometers or more during a single foraging trip. These extensive movements have important implications for seabird management and conservation because individual birds traversing that large of an area can encounter an array of environmental conditions and anthropogenic activities, some of which may pose threats to their survival. Advances in tracking technology allow biologists to now consider, for example, how a bycatch threat or the presence of a marine protected area thousands of kilometers distant from a colony may affect seabird ecology (Hyrenbach et al. 2006; Prince et al. 1992).

8.3.3   *The Transboundary Nature of Contaminants for Seabirds*

Seabirds have often been used as biosentinels for contaminants, pollution, and other chemical stressors in the marine and coastal environment (Braune et al. 2001; Vander Pol and Becker 2007). Their position at the apex of trophic webs exposes them to biomagnification effects of contaminants. Strong site fidelity to breeding and foraging areas exposes them to persistent point-source contaminants, and wide-ranging foraging habits expose them to spatially diverse contaminant sources and politically inconsistent regulatory policies in both marine and terrestrial environments (Burger and Gochfeld 2002). Here we describe the transboundary nature by which seabirds encounter contaminants.

Nearshore species are exposed to contaminants during the breeding season in much the same way as a raptor or songbird. A parent forages within a relatively local area while provisioning itself and its chicks and acquires some contaminants from their prey. For example, Wenzel et al. (1996) examined the distribution of five trace elements in nestling Black-legged Kittiwakes in the North Sea and attributed elevated concentrations to local food sources. Becker (1989) and Becker et al. (1991) also attributed mercury contamination in eggs of nearshore Common Terns within the Elbe estuary to local sources of mercury. The transboundary nature of contaminants for nearshore seabirds often arises because the contaminants of concern, although produced at a single point source, are either transported across ecosystem boundaries (e.g., from agricultural to marine systems) or obtained as birds forage in agricultural or urban systems (Cifuentes et al. 2003).

Pelagic seabirds provide examples of transboundary contamination at extensive spatial scales. Breeding adults, in particular, may regularly travel a significant distance from an uncontaminated breeding site to a contaminated site to forage. For example, Finkelstein et al. (2006) examined organochlorine and mercury contamination in the Black-footed Albatross (*Phoebastria nigripes*) and Laysan Albatross (*Phoebastria immutabilis*), two sympatrically breeding species in the Northwestern Hawaiian Islands. Contamination levels were about 400% higher in Black-footed Albatrosses compared to Laysan Albatrosses despite similarities in diets, breeding behavior, and nesting locations. The difference in contaminant loads was attributed to a difference in foraging locations between the two species. Black-footed Albatross were foraging northeast of the Islands toward and along the west coast of North America where contamination history is strong, while Laysan Albatrosses were foraging north and west of the Hawaiian Islands in areas without a strong contamination history. While this example clearly demonstrates the transboundary nature of contamination for both species, it also demonstrates that species that breed sympatrically may be exposed differentially to contamination depending on the location and extent of the foraging range. Therefore, not only can it be difficult to predict contamination effects on pelagic seabirds due to their extensive foraging habits, but it cannot be assumed that the intensity or type of contamination will be consistent among species breeding in a single location due to the variability that may occur in foraging ranges.

Both pelagic and nearshore seabirds also may be exposed to contaminants that are being transported. For example, seabirds may forage relatively locally in an area that does not contain a contaminant source, but oceanic and atmospheric currents may move contaminants across boundaries and hence affect seabirds. Ricca et al. (2008) found elevated levels of contaminants in a suite of seabirds from the Aleutian Islands that were nesting and foraging in locations that were not associated with point sources of contamination. The species sampled represented multiple trophic positions and the authors suggested that the contaminants were being transported from the Western Pacific through oceanic and atmospheric processes.

Another potential mechanism for contaminant transport considers the transboundary nature of seabirds foraging upon discarded bycatch from commercial fishing vessels. Many seabirds attend commercial fishing vessels where they scavenge for discarded bycatch (Furness et al. 1988; Garthe and Hüppop 1994). In many cases the discarded bycatch items are demersal (bottom-dwelling) fish while the seabirds themselves are surface feeders (Walter and Becker 1997; Wickliffe and Jodice in press). Demersal prey often contain higher levels of contaminants such as mercury due to biomagnification and bioavailability in deeper waters (Monteiro et al. 1996). As these demersal fish are brought to the surface and hence made available to seabirds during the discarding process, any contaminants they may contain are effectively transported across depth boundaries. For example, Arcos et al. (2002) suggested that levels of mercury in Audouin's Gulls (*Larus audouinii*) that foraged upon discarded benthic prey were elevated compared to levels in Common Terns that did not forage upon discarded prey.

Oil spills also represent a transboundary contamination source for marine wildlife. Seabirds are exposed to oil primarily through direct contact and contamination of their prey base. Effects may be lethal or sublethal, occur proximate to or distant from seabird colonies, and be persistent. For example, the *Exxon Valdez* oil spill in Northern Prince William Sound (PWS), Alaska, created a sudden and severe point source of contamination. Ultimately the oil spread 750 km to the southwest and eventually contaminated more than 2,000 km of shoreline (Peterson et al. 2003). Approximately 878,000 seabirds were breeding at colonies within the ultimate path of the spill and many colonies experienced direct and immediate oiling (Piatt et al. 1990). Pigeon Guillemot (*Cepphus columba*) colonies on the Naked Islands in Central PWS about 35 km from the spill were directly in the path of the prevailing currents and were oiled within 3–4 days of the spill. In contrast, seabird colonies 300–400 km from the spill site in lower Cook Inlet and at the mouth of the Gulf of Alaska in the Barren Islands group did not originally appear to be in the direct path of the spill but were oiled within 3 weeks. Hence the oil spilled in Northern PWS acted as a proximate and somewhat predictable source of contamination at colonies in Central PWS but also acted as a distant and somewhat less predictable source of contamination at colonies elsewhere.

Seabirds also may be exposed to oil indirectly at the foraging grounds when they ingest prey (e.g., fish) that have been exposed to oil, and these effects may be quite persistent over time (Jewett et al. 2002). For example, Yellow-legged Gulls (*Larus michahellis*) experienced changes in plasma biochemistry and elevated levels of polycyclic aromatic hydrocarbons that were consistent with the ingestion of fuel oil 17 months after the *Prestige* oil spill occurred off the coast of Spain (Alonso-Alvarez et al. 2007a, b). Similarly, adult Pigeon Guillemots in PWS, which forage on fish and benthic invertebrates, showed elevated levels of CYP1A, a detoxification enzyme associated with exposure to oil, 9 years post-spill (Golet et al. 2002). These examples demonstrate that seabirds may be affected by oil both at and away from the colony, and that these effects can span temporal scales of months to years.

The abundance of plastics in the marine environment has become well-documented and presents another type of transboundary contaminant source for seabirds. Seabirds ingest plastic while foraging, and plastics also are brought back to the nest and ingested by chicks. Robards et al. (1995) and Blight and Burger (1997) noted plastic was very common in 11 species of seabirds from an area in the Eastern North Pacific where both regionally breeding species (e.g., Tufted Puffins [*Fratercula cirrhata*] and Rhinoceros Auklets [*Cerorhinca monocerata*]) and species from the West and South Pacific foraged (e.g Black-footed Albatross and Sooty Shearwater). Similarly, eight species of shearwaters, albatross, and petrels captured incidentally in drift-net fisheries or gathered from beached-bird surveys off the coast of Brazil frequently had plastics in their systems (Colabuono et al. 2009). Young et al. (2009) found that Laysan Albatrosses nesting on Kure Atoll spent more time foraging within the range of the 'Western Pacific garbage patch' compared to Laysan Albatrosses nesting on Oahu, and they also found a higher incidence of plastics in boluses regurgitated from chicks at Kure Atoll colony.

As with contaminants, plastics may be transported to pristine colony sites. Morishige et al. (2007) examined the amount and type of plastic debris on beaches of the Hawaiian Islands NWR. In a 16-year assessment, they found over 52,000 pieces of plastic washed up on the beaches of some of the most remote uninhabited atolls on the planet. Interestingly, they also found a positive correlation between deposition rates and the occurrence of El Niño events, suggesting that the amount of plastic appearing on these remote beaches may vary as changes in wind patterns cause a shift in ocean currents.

These examples demonstrate that seabirds are exposed to contaminants at a wide range of spatial and temporal scales, and that relatively pristine systems can accumulate contaminants. Based on their abundance and global distribution, seabirds also represent a significant biovector of nutrients and contaminants from the ocean to the land. Blais et al. (2005) clearly demonstrated that Arctic ponds subjected to deposition of seabird guano had 10, 25, and 60 times the level of hexachlorobenzene, mercury, and DDT, respectively, compared to ponds that were not exposed to seabird guano. Therefore, not only does seabird guano stimulate productivity via the addition of nutrients, it also provides a transport mechanism for industrial and agricultural pollutants in high-latitude systems where these contaminants are not native (Blais et al. 2005).

8.3.4 Environmental Variability/Climate Forcing

Seabirds are strongly affected by environmental variables, including climate forcing, operating at multiple temporal and spatial scales. This may range from a localized storm event that causes nest loss at one colony to a hemispheric shift in weather patterns or ocean currents that affect the entire breeding range of a species. Furthermore, climate conditions affecting prey availability on the foraging grounds may reduce chick survival thousands of kilometers away. Environmental variables such as these affect seabirds at various spatial and temporal scales, and changes in seabird populations will likely play a role in restructuring coastal ecosystems.

Seabird life histories expose individuals and populations to environmental conditions affecting both terrestrial and marine habitats. Environmental effects on terrestrial nesting or resting habitat (excluding anthropogenic habitat alteration or predator introductions for this discussion) can be unique for seabirds at times, but in general are mostly similar to those affecting other terrestrial organisms, including severe weather events that generally have localized, short-term consequences. Changes in the marine environment, however, often have the most dramatic, widespread, or longest-lasting consequences to seabird populations. The fluid and dynamic nature of marine systems, however, requires seabirds to adapt to environmental fluctuations in ways drastically different than wholly terrestrial species. This is particularly true during the breeding season when most seabird species are constrained to central-place foraging from their terrestrial nesting habitat yet are required to constantly adapt to their marine foraging habitat, which is in constant three-dimensional motion via horizontal currents and vertical mixing that affects the distribution of prey, themselves often highly mobile organisms.

A dominant force driving horizontal currents and vertical mixing are ocean-atmosphere interactions. Atmospheric winds and temperature affect ocean currents, mixing, and the distribution of seabird prey both locally on time scales of hours or days and regionally on time scales of years to decades. Not surprisingly, these ocean-atmosphere interactions are often themselves transboundary in nature. For example, changes in wind patterns over the equatorial region (e.g., El Niño-Southern Oscillation) affect currents, temperature, and prey availability (i.e., distribution and abundance) thousands of kilometers away in the North and South Pacific and beyond (including terrestrial habitats over the Americas and Asia; Black et al. 2009; Chavez et al. 1999, 2003).

One of the most clear and dramatic examples of ocean-atmosphere interaction and transboundary connections affecting seabird populations is that of Cassin's Auklets breeding in the California Current System off the west coasts of Canada and the U.S. During the 2005 breeding season, Sydeman et al. (2006) reported that unusual atmospheric blocking in the Gulf of Alaska caused the jet stream, which affects coastal winds, to shift southward and cause anomalously warm sea-surface temperatures and unfavorable conditions for auklet prey (zooplankton) in the Northern California Current but not further south. Northern colonies of auklets off Canada and Central California experienced unprecedented (within a 35-year time series) reproductive failure and colony abandonment. In contrast, the abundance of auklet and their prey to the south, off Southern California, was anomalously high. Other examples of changes in ocean conditions affecting seabird prey availability, and hence reproductive success or population abundance, include cool water temperature delaying the inshore migration of key forage fish prey for the Common Murre off Newfoundland (Davoren and Montevecchi 2003), and the opposing effects of cold ocean temperatures benefitting planktivorous seabirds and warm ocean temperatures benefiting piscivorous seabirds in Tauyskaya Bay, Russia (Kitaysky and Golubova 2000).

Several well-documented climate signals that affect terrestrial and marine eco-systems over entire ocean basins have profound effects on seabird populations. In fact, seabirds often provide early warning signs of these large-scale climate changes, even though the actual physical drivers are thousands of kilometers away. One example is the El Niño-Southern Oscillation, which results from changes in atmospheric pressure over the South Pacific and Indian Oceans. Changes in pressure affect equatorial winds (and therefore ocean currents), ocean mixed-layer depth, overall productivity, and consequently food for seabirds. While the extent of El Niño is global, the effects are strongest in the equatorial Pacific. The 1982–1983 El Niño, one of the strongest recorded, resulted in the death of millions of seabirds in the equatorial Pacific due to starvation and also affected reproductive success of some species globally (Schreiber and Schreiber 1989). El Niño events occur relatively frequently, every 2–7 years; however, they are generally short-lived, lasting a year or less. Other well-documented, longer-lasting climate forcing that affect seabirds in the North Atlantic and North Pacific Oceans include the North Atlantic Oscillation, the Pacific Decadal Oscillation, and the Arctic Oscillation. These climate oscillations switch between alternate states lasting decades and, like El Niño, affect entire ecosystems from zooplankton to seabirds at ocean-basin scales. Effects of these oscillations have been shown to influence seabirds and their prey in the North Atlantic (Aebischer et al. 1990) and North Pacific (Anderson and Piatt 1999), sometimes alternating effects between these two regions (Irons et al. 2008). The effects can also vary by species, and studies in the North Atlantic demonstrate that these broader scale (hemispheric) climate shifts can have great effects on wider ranging species (i.e., more broadly dispersive or migratory during the non-breeding season) but have little or no effect on more locally residing species (Frederiksen et al. 2004).

In addition to these cyclical climate patterns, linearly changing or non-periodic trends also affect seabird populations through a wide variety of mechanisms. One potential mechanism is warming trends that affect wind patterns over the ocean, which, in turn, affect currents, water column mixing, and seabird food supply. For example, Bakun (1990) postulated that greenhouse gas-induced warming could, by warming coastal land masses more than water masses, create greater pressure differences between land and sea and thereby intensify coastal winds and water column mixing, with potentially dramatic effects on marine ecosystems (Bakun and Weeks 2004). In the California Current System off Western North America, long-term ocean warming has affected the community composition and abundance of seabirds in offshore waters (Veit et al. 1996, 1997), with an overall decline in numbers resulting from fewer cold-water associated pursuit-diving seabirds despite the increase in warm-water associated near-surface feeding species (Hyrenbach and Veit 2003). In the Northern California Current, warming ocean temperatures were correlated with declines in reproductive success of Tufted Puffins, a cold-water associated pursuit-diving seabird (Gjerdrum et al. 2003). More extreme, anomalous weather events may occur if climate change occurs, which may affect seabird species as well (Frederiksen et al. 2008). Likewise, changes in sea-level rise of even one meter could greatly impact seabird breeding habitat on low-lying beaches, atolls, and rocks (Baker et al. 2006) and in coastal estuarine habitat (Daniels et al. 1993).

8.3.5 *Seabirds and Commercial Fisheries: Efforts to Reduce Bycatch Mortality*

Seabirds provide many examples of research, management, conservation, and policy actions that require transboundary efforts for success and implementation (e.g., Wolf et al. 2006). Here we briefly examine the case of seabird mortality that occurs as bycatch within commercial fisheries.

Procellariids (albatrosses, petrels, and shearwaters) are the epitome of ocean wanderers, regularly traversing ocean basins within breeding seasons or crossing hemispheres and circumnavigating the globe during the non-breeding seasons (Croxall et al. 2005; Felicísimo et al. 2008; Fernández et al. 2001). Albatrosses (Diomedeidae), which range over long distances and often forage opportunistically, are particularly prone to incidental mortality in industrial longline fishing operations. Birds are most often hooked when longlines are being deployed and baited hooks are accessible at the surface near the vessel. Due in large part to this bycatch mortality, the Diomediadae are now one of the most endangered families of birds with 19 of 21 species on the International Union for the Conservation of Nature Red List (Croxall et al. 2005). Not only might individual albatrosses forage within the exclusive economic zones (200 nautical mile limit) of different nations, but also within international, high-seas regions outside national jurisdictions where vessels from many nations fish unregulated at times.

While this conservation challenge is far from solved, significant progress has been made during the past decade. Researchers have worked with the fishing industry to develop methods to prevent seabirds from attacking baited hooks while being deployed near the vessel. These include streamer lines that scare birds away from the baited hooks when they are near the water surface and additional weight added to lines that causes them to sink more rapidly (Dietrich et al. 2008; Melvin et al. 2001; Robertson et al. 2006).

Because methodologies for various fisheries are so diverse, no one solution works in all situations; therefore, it is important that a 'toolbox' of options are available to the fishing industry (Melvin and Parrish 2001). For example, national governments and regional fishery management organizations have enacted, through binding agreements such as ACAP (Agreement on the Conservation of Albatrosses and Petrels) and CCAMLR (Convention on the Conservation of Antarctic Marine Living Resources), (1) regulations on the discharge of fish bycatch and fish waste that attracts birds to fishing vessels, (2) area or seasonal closures, and (3) regulations that limit vessels to fishing only at night when some seabird species are less active. Night-setting, however, can increase the undesirable bycatch of other marine life, including sharks, thereby having unintended ecological consequences. Non-governmental organizations, such as BirdLife International, also have initiated multinational, grassroots programs (e.g., Save the Albatross Campaign) to work with fishers to implement measures proven to reduce seabird bycatch across a range of fisheries from local and artisanal to regional and industrial.

8.4 Lessons Learned

Throughout this chapter we have provided numerous examples of the transboundary nature of seabird ecology. In a basic sense, seabirds exemplify the transboundary concept because they require both terrestrial and marine habitats. Therefore, wherever conservation of seabirds or the management of their populations is the goal, consideration must be given to ecosystem dynamics on land and at sea. Because the jurisdiction of agencies does not cross the land-sea boundary in the same manner as the seabirds they are managing, these efforts are facilitated by multi-agency communication and collaboration.

From coastal species to ocean wanderers, seabirds traverse ecological and political boundaries on a regular basis and with a frequency and magnitude that is relatively unique among wildlife. Research and the technology underlying these efforts have evolved over the decades to address this unique aspect of seabird ecology as has the thinking of scientists. Many of the examples provided above have benefitted from an interdisciplinary approach to research that includes team members with expertise not just in wildlife but from a wide range of other disciplines, including chemists who understand contaminant transport, fisheries biologists who understand population dynamics of seabird prey, oceanographers and atmospheric scientists who understand ocean circulation and wind patterns, and engineers who can design microelectronic devices that allow the movements of individuals to be tracked across ocean basins for years at a time. Addressing complex ecological questions and improving our understanding of the complex systems we study are benefited by collaborative, cross-disciplinary research teams including such expertise as seabird biologists, fisheries biologists, and oceanographers.

Many seabirds live in remote places that are difficult for researchers to access. As such, knowledge of even basic distributions and status can be lacking, although the need for such data can be critical when attempting to understand seabird ecology and how changes to the land or sea environment might affect a species or site. Therefore, basic inventories of the occurrence and distribution of seabirds both at sea and at breeding sites continue to be important undertakings. For example, a recent inventory of breeding seabirds in the Caribbean makes available, for the first time, a comprehensive, island-by-island review of seabird occurrence in that region (Bradley and Norton 2009).

Seabirds exemplify a suite of wildlife that, throughout their daily, seasonal, and annual cycles, cross multiple ecological and political boundaries. The examples we have provided demonstrate that research, management, conservation, and policy efforts focused on these species often include a transboundary approach and often consider natural and anthropogenic stressors in marine and terrestrial systems that function at multiple scales in both time and three-dimensional space. Many other examples of the ecoregional and transboundary nature of seabird ecology exist that we did not cover here, including eradicating and preventing the reintroduction of exotic predators on terrestrial breeding areas (Keitt et al. 2002; VanderWerf et al. 2007) and managing direct competition for prey species between seabirds and humans via commercial fisheries extraction (Wanless et al. 2007). These and all of the examples we have discussed demonstrate that, by their very nature and by the nature of the systems that they must function within, seabirds embody the complexity of wildlife ecology and conservation in the twenty-first century.

Acknowledgements This manuscript benefited from reviews by W. Mackin and S. Vander Pol. The South Carolina Cooperative Fish and Wildlife Research Unit is supported jointly by the U.S. Geological Survey, South Carolina Department of Natural Resources, and Clemson University.

References

Adams, J., Takekawa, J., & Carter, H. R. (2004). Foraging distance and home range of Cassin's Auklets nesting at two colonies in the California Channel Islands. *The Condor, 106*, 618–637.

Aebischer, N. J., Coulson, J. C., & Colebrookl, J. M. (1990). Parallel long-term trends across four marine trophic levels and weather. *Nature, 347*, 753–755.

Ainley, D. G., Podolsky, R., Deforest, L., Spencer, G., & Nur, N. (2001). The status and population trends of the Newell's Shearwater on Kaua'i: Insights from modeling. *Studies in Avian Biology, 22*, 108–123.

Alonso-Alvarez, C., Munilla, I., López-Alonso, M., & Velando, A. (2007a). Sublethal toxicity of the Prestige oil spill on yellow-legged gulls. *Environment International, 33*, 773–781.

Alonso-Alvarez, C., Pérez, C., & Velando, A. (2007b). Effects of acute exposure to heavy fuel oil from the Prestige spill on a seabird. *Aquatic Toxicology, 84*, 103–110.

Anderson, P. J., & Piatt, J. F. (1999). Community reorganization in the Gulf of Alaska following ocean climate regime shift. *Marine Ecology Progress Series, 189*, 117–123.

Arcos, J. M., Ruiz, X., Bearhop, S., & Furness, R. W. (2002). Mercury levels in seabirds and their fish prey at the Ebro Delta (NW Mediterranean): The role of trawler discards as a source of contamination. *Marine Ecology Progress Series, 232*, 281–290.

Ashmole, N. P. (1963). The regulation of numbers of tropical oceanic birds. *Ibis, 103b*, 458–473.

Ashmole, N. P. (Ed.). (1971). *Seabird ecology and the marine environment*. New York: Academic.

Baker, J. D., Littnan, C. L., & Johnston, D. W. (2006). Potential effects of sea level rise on the terrestrial habitats of endangered and endemic megafauna in the Northwestern Hawaiian Islands. *Endangered Species Research, 4*, 1–10.

Bakun, A. (1990). Coastal ocean upwelling. *Science, 247*, 198–201.

Bakun, A., & Weeks, S. J. (2004). Greenhouse gas buildup, sardines, submarine eruptions and the possibility of abrupt degradation of intense marine upwelling ecosystems. *Ecology Letters, 7*, 1015–1023.

Ballance, L. T., Pitman, R. L., & Reilly, S. B. (1997). Seabird community structure along a productivity gradient: Importance of competition and energetic constraint. *Ecology, 78*, 1502–1518.

Becker, P. H. (1989). Seabirds as monitor organisms of contaminants along the German North Sea Coast. *Helgolaender Marine Research, 43*, 395–403.

Becker, P. H., Perrins, C. M., Lebreton, J. D., & Hirons, G. J. M. (1991). Population and contamination studies in coastal birds: The Common Tern (*Sterna hirundo*). In C. M. Perrins, J. D. Lebreton, & G. J. M. Hirons (Eds.), *Bird Population Studies: Relevance to conservation and management* (pp. 433–460). Oxford, UK: Oxford University Press.

BirdLife International. (2008). *Critically endangered birds: A global audit*. Cambridge, UK: BirdLife International.

BirdLife International. (2009a). *Species factsheet*: Puffinus huttoni. Retrieved November 1, 2009, from http://www.birdlife.org

BirdLife International. (2009b). *Species factsheet*: Puffinus newelli. Retrieved November 1, 2009, from http://www.birdlife.org

Birt, V. L., Birt, T. P., Goulet, D., Cairns, D. K., & Montevecchi, W. A. (1987). Ashmole's halo: Direct evidence for prey depletion by a seabird. *Marine Ecology Progress Series, 40*, 205–208.

Black, B. A., Copenheaver, C. A., Frank, D. C., Stuckey, M. J., & Kormanyos, R. E. (2009). Multiproxy reconstructions of northeastern Pacific sea surface temperature data from trees and Pacific geoduck. *Palaeoclimatology, Palaeogeography, Palaeoecology, 278*, 40–47.

Blais, J. M., Kimpe, L. E., McMahon, D., Keatley, B. E., Mallory, M. L., Douglas, M. S. V., et al. (2005). Arctic seabirds transport marine-derived contaminants. *Science, 309*, 445.

Blight, L. K., & Burger, A. E. (1997). Occurrence of plastic particles in seabirds from the eastern North Pacific. *Marine Pollution Bulletin, 34*, 323–325.

Bradley, P. E., & Norton, R. L. (Eds.). (2009). *An inventory of breeding seabirds of the Caribbean*. Gainesville, FL: University Press of Florida.

Braune, B. M., Donaldson, G. M., & Hobson, K. A. (2001). Contaminant residues in seabird eggs from the Canadian Arctic. Part I. Temporal trends 1975–1998. *Environmental Pollution, 114*, 39–54.

Burger, J., & Gochfeld, M. (2002). Effects of chemicals and pollution on seabirds. In E. A. Schreiber & J. Burger (Eds.), *Biology of marine birds* (pp. 485–525). Boca Raton, FL: CRC.

Chape, S., Harrison, J., Spalding, M., & Lysenko, I. (2005). Measuring the extent and effectiveness of protected areas as an indicator for meeting global biodiversity targets. *Philosophical Transactions of the Royal Society B-Biological Sciences, 360*, 443–455.

Chavez, F. P., Strutton, P. G., Friederich, G. E., Feely, R. A., Feldman, G. C., Foley, D. G., et al. (1999). Biological and chemical response of the equatorial Pacific Ocean to the 1997–98 El Niño. *Science, 286*, 2126–2131.

Chavez, F. P., Ryan, J., Lluch-Cota, S. E., & Ñiquen, M. (2003). From anchovies to sardines and back: multidecadal change in the Pacific Ocean. *Science, 299*, 217–221.

Cherel, Y., Stahl, J.-C., & Le Maho, Y. (1987). Ecology and physiology of fasting in king penguin chicks. *Auk, 104*, 254–262.

Cifuentes, J. M., Becker, P. H., Sommer, U., Pacheco, P., & Schlatter, R. (2003). Seabird eggs as bioindicators of chemical contamination in Chile. *Environmental Pollution, 126*, 123–137.

Colabuono, F. I., Barquete, V., Domingues, B. S., & Montone, R. C. (2009). Plastic ingestion by Procellariiformes in Southern Brazil. *Marine Pollution Bulletin, 58*, 93–96.

Coulson, J. C. (1968). Differences in the quality of birds nesting in the centre and on the edges of a colony. *Nature, 217*, 478–479.

Croxall, J. P., Silk, J. R. D., Phillips, R. A., Afanasyev, V., & Briggs, D. R. (2005). Global circumnavigations: tracking year-round ranges of nonbreeding albatrosses. *Science, 307*, 249–250.

Cuthbert, R., & Davis, L. S. (2002). The breeding biology of Hutton's Shearwater. *Emu, 102*, 323–329.

Cuthbert, R., Fletcher, D., & Davis, L. S. (2001). A sensitivity analysis of Hutton's shearwater: prioritizing conservation research and management. *Biological Conservation, 100*, 163–172.

Daniels, R. C., White, T. W., & Chapman, K. K. (1993). Sea-level rise – destruction of threatened and endangered species habitat in South Carolina. *Environmental Management, 17*, 373–385.

Davoren, G. K., & Montevecchi, W. A. (2003). Signals from seabirds indicate changing biology of capelin stocks. *Marine Ecology Progress Series, 258*, 253–261.

Day, R. H., & Cooper, B. A. (1995). Patterns of movement of Dark-Rumped Petrels and Newell's Shearwaters on Kauai, Hawaii. *Condor, 97*, 1011–1027.

Dietrich, K. S., Melvin, E. F., & Conquest, L. (2008). Integrated weight longlines with paired streamer lines – best practice to prevent seabird bycatch in demersal longline fisheries. *Biological Conservation, 141*, 1793–1805.

Egevang, C., Stenhouse, I. J., Phillips, R. A., Petersen, A., Fox, J. W., Silk, J. R. D. (2010). Tracking of Arctic terns *Sterna paradisaea* reveals longest animal migration. Proceedings of the National Academy of Sciences *107*, 2072–2081.

Felicísimo, Á. M., Muñoz, J., González-Solis, J. (2008). Ocean surface winds drive dynamics of transoceanic aerial movements. *PLoS ONE, 3*, e2928.

Fernández, P., Anderson, D. J., Sievert, P. R., & Huyvaert, K. P. (2001). Foraging destinations of three low-latitude albatross (*Phoebastria*) species. *Journal of Zoology, 254*, 391–404.

Finkelstein, M., Keitt, B. S., Croll, D. A., Tershy, B., Jarman, W. M., Rodriguez-Pastor, S., et al. (2006). Albatross species demonstrate regional differences in North Pacific marine contamination. *Ecological Applications, 16*, 678–686.

Frederiksen, M., Harris, M. P., Daunt, F., Rothery, P., & Wanless, S. (2004). Scale-dependent climate signals drive breeding phenology of three seabird species. *Global Change Biology, 10*, 1214–1221.

Frederiksen, M., Daunt, F., Harris, M. P., & Wanless, S. (2008). The demographic impact of extreme events: stochastic weather drives survival and population dynamics in a long-lived seabird. *Journal of Animal Ecology, 77*, 1020–1029.

Furness, R. W. (2003). Impacts of fisheries on seabird communities. *Scientia Marina, 67*, 33–45.

Furness, R. W., & Monaghan, P. (1987). *Seabird ecology*. New York: Chapman & Hall.

Furness, R. W., Hudson, A. V., & Ensor, K. (1988). Interactions between scavenging seabirds and commercial fisheries around the British Isles. In J. Burger (Ed.), *Seabirds and other marine vertebrates* (pp. 240–268). New York: Columbia University Press.

Garthe, S., & Hüppop, O. (1994). Distribution of ship-following seabirds and their utilization of discards in the North Sea in summer. *Marine Ecology Progress Series, 106*, 1–9.

Garthe, S., Camphuysen, C. J., & Furness, R. W. (1996). Amounts of discards by commercial fisheries and their significance as food for seabirds in the North Sea. *Marine Ecology Progress Series, 136*, 1–11.

Garthe, S., Montevecchi, W. A., Chapdelaine, G., Rail, J. -F., & Hedd, A. (2007). Contrasting foraging tactics by northern gannets (*Sula bassana*) breeding in different oceanographic domains with different prey fields. *Marine Biology, 151*, 687–694.

Gaston, A. J. (2004). *Seabirds: A natural history*. New Haven, CT: Yale University Press.

Gaston, A. J., & Jones, I. L. (1998). *The Auks: Alcidae (Bird families of the world)*. Oxford, UK: Oxford University Press.

Gaston, A. J., Ydenberg, R. C., & Smith, G. E. J. (2007). Ashmole's halo and population regulation in seabirds. *Marine Ornithology, 35*, 119–126.

Gill, V. A., Hatch, S. A., & Lanctot, R. B. (2002). Sensitivity of breeding parameters to food supply in Black-legged Kittiwakes *Rissa tridactyla*. *Ibis, 144*, 268–283.

Gjerdrum, C., Vallée, A. M. J., St. Clair, C. C., Bertram, D. F., Ryder, J. L., & Blackburn, G. S. (2003). Tufted puffin reproduction reveals ocean climate variability. *Proceedings of the National Academy of Sciences, 100*, 9377–9382.

Gochfeld, M., Burger, J., Haynes-Sutton, A., Halewyn, R. V., & Saliva, J. E. (1994). Successful approaches to seabird protection in the West Indies. In D. N. Nettleship, J. Burger, & M. Gochfeld (Eds.), *Seabirds on islands: threats, case studies and action plans* (pp. 186–209). Cambridge, UK: Birdlife International.

Golet, G. H., Irons, D. B., & Estes, J. A. (1998). Survival costs of chick-rearing in Black-legged Kittiwakes. *Journal of Animal Ecology, 67*, 827–841.

Golet, G. H., Seiser, P. E., McGuire, A. D., Roby, D. D., Fischer, J. B., Kuletz, K. J., et al. (2002). Long-term direct and indirect effects of the 'Exxon Valdez' oil spill on pigeon guillemots in Prince William Sound, Alaska. *Marine Ecology Progress Series, 241*, 287–304.

González-Solís, J., Croxall, J. P., Oro, D., & Ruiz, X. (2007). Trans-equatorial migration and mixing in the wintering areas of a pelagic seabird. *Frontiers in Ecology and the Environment, 5*, 297–301.

Gore, J. A., & Kinnison, M. J. (1991). Hatching success in roof and ground colonies of least terns. *Condor, 93*, 759–762.

Hamer, K. C., Phillips, R. A., Hill, J. K., Wanless, S., & Wood, A. G. (2001). Contrasting foraging strategies of gannets *Morus bassanus* at two North Atlantic colonies: Foraging trip duration and foraging area fidelity. *Marine Ecology Progress Series, 224*, 283–290.

Hatch, S. A., Meyers, P. M., Mulcahy, D. M., & Douglas, D. C. (2000). Seasonal movements and pelagic habitat use of murres and puffins determined by satellite telemetry. *Condor, 102*, 145–154.

Hebshi, A. J., Duffy, D. C., & Hyrenbach, K. D. (2008). Associations between seabirds and subsurface predators around Oahu, Hawaii. *Aquatic Biology, 4*, 89–98.

Hyrenbach, K. D., & Veit, R. R. (2003). Ocean warming and seabird communities of the southern California Current System (1987–98): Response at multiple temporal scales. *Deep-Sea Research, Part II, 50*, 2537–2565.

Hyrenbach, K. D., Keiper, C., Allen, S. G., Ainley, D. G., & Anderson, D. J. (2006). Use of marine sanctuaries by far-ranging predators: Commuting flights to the California Current System by breeding Hawaiian albatrosses. *Fisheries Oceanography, 15*, 95–103.

Irons, D. B., Anker-Nilssen, T., Gaston, A. J., Byrd, G. V., Falk, K., Gilchrist, G., et al. (2008). Fluctuations in circumpolar seabird populations linked to climate oscillations. *Global Change Biology, 14*, 1455–1463.

Jewett, S. C., Dean, T. A., Woodin, B. R., Hoberg, M. K., & Stegeman, J. J. (2002). Exposure to hydrocarbons 10 years after the Exxon Valdez oil spill: Evidence from cytochrome P4501A expression and biliary FACs in nearshore demersal fishes. *Marine Environmental Research, 54,* 21–48.

Jodice, P. G. R., Roby, D. D., Turco, K. R., Suryan, R. M., Irons, D. B., Piatt, J. F., et al. (2006). Assessing the nutritional stress hypothesis: Relative influence of diet quantity and quality on seabird productivity. *Marine Ecology Progress Series, 325,* 267–279.

Jodice, P. G. R., Murphy, T. M., Sanders, F. J., & Ferguson, L. M. (2007). Longterm trends in nest counts of colonial seabirds in South Carolina, USA. *Waterbirds, 30,* 40–51.

Jouventin, P., & Weimerskirch, H. (1990). Satellite tracking of wandering albatrosses. *Nature, 343,* 746–748.

Keitt, B. S., Wilcox, C., Tershey, B. R., Croll, D. A., & Donlan, C. J. (2002). The effect of feral cats on the population viability of Black-vented Shearwaters (*Puffinus opisthomelas*) on Natividad Island, Mexico. *Animal Conservation, 5,* 217–223.

Kitaysky, A. S., & Golubova, E. G. (2000). Climate change causes contrasting trends in reproductive performance of planktivorous and piscivorous alcids. *Journal of Animal Ecology, 69,* 248–262.

Konarzewski, M., & Taylor, J. R. E. (1989). The influence of weather conditions on the growth of Little Auk *Alle alle* chicks. *Ornis Scandinavica, 20,* 112–116.

Krogh, M. G., & Schweitzer, S. H. (1999). Least Terns nesting on natural and artificial habitats in Georgia, USA. *Waterbirds, 22,* 290–296.

Lee, D. S. (2000a). Status and conservation priorities for Audubon's Shearwaters. In E. A. Schreiber & D. S. Less (Eds.), *Status and conservation of West Indian seabirds* (pp. 25–30). Ruston, LA: Society of Caribbean Ornithology.

Lee, D. S. (2000b). Status and conservation priorities for Black-capped Petrels in the West Indies. In E. A. Schreiber & D. S. Less (Eds.), *Status and conservation of West Indian seabirds* (pp. 11–18). Ruston, LA: Society of Caribbean Ornithology.

Longhurst, A. R. (2007). *Ecological geography of the sea.* Boston, MA: Academic.

McGehee, M. W. (2000). Status and conservation priorities for White-tailed Tropicbirds and Red-billed Tropicbirds in the West Indies. In E. A. Schreiber & D. S. Less (Eds.), *Status and conservation of West Indian seabirds* (pp. 31–38). Ruston, LA: Society of Caribbean Ornithology.

Melvin, E. F., & Parrish, J. K. (2001). *Seabird bycatch: Trends, roadblocks, and solutions.* Fairbanks, AK: University of Alaska Sea Grant Program.

Melvin, E. F., Parrish, J. K., Dietrich, K. S., & Hamel, O. S. (2001). *Solutions to seabird bycatch in Alaska's demersal longline fisheries.* Seattle, WA: Washington Sea Grant Program.

Miskelly, C. M., Taylor, G. A., Gummer, H., & Williams, R. (2009). Translocations of eight species of burrow-nesting seabirds (genera *Pterodroma*, *Pelecanoides*, *Pachyptila* and *Puffinus*: family Procellariidae). *Biological Conservation, 142,* 1965–1980.

Monteiro, L. R., Costa, V., Furness, R. W., & Santos, R. S. (1996). Mercury concentrations in prey fish indicate enhanced bioaccumulation in mesopelagic environments. *Marine Ecology Progress Series, 141,* 21–25.

Morishige, C., Donohue, M. J., Flint, E., Swenson, C., & Woolaway, C. (2007). Factors affecting marine debris deposition at French Frigate Shoals, Northwestern Hawaiian Islands Marine National Monument, 1990–2006. *Marine Pollution Bulletin, 54,* 1162–1169.

Mowbray, T. B. (2002). Northern Gannet (*Morus bassanus*) (No. 693). In A. Poole & F. Gill (Eds.), *The birds of North America (No. 693).* Philadelphia, PA: Academy of Natural Sciences.

Nelson, J. B. (2005). *Pelicans, cormorants and their relatives: the Pelecaniformes.* Oxford, UK: Oxford University Press.

Nevitt, G. A., Losekoot, M., & Weimerskirch, H. (2008). Evidence for olfactory search in wandering albatross, *Diomedea exulans. Proceedings of the National Academy of Sciences, 105,* 4576–4581.

Perrow, M. R., Skeate, E. R., Lines, P., Brown, D., & Tomlinson, M. L. (2006). Radio telemetry as a tool for impact assessment of wind farms: The case of Little Terns *Sterna albifrons* at Scroby Sands, Norfolk, UK. *Ibis, 148*, 57–75.

Peterson, C. H., Rice, S. D., Short, J. W., Esler, D., Bodkin, J. L., Ballachey, J. L., et al. (2003). Long-term ecosystem response to the Exxon Valdez oil spill. *Science, 302*, 2082–2086.

Piatt, J. F., Lensink, C. J., Butler, W., Kendziorek, M., & Nysewander, D. R. (1990). Immediate impact of the Exxon Valdez oil-spill on marine birds. *Auk, 107*, 387–397.

Pinson, D., & Drummond, H. (1993). Brown pelican siblicide and the prey-size hypothesis. *Behavioral Ecology and Sociobiology, 32*, 111–118.

Prince, P. A., Wood, A. G., Barton, T., & Croxall, J. P. (1992). Satellite tracking of wandering albatrosses (*Diomedea exulans*) in the South Atlantic. *Antarctic Science, 4*, 31–36.

Ricca, M. A., Miles, A. K., & Anthony, R. G. (2008). Sources of organochlorine contaminants and mercury in seabirds from the Aleutian archipelago of Alaska: Inferences from spatial and trophic variation. *Science of the Total Environment, 406*, 308–323.

Robards, M. D., Piatt, J. F., Wohl, K. D. (1995). Increasing frequency of plastic particles ingested by seabirds in the subarctic North Pacific. *Marine Pollution Bulletin, 30*, 151.

Robertson, G., McNeill, M., Smith, N., Wienecke, B., Candy, S., & Olivier, F. (2006). Fast sinking (integrated weight) longlines reduce mortality of white-chinned petrels (*Procellaria aequinoctialis*) and sooty shearwaters (*Puffinus griseus*) in demersal longline fisheries. *Biological Conservation, 132*, 458–471.

Sachs, E. B., & Jodice, P. G. R. (2009). Behavior of parent and nestling Brown Pelicans during early brood rearing. *Waterbirds, 32*, 276–281.

Schreiber, E. A., & Burger, J. (2001). *Biology of marine birds*. Boca Raton, FL: CRC.

Schreiber, E. A., & Schreiber, R. W. (1989). Insights into seabird ecology from a global "natural experiment". *National Geographic Research, 5*, 64–81.

Shaffer, S. A., Tremblay, Y., Weimerskirch, H., Scott, D., Thompson, D. R., Sagar, P. M., et al. (2006). Migratory shearwaters integrate oceanic resources across the Pacific Ocean in an endless summer. *Proceedings of the National Academy of Sciences, 103*, 12799–12802.

Spalding, M. D., Fox, H. E., Allen, G. R., Davidson, N., Ferdaña, Z. A., Finlayson, M., et al. (2007). Marine ecoregions of the world: A bioregionalization of coastal and shelf areas. *BioScience, 57*, 573–583.

Suryan, R. M., Irons, D. B., Kaufman, M., Benson, J., Jodice, P. G. R., Roby, D. D., et al. (2002). Short-term fluctuations in forage fish availability and the effect on prey selection and brood-rearing in the black-legged kittiwake *Rissa tridactyla*. *Marine Ecology Progress Series, 236*, 273–287.

Suryan, R. M., Sato, F., Balogh, G. R., Hyrenbach, K. D., Sievert, P. R., & Ozaki, K. (2006). Foraging destinations and marine habitat use of short-tailed albatrosses: A multi-scale approach using first-passage time analysis. *Deep-Sea Research, Part II, 53*, 370–386.

Sydeman, W. J., Bradley, R. W., Warzybok, P., Abraham, C. L., Jahncke, J., Hyrenbach, K. D., et al. (2006). Planktivorous auklet *Ptychoramphus aleuticus* responses to ocean climate, 2005: Unusual atmospheric blocking? *Geophysical Research Letters, 33*. doi: 10.1029/2006GL026736.

Tickell, W. L. N. (2000). *Albatrosses*. New Haven, CT: Yale University Press.

Trivelpiece, W. Z., Trivelpiece, S. G., & Volkman, N. J. (1987). Ecological segregation of adelie, gentoo, and chinstrap penguins at King George Island, Antarctica. *Ecology, 68*, 351–361.

UNEP [United Nations Environment Programme]. (2006). *Marine and coastal ecosystems and human well-being: A synthesis report based on the findings of the Millennium Ecosystem Assessment*. Nairobi: UNEP.

Vander Pol, S. S., & Becker, P. M. (2007). Monitoring contaminants in seabirds: The importance of specimen banking. *Marine Ornithology, 35*, 113–118.

VanderWerf, E. A., Wood, K. R., Swenson, C., LeGrande, M., Eijzenga, H., & Walker, R. L. (2007). Avifauna of Lehua Islet, Hawai'i: Conservation value and management needs. *Pacific Science, 61*, 39–52.

8 The Transboundary Nature of Seabird Ecology

Veit, R. R., Pyle, P., & McGowan, J. A. (1996). Ocean warming and long-term change in pelagic bird abundance within the California current system. *Marine Ecology Progress Series, 139,* 11–18.

Veit, R. R., McGowan, J. A., Ainley, D. G., Wahls, T. R., & Pyle, P. (1997). Apex marine predator declines ninety percent in association with changing oceanic climate. *Global Change Biology, 3,* 23–28.

Velando, A., Ortega-Ruano, J. E., & Freire, J. (1999). Chick mortality in European Shag *Stictocarbo aristotelis* related to food limitation during adverse weather events. *Ardea, 87,* 51–59.

Velarde, E., Ezcurra, E., Cisneros-Mata, M. A., & Lavin, M. F. (2004). Seabird ecology, El Nino anomalies, and prediction of sardine fisheries in the Gulf of California. *Ecological Applications, 14,* 607–615.

Walter, U., & Becker, P. H. (1997). Occurrence and consumption of seabirds scavenging on shrimp trawler discards in the Wadden Sea. *ICES Journal of Marine Science, 54,* 684–694.

Wanless, S., Frederiksen, M., Daunt, F., Scott, B. E., & Harris, M. P. (2007). Black-legged kittiwakes as indicators of environmental change in the North Sea: Evidence from long-term studies. *Progress in Oceanography, 72,* 30–38.

Warham, J. (1990). *The petrels: Their ecology and breeding systems.* London: Academic.

Weimerskirch, H., Le Corre, M., Marsac, F., Barbraud, C., Tostain, O., & Chastel, O. (2006). Postbreeding movements of frigatebirds tracked with satellite telemetry. *Condor, 108,* 220–225.

Wenzel, C., Adelung, D., & Theede, H. (1996). Distribution and age-related changes of trace elements in kittiwake *Rissa tridactyla* nestlings from an isolated colony in the German Bight, North Sea. *Science of the Total Environment, 193,* 13–26.

Wickliffe, L. C., & Jodice, P. G. R. (in press). Abundance of nearshore seabirds at shrimp trawlers in South Carolina. *Marine Ornithology.*

Wittenberger, J. F., & Hunt, G. L. J. (1985). The adaptive significance of coloniality in birds. In D. S. Farner, J. R. King, & K. C. Parkes (Eds.), *Avian biology, 8* (pp. 1–75). London: Academic.

Wolf, S., Keitt, B., Aguirre-Muñoz, A., Tershey, B., Palacios, E., & Croll, D. (2006). Transboundary seabird conservation in an important North American marine ecoregion. *Environmental Conservation, 33,* 294–305.

Young, L. C., Vanderlip, C., Duffy, D. C., Afanasyev, V., Shaffer, S. A. (2009). Bringing home the trash: do colony-based differences in foraging distribution lead to increased plastic ingestion in Laysan albatrosses? *PLoS ONE, 4,* e7623.

Chapter 9
Conservation Planning with Large Carnivores and Ungulates in Eastern North America: Learning from the Past to Plan for the Future

Justina C. Ray

Abstract While large mammals are often important targets of conservation activities in their own right, they can serve as effective tools for designing conservation landscapes and management measures at the human–wildlife interface. This chapter explores the potential role of large mammals in conservation planning in the Northern Appalachians/Acadian ecoregion, exploring two major questions: What can we learn from the past about the status of large mammals and the drivers of change, and what can this knowledge tell us about how both to plan for their continued persistence or recovery and to deploy them to help cover at least some of the needs of other, less visible components of biological diversity? An analysis of the individual trajectories of 10 large mammal species over the past four centuries of landscape and climate changes in the Northern Appalachian/Acadian ecoregion reveals several patterns of decline and recovery having occurred against a backdrop of variable environmental conditions such as land-use change, climate shifts, prevailing human attitudes, and interspecific relationships. Deploying large mammals as conservation planning tools can range from expanding the scale of conservation ambition to guiding the identification of core conservation lands, connectivity within the overall landscape, and thresholds of development intensity.

Keywords Carnivores • Conservation planning • Mammals • Population trends • Ungulates

---

J.C. Ray (✉)
Wildlife Conservation Society Canada, 720 Spadina Avenue, #600,
Toronto ON M5S 2T9, Canada
e-mail: jray@wcs.org

S.C. Trombulak and R.F. Baldwin (eds.), *Landscape-scale Conservation Planning*,
DOI 10.1007/978-90-481-9575-6_9, © Springer Science+Business Media B.V. 2010

9.1 Introduction

In keeping with the principle that an essential goal of conservation planning is safeguarding biological diversity, it is often assumed that the conservation of large-bodied mammals represents the pinnacle of achievement in this endeavor. This is because ensuring for the persistence of such species often constitutes formidable challenges, as judged by the history of large faunal change in Europe and North America over the past 500 years driven by human-mediated stressors. Members of these species can have high demands for space that collide easily with human interests, and their often low reproductive capacity makes it difficult for populations to recover once in decline (Weaver et al. 1996). To add to this natural vulnerability, such species are also highly valued for meat and other products, or maligned as a source of real or perceived threat to human inhabitants. Through force of gun and plow amidst rising industrial societies, large mammal populations, such as ungulates and carnivores, have been generally among those that are the first to dwindle or disappear worldwide (Laliberte and Ripple 2004; Morrison et al. 2007).

The Northern Appalachian/Acadian ecoregion in Northeastern North America has the continent's longest history of European settlement (Whitney 1994). The shifting distributions of large mammal species occupying the same region mirror the trajectory of land conversion and recovery over the past several centuries. Dwelling in a region characterized by topographic and ecological diversity, large mammal fauna have included at one time or another several ungulate species (caribou [*Rangifer tarandus*], moose [*Alces alces*], and white-tailed deer [*Odocoileus virginianus*], with elk [*Cervus elaphus*] and bison [*Bison bison*] on the outskirts) and carnivores (wolves [*Canis lupus* or *lycaon*], black bears [*Ursus americanus*], cougars [*Puma concolor*], wolverines [*Gulo gulo*], lynx [*Lynx canadensis*], bobcat [*Lynx rufus*], and coyotes [*Canis latrans*]). Each species has individual histories and has responded independently to changing climates and landscapes; not one has enjoyed stable population levels or distributions over the past few hundred years. While ultimate factors lie in land shifts stimulated by humans or more natural climatic changes, in many cases it is the biology of the animals themselves and the strength of their interaction with closely related species that have dictated their status at any given time.

A "focal species" approach to conservation planning refers to the process whereby conservation planning is designed in whole or part on the needs of selected species (Ray 2005a). This is accomplished through assessing their potential for recovery and/or continued persistence under various planning designs (Wilson et al. 2009). Deploying species as both targets and tools in conservation planning can be a logical investment of resources as long as the right species are chosen and perceived relationships with ecological processes are tested (Lindenmayer et al. 2002; Chap. 17). Large mammals can offer particular advantages in this regard because their decline or disappearance from an area says a lot about the state of biological diversity in that region (Morrison et al. 2007; Ray 2005a). A retrospective view is helpful in documenting the range of processes that have affected species persistence within a region over time. Information on historical trends can

9 Conservation Planning with Large Carnivores and Ungulates in Eastern North America   169

highlight the drivers behind species-specific relative abundance and distribution patterns. This can help in efforts to select desired and practical conservation goals as well as effective approaches to achieve them (Motzkin and Foster 2004).

In this chapter, I explore the potential role of large mammals in conservation planning in the Northern Appalachians/Acadian ecoregion. I address two major questions: What can we learn from the past about the status of large mammals and the drivers of change? And what can this knowledge tell us about how both to plan for their continued persistence or recovery and to deploy them to help cover at least some of the needs of other, less visible components of biological diversity? This analysis begins with an exploration of the individual histories of ten species of ungulates and carnivores resident in the ecoregion over the past four centuries. After exploring the drivers behind their declines and/or recoveries, the latter half of the chapter discusses lessons learned that are relevant for conservation planning today in this region and beyond.

9.2   Historical Trends and Limiting Factors

This section contains a review what is known about the historical trajectories of relative abundances and distributions of ten ungulates and carnivores in the ecoregion since the time of European settlement (late 1600s). With each species, I explore what is known about the drivers behind their declines and/or recoveries at particular points of time. I also provide information on their current status and discuss the abiotic and biotic factors that are known to limit their distribution.

9.2.1   Caribou (Rangifer tarandus)

Caribou have lost almost one-third of their historical North American range from Southern Canada and the lower 48 states of the U.S. (Hummel and Ray 2008). In the mid-1800s, the coniferous forests of Maine, Northern Vermont, New Hampshire, the Gaspé Peninsula and the Atlantic provinces of Canada were all home to caribou populations (Bergerud and Mercer 1989); Grant (1902) claimed caribou never occurred in Northern New York. This species disappeared from the St. Lawrence Valley of Québec, New Hampshire, and Vermont in the middle part of the nineteenth century but was still hunted in Maine in the late 1890s, with the last native caribou recorded in the state in 1905 (Martin 1980, cited in Courtois et al. 2003). Historians in the region speak of the coming and going of caribou in local areas (Krohn 2005; Parker 2004) due to their propensity to move around in the landscape in an unpredictable fashion; there were also known periods of localized abundance. For example, caribou were described as highly abundant in Northern New Brunswick in the 1880s (Parker 2004), and they occurred throughout the southern shore of the St. Lawrence River through the Gaspé Peninsula (Courtois et al. 2003).

Reasons for the precipitous declines of caribou in the Northern Appalachian/Acadian ecoregion, as elsewhere, stemmed from overhunting, the rising tide of white-tailed deer that took advantage of converted landscapes and milder winters in the region, and increased levels of predation that likewise accompanied the extensive habitat changes (Miller et al. 2003). Evidence suggests that caribou distribution extended further south prior to or following the Little Ice Age (1300–1850) (Telfer and Kelsall 1984), such that the range reflected in our historical record most likely represents a maximum range for this northern species (Bergerud and Mercer 1989). In turn, white-tailed deer responded favorably to the warming trend that took place after this period. Although early authors remarked on the ecological incompatibility between caribou and white-tailed deer (e.g., Palmer 1938), the basis for this was not known until the transmission of the meningeal worm (*Paralastrongylus tenius*, a parasite that when transmitted to caribou is 100% fatal [Anderson 1972]), was discovered.

Following localized declines or population fluctuations during the late nineteenth century, caribou largely disappeared from the ecoregion by the early 1900s, persisting only in the Cape Breton highlands of Nova Scotia until 1925 (Kelsall 1984). One remnant population remains today in Gaspésie National Park in Québec, which numbers just 140 individuals (Courtois et al. 2003). Several translocations into the ecoregion took place since the 1920s, but none was successful, primarily because of transmission of meningeal worms from white-tailed deer (Bergerud and Mercer 1989). Although caribou have never been recorded to reoccupy a range from which they have been extirpated, small, isolated populations have been able to persist for some time, often by means of predator control efforts conducted annually (Festa-Bianchet et al. 2010; Hummel and Ray 2008).

While overharvesting was the primary factor for the original decline of caribou on the Gaspé Peninsula, the cause has shifted in recent decades to predation by coyotes and black bears. Since the 1980s, members of the herd have been increasingly confined to high-elevation areas, nearly coincident with the time when coyotes moved into the region (Mosnier et al. 2003). Repeated predator control efforts were made from 1990 to 1996 and again since 2003 and appear to be necessary in order to ensure for the continued survival of this population (Mosnier et al. 2008).

9.2.2 Moose (Alces alces)

When European settlers first arrived on the North American continent, moose were abundant throughout most of the East (Alexander 1993; Bontaites and Gustafson 1993; Courtois and Lamontagne 1997). This was in sharp contrast to the status of the same species that once persisted all over Europe but had already experienced widespread declines in the southern half of that continent by the time the first European settlers arrived in North America. Moose are a favored game species wherever they occur, and were relied on by early aboriginal peoples for food, clothing, and leather products – traditionally hunted in winter when the hide properties and body condition were most favorable (Parker 2004). During the period of unrelenting exploitation, moose were among the principal quarry of the new inhabitants of the New World such that even prior to the emergence of markets, unregulated hunting not only drove population numbers down in some areas but was responsible for the slow retraction of the species' range. Moose were extirpated from Pennsylvania by 1790 (Karns 1997), Massachusetts in the early 1800s (Vecellio et al. 1993), and the Adirondack Mountains in 1861 (Terrie 1993). By the late nineteenth century, numbers of this species were low throughout New England, the Gaspé Peninsula of Québec, and the Maritime provinces (Alexander 1993; Bontaites and Gustafson 1993; Courtois and Lamontagne 1997; Parker 2004). Even in relatively wild areas like New Brunswick, it was increasingly rare to find moose within easy reach of river shores (Parker 2004). In Cape Breton, Nova Scotia, moose were extirpated altogether by the 1800s and never became re-established until introductions in the late 1940s from Alberta (Beazley et al. 2006).

In Nova Scotia and New Brunswick, hunting restrictions allowed for some recovery of moose populations, but this did not last long. By the late 1930s, hunting seasons in both provinces were closed once again (Beazley et al. 2006; Parker 2004). In New England, moose populations remained in a depressed state and confined to a small portion of their former range, and it took almost a century before they demonstrated signs of recovery beginning in the 1970s (Karns 1997).

Although overhunting had for the most part been the principal driver in bringing down moose populations, ending hunting was not generally sufficient for recovery. While the reason for this was unknown at the time, some observers noted that moose declines were coincident with expansion of white-tailed deer (Parker 2004). It was later determined that the same meningeal worm that limited recovery by caribou were also affecting moose. Indeed, moose populations were able to begin to recover again beginning in the 1900s when three conditions were in place: (1) harvest regulations were instated, (2) forest cover returned, and (3) white-tailed deer declined due to occasional severe winters and return of mature forest. Other factors promoting moose recovery in localized areas included forest clearing in patches. This provided young browse and the increase of wetland habitats following the recovery of beaver (*Castor canadensis*) populations, which themselves had been decimated by overharvesting during the preceding centuries (Alexander 1993; Parker 2004).

Currently, moose in the Northern Appalachians are on a trajectory of recovery, with population increases and range reclamation occurring throughout the ecoregion. For example, in Vermont, moose populations were estimated at 200 in 1980 and grew to over 1,500 by 1993 (Alexander 1993). Today, it is estimated at 4,700 in Vermont, covering the majority of the state (Vermont Department of Fish and Wildlife 2009). In New Hampshire, moose were estimated at 500 in 1977, jumped to 1,600 in 1982, 5,000 in 1993 (Bontaites and Gustafson 1993), growing to approximately 7,000 in the state by 2000 (Aldrich and Phippen 2000). In New York, it appears that moose began immigrating into the state in the 1970s from Vermont (Jenkins 2004). Since then, the population has been increasing steadily and is officially estimated at 300–500 animals (NYDEC, New York State

Department of Environmental Conservation 2009). A small population has even become established in Massachusetts (Vecellio et al. 1993).

Within the ecoregion, two areas stand out where moose populations are not recovering: southern Québec and mainland Nova Scotia. In Nova Scotia, moose are by and large confined to the most remote areas in small populations of questionable viability and appear to be functionally isolated from one another with little evidence of genetic exchange (Beazley et al. 2006). The eastern moose from mainland Nova Scotia has been classified as 'endangered' under the Nova Scotia Endangered Species Act since 2003.

Residing in boreal forests in North America and conifer-dominated forest systems of the Northern Appalachian/Acadian ecoregion, the primary factors that determine the northern limit of moose distribution are availability of food and cover, while the southern extent of their range is predominantly limited by climate (Karns 1997). Heat is the most critical of factors, with maximum temperatures leading to stress in summer (Karns 1997). Also, the meningeal worm can be a significant mortality factor where the ranges of moose and white-tailed deer overlap. This disease has been responsible for moose population declines where white-tailed deer are able to exist at high densities (e.g., where snow is shallow enough for them to persist during winter [Beazley et al. 2006]). Unlike caribou, however, moose can co-exist with low-density white-tailed deer populations.

In the northern stronghold of their range, moose are not particularly averse to humans, living in areas that have been settled by humans for long periods of time and flourishing where heavy and repeated logging activity has occurred. However, towards the southern limit of their distribution, such as the Northern Appalachians, it appears that this species has a lower threshold of human disturbance. Most mortality among moose populations in the region is due to vehicle collisions, followed by hunting and meningeal worm (Alexander 1993). Radio-collared moose have been shown to avoid highways at coarse scales of habitat selection (Laurian et al. 2008). Although they are able to adapt to disturbances that are predictable and do not pose any particular threat to individuals, moose tend to avoid areas that are used regularly by cross-country skiers, snowmobilers, and hunters (Forman et al. 1997). Where such uses are squeezed into a relatively small area, such as Nova Scotia, the St. Lawrence Valley of southern Québec, or Southern New England and New York outside of the Adirondack Mountains (Hicks 1986), moose populations have not flourished. Another factor that appears to limit the recovery of moose populations when deer are not present is their 'social carrying capacity' (Bontaites and Gustafson 1993). In urbanized areas, moose begin to pose a real or perceived threat to human life and property and increasingly become victims of collisions with cars and trains (Karns 1997).

9.2.3 White-Tailed Deer (Odocoileus virginianus)

White-tailed deer are among the animals in New England forests most used by humans during the last 5,000 years. Their hides were shipped to Europe in a vibrant commercial trade that peaked around 1700 (McCabe and McCabe 1984).

Although it is difficult to imagine given today's overabundance of deer in suburban and rural areas alike, deer were overhunted to near extirpation in all but the unsettled portions of the ecoregion by the late 1800s. In some places, concerns for white-tailed deer populations prompted the closure of hunting seasons in Massachusetts as early as 1698 (Bernardos et al. 2004). Throughout the 1800s, the combined influence of market hunting and deforestation acted to depress white-tailed deer populations to a fraction of their original abundance (Miller et al. 2003). In the U.S., available funding stimulated by the passage of the Pittman-Robertson Wildlife Restoration Act of 1937 enabled organized restoration efforts to begin in the late 1930s with almost immediate success (Miller et al. 2003).

White-tailed deer have experienced shifts in their distribution in response to changes in the regional climate and habitat conditions. The recent history of deer in Northern New York serves as a fascinating illustration of the ebbs and flows of deer populations in the region (Jenkins 2004). Deer were common throughout the Adirondack Mountains in the middle of the twentieth century – a time when the region was dominated by young forests with excellent winter browse and large predators had mostly disappeared. Numerous hunt clubs within the Adirondack Mountains were created with deer as the principal quarry. The designation of wilderness areas in the 1970s commenced a trend of aging forests. This, in concert with several winters in a row in the 1970s where snow was exceptionally deep, triggered sharp reductions of deer populations in the region. Currently, deer are more common outside the Adirondack Park than within, promoted by the widespread abandonment of farms and decreasing numbers of hunters since the 1950s. In New Brunswick, white-tailed deer were actually absent when settlers first arrived, but archaeological evidence from middens indicated that they did reside in the region before the onset of long and cold winters and deep snows. Deer eventually reached New Brunswick again during the period of moderating climate following the Little Ice Age (Parker 2004).

White-tailed deer are one of the most successful North American mammals in modern times, with populations readily thriving in conditions created by human settlement. Overabundance of white-tailed deer is in fact a challenging issue for wildlife managers from the perspectives of both human conflict and ecosystem health (Warren 1997). Regarding the latter, overbrowsing by deer populations have well-documented cascading impacts on forest ecosystems (Côté et al. 2004). Along with the decline of hunting (Bernardos et al. 2004), the cultural perspective of deer 'is undergoing a remarkable shift in recent decades from a noble and wild game animal to neighborhood pest' (Foster 2002).

The most important limiting factor for white-tailed deer, determining the northern extent of its distribution at any given time, is snow accumulation in winter, which can increase mortality (de Vos 1964). White-tailed deer also require an abundant supply of relative young hardwood trees and shrubs and, therefore, tend to be absent from older-aged forests. In contrast to historical times, the average deer today is in minimal danger of being shot and killed other than at certain times of the year, especially close to human settlement. Deer populations have responded favorably to year-round subsidization of food sources in agricultural fields and suburban gardens and population densities reflect this, particularly in the absence of predators. Strongholds of white-tailed deer across the Northern Appalachians are almost a mirror image of that of moose, although zones of overlap are increasing. Overall population numbers are probably quite close to pre-exploitation levels, speaking to the propensity of this species to rebound from population lows under the right conditions (McCabe and McCabe 1984).

9.2.4 Wolf (Canis lupus or lycaon)

At one time widespread across North America and Eurasia, wolves persist in just a fraction of their historic global range today. Their story in the Northern Appalachians is one of abundance at the time of European colonization followed by sharp declines, with little sign of recovery at present. Owing to their generalist habitat tendencies, wolves were at one time widespread throughout the ecoregion. Population declines were largely driven by direct persecution, chiefly aided by the bounty system. The first wolf bounty in the region was established by the colony of Massachusetts in 1630 while the last wolf in the state was shot 200 years later (Bernardos et al. 2004). Wolves stood in the way of the new life of European settlers who had become accustomed to keeping free-ranging hogs and sheep in their predator-free lives back home (Conover 2002; Whitney 1994). While direct persecution was responsible for most mortalities, changes in land cover and declining prey levels likely dealt the final blow (Foster 2002). Wolves disappeared from Nova Scotia by 1845–1847 (Scott and Hebda 2004), the Adirondack Mountains by the mid-1890s (Kays and Daniels 2009), Northern New England by the early 1900s (Whitney 1994), and New Brunswick by the early 1920s (Lohr and Ballard 1996).

In contrast to Northern Appalachians where wolf populations have never rebounded, the process of wolf recovery in the Western U.S. began in the 1970s under the umbrella of the Endangered Species Act to considerable success. Not only did the species begin receiving protection from exploitation and farming, but sources of conflict with domestic animals declined at the same time as ungulate prey populations in the region increased. Recovery in the Great Lakes and the U.S. Rockies was kick-started by a handful of active reintroductions and aided by the natural expansion of Canadian wolves (Paquet and Carbyn 2003).

Today, wolves are traditionally regarded as synonymous with western notions of wilderness. As one of the most maligned animals in human history, however, the principal limiting factor for their occurrence has not been habitat disturbance but rather direct persecution. In human-dominated landscapes, road density offers an excellent proxy for this threat (Mladenoff et al. 1995). However, roads fail to predict wolf presence in environments where human settlements and agriculture do not prevail (e.g., areas with forestry) and thus where encounters with humans and conflict with livestock are infrequent (Musiani and Paquet 2004). Judging by the successful expansion and reintroductions elsewhere in North America, wolves have an inherent ability, by virtue of their exceptional adaptability and favorable life-history traits (Weaver et al. 1996), to withstand high levels of mortality and rebuild following population declines when provided the opportunity.

Potential habitat for wolf populations in the Northern Appalachian/Acadian ecoregion have been identified in several independent analyses (Carroll 2003; Harrison and Chapin 1998; Paquet et al. 1999) with all authors commenting that the likelihood that members of Ontario or Québec populations would arrive into the region on their own accord is seriously limited by human barriers (Wydeven et al. 1998). Nevertheless, in 2002, one canid confirmed through genetic analyses to be an 'eastern wolf' was trapped in the eastern townships of southern Québec (Villemure and Jolicoeur 2004). While no crossing of the St. Lawrence River has ever been documented, wolves, particularly subadults, can be highly mobile, and elsewhere they have colonized previously occupied habitat decades after their extirpation (Gehring and Potter 2005). The extent to which neighboring populations in Québec can serve as a source for population recovery in the Northern Appalachians is likely to be highly limited, given the high levels of hunting and trapping pressures outside reserves in the St. Lawrence Valley (Carroll 2003), not to mention the open-water barrier of the river itself, which is an active, year-round shipping channel with 4-lane highways and associated human settlement that parallel the river (Harrison and Chapin 1998).

A complicating factor in re-constructing historical distributional trends of the wolf in Eastern North America is the taxonomic confusion characterizing this species. A prevailing hypothesis is that wolves currently residing in Eastern Canada belong to *C. lycaon* – a species distinct from northern timber wolves (*C. lupus*; Wilson et al. 2003). However, since the eastward spread of coyotes, interbreeding is common between wolves and coyotes, such that large canids inhabiting the region today are actually hydrids of several forms (Kyle et al. 2006; Leonard and Wayne 2008). Jenkins (2004) has most aptly described wolves and coyotes as 'endpoints of a genetic continuum.' Still unclear is which wolf species occupied the ecoregion historically where they are now absent (Forbes et al. in press). With the smaller *C. lycaon* being primarily a predator of white-tailed deer, it is an open question as to what predators exploited once abundant moose, caribou, and elk populations.

9.2.5 Coyote (Canis latrans)

The most successful colonizing mammal in recent history, coyotes were not present in the Northern Appalachian/Acadian ecoregion in historical times (Parker 1995; Forbes et al. in press). At the time of European settlement in the Western U.S. (ca. 1830), coyotes were limited in their distribution to the prairies and grasslands of the midwestern portion of the continent. Beginning in the early 1900s, they expanded rapidly eastward through both natural means and casual transplantations (Parker 1995; Voigt and Berg 1987). Coyotes first penetrated Ontario in 1919, colonized New York in the 1950s, reached the south shore of the St. Lawrence River in 1963, and arrived in Newfoundland in the 1990s (Fener et al. 2005; Larivière and Crête 1992; Parker 1995). They were firmly established in the ecoregion by the 1980s (Parker 1995).

From a mammalian perspective, the speed of their colonization in the East was unusually swift: in Maine, it proceeded at a rate of 1,867 km^2/year and in New York, at 2,240 km^2/year (Richens and Hugie 1974).

Prior to the twentieth century, coyotes did not venture far from grassland habitats (Voigt and Berg 1987). Their eastward expansion coincided with landscape alteration through intensive logging and agricultural development and the local extermination of gray wolves – a chief competitor in forested habitats (Larivière and Crête 1992; Parker 1995). At the same time, improved habitat conditions for white-tailed deer, together with a more favorable climate and the disappearance of wolf, facilitated coyote range expansion northward (Parker 1995). Bounties were immediately set up upon first discovery of coyotes in Maine in the early 1930s 'to concentrate efforts towards their extermination' (Aldous 1939). However, efforts to control coyotes in this manner throughout their North American range have been largely ineffective (Bekoff and Gese 2003).

Coyotes occupy a great range of habitats but are not as abundant in dense forest as in more open or disturbed habitats (Kays et al. 2008). Similar to white-tailed deer, the range of the coyote appears to be limited by snow cover and food resources. The disappearance of the wolf in the ecoregion not only paved the way for the entrance of coyotes but enabled it to secure a spot as the top terrestrial predator in the region (Gompper 2002). Having grown larger in size than their grassland progenitors and evolved into efficient predators of deer, they appear to serve as a partial ecological replacement for wolves in the ecoregion (Ballard et al. 1999). Coyotes do not coexist well with wolves; studies have documented direct and indirect competition with larger carnivores, involving outright killing and other behavioral shifts (Bekoff and Gese 2003).

9.2.6 Black Bear (Ursus americanus)

At one time occurring in all forested habitats across North America, black bears lost substantial portions of the southern part of their original range in the 1800s through the twin threats of overexploitation and habitat loss. Together with white-tailed deer, they have been the most used among large mammals in the Northeastern North America during the past 5,000 years and the most common food animal in Indian middens (Loskiel 1794, cited in Whitney 1994). Black bears were hunted to near expiration in the nineteenth century. Along with wolves, they were subject to bounties in Northeastern North America upon the arrival of European settlers, and other indiscriminate hunting in defense of people, livestock, and crops (Parker 2004). Once this pressure relaxed in the age of conservation and management, and forest cover returned in Northeastern North America, the conditions for black bear population growth were reinstated (Foster 2002). Populations began recovering virtually without interruption everywhere in the range where human settlement and accompanying road densities were not too intense (Pelton et al. 1999).

Currently, black bears can be found in most forested regions away from heavily settled areas in the ecoregion. In areas with substantial human development, black bear habitat has become increasingly fragmented or has disappeared. For example, clearing of forest for agriculture and human settlement along the St. Lawrence River between Montreal and Québec City, forest clearing through human development in Central New Brunswick, and suburbanization in Southern New England has led to loss of black bear habitat even as populations were recovering elsewhere in the region. They were extirpated from Prince Edward Island by 1937 and have not reclaimed this range (Williamson 2002). Most black bear mortalities are human-related, through hunting, poaching (for a limited global trade in bear body parts), killing of nuisance bears, and vehicle accidents (Williamson 2002).

Black bear populations are thought to be stable in the Northern Appalachian/ Acadian ecoregion and are even increasing in some areas (Pelton et al. 1999; Williamson 2002). Like with wolves, the life-history traits of black bears – age of first reproduction, life span, and litter size – bestow populations of this species with an ability to recover quickly from declines. Black bears are not necessarily limited by human disturbance and are able to adapt within short time periods to human settlement (Beckmann and Berger 2003). However, large home range sizes of bears mean that road crossings are generally inevitable, which become more frequent, with associated higher mortality risk, with increased density of the road network (Brody and Pelton 1989).

With the simultaneous expansion of human and bear populations, these two species have come into increasing contact in recent decades. In such cases of coexistence, the documented sightings and incidents of 'nuisance bears' have been growing, with individuals (particularly subadult males and females) increasingly engaged in crop and livestock depredation, apiary damage, and garbage raiding. Their propensity for this behavior increases in years when failures in berry production during drought periods necessitate wide-ranging searches for alternative food sources. Indeed, bears are readily attracted to year-round predictable food sources in areas of human settlement, and the physical movement of populations from wildlands has been followed by shifts in behavior, morphology, and ecology of black bears (Beckmann and Berger 2003). As such, over the course of the past century, the mandate of wildlife management departments throughout the region has transferred from black bear harvest control and population restoration to management of human conflict (Hristienko and McDonald 2007). The prevailing strategy by jurisdictions is to kill or relocate nuisance individuals and rely on areas that are protected from human intrusion to maintain populations (Mattson 1990). Under such conditions, when individual bears can become habituated to humans, their increasing visibility to humans can trigger perceptions that overall bear populations are thriving or even overabundant. In fact, however, increases in mortality due to such conflicts and the draining of populations from wildland areas can have the opposite results (Beckmann et al. 2008).

9.2.7 Wolverine (Gulo gulo)

As a circumpolar species, wolverines are largely confined to boreal and tundra regions of North America, thriving most in landscapes that have been largely unaltered by humans (Copeland and Whitman 2003). This is another species that has lost a large part of its original range, although the extent of this loss is not well-understood due to incomplete knowledge of its historical distribution (Aubry et al. 2007). However, the wolverine is commonly identified as an extirpated species in the Northern Appalachians (e.g., Bernardos et al. 2004; Foster et al. 2002) and on state and provincial endangered species lists (Table 9.1). Yet it is unknown the extent to which southern historical records reflect extreme dispersal events and occurrences that are in fact extra-limital (Copeland and Whitman 2003). Once researchers finally made a concerted effort to assemble and verify historical observation records, the conclusion for the Northeastern U.S. was that wolverine occurrence was, at best, 'sparse and haphazard' (Aubry et al. 2007). Of the 11 wolverine records that extended down to Pennsylvania since the early 1800s, only two, both from New Hampshire, were considered verifiable. The latest one was from 1811 (Aubry et al. 2007). Either wolverines disappeared prior to that time or they were never very abundant. No parallel verification effort has been conducted for the Eastern Canadian provinces. Wolverines are assumed to have occurred historically in Québec south of the St. Lawrence River and Northwestern New Brunswick (Slough 2007; Wrigley 1967 in Forbes et al. in press) but it is unknown whether such records have been verified. The species never occurred in Nova Scotia or Prince Edward Island (Scott and Hebda 2004; Slough 2007). de Vos (1964) identified their distribution in the neighboring Great Lakes region as widespread but 'nowhere common.' The northern stronghold of the species has in fact receded or disappeared from Québec and Ontario, with fewer possibilities for dispersing individuals to find their way north of the St. Lawrence River (COSEWIC, 2003; Dawson 2000).

Currently confined to northern environments, wolverine distribution in North America is closely associated with persistence of snow cover through the spring denning period (May–June; Aubry et al. 2007). Such conditions no longer exist in the Northern Appalachian/Acadian ecoregion except at the highest elevations and are expected to disappear altogether under most scenarios of climate change (Chap. 15). Like many larger-bodied carnivores, protection from overexploitation and the availability of large-ungulate prey biomass serve as additional requirements for this species. Wolverines are identified in the popular imagination as creatures of wilderness, although whether humans have caused the retreat of wolverines into such areas or their required conditions are simply confined to areas where human development has not yet extended is still an open question. As noted by Copeland and Whitman (2003), however, 'large tracts of pristine habitat may be the only assurance of their continued existence.'

9.2.8 Cougar (Puma concolor)

Cougars, or mountain lions, were at one time the most widely distributed animal in the New World, after humans, occurring in forested habitats all the way to the

Table 9.1 Explanatory factors driving relative abundance and distributional dynamics of large mammal species in Eastern North America

Species	Human landscape change	Climate limitations	Directly interacting species	Official status
Caribou	Mature coniferous forests critical; tolerate low levels of disturbance	Northern, deep snow	White-tailed deer (disease transmission), wolves, coyotes, bears (predators)	COSEWIC (CA): end NB: ext; NS: ext; QC: thr
Moose	Favor early successional forests; upper limit of settlement and road density	Northern, temperature	White-tailed deer (disease transmission)	NS: end
White-tailed deer	Thrive in early successional forests, agricultural and suburban edges	Southern, limited by deep snow, cold winters	Wolves, coyotes, bears (localized predation)	–
Wolf	Generalist habitat associations; upper limit of road density and settlement dictated by human conflict	No climate limitations	–	ESA (US): thrNS: ext; ME: sc; NH: end; NY: ext
Coyote	Generalist habitat associations;	Southern, limited by deep snow, cold winters	Wolves (competitor/predator)	–
Black bear	Favor early successional forests; upper limit of road density and settlement	No climate limitations	–	–
Wolverine	Low tolerance for human conflict	Northern, require deep spring snow for denning	Wolves (competitor)	NB: ext; QC: end

(continued)

Table 9.1 (continued)

Species	Human landscape change	Climate limitations	Directly interacting species	Official status
Cougar	Generalist habitat associations; upper limit of road density and settlement dictated by human conflict	No climate limitations	–	ESA (US): endNB: ext; VT: end; NY: ext
Canada lynx	–	Morphologically adapted for deep snow	Bobcat (competitor)	ESA (US): thrNS: end; NB: end; ME: SC; NH: end; VT: end; NY: thr
Bobcat	–	Limited by deep snow	Lynx (competitor)	QC: SC

Note that overexploitation is a potential significant driver for all species. See individual species accounts for references. Status codes: end, endangered; thr, threatened; SC, special concern; ext, extirpated. Categories are defined and assessed at the jurisdictional level southern tip of South America (Young and Goldman 1946). In Northeastern North America, the range of this carnivore likely extended as far north as that of white-tailed deer, with a northern limit of Southern Québec (south of the St. Lawrence River), Central Maine, and Northern New Hampshire (Parker 1998). They were most common in the rugged portions of the ecoregion, such as the Adirondack, Green, and White Mountains. As with other large carnivores, the eighteenth and nineteenth centuries brought about rampant persecution and targeted bounties. The consequence was widespread loss of range and population declines on the continent. In Vermont, for example, the bounty on cougars was one of the first acts instated in the very first session of the legislature, and was not discontinued until 1904 (Parker 1998). Bounty records and those from the famous 'circle hunts' provide some evidence that cougars were captured in large numbers until the early decades of the nineteenth century and continued to appear in the records until the late 1800s (Parker 1998). When combined with large-scale habitat loss and depletion of large ungulate prey in the Northern Appalachians, cougars were considered to be extirpated from the entire ecoregion by the late 1800s, having disappeared from strongholds to the south, such as Massachusetts and Connecticut even earlier that century. Goodwin (1936) claimed the Adirondack Mountains to be the final stronghold of cougar in Eastern U.S., last recorded in New York in 1894. However, Brocke (2009) contended that the mountain ranges and surrounding areas of the Adirondacks were never rugged or vast enough to support populations of this wide-ranging carnivore and that the false impression of their abundance was generated by fraudulent bounty collections.

Cougar have lost as much as two-thirds of their historical range in North America (Pierce and Bleich 2003); the process of natural recovery since the 1960s has been aided by active protection measures, evolving human attitudes, and increasing ungulate populations. Moreover, pockets in the West largely devoid of human influence acted as source populations for range expansion, which has occurred to a modest extent in Midwestern U.S. and Canada through subadult dispersal (Nielsen et al. 2006). The cessation of active hunting of cougars and the resurgence of forest cover in Northeastern North America has not, however, resulted in the recovery of cougars, as it has in the case of other large mammals such as black bear and moose. The most likely reason is that source populations of extant cougars in Western North America are separated at too far distances over highly settled, and therefore inhospitable, terrain.

No animal in the ecoregion sparks as much passionate debate regarding its continued existence. Indeed, the controversy has steadily heightened since the 1980s with hundreds of sightings reported in Northeastern North America (Jenkins 2004; Stocek 1995). Logs of such sightings and other evidence such as tracks and hairs are kept faithfully in most jurisdictions or by independent enterprises, but few are verifiable. Confirmed records verified and collated by The Cougar Network, for example, indicate that the preponderance of these in North America are located in the vicinity of the contiguous western range of this species, with only a smattering in the Northern Appalachians (The Cougar Network 2009; Nielsen et al. 2006). In New Brunswick and Québec, several hair and track samples collected since 2002 have been confirmed through DNA analyses to belong to cougar, and in 1996 a cougar was killed by a truck in the Eastern Townships region of Estrie, Québec, not far from the New Hampshire border (Forbes et al. in press). Of seven samples, three are of South or Central American origin, indicating that these individuals are escaped or released animals or the offspring of once captive animals. DNA analysis on the other four samples could not reject a similar identity (Forbes et al. in press).

This chapter does not seek to resolve this mystery, and the pursuit of evidence of the existence of the 'eastern' cougar will undoubtedly continue unabated. Even if definitive conclusions about the existence and origin of this species in the ecoregion cannot be made at this time, it is safe to say that the species exists nowhere near the densities approaching the ecological functional role that it plays in western environments or that it supposedly did in historical times in this part of the world. Whether what is currently present represents the seeds of a future population remains to be seen and should be closely monitored. It should also be noted that where populations of cougars are established in Western North America, their presence is readily confirmable through roadkills, incidental take through trapping and hunting, and track surveys (Kurta et al. 2007; Parker 1998).

Cougars are catholic in their habitat requirements, with prey availability, vegetation structure, and topography determining habitat use (Pierce and Bleich 2003). Human fear and conflict were the root causes of cougar extirpation in the ecoregion, and while this may not translate into the same likelihood of mortality, the potential for conflict still exists. In suburban areas where they have been studied, cougars are known to navigate through remaining natural areas, adjust their activity patterns, and for the most part move about unnoticed (Beier 1995). While individuals from established populations tend to avoid roads and associated human development, they can be less discerning when traveling short distances through inhospitable terrain (Beier 1995; Logan and Sweanor 2001).

9.2.9 Canada lynx (Lynx canadensis)

Because early lynx records are impossible to distinguish from bobcat (Hoving et al. 2003), it is not possible to determine the relative abundance of lynx when European settlers arrived. What is known about the conditions of the Little Ice Age, however, suggests a favorable environment for this northern species. It is known that the southern range of Canada lynx once extended as far south as Connecticut and Pennsylvania, but has receded during the past century (Hoving et al. 2003). This species was also extirpated from Prince Edward Island in the early 1800s and mainland Nova Scotia by the 1920s (Parker 2001). Lynx are currently restricted to Cape Breton Island, Northwestern New Brunswick, the Gaspé Peninsula, and Central and Northern Maine (Hoving et al. 2003; Ray et al. 2002). Sightings continue to be reported from high elevation areas in Vermont and New Hampshire where suitable, albeit fragmented, habitat conditions occur (Ray et al. 2002). Jenkins (2004) referred to the Adirondack Mountains as 'borderline lynx country at best.' While reliable historical records do exist, whether they

9 Conservation Planning with Large Carnivores and Ungulates in Eastern North America 183

represent occasional wanderers or residents has never been properly resolved. An unsuccessful reintroduction attempt in the Adirondacks took place from 1989 to 1991, with high mortality rates through vehicle collisions (Brocke 2009) and no evidence that the species resides in the area today (Weaver 1999). Certainly, the lack of continuous expanses of coniferous-dominated forests coupled with a less-than-robust snowshoe hare (*Lepus americanus*) population would preclude this possibility (Jenkins 2004).

The main reduction in lynx range in the ecoregion occurred around the turn of the twentieth century; the species was no longer detected in Southern Maine after 1904. Records of lynx in this part of the world were essentially absent for the first half of the 1900s and then began to increase steadily after 1973 (Hoving et al. 2003). Unlike wolverine, once researchers began to scour historical observation records, it became clear that lynx do enjoy a population stronghold in Northeastern North America, contrary to previous impressions of the species being primarily confined to the western part of the continent currently and in the past (e.g., McKelvey et al. 2000).

Lynx are primarily restricted to boreal forest habitats, which in the Northern Appalachian/Acadian ecoregion extend southward only into cool and mesic high-elevation areas (Hoving et al. 2005). They are particularly limited by the availability of snowshoe hare, their preferred prey throughout their distribution. Habitat conditions for lynx in the southern periphery of its range tends to be highly variable in distribution and quality, where hares do not tend to achieve the same peak abundances as they do in northern boreal forests (Murray et al. 2008). The Québec portion of the Northern Appalachians contains the most robust lynx populations and likely serves as a source for the rest of the ecoregion (Ray et al. 2002). Genetic analyses have confirmed the relative isolation of Northern Appalachian lynx from populations that occur north of the St. Lawrence River, which is unlikely to be well-connected due to tremendous development activity alongside the river and year-round open waters due to shipping (USFWS, 2000).

The southern distribution of lynx appears to be limited by deep snow conditions (Hoving et al. 2005), which confer the lynx with a competitive advantage over bobcat and other potential competitors such as coyote (Parker et al. 1983). This notion is supported by (1) the expansion of bobcat range following the contraction of lynx range in the early 1900s (de Vos 1964; Hoving et al. 2003; Lariviere and Walton 1997); (2) the recent incidence of naturally occurring hybridization between bobcats and lynx at the southern edge of lynx range in Maine, New Brunswick, and Minnesota (Homyack et al. 2008); (3) the retraction of lynx range corresponding with changes in climate in the region, including recent warming trends with less snowfall (Hoving et al. 2003); and (4) reduction in lynx range in Southern Alberta associated with the interactive effect of roads and coyotes (Bayne et al. 2008).

9.2.10 Bobcat (Lynx rufus)

Bobcats are currently the most widely distributed among North American native felids (Anderson and Lovallo 2003). This situation, however, has been by no means static over the past two centuries, with the Northern Appalachian/Acadian ecoregion a case in point. Further confusing the historical record is the fact that bobcat and lynx were not distinguished in the literature with any consistency until the mid 1800s (Hoving et al. 2003). Some evidence suggests that bobcats were not very abundant when the early European colonists arrived but that they benefitted from early land clearing associated with human settlement (Litvaitis et al. 2006). There were bounty programs within bobcat range beginning in the early 1800s, but records did not generally differentiate lynx and bobcat (Litvaitis et al. 2006). Beginning in the early 1900s, a northward expansion of the distribution of this species took place concomitant with land clearing for agriculture, ameliorating climate and snow conditions following the end of the Little Ice Age, and a corresponding northward retreat of the southern limit of lynx (de Vos 1964; Hoving et al. 2003; Lariviere and Walton 1997; Litvaitis et al. 2006). Throughout the first half of the twentieth century, bobcat records began to appear in localities where they were hitherto unknown (e.g., Cape Breton [Parker et al. 1983] and Southern Maine [Hoving et al. 2003]).

Although bobcats have never been present on Prince Edward Island, the current distribution of this species extends across the rest of the ecoregion. In the past four decades, however, its upward trend appears to have reversed somewhat; population declines have provoked the listing of this species in some jurisdictions (Litvaitis et al. 2006; Table 9.1). Its broad distribution belies this felid's specialist tendencies as an early successional habitat 'obligate' (Litvaitis 2001); population declines have been associated with overall decline of such habitats in the region (Litvaitis 2003). Furthermore, because bobcat declines occurred at the same time that coyotes began to secure their foothold in the region, some authors have suggested a competitive relationship between the two species (Brocke 2009; Hoving et al. 2003; Litvaitis and Harrison 1989). On the other hand, bobcats have continued to move into areas formerly occupied by lynx (Hoving et al. 2003), and hybridization between the two species has been recorded at this frontier (Homyack et al. 2008). Snow depth plays a likely role in limiting the northern range of bobcats (Hoving et al. 2003; McCord 1974). They are not morphologically adapted for travel in deep snow, and their winter habitat use appears to be governed by avoidance of such conditions. As a result, they are inferior competitors with lynx in areas characterized by severe winters (McCord 1974). Some areas of prime habitat have been consistently occupied by bobcats for four centuries (Litvaitis et al. 2006). Where bobcat populations occur at modest levels, their persistence is further challenged by high road densities and otherwise modified landscapes, incurring heightened levels of mortality (Litvaitis et al. 2006).

9.3 Learning from the Past to Plan for the Future

Wildlife populations and hence communities have been highly dynamic in Northeastern North America over the past 300 years (Bernardos et al. 2004; Foster et al. 2002; Whitney 1994), as illustrated by the patterns of the ten large mammals discussed here (Fig. 9.1). Most of these species have experienced relatively dramatic declines and recoveries in their relative abundances and distributional shifts in the region since colonization by European settlers. The trajectory of each species has been unique, affirming the tendency of responses to the changing ecosystem context to be highly species-specific. Relative abundance and distributional limits reflect a legacy of species-specific habitat associations, sensitivities to human disturbance, levels of conflict with human residents, and adaptations to climate (Table 9.1).

The exhibited patterns do, however, fall into four general categories that represent distinct trends in dynamics characteristic of large mammals in Northeastern North America (modified from Foster et al. 2002; Bernardos et al. 2004):

1. Species declining historically with recent increases and recovery: moose, white-tailed deer, and black bear
2. Species declining historically with little or no recovery: wolf, wolverine, caribou, cougar, and lynx
3. Species expanding their range: coyote
4. Species increasing with forest clearance and agriculture and decreasing with forest maturation: bobcat and white-tailed deer

Inherent in shifting ecological conditions over the past 400 years has been the changing role of the human footprint (Cronon 1983; Fuller et al. 2004; Whitney 1994). The history of land-use change in the Northern Appalachian/Acadian

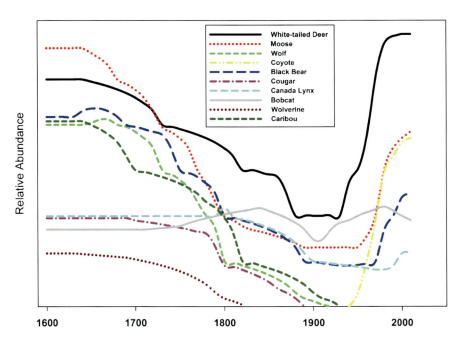

Fig. 9.1 Historical trends of ten large mammal species profiled in this chapter. Each species is represented by a generalized pattern averaged over the ecoregion; local dynamics are not captured

ecoregion has taken place in several discrete stages and provides the background behind the faunal changes discussed here. From the time of arrival of European colonists in the late fifteenth century, the region was dominated by an agrarian economy where individual pursuits of survival and subsistence dominated land use patterns. Industrial growth commenced in the late 1700s, accompanied by the transportation revolution, which signaled the emergence of canals, road networks, and railways. Of relevance to wildlife, market hunting was facilitated by this development, in addition to the invention of the repeating rifle and the refrigerated railroad car. As such, this period was defined by peak levels of exploitation, through hunting and trapping, of the region's fauna and the first extirpations and even extinctions. All this coincided with the highest levels of land conversion from logging and human settlement during the latter half of the nineteenth century. It was at this time that the opening of the American West began to promote a geographic shift of commercial-scale agriculture in that direction. Accordingly, widespread cropland abandonment commenced in 1860s in Eastern North America, setting the scene for the gradual re-growth of forest cover from a virtually denuded state at that time.

The turn of the twentieth century also brought in a new era of conservation and management that allowed many populations of large mammals to climb from historic lows. While forest cover in the region has by now returned almost to its earlier extent of coverage, land uses of the past have been supplanted by extensive suburbanization, increasing industrialization, and fragmentation of natural areas (Woolmer et al. 2008). This character of disturbance presents an altogether novel challenge for biological diversity: full conversion from natural vegetation to human infrastructure for dwelling and transportation purposes, resulting in long-term or permanent habitat loss with little chance for successional recovery (Fuller et al. 2004; McKinney 2006).

Ensuring that planning and regulatory tools provide adequately for the persistence of biological diversity requires a comprehensive understanding of the factors that threaten the conservation of individual species. Centuries of changing circumstances provide a good opportunity to gain understanding of the relative vulnerability of large mammals to threatening processes – information that is vital for effective conservation planning (Wilson et al. 2005).

With habitat loss being the overwhelming cause of species endangerment in modern times (Sala et al. 2000), it is natural to view this as the dominant historical driver of species population and distributional dynamics in Northeastern North America, given the known changes in land use that have occurred. However, overharvest, climatic conditions, species interactions, and societal attitudes have also played important roles. The interacting and cumulative nature of all four factors, in addition to time lags in ecological response (Bernardos et al. 2004), complicate our ability to understand the strength of the role of each driver.

First, although the reasons differed according to the species, all 10 large mammals were subject to uncontrolled killing, an activity that began as soon as the first settlers arrived (Cronon 1983). As a corollary, concerted efforts to manage populations in the form of harvest control and some augmentation promoted the gradual recovery of many, although not all, these species beginning in the 1900s (Conover 2002).

Second, climate conditions also underwent broad-scale shifts during the same period. The Little Ice Age between the 1300s and mid-1800s was characterized by unusually cold temperatures, with the coldest temperatures in Northeastern North America recorded in 1776 (Fagan 2000). A gradual warming trend occurred in the region through the 1800s.

Third, direct and indirect interactions with sympatric species through killing, predation risk, or competition can be key determinants of wildlife abundance and distribution patterns on the landscape (Ritchie and Johnson 2009). Range shifts for both Canada lynx and bobcat have occurred in opposite patterns from one another over the past century, likely reflecting a competitive relationship interacting with snow depth (Krohn et al. 2004). Coyotes would likely never have been so successful in invading the region without the preceding demise of wolf populations and may likewise have influenced present-day patterns of bobcat distribution and abundance (Brocke 2009). Predation can also play a limiting role for ungulates in particular, although this is most likely to occur in localized situations on small populations (Ray et al. 2005).

Finally, social attitudes governing individual human behavior as it relates to coexisting with ungulates and carnivores have also undergone substantial shifts in this region. Broadly speaking, the historical 'shoot-on-sight' philosophy held by every settler who feared for his life or livelihood or needed food on the table has palpably changed to today's level of tolerance by comparison (Conover 2002). This means that direct mortality at the hands of human co-inhabitants has declined. This has, of course, been aided by the fact that most of the large predators, particularly wolves, are no longer present. People in the Northern Appalachians are less likely to be relying on a subsistence existence, so conflicts are less deleterious to overall livelihoods, and the wildlife management policies (including some control programs) are now in place to address conflicts (people may be more apt to call an official to address a conflict rather than deal with the situation themselves). Many of the species discussed here that survived last century's era of exploitation have also become, at least to some extent, 'culturally conditioned,' having developed strategies to co-exist amid human infrastructure, and displaying a flexibility in their behavior and ecology (e.g., Beckmann and Berger 2003). Nevertheless, current increases in development and low-density housing patterns are bringing about heightened opportunities for conflicts; for example, suburban and exurban developments (Kretser et al. 2008) can increase population mortality rates (Beckmann and Berger 2003).

9.4 Planning for Ecoregional Conservation Through the Lens of Large Mammals

The discipline of conservation planning has emerged as a systematic process to plan for biological diversity conservation in response to its continued loss in the face of an expanding human footprint (Groves 2003; Margules and Pressey 2000). In the Northern Appalachian/Acadian ecoregion in particular, the need to ensure that the remaining natural areas and native biological diversity are protected and adequate management measures at the human–wildlife interface are in place against the backdrop of a continually changing environment has led to the mounting imperative for comprehensive conservation planning at multiple scales (Trombulak et al. 2008).

Species should figure prominently in such exercises both as targets of conservation interventions but also as tools to guide conservation area design (Groves 2003; Ray 2005a). It stands to reason that large mammals in particular, by virtue of their life history requirements, vulnerability to certain threatening processes, and cultural connections (both negative and positive), should be of particular use in this regard. For example, such species may be useful for informing both the goals and broader vision of a conservation planning exercise. They may also be used to define high-level aspects of a conservation 'blueprint' itself. When it comes to real-world implementation, however, using a surrogate or focal species approach to guide land-use planning decisions is fraught with challenges (Favreau et al. 2006). The most formidable among these are that (1) the cost and length of time required to collect suitable data on such species to guide decision-making is often in conflict with budgets and timeframes for decision-making, and (2) the extent to which chosen species can fulfill any promise to serve as adequate surrogates for myriad other components of biological diversity is often too readily assumed and seldom tested. Understanding the promises and limitations of the role of large mammals in such exercises, therefore, is paramount for effective conservation planning. In the discussion that follows, I will address both of these in concert by outlining both the manner in which large mammals can play useful roles in conservation planning and the limitations of such an approach.

9.4.1 Large Mammals as Conservation Targets

At the root of any conservation planning exercise is the vision that drives it and a set of goals or targets that describe its ultimate purpose. Desired outcomes often include biological diversity targets, or those set for individual species (Groves 2003; Tear et al. 2005). These can represent a desire either to maintain or restore the status of individual species (or even their elimination in the case of exotics). Planners must make decisions about two particular aspects of their stated objectives as they relate to large mammals. The first is which species to focus on, and the second is the state of the target, spelled out in qualitative or quantitative terms, to be addressed in conservation planning (Chap. 17).

A list of species that serve as conservation targets in a planning exercise will include those on federal, provincial, or state species-at-risk lists or other extinction-prone species, recognizing that focused attention is likely required to ensure their recovery. Others will be selected as targets under the assumption that attention to their conservation will bring about benefits for numerous co-occurring species (Groves 2003; Ray 2005a; Sanderson 2006). Large mammals of the Northern Appalachians serve as examples of both categories.

Inherent in choosing species that will receive special attention are decisions about restoration targets that have always occupied the ambitions of some of the more visionary conservationists in the region. Whether restoration is envisioned through active reintroduction attempts or through more passive means of natural recovery (as may currently be true for cougars and is certainly true for moose), room must be made for such species in the planning context. Otherwise, if ever-changing circumstances, such as an expanding human footprint or warming climate, compromise overall habitat quality and leave little room for space-demanding animals, formulating plans for their reintroduction is simply irresponsible. Success is unlikely if the same factors that drove the disappearance of such creatures are left undiminished. The failed introductions of Canada lynx and caribou in the region demonstrated some of these pitfalls. With the science of reintroductions having increased markedly in sophistication since then (Seddon et al. 2007), such misjudgments are less likely to occur in the future.

The oft-cited desire to restore the full complement of large mammals that were present when colonists first arrived must be tempered by reality, including the increasingly unfavorable climatic and associated habitat conditions in the ecoregion for northern species like caribou and wolverine. By contrast, the ambition to restore wolves in the region has stimulated multiple serious analyses of the ecological potential of such an enterprise, indicating that the notion is not so far-fetched (Carroll 2003; Harrison and Chapin 1998; Paquet et al. 1999; Wydeven et al. 1998). During the time it takes geneticists and ecologists to sort out what taxonomic form of wolf is truly native to the region (Rutledge et al. 2009), it is possible that they will make their way into the ecoregion from populations to the north and west and establish themselves on their own. Finally, it must be stated that while accounting for the restorative potential of large mammals in conservation planning allows for the expression of ambitious dreams and rallying points for optimism, the trade-offs in terms of time, energy, and resources must be evaluated.

The second decision relevant to conservation planning concerns the state of species targets, or the specific desired outcomes. The key role of population size and trend as a determinant of vertebrate extinction risk (O'Grady et al. 2004) underscores the importance of this step. The ultimate planning objective for a species should at the very least strive for demographic sustainability of identified populations, but it can be as ambitious as achieving historic population levels or ecological functionality or as modest as merely ensuring the presence of the species (Sanderson 2006). Our ability to predict the capacity of the Northern Appalachian/Acadian ecoregion to support large mammal densities makes accurate 'reference levels' for resident wildlife populations challenging: few would have predicted even 20 years ago that moose populations would achieve the heights they have in some areas of the region or how rapidly coyotes would colonize new areas. Likewise, the more effort that has been put into radio-collaring and surveying Canada lynx in Maine, the higher the population estimates rise. Minimum viable population estimates are readily attainable (Sanderson 2006), as are general rules of thumb for population levels (7,000 adults; Reed et al. 2003). One challenge is the difficulty of evaluating population boundaries for wide-ranging species without demographic or genetic data or unless they are isolated (e.g., Gaspé caribou), or alternatively, assessing the amount and quality of habitat that is required to support viable populations. The more heterogeneous the habitat or the broader the scale of planning, the more difficult it will be to set and measure progress against specific population-level targets.

An attractive option is to set historical 'baseline' conditions as population-level targets. The many changes that have characterized the ecoregion, however, render this a somewhat futile exercise (Motzkin and Foster 2004). It is difficult enough to ascertain which species were present at what time in the past or even the population levels of species in the region today. It is also not possible to establish what represents the true baseline conditions or those within the 'natural range of variability' (Landres et al. 1999). For example, conservationists are often wooed by the historical accounts of large-mammal abundances when the settlers first landed in Northeastern North America, but it is difficult to know the extent to which erstwhile perceptions of such great quantities were merely influenced by their comparisons with the relative impoverishment of large mammals back home, or even whether predator populations here became artificially high in response to the abundance of livestock animals that reached the shores at the same time (Anderson 2004). Against this backdrop, current cultural constraints and extent of permanent conversion of the natural landscape may be the ultimate determinant of population-level targets.

9.4.2 Large Mammals as Conservation Tools

While much has been written about the validity of assumptions that underlie the role of focal species in conservation planning (e.g., Favreau et al. 2006; Ray 2005a), the following discussion centers on the manner in which large mammals can be of use in this particular geography. In Western North America, where mammal communities are represented by a diverse array of large-bodied ungulates and carnivores, their role in conservation planning can more easily be justified (Carroll et al. 2001; Noss et al. 2002). In the Northern Appalachian/Acadian ecoregion where human settlement prevails, the community is somewhat a relic of the past, with the species most sensitive to prevailing threats having disappeared. Those that remain are for the most part habitat generalists that exhibit sometimes remarkable degrees of flexibility and adaptability to the human-dominated landscape. So how can large mammals in this landscape aid us in conservation planning? Keeping in mind not only that large mammals must be joined by other species to make a full complement of focal species (Coppolillo et al. 2004), or that many other conservation tools than use of focal species exist (Groves 2003; Tear et al. 2005; Trombulak et al. 2008), large mammals can serve as guides for several elements of conservation plans, including the location of core conservation areas and priorities for connectivity, ecological thresholds of development, and management options outside conservation areas.

Core Areas With body size directly related to home range size (Lindstedt et al. 1986), large mammals are generally among the most space-demanding representatives of regional biological diversity. Accordingly, they should be of use in guiding the extent and selection of conservation areas, or those that are designated as off-limits to major development and associated roads. For most large mammals considered here, past and present strongholds are the relatively unsettled areas of the ecoregion at the far end of the continuum of human impact. This statement may appear contradictory from a local perspective since many large mammals in this region exhibit surprising abilities to adapt to the human landscape. Nevertheless, the importance of such areas to population persistence of most such species is quite clear when viewed from a broad scale. It is highly probable, for example, that the recovery of black bear and moose in the region would not have been possible without the persistence of these wildlands, although this is difficult to quantify. As has been demonstrated with black bear elsewhere, at the same time individuals driven by an opportunistic and curious nature have been attracted to anthropogenic food sources, the high-enough mortality they suffer renders such areas population sinks (Beckmann and Lackey 2008). Species such as cougar apparently have been limited in their ability to recover in this area as they have in the West due to the long distances from source populations.

Developing recommendations for the size of such wildland patches is risky since those who implement such guidelines have a tendency to regard these as minimum sizes. In addition, none of the species under discussion here requires wilderness areas to the exclusion of the anthropogenic landscape, making it that much more difficult to develop recommendations for required size of core areas. Clearly, accommodating the needs of large mammals generally requires areas of natural habitat several orders of magnitude above those required to meet those of bird, plant, or invertebrate communities (ELI, 2003). Because minimum areas can be calculated in a variety of ways (e.g., occupancy analyses, species-area relationships, and minimum viable population estimates), it is easy to be confused by scientifically-derived recommendations.

Ultimately, the landscape context in which such fragments are embedded has as much or more importance than the size of the patches themselves in contributing to persistence of large mammals (Franklin and Lindenmayer 2009). Designing landscapes to minimize fragmentation and maximize overall extent of natural cover over scales that are meaningful for populations may have more dividends than a focus on securing large patches which, if embedded in hostile landscapes, may be relatively worthless (Ray 2005b). Specifically, minimum patch size will be less important than the overall amount and dispersion of habitat and the intensity of the human footprint in the intervening landscape. Cookbook remedies or rules of thumb regarding size of core areas to receive protection are therefore generally inappropriate (Ray 2005b).

At present in the Northern Appalachians/Acadia ecoregion, a number of sizeable blocks of core unroaded area serve undetermined functions for the persistence of large mammals. Depending on the scale of planning, one option is to build targets for individual fragment sizes around calculations for the area needed to support a certain number of individuals of carnivore or ungulate, or alternatively, identify those that are too small to be able to reliably contribute to mammalian conservation (e.g., Kerley et al. 2003). Another is to identify source areas through spatially explicit population modeling (Carroll 2007) or grid-based landscape population modeling (Fahrig 2001) and keep such areas as intact as possible. Even in the absence of precise formulation of minimum areas required, the scale and ambition of a planning exercise will be ratcheted up a notch or two if the continued viability or recovery of wide-ranging species like black bear, wolves, moose, Canada lynx, and bobcat is accommodated.

Connectivity Designing for connectivity between natural areas is also a critical feature of conservation planning for large mammals as well as other species (Hilty et al. 2006; Chap. 16). Connectivity is a species-specific and scale-dependent emergent property of landscapes, facilitating movement of organisms across heterogeneous landscapes (Schmiegelow 2007). It enhances the value of, but cannot replace, the role of core areas (Noss and Daly 2006). Connectivity becomes an issue for these animals in landscapes that have undergone extensive development where pinch points that impede species movements begin appearing. Providing for connectivity becomes imperative to prevent the isolation of populations or subpopulations. For example, moose habitat on mainland Nova Scotia is already isolated, which is one assumed driver of their population decline and genetic impoverishment in this area (Beazley et al. 2006).

Given the sometimes extensive movements of large mammals, accommodating their needs in this regard can be another way to scale up the ambition of a conservation plan. However, although the concept of corridor placement in land-use planning is generally widely embraced, in practice, the objectives of the corridor are often disconnected from the manner in which animals actually move through the landscape (Chetkiewicz et al. 2006). It is not always possible to view the relative permeability of a fragmented landscape through our own eyes, yet most corridors are designed on the basis of how humans define connectivity. As such, it is vital to distinguish structural connectivity with functional connectivity from the viewpoint of individual species (Hilty et al. 2006; Chap. 16). Identifying habitats suitable for linking core areas should be based on knowledge of the species' response to vegetation, land use, road density, and topography (Beier et al. 2006) and to behavioral decisions regarding individual movements (Chetkiewicz et al. 2006).

The species-specific nature of planning for linkages becomes most important in landscapes where multiple options for connectivity still exist. Otherwise, 'path of least resistance' approaches will be easiest to spot in circumstances where these are vanishing, and it is a matter of saving what is left before it vanishes (Noss and Daly 2006). Ideally, the latter approach would be accompanied by careful monitoring of selected species to understand their movement decisions in such a context and to evaluate the effectiveness of designed linkages. We should, however, be careful to remind ourselves of the habitat generalist nature of many of the remaining large mammals considered here. Because many can and do move through marginal and degraded habitats, a corridor designed for any of them does not serve most habitat specialists with limited mobility (Noss and Daly 2006). Nevertheless, ensuring that conservation designs accommodate the long-range movements of large mammals that have high dispersal requirements and for which some anthropogenic environments, such as roads, constitute barriers will add value and ambition to planning at various scales. The more modified the landscape and the more connectivity is compromised, the more important this step becomes, as long as such species remain residents.

Thresholds of Landscape Change Large mammals should serve as a useful lens to tackle questions of conservation design from yet a different direction. Rather than wring our hands over 'how much is enough,' we can ask 'how much is too much?' In other words, we can draw from the science investigating impacts of land-use change on wildlife populations to guide conservation planning efforts that place limits on the extent and intensity of our human footprint (Schmiegelow et al. 2008). This obviously depends on the extent to which land-use changes represent threatening processes to a particular species and whether an empirically-derived relationship is known or can be evaluated between population status and degree of habitat loss, fragmentation, or degradation. While most of the species profiled in this chapter are not habitat specialists and enjoy wide distributions when not overharvested, the reviews provided here demonstrate that all show some upper limit of human development that can be tolerated lying somewhere on the continuum between unroaded wilderness and a parking lot (Table 9.1).

The science of 'thresholds' is in its relative infancy and again will generally be both species- and context-specific. One rich area relevant for this discussion is the evolving field of road ecology. A recent review of the population-level effects of roads and traffic on animals confirmed that for large mammals, the impacts are predominantly negative (Fahrig and Rytwinski 2009). Most affected are those species that have large movement ranges and do not avoid roads or traffic and for which increased mortality leads to population declines because of low reproductive rates (Forman et al. 2003). Species like cougar, black bear, and wolf appear to be absent from areas with relatively high road densities where they occur, either because of behavioral avoidance or increased mortality (Dickson and Beier 2002; Fahrig and Rytwinski 2009; Mladenoff et al. 1995; Morrison et al. 2007). For wolves, multiple studies have even identified road-density thresholds beyond which wolves have a high probability of extirpation in settled landscape (Fuller et al. 1992; Mladenoff et al. 1995; Wydeven et al. 2001). Similarly, bobcats in New Hampshire do not appear to occupy potential home ranges with more than one major road (Litvaitis et al. 2006). For some species in this region, such as moose, high road densities may not have particularly deleterious impacts on population size, but may nevertheless limit the extent to which populations will increase or expand their range.

Because roads are an integral part of the human footprint in that the building of human infrastructure is always accompanied by roads, this is a fruitful area on which to focus planning efforts. In this regard, rather than planning for roads in a piecemeal fashion, conservation design would contain some notion of an upper limit to road density before build-out occurs. Future scenarios of road development can occur in a predictable fashion, particularly as it relates to exurban settlement (Baldwin et al. 2007). Combining this insight with knowledge of road density

194   J.C. Ray

thresholds for large mammals would add a refreshingly proactive dimension to land-use planning.

Thresholds of overall habitat conversion and extent of fragmentation may provide additional guidance for conservation planning (With and Crist 1995). However, the relationship between such parameters and the occurrence or population status of large mammals has not been explored to the extent it has with other species (e.g., birds [Radford et al. 2005; Trollope et al. 2009]; American marten [*Martes americana*; Hargis et al. 1999]). Context, particularly scale, will likely complicate our ability to generate rules of thumb for this as it relates to large mammals. Hence, concentrating on core area for design purposes would, for the time being at least, constitute a more prudent approach.

9.4.3   Other Considerations for Using Large Mammals in Conservation Planning

When large mammals are among the residents of a landscape, conservation planning must incorporate several additional factors that transcend traditional conservation design. While not necessarily unique to this wildlife group, three factors – planning for change, management measures, and monitoring – must be integral to any conservation plan that explicitly considers this group of species.

Planning for Change   Most conservation designs are inherently static entities. Patterns of biological diversity and the processes that generate them, including human threats, tend to be evaluated for one point in time yet serve as the basis of plans for which the goal is to maintain diversity in the future. It is, however, increasingly clear that, to be effective, plans must take inevitable ecological changes into account. This includes not only natural processes that influence patterns of biological diversity, but the cumulative effects of human modification of landscapes and a rapidly changing climate (Pressey et al. 2007; Chap. 15).

While thresholds of tolerance to modified landscapes will probably remain the same for a given large mammal species, the locations and size of core conservation areas and key linkages almost certainly will not. Accordingly, designs that designate core areas but particularly those that aim to maximize connectivity must provide for anticipated range shifts in response to changing conditions in addition to movements within home ranges or dispersal among populations (Noss and Daly 2006). Emerging analyses that predict future habitat conversion (e.g., Baldwin et al. 2007, 2009; Trombulak et al. 2008) and ecological consequences of a changing climate (Rodenhouse et al. 2009) should be used to forecast future scenarios of landscape condition. These can in turn be related to the status of large mammals (e.g., Canada lynx; [Carroll 2007]) to make necessary adjustments to elements of conservation design that have a better chance of accommodating future change and the considerable uncertainty associated with it.

Management Measures to Mitigate Impacts on Large Mammals   No matter how robust a conservation design, some degree of management intervention above and beyond the conservation plan will be required to better ensure large mammal populations are conserved. Wildlife management includes active interventions beyond land-use designations that range from habitat manipulation, to harvest regulations, to mitigation strategies for development infrastructure (e.g., roads), to strategies aimed at dealing with nuisance animals, including culling of populations. Because protected areas are not likely to surpass a mere fraction of most ecoregions, management practices in the intervening 'matrix' habitats outside such areas are required to help ensure that these do not become hostile environments and instead actively contribute to the conservation landscape as a whole (Franklin and Lindenmayer 2009).

This review makes clear that the changing status of all 10 profiled large mammals has occurred against a backdrop of shifting human attitudes (Conover 2002). Indeed, particularly when it comes to large carnivores, only adjustments in human behavior will enable some wildlife populations to achieve some level of security. This factor will only increase in importance the more human populations encroach on wildlife habitat, as with low-density exurban sprawl (Kretser et al. 2008). In light of the reality that many present-day large predator populations do not likely exist at densities that allow for an expression of their functional ecological roles, special strategies must be devised that not only seek to replace or augment the role of such animals (e.g., in highly settled areas) but foster a wider expression of acceptance for their continued presence.

Monitoring A third obligatory feature of conservation planning relates to long-term monitoring of the status of large mammal populations. This is important for testing how they are faring in response to land-use designations or management interventions that have been put in place for their welfare. Monitoring the effectiveness of such activities is unfortunately often among the first line items to be removed from budgets during periods of financial restraint but is critical for achieving adequate understanding of how interventions ultimately relate to large mammal conservation status. While many strategies detailed in this chapter have been devised in the name of conserving large mammals, it is rare for them to come with demonstrated success since prescriptions are seldom accompanied by long-term monitoring information.

Budgetary limitations are not the sole challenge that stands in the way of this. The life-history characteristics of the animals discussed here (large-bodied, long-lived, wide-ranging, and often elusive) mean that designing robust survey protocols over sufficiently large areas that will have the statistical power to be able to detect true trends over time is logistically challenging, to say the least (MacKay et al. 2008). Arguably for many species, present-day techniques render this unfeasible. Responses of large-mammal populations to both management and changing ecological conditions are difficult to determine either because of confounding and interacting factors or because of the reality of time lags in response. Nevertheless, a commitment to monitor large-mammal populations will not only help to inform the success of conservation planning in that landscape (as long as management responses are nimble enough to change course in response to the weight of evidence), but those with similar conditions elsewhere.

9.5 Lessons Learned

In this chapter, I have pieced together existing information on the individual trajectories of 10 large mammal species over the past four centuries of change in the Northern Appalachian/Acadian ecoregion. Several broad categories of patterns emerge: species that have declined historically with recent increases and recovery, those that have declined historically with little or no recovery, those that have expanded their range into the ecoregion in recent history, and those already resident in the area that have fared particularly well with forest clearance and expanding agriculture but have then decreased in localized areas with forest maturation. Nevertheless, the path of each individual carnivore and ungulate species has been unique, demonstrating the interplay between background conditions and the particular inherent characteristics of each. Because the past has been defined by alternate periods of decline and recovery of environmental conditions, it has provided an interesting perspective on the interactions between drivers such as land-use change, climate shifts, prevailing human attitudes, and interspecific relationships. In any case, this exploration underscores the reality that in any given place in the ecoregion the large-mammal community structure (as defined by the relative abundances of component species) has undergone significant shifts over this time period.

This retrospective analysis has some clear usefulness for ongoing and future land-use planning, in that it has provided insight into both the limitations and conservation prospects of individual species, many of which serve as conservation focal points, by virtue of their high-profile status. As such, these species can be important not merely as targets of conservation activities in their own right, but can effectively serve as planning tools. This can be as simple as expanding the scale of ambition for conservation by wrapping their fates in the success of the plan, but also as a means by which to guide the identification of core conservation lands, connectivity within the overall landscape, and thresholds of development intensity.

Acknowledgements I extend my sincere appreciation to Graham Forbes and Michaele Glennon, who reviewed various drafts of this manuscript and provided insightful comments that helped improve the final product. I also take this occasion to acknowledge Bill Krohn and Jerry Jenkins, who have both influenced my thinking on the historical ecology of carnivores in the Northeast through their important work and stimulating discussions.

References

Aldous, C. M. (1939). Coyotes in Maine. *Journal of Mammalogy, 20*, 104–106.
Aldrich, E., & Phippen, W. (2000). Sky high for New Hampshire's moose. *New Hampshire Wildlife Journal*, November/December, 8–11. Retrieved January 31, 2010, from http://www.wildlife.state.nh.us/Wildlife_Journal/WJ_sample_stories/WJ_f00_Moose_research.pdf
Alexander, C. E. (1993). The status and management of moose in Vermont. *Alces, 29*, 187–195.
Anderson, R. C. (1972). The ecological relationships of meningeal worm and native cervids in North America. *Journal of Wildlife Diseases, 8*, 304–310.

Anderson, V. D. (2004). *Creatures of empire: How domestic animals transformed early America*. New York: Oxford University Press.

Anderson, E. M., & Lovallo, M. J. (2003). Bobcat and lynx. In G. A. Feldhamer, B. C. Thompson, & J. A. Chapman (Eds.), *Wild mammals of North America: Biology, management, and conservation* (pp. 758–786). Baltimore, MD: Johns Hopkins University Press.

Aubry, K. B., McKelvey, K. S., & Copeland, J. P. (2007). Distribution and broadscale habitat relations of the wolverine in the contiguous United States. *Journal of Wildlife Management, 71*, 2147–2158.

Baldwin, R. F., Trombulak, S. C., Anderson, M. G., & Woolmer, G. (2007). Projecting transition probabilities for regular public roads at the ecoregion scale: A Northern Appalachian/Acadian case study. *Landscape and Urban Planning, 80*, 404–411.

Baldwin, R. F., Trombulak, S. C., & Baldwin, E. D. (2009). Assessing risk of large-scale habitat conversion in lightly settled landscapes. *Landscape and Urban Planning, 91*, 219–225.

Ballard, W. B., Whitlaw, H. A., Young, S. J., Jenkins, R. A., & Forbes, G. J. (1999). Predation and survival of white-tailed deer fawns in northcentral New Brunswick. *Journal of Wildlife Management, 63*, 574–579.

Bayne, E. M., Boutin, S. R., & Moses, R. A. (2008). Ecological factors influencing the spatial pattern of Canada lynx relative to its southern range edge in Alberta, Canada. *Canadian Journal of Zoology, 86*, 1189–1197.

Beazley, K., Ball, M., Isaacman, L., McBurney, S., Wilson, P., & Nette, T. (2006). Complexity and information gaps in recovery planning for moose (*Alces alces americana*) in Nova Scotia, Canada. *Alces, 42*, 89–109.

Beckmann, J. P., & Berger, J. (2003). Rapid ecological and behavioural changes in carnivores: The responses of black bears (*Ursus americanus*) to altered food. *Journal of Zoology, 261*, 207–212.

Beckmann, J. P., Karasin, L., Costello, C., Matthews, S., & Smith, Z. (2008). *Coexisting with black bears: Perspectives from four case studies across North America (WCS Working Paper)*. Bronx, NY: Wildlife Conservation Society.

Beckmann, J. P., & Lackey, C. W. (2008). Carnivores, urban landscapes, and longitudinal studies: A case history of black bears. *Human – Wildlife Conflicts, 2*, 77–83.

Beier, P. (1995). Dispersal of juvenile cougars in fragmented habitat. *Journal of Wildlife Management, 59*, 228–237.

Beier, P., Penrod, K. L., Luke, C., Spencer, W. D., & Cabanero, C. (2006). South Coast missing linkages: Restoring connectivity to wildlands in the largest metropolitan area in the USA. In K. R. Crooks & M. Sanjayan (Eds.), *Connectivity conservation* (pp. 555–586). Cambridge, UK: Cambridge University Press.

Bekoff, M., & Gese, E. M. (2003). Coyote (*Canis latrans*). In G. A. Feldhamer, B. C. Thompson, & J. A. Chapman (Eds.), *Wild mammals of North America: Biology, management, and conservation* (pp. 467–481). Baltimore, MD: Johns Hopkins University Press.

Bergerud, A. T., & Mercer, W. E. (1989). Caribou introductions in eastern North America. *Wildlife Society Bulletin, 17*, 111–120.

Bernardos, D., Foster, D., Motzkin, G., & Cardoza, J. (2004). Wildlife dynamics in the changing New England landscape. In D. R. Foster & J. D. Aber (Eds.), *Forests in time: The environmental consequences of 1,000 years of change in New England* (pp. 142–168). New Haven, CT: Yale University Press.

Bontaites, K. M., & Gustafson, K. (1993). The history and status of moose and moose management in New Hampshire. *Alces, 29*, 163–167.

Brocke, R. H. (2009). Wildlife for a wilderness: Restoring large predators in the Adirondacks. In W. F. Porter, J. D. Erickson, & R. S. Whaley (Eds.), *The great experiment in conservation: Voices from the Adirondack Park* (pp. 169–190). Syracuse, NY: Syracuse University Press.

Brody, A. J., & Pelton, M. R. (1989). Effects of roads on black bear movements in western North Carolina. *Wildlife Society Bulletin, 17*, 5–10.

Carroll, C. (2003). *Impacts of landscape change on wolf viability in the northeastern U.S. and southeastern Canada: Implications for wolf recovery (Special Report 5)*. Richmond, VT: Wildlands Project.

Carroll, C. (2007). Interacting effects of climate change, landscape conversion, and harvest on carnivore populations at the range margin: Marten and lynx in the northern Appalachians. *Conservation Biology, 21*, 1092–1104.

Carroll, C., Noss, R. F., & Paquet, P. C. (2001). Carnivores as focal species for conservation planning in the Rocky Mountain region. *Ecological Applications, 11*, 961–980.

Chetkiewicz, C. L. B., Clair, C. C. S., & Boyce, M. S. (2006). Corridors for conservation: Integrating pattern and process. *Annual Review of Ecology, Evolution, and Systematics, 37*, 317–342.

Conover, M. R. (2002). *Resolving human–wildlife conflicts: The science of wildlife damage management*. Boca Raton, FL: Lewis Publishers.

Copeland, J. P., & Whitman, J. S. (2003). Wolverine. In G. A. Feldhamer, B. C. Thompson, & J. A. Chapman (Eds.), *Wild mammals of North America: Biology, management, and conservation* (pp. 672–682). Baltimore, MD: Johns Hopkins University Press.

Coppolillo, P., Gomez, H., Maisels, F., & Wallace, R. (2004). Selection criteria for suites of landscape species as a basis for site-based conservation. *Biological Conservation, 115*, 419–430.

COSEWIC (Committee on the Status of Endangered Wildlife in Canada). (2003). *COSEWIC assessment and update status report on the wolverine* Gulo gulo *in Canada*. Ottawa, ON: Committee on the Status of Endangered Wildlife in Canada.

Côté, S. D., Rooney, T. P., Tremblay, J. P., Dussault, C., & Waller, D. M. (2004). Ecological impacts of deer overabundance. *Annual Review of Ecology, Evolution, and Systematics, 35*, 113–147.

Courtois, R., & Lamontagne, G. (1997). Management system and current status of moose in Quebec. *Alces, 33*, 97–114.

Courtois, R., Ouellet, J. P., Gingras, A., Dussault, C., Breton, L., & Maltais, J. (2003). Historical changes and current distribution of Caribou, *Rangifer tarandus*, in Quebec. *Canadian Field-Naturalist, 117*, 399–414.

Cronon, W. (1983). *Changes in the land: Indians, colonists, and the ecology of New England*. New York: Hill and Wang.

Dawson, F. N. (2000). *Report on the status of the wolverine* (Gulo gulo) *in Ontario. Report for the Committee on the Status of Species at Risk in Ontario*. Thunder Bay, ON: Ontario Ministry of Natural Resources.

de Vos, A. (1964). Range changes of mammals in the Great Lakes region. *American Midland Naturalist, 71*, 210–231.

Dickson, B. G., & Beier, P. (2002). Home-range and habitat selection by adult cougars in southern California. *Journal of Wildlife Management, 66*, 1235–1245.

ELI (Environmental Law Institute). (2003). *Conservation thresholds for land use planners*. Washington, DC: Environmental Law Institute.

Fagan, B. M. (2000). *The Little Ice Age: How climate made history, 1300–1850*. New York: Basic Books.

Fahrig, L. (2001). How much habitat is enough? *Biological Conservation, 100*, 65–74.

Fahrig, L., & Rytwinski, T. (2009). Effects of roads on animal abundance: An empirical review and synthesis. *Ecology and Society, 14*, Article 21. Retrieved on January 31, 2010, from http://www.ecologyandsociety.org/vol14/iss1/art21/

Favreau, J. M., Drew, C. A., Hess, G. R., Rubino, M. J., Koch, F. H., & Eschelbach, K. A. (2006). Recommendations for assessing the effectiveness of surrogate species approaches. *Biodiversity and Conservation, 15*, 3949–3969.

Fener, H. M., Ginsberg, J. R., Sanderson, E. W., & Gompper, M. E. (2005). Chronology of range expansion of the Coyote, *Canis latrans*, in New York. *Canadian Field-Naturalist, 119*, 1–5.

Festa-Bianchet, M., Ray, J. C., Boutin, S., Côté, S. D., & Gunn, A. (2010). Caribou conservation in Canada: An uncertain future. *Canadian Journal of Zoology* (in press).

Forbes, G. J., McAlpine, D. F., & Scott, F. W. (2010). Mammals of the atlantic maritime ecozone. In D. F. McAlpine & I. M. Smith (Eds.), *Assessment of species diversity in the Atlantic Maritime Ecozone*. Ottawa, ON: NRC Research Press, National Research Council Canada (in press).

9 Conservation Planning with Large Carnivores and Ungulates in Eastern North America 199

Forman, R. T. T., Friedman, D. S., Fitzhenry, D., Martin, J. D., Chen, A. S., & Alexander, L. E. (1997). Ecological effects of roads: Toward three summary indices and an overview for North America. In K. Canters, A. Piepers, & D. Hendriks-Heersma (Eds.), *Habitat fragmentation and infrastructure* (pp. 40–54). Maastricht and The Hague, The Netherlands: Ministry of Transport, Public Works and Water Management.

Forman, R. T. T., Sperling, D., Bissonette, J. A., Clevenger, A. P., Cutshall, C. D., Dale, V. H., et al. (2003). *Road ecology: Science and solutions*. Washington, DC: Island Press.

Foster, D. R. (2002). Insights from historical geography to ecology and conservation: Lessons from the New England landscape. *Journal of Biogeography, 29*, 1269–1275.

Foster, D. R., Motzkin, G., Bernardos, D., & Cardoza, J. (2002). Wildlife dynamics in the changing New England landscape. *Journal of Biogeography, 29*, 1337–1357.

Franklin, J. F., & Lindenmayer, D. B. (2009). Importance of matrix habitats in maintaining biological diversity. *Proceedings of the National Academy of Sciences, 106*, 349–350.

Fuller, T. K., Berg, W. E., Radde, G. L., Lenarz, M. S., & Joselyn, G. B. (1992). A history and current estimate of wolf distribution and numbers in Minnesota. *Wildlife Society Bulletin, 20*, 42–55.

Fuller, J. L., Foster, D. R., Motzkin, G., McLachlan, J., & Barry, S. (2004). Broad-scale forest response to land-use and climate change. In D. R. Foster & J. D. Aber (Eds.), *Forests in time: The environmental consequences of 1, 000 years of change in New England* (pp. 101–124). New Haven, CT: Yale University Press.

Gehring, T. M., & Potter, B. A. (2005). Wolf habitat analysis in Michigan: An example of the need for proactive land management for carnivore species. *Wildlife Society Bulletin, 33*, 1237–1244.

Gompper, M. E. (2002). Top carnivores in the suburbs? Ecological and conservation issues raised by colonization of north-eastern North America by coyotes. *BioScience, 52*, 185–190.

Goodwin, G. C. (1936). Big game animals in the northeastern United States. *Journal of Mammalogy, 17*, 48–50.

Grant, M. (1902). *The caribou*. Bronx, NY: Office of the New York Zoological Society.

Groves, C. R. (2003). *Drafting a conservation blueprint: A practitioner's guide to planning for biodiversity*. Washington, DC: Island Press.

Hargis, C. D., Bissonette, J. A., & Turner, D. L. (1999). The influence of forest fragmentation and landscape pattern on American martens. *Journal of Applied Ecology, 36*, 157–172.

Harrison, D. J., & Chapin, T. G. (1998). Extent and connectivity of habitat for wolves in eastern North America. *Wildlife Society Bulletin, 26*, 767–775.

Hicks, A. (1986). The history and current status of moose in New York. *Alces, 22*, 245–252.

Hilty, J. A., Lidicker, W. Z., & Merenlender, A. M. (2006). *Corridor ecology: The science and practice of linking landscapes for biodiversity conservation*. Washington, DC: Island Press.

Homyack, J. A., Vashon, J. H., Libby, C., Lindquist, E. L., Loch, S., McAlpine, D. F., et al. (2008). Canada lynx-bobcat (*Lynx canadensis* x L. *rufus*) hybrids at the southern periphery of lynx range in Maine, Minnesota and New Brunswick. *American Midland Naturalist, 159*, 504–508.

Hoving, C. L., Harrison, D. J., Krohn, W. B., Joseph, R. A., & O'Brien, M. (2005). Broad-scale predictors of Canada lynx occurrence in eastern North America. *Journal of Wildlife Management, 69*, 739–751.

Hoving, C. L., Joseph, R. A., & Krohn, W. B. (2003). Recent and historical distributions of Canada lynx in Maine and the Northeast. *Northeastern Naturalist, 10*, 363–382.

Hristienko, H., & McDonald, J. E. (2007). Going into the 21st century: A perspective on trends and controversies in the management of the American black bear. *Ursus, 18*, 72–88.

Hummel, M., & Ray, J. C. (2008). *Caribou and the North: A shared future*. Toronto, ON: Dundurn Press.

Jenkins, J. C. (2004). *The Adirondack atlas: A geographic portrait of the Adirondack Park*. Syracuse, NY: Syracuse University Press.

Karns, P. D. (1997). Population distribution, density and trends. In A. W. Franzmann & C. C. Schwartz (Eds.), *Ecology and management of the North American moose* (pp. 125–139). Washington, DC: Wildlife Management Institute.

Kays, R. W., & Daniels, R. A. (2009). Fish and wildlife communities of the Adirondacks. In W. F. Porter, J. D. Erickson, & R. S. Whaley (Eds.), *The great experiment in conservation: Voices from the Adirondack Park* (pp. 71–86). Syracuse, NY: Syracuse University Press.

Kays, R. W., Gompper, M. E., & Ray, J. C. (2008). Landscape ecology of eastern coyotes based on large-scale estimates of abundance. *Ecological Applications, 18*, 1014–1027.

Kelsall, J. P. (1984). *Status report on woodland caribou*. Ottawa, ON: Committee on the Status of Endangered Wildlife in Canada.

Kerley, G. I. H., Pressey, R. L., Cowling, R. M., Boshoff, A. F., & Sims-Castley, R. (2003). Options for the conservation of large and medium-sized mammals in the Cape Floristic Region hotspot, South Africa. *Biological Conservation, 112*, 169–190.

Kretser, H. E., Sullivan, P. J., & Knuth, B. A. (2008). Housing density as an indicator of spatial patterns of reported human–wildlife interactions in Northern New York. *Landscape and Urban Planning, 84*, 282–292.

Krohn, W. B. (2005). *Manly Hardy (1832–1910): The life and writing of a Maine fur-buyer, hunter, and naturalist*. Orono, ME: Maine Folklife Center, University of Maine.

Krohn, W., Hoving, C., Harrison, D., Phillips, D., & Frost, H. (2004). *Martes* footloading and snowfall distribution in eastern North America: Implications to broad-scale distributions and interactions of mesocarnivores. In D. Harrison, A. Fuller, & G. Proulx (Eds.), *Martens and fishers (Martes) in human altered environments: An international perspective* (pp. 115–131). New York: Springer.

Kurta, A., Schwartz, M. K., & Anderson, C. R., Jr. (2007). Does a population of cougars exist in Michigan? *American Midland Naturalist, 158*, 467–471.

Kyle, C. J., Johnson, A. R., Patterson, B. R., Wilson, P. J., Shami, K., Grewal, S. K., et al. (2006). Genetic nature of eastern wolves: Past, present and future. *Conservation Genetics, 7*, 273–287.

Laliberte, A. S., & Ripple, W. J. (2004). Range contractions of North American carnivores and ungulates. *BioScience, 54*, 123–138.

Landres, P. B., Morgan, P., & Swanson, F. J. (1999). Overview of the use of natural variability concepts in managing ecological systems. *Ecological Applications, 9*, 1179–1188.

Larivière, S., & Crête, M. (1992). *Causes et conséquences de la colonisation du Québec par le coyote (Canis latrans)*. Québec, QC: Ministère du Loisir, de la Chasse et de la Pêche, Direction de la gestion des espèces et des habitats.

Lariviere, S., & Walton, L. R. (1997). *Lynx rufus. Mammalian Species, 563*, 1–8.

Laurian, C., Dussault, C., Ouellet, J. P., Courtois, R., Poulin, M., & Breton, L. (2008). Behavior of moose relative to a road network. *Journal of Wildlife Management, 72*, 1550–1557.

Leonard, J. A., & Wayne, R. K. (2008). Native Great Lakes wolves were not restored. *Biology Letters, 4*, 95–98.

Lindenmayer, D. B., Manning, A. D., Smith, P. L., Possingham, H. P., Fischer, J., Oliver, I., et al. (2002). The focal-species approach and landscape restoration: A critique. *Conservation Biology, 16*, 338–345.

Lindstedt, S. L., Miller, B. J., & Buskirk, S. W. (1986). Home range, time, and body size in mammals. *Ecology, 67*, 413–418.

Litvaitis, J. A. (2001). Importance of early successional habitats to mammals in eastern forests. *Wildlife Society Bulletin, 29*, 466–473.

Litvaitis, J. A. (2003). Are pre-Columbian conditions relevant baselines for managed forests in the northeastern United States? *Forest Ecology and Management, 185*, 113–126.

Litvaitis, J. A., & Harrison, D. J. (1989). Bobcat-coyote niche relationships during a period of coyote population increase. *Canadian Journal of Zoology, 67*, 1180–1188.

Litvaitis, J. A., Tash, J. P., & Stevens, C. L. (2006). The rise and fall of bobcat populations in New Hampshire: Relevance of historical harvests to understanding current patterns of abundance and distribution. *Biological Conservation, 128*, 517–528.

Logan, K. A., & Sweanor, L. L. (2001). *Desert puma: Evolutionary ecology and conservation of an enduring carnivore*. Washington, DC: Island Press.

Lohr, C., & Ballard, W. B. (1996). Historical occurrence of Wolves, *Canis lupus*, in the maritime provinces. *Canadian Field-Naturalist, 110*, 607–610.

MacKay, P., Zielinski, W. J., Long, R. A., & Ray, J. C. (2008). Noninvasive research and carnivore conservation. In R. Long, P. MacKay, W. J. Zielinski, & J. C. Ray (Eds.), *Noninvasive survey methods for carnivores* (pp. 1–7). Washington, DC: Island Press.

Margules, C. R., & Pressey, R. L. (2000). Systematic conservation planning. *Nature, 405*, 243–253.

Mattson, D. J. (1990). Human impacts on bear habitat use. In L. M. Darling & W. R. Archibald (Eds.), *Bears – their biology and management: Proceedings of the eighth international conference on bear research and management, Victoria, BC, 20–25 February 1989* (pp. 33–56). Washington, DC: International Association for Bear Research and Management.

McCabe, R. E., & McCabe, T. R. (1984). Of slings and arrows: An historical retrospection. In L. K. Halls (Ed.), *White-tailed deer: Ecology and management* (pp. 19–72). Harrisburg, PA: Stackpole.

McCord, C. M. (1974). Selection of winter habitat by bobcats (*Lynx rufus*) on the Quabbin Reservation, Massachusetts. *Journal of Mammalogy, 55*, 428–437.

McKelvey, K. S., Aubry, K. B., & Ortega, Y. K. (2000). History and distribution of lynx in the contiguous United States. In L. F. Ruggiero, K. B. Aubry, S. W. Buskirk, G. M. Koehler, C. J. Krebs, K. S. McKelvey, et al. (Eds.), *Ecology and conservation of lynx in the United States* (pp. 207–264). Boulder, CO: University Press of Colorado.

McKinney, M. L. (2006). Urbanization as a major cause of biotic homogenization. *Biological Conservation, 127*, 247–260.

Miller, K. V., Muller, L. I., & Demarais, S. (2003). White-tailed deer (*Odocoileus hemionus*). In G. A. Feldhamer, B. C. Thompson, & J. A. Chapman (Eds.), *Wild mammals of North America: Biology, management, and conservation* (pp. 906–930). Baltimore, MD: The Johns Hopkins University Press.

Mladenoff, D. J., Sickley, T. A., Haight, R. G., & Wydeven, A. P. (1995). A regional landscape analysis and prediction of favorable gray wolf habitat in the northern Great-Lakes region. *Conservation Biology, 9*, 279–294.

Morrison, J. C., Sechrest, W., Dinerstein, E., Wilcove, D. S., & Lamoreux, J. F. (2007). Persistence of large mammal faunas as indicators of global human impacts. *Journal of Mammalogy, 88*, 1363–1380.

Mosnier, A., Boisjoly, D., Courtois, R., & Ouellet, J. P. (2008). Extensive predator space use can limit the efficacy of a control program. *Journal of Wildlife Management, 72*, 483–491.

Mosnier, A., Ouellet, J. P., Sirois, L., & Fournier, N. (2003). Habitat selection and home-range dynamics of the Gaspe caribou: A hierarchical analysis. *Canadian Journal of Zoology, 81*, 1174–1184.

Motzkin, G., & Foster, D. R. (2004). Insights for ecology and conservation. In D. R. Foster & J. Aber (Eds.), *Forests in time: The environmental consequences of 1,000 years of change in New England* (pp. 367–379). New Haven, CT: Yale University Press.

Murray, D. L., Steury, T. D., & Roth, J. D. (2008). Assessment of Canada lynx research and conservation needs in the southern range: Another kick at the cat. *Journal of Wildlife Management, 72*, 1463–1472.

Musiani, M., & Paquet, P. C. (2004). The practices of wolf persecution, protection, and restoration in Canada and the United States. *BioScience, 54*, 50–60.

Nielsen, C. K., Dowling, M., Miller, K., & Wilson, B. (2006). The Cougar Network: Using science to assess the status of cougars in eastern North America. In H. J. McGinnis, J. W. Tischendorf, & S. J. Ropski (Eds.), *Proceedings of the eastern cougar conference 2004* (pp. 82–86). Morgantown, WV. American Ecological Research Institute, Fort Collins, Colorado.

Noss, R. F., Carroll, C., Vance-Borland, K., & Wuerthner, G. (2002). A multicriteria assessment of the irreplaceability and vulnerability of sites in the Greater Yellowstone Ecosystem. *Conservation Biology, 16*, 895–908.

Noss, R. F., & Daly, K. M. (2006). Incorporating connectivity into broad-scale conservation planning. In K. R. Crooks & M. Sanjayan (Eds.), *Connectivity conservation* (pp. 587–619). Cambridge: Cambridge University Press.

NYDEC (New York State Department of Environmental Conservation). (2009). Moose fact sheet. New York State Department of Environmental Conservation. Retrieved December 5, 2009, from http://www.dec.ny.gov/animals/6964.html.

O'Grady, J. J., Reed, D. H., Brook, B. W., & Frankham, R. (2004). What are the best correlates of predicted extinction risk? *Biological Conservation, 118*, 513–520.

Palmer, R. S. (1938). Late records of caribou in Maine. *Journal of Mammalogy, 19*, 37–43.

Paquet, P. C., & Carbyn, L. N. (2003). Gray wolf (*Canis lupus* and allies). In G. A. Feldhamer, B. C. Thompson, & J. A. Chapman (Eds.), *Wild mammals of North America: Biology, management, and conservation* (pp. 482–510). Baltimore, MD: Johns Hopkins University Press.

Paquet, P. C., Strittholt, J. R., & Staus, N. L. (1999). *Wolf reintroduction feasibility in the Adirondack Park.* Corvallis, OR: Conservation Biology Institute.

Parker, G. (1995). *Eastern coyote: Story of its success.* Halifax, NS: Nimbus Publishing.

Parker, G. (1998). *The eastern panther: Mystery cat of the Appalachians.* Halifax, NS: Nimbus Books.

Parker, G. (2001). *Status report on the Canada lynx in Nova Scotia.* Halifax, NS: Nova Scotia Department of Natural Resources.

Parker, G. (2004). *Men of the autumn woods: Non-resident big game hunting in New Brunswick.* Sackville, NB: Gerry Parker.

Parker, G. R., Maxwell, J. W., Morton, L. D., & Smith, G. E. J. (1983). The ecology of the lynx (*Lynx canadensis*) on Cape Breton Island. *Canadian Journal of Zoology, 61*, 770–786.

Pelton, M. R., Coley, A. B., Eason, T. H., Doan Martinez, D. L., Pederson, J. A., van Manen, F. T., et al. (1999). American black bear conservation action plan. In C. Servheen, S. Herrero, & B. Peyton (Eds.), *Bears: Status survey and conservation action plan* (pp. 144–146). Gland, Switzerland: IUCN/SSC Bear and Polar Bear Specialist Groups.

Pierce, B. M., & Bleich, V. C. (2003). Mountain lion (*Puma concolor*). In G. A. Feldhamer, B. C. Thompson, & J. A. Chapman (Eds.), *Wild mammals of North America: Biology, management, and conservation* (pp. 744–757). Baltimore, MD: Johns Hopkins University Press.

Pressey, R. L., Cabeza, M., Watts, M. E., Cowling, R. M., & Wilson, K. A. (2007). Conservation planning in a changing world. *Trends in Ecology and Evolution, 22*, 583–592.

Radford, J. Q., Bennett, A. F., & Cheers, G. J. (2005). Landscape-level thresholds of habitat cover for woodland-dependent birds. *Biological Conservation, 124*, 317–337.

Ray, J. C. (2005a). Large carnivorous animals as tools for conserving biodiversity: Assumptions and uncertainties. In J. C. Ray, K. H. Redford, R. S. Steneck, & J. Berger (Eds.), *Large carnivores and the conservation of biodiversity* (pp. 34–56). Washington, DC: Island Press.

Ray, J. C. (2005b). Sprawl or highly mobile or wide-ranging species. In E. A. Johnson & M. W. Klemens (Eds.), *Nature in fragments: Urban sprawl's effects on biodiversity* (pp. 181–205). New York: Columbia University Press.

Ray, J. C., Organ, J. F., & O'Brien, M. S. (2002). *Canada lynx (Lynx canadensis) in the Northern Appalachians: Current knowledge, research priorities, and a call for regional cooperation and action.* Toronto, ON: Wildlife Conservation Society.

Ray, J. C., Redford, K. H., Berger, J., & Steneck, R. S. (2005). Is large carnivore conservation equivalent to biodiversity conservation and how can we achieve both? In J. C. Ray, K. H. Redford, R. S. Steneck, & J. Berger (Eds.), *Large carnivores and the conservation of biodiversity* (pp. 400–427). Washington, DC: Island Press.

Reed, D. H., O'Grady, J. J., Brook, B. W., Ballou, J. D., & Frankham, R. (2003). Estimates of minimum viable population sizes for vertebrates and factors influencing those estimates. *Biological Conservation, 113*, 23–34.

Richens, V. B., & Hugie, R. D. (1974). Distribution, taxonomic status, and characteristics of coyotes in Maine. *Journal of Wildlife Management, 38*, 447–454.

Ritchie, E. G., & Johnson, C. N. (2009). Predator interactions, mesopredator release and biodiversity conservation. *Ecology Letters, 12*, 982–998.

Rodenhouse, N. L., Christenson, L. M., Parry, D., & Green, L. E. (2009). Climate change effects on native fauna of northeastern forests. *Canadian Journal of Forest Research, 39*, 249–263.

Rutledge, L. Y., Bos, K. I., Pearce, R. J., & White, B. N. (2009). Genetic and morphometric analysis of sixteenth century *Canis* skull fragments: Implications for historic eastern and gray wolf distribution in North America. *Conservation Genetics, 2009*, 1–9.

Sala, O. E., Chapin, F. S., Armesto, J. J., Berlow, E., Bloomfield, J., Dirzo, R., et al. (2000). Biodiversity: Global biodiversity scenarios for the year 2100. *Science, 287*, 1770–1774.

Sanderson, E. W. (2006). How many animals do we want to save? The many ways of setting population target levels for conservation. *BioScience, 56*, 911–922.

Schmiegelow, F. K. A. (2007). Corridors, connectivity, and biological conservation. In D. B. Lindemayer & R. J. Hobbs (Eds.), *Managing and designing landscapes for conservation: Moving from perspectives to principles* (pp. 251–262). Oxford: Blackwell.

Schmiegelow, F. K. A., Cumming, S. G., Lisgo, K. A., Leroux, S. J., Anderson, L. G., & Krawchuk, M. (2008). *A science-based framework for identifying system-level benchmarks in boreal regions of Canada. Canadian BEACONs Project Report #5*. Edmonton, AB: The Canadian BEACONs Project, University of Alberta.

Scott, F. W., & Hebda, A. J. (2004). Annotated list of the mammals of Nova Scotia. *Proceedings of the Nova Scotia Institute of Science, 42*, 189–208.

Seddon, P. J., Armstrong, D. P., & Maloney, R. F. (2007). Developing the science of reintroduction biology. *Conservation Biology, 21*, 303–312.

Slough, B. G. (2007). Status of the wolverine *Gulo gulo* in Canada. *Wildlife Biology, 13*, 76–82.

Stocek, R. F. (1995). The cougar, *Felis concolor*, in the Maritime provinces. *Canadian Field-Naturalist, 109*, 19–22.

Tear, T. H., Kareiva, P., Angermeier, P. L., Comer, P., Czech, B., Kautz, R., et al. (2005). How much is enough? The recurrent problem of setting measurable objectives in conservation. *BioScience, 55*, 835–849.

Telfer, E. S., & Kelsall, J. P. (1984). Adaptation of some large North American mammals for survival in snow. *Ecology, 65*, 1828–1834.

Terrie, P. G. (1993). *Wildlife and wilderness: A history of Adirondack mammals*. Fleischmanns, NY: Purple Mountain Press.

The Cougar Network. (2009). *The Cougar network: Using science to understand cougar ecology*. Retrieved December 5, 2009, from http://www.easterncougarnet.org/.

Trollope, S. T., White, J. G., & Cooke, R. (2009). The response of ground and bark foraging insectivorous birds across an urban-forest gradient. *Landscape and Urban Planning, 93*, 142–150.

Trombulak, S. C., Anderson, M. G., Baldwin, R. F., Beazley, K., Ray, J. C., Reining, C., et al. (2008). *The Northern Appalachian/Acadian ecoregion: Priority locations for conservation action* (Special Report 1). Warner, NH: Two Countries, One Forest.

USFWS (U.S. Fish and Wildlife Service). (2000). Endangered and threatened wildlife and plants; determination of threatened status for the contiguous U.S. distinct population segment of the Canada lynx and related rule. *Federal Register, 65*, 16052–16085.

Vecellio, G. M., Deblinger, R. D., & Cardoza, J. E. (1993). Status and management of moose in Massachusetts. *Alces, 29*, 1–7.

Vermont Department of Fish and Wildlife. (2009). *Wildlife programs: Hunting opportunities: Moose*. Retrieved on December 4, 2009, from http://www.vtfishandwildlife.com/moose_hunt_opps.cfm.

Villemure, M., & Jolicoeur, H. (2004). First confirmed occurrence of a Wolf, *Canis lupus*, south of the St. Lawrence River in over 100 years. *Canadian Field-Naturalist, 118*, 608–610.

Voigt, D. R., & Berg, W. E. (1987). Coyote. In M. Novak, J. A. Baker, M. E. Obbard, & B. Malloch (Eds.), *Wild furbearer management and conservation in North America* (pp. 345–357). Toronto, ON: The Ontario Trappers Association and Ontario Ministry of Natural Resources.

Warren, R. J. (1997). The challenge of deer overabundance in the 21st century. *Wildlife Society Bulletin, 25*, 213–214.

Weaver, J. L. (1999). *Lynx survey in the Adirondack Park*. Bronx, NY: Wildlife Conservation Society.

Weaver, J. L., Paquet, P. C., & Ruggiero, L. F. (1996). Resilience and conservation of large carnivores in the Rocky Mountains. *Conservation Biology, 10*, 964–976.

Whitney, G. G. (1994). *From coastal wilderness to fruited plain: A history of environmental change in temperate North America 1500 to the present*. Cambridge: Cambridge University Press.

Williamson, D. F. (2002). *In the black: Status, management, and trade of the American black bear (Ursus americanus) in North America*. Washington, DC: TRAFFIC North America, World Wildlife Fund.

Wilson, K. A., Carwardine, J., & Possingham, H. P. (2009). Setting conservation priorities. *Annals of the New York Academy of Sciences, 1162*, 237–264.

Wilson, P. J., Grewal, S., McFadden, T., Chambers, R. C., & White, B. N. (2003). Mitochondrial DNA extracted from eastern North American wolves killed in the 1800s is not of gray wolf origin. *Canadian Journal of Zoology, 81*, 936–940.

Wilson, K., Pressey, R. L., Newton, A., Burgman, M., Possingham, H., & Weston, C. (2005). Measuring and incorporating vulnerability into conservation planning. *Environmental Management, 35*, 527–543.

With, K. A., & Crist, T. O. (1995). Critical thresholds in species' responses to landscape structure. *Ecology, 76*, 2446–2459.

Woolmer, G., Trombulak, S. C., Ray, J. C., Doran, P. J., Anderson, M. G., Baldwin, R. F., et al. (2008). Rescaling the Human Footprint: A tool for conservation planning at an ecoregional scale. *Landscape and Urban Planning, 87*, 42–53.

Wydeven, A. P., Fuller, T. K., Weber, W., & MacDonald, K. (1998). The potential for wolf recovery in the northeastern United States via dispersal from southeastern Canada. *Wildlife Society Bulletin, 26*, 776–784.

Wydeven, A. P., Mladenoff, D. J., Sickley, T. A., Kohn, B. E., Thiel, R. P., & Hansen, J. L. (2001). Road density as a factor in habitat selection by wolves and other carnivores in the Great Lakes Region. *Endangered Species Update, 18*, 110–114.

Young, S. P., & Goldman, E. A. (1946). *The puma: Mysterious American cat*. New York: Dover.

Chapter 10
Protecting Natural Resources on Private Lands: The Role of Collaboration in Land-Use Planning

Jessica Spelke Jansujwicz and Aram J.K. Calhoun

Abstract Private lands are important for managing biological diversity, but tensions between a landowner's perceived property rights and conservation interests make landscape-scale conservation a challenge. To reconcile this conflict, there is a growing trend toward more inclusive, collaborative efforts to involve multiple stakeholders in land-use policy decisions. In theory, a collaborative approach is a logical framework for decision-making and action, and the benefits of collaboration are touted in the academic literature and popular press. This strategy is not without critics, however, and the merits of collaboration are at the center of debate. This chapter reviews the rhetorical and theoretical debate over collaboration; identifies the limitations of past and current approaches to measure the success of collaboration in practice; and applies a performance evaluation framework to investigate and link the process and outputs of a multi-stakeholder, conservation planning process in Maine to social and environmental outcomes. While this analysis focuses on the Vernal Pool Working Group, a state-initiated and led collaborative planning process, it offers noteworthy lessons about the possibilities and limits of using collaboration as a tool to manage natural resources on private lands. By offering an example of progressive collaborative conservation, this chapter illustrates the central role collaborative communication can play in shaping the character of local-level planning efforts and, by extension, planning at larger spatial scales.

Keywords Collaboration · Conservation planning · Private lands · Property rights · Vernal pools

---

J.S. Jansujwicz (✉)
University of Maine, 5755 Nutting Hall, Room 232A, Orono, ME 04469-5755, USA
e-mail: jessica.jansujwicz@maine.edu

A.J.K. Calhoun
Associate Professor of Wetlands Ecology, University of Maine, 5755 Nutting Hall, Room 222, Orono, ME 04469-5775, USA
e-mail: calhoun@maine.edu

S.C. Trombulak and R.F. Baldwin (eds.), *Landscape-scale Conservation Planning*,
DOI 10.1007/978-90-481-9575-6_10, © Springer Science+Business Media B.V. 2010

10.1 Introduction

Managing natural resources for the common good is a complex issue, particularly when achieving conservation goals requires management of private lands. Over 60% of the land in the United States is privately owned (USDA 2002), making private lands an essential component of any comprehensive natural resource management strategy. Yet, while private lands are important for managing biological diversity, tensions between a landowner's perceived property rights and conservation interests make landscape-scale conservation a challenge. Landowners are often reluctant to cooperate in resource management strategies that may incur a personal cost, lower the value of their land, or impose restrictions on land use. Many also resent the layers of regulation affecting their property, questioning the personal benefits of protecting or even identifying individual species or natural habitats on their land.

Government restrictions designed to protect wildlife and other significant natural resources on private land are often controversial. Whether land-use restrictions interfere with individual private property rights to an extent requiring compensation to the property owner has been litigated frequently in both federal and state courts (Bean and Rowland 1997; Dwyer et al. 1995; Shogren 1998). As the spiraling number of so-called 'takings' lawsuits suggests, citizen resistance to environmental regulations has significant political implications (Jansujwicz 1999). An expanded regulatory takings doctrine that redefines when a government action requires landowner compensation may effectively chill the predisposition and ability of environmental managers to implement environmental regulations (Wise 2004). Environmental managers may shy away from controversy, avoiding stringent enforcement in cases that may later be subject to intense scrutiny by the courts. Government reluctance to enforce strict regulatory limits such as is embodied in the Endangered Species Act impedes the protection of significant natural resources on private lands. This inevitably begs the question, 'Can private property and conservation coexist?' (Freyfogle 2003). Because agency mandates to protect natural resources often clash with property-rights interests, environmental regulators will continue to face the difficult task of designing resource management strategies that effectively balance property rights and economic development with environmental and natural resource protection in a manner acceptable to state legislatures and their constituents.

To reconcile these differences, there is a growing trend toward more inclusive, collaborative efforts to involve multiple stakeholders in land-use policy decisions. Called many things – public-private partnerships, collaborative conservation planning, cooperative ecosystem management, consensus decision making, and alternative dispute resolution models – these new approaches to multi-stakeholder participation in environmental decision-making are emerging in hundreds of communities across the country as citizens, environmentalists, business leaders, and public officials are meeting face-to-face to work through their differences, resolve conflicts, and design new strategies to address resource-related issues (Chap. 4).

10 Protecting Natural Resources on Private Lands

Today, the rhetoric of collaboration is commonplace and multi-stakeholder planning processes are an important cornerstone for a rapidly increasing number of federal, state, and local natural resource and environmental programs addressing wetlands, wildlife, endangered species, water quality, and other watershed management concerns (Carr et al. 1998; EPA 1996, 1998; USDA and U.S. DOC 2000; U.S. GAO 2008; Wondolleck and Yaffee 2000). Increasingly, the term 'collaboration' is used to represent a broad array of strategies from collaborative engagement processes and informal organizations, to more formalized partnerships or super-agencies (e.g., CALFED Bay-Delta Program, a collaborative effort of 25 state and federal agencies with management or regulatory responsibilities for the San Francisco Bay-Delta system) (Sabatier et al. 2005). Under the umbrella of collaborative resource management, for example, are interagency task forces and work groups as well as many examples of local initiatives that involve the community planning process, including habitat conservation planning (Noss et al. 1997; Thomas 2001, 2003), watershed partnerships (Born and Genskow 1999; Kenney et al. 2000), community-based forestry (Carr et al. 1998; Danks 2008), and citizen-science programs (Calhoun and Reilly 2008).

Collaborative planning processes can be government-driven ('top-down') or citizen-initiated ('bottom-up'), but all share common organizing principles and theoretical underpinnings. By encouraging stakeholder participation early in the planning process, advocates claim that collaboration can temper the confrontational politics of conventional regulatory approaches and overcome inefficiencies inherent in traditional models of environmental governance, thereby offering an alternative strategy to achieve a widening array of government-mandated environmental objectives (Beierle and Cayford 2002; Busenberg 1999; Kemmis 1990; Susskind and Cruikshank 1987; Wondolleck and Yaffee 2000). While many tout the benefits of collaborative processes, others raise important concerns of accountability and legitimacy (McCloskey 1996; Moote 2008; Weber 2003; Wondolleck and Yaffee 2000), representation (McCloskey 2004–2005; Weber 2003), and scientific credibility (Coglianese 1999; Weber 2003).

Over the past 2 decades, the debate over the merits of collaboration has been largely rhetorical and theoretical, and little empirical evidence suggests whether collaboration has positive or negative impacts on the environment (Layzer 2008; Thomas 2008), the community, government officials, and future policy decisions. For the most part, existing research on collaboration has focused on process (e.g., Kenney et al. 2000; Leach and Pelkey 2001; Wondolleck and Yaffee 2000), policy outputs (e.g., Koontz 2005), and more recently on social outcomes (e.g., Sabatier et al. 2005), but very little is known about environmental outcomes (Koontz and Thomas 2006; Thomas 2008). Moreover, few empirical studies link the process and outputs of collaboration with both social and environmental outcomes (Mandarano 2008). Such evaluation is necessary to support collaborative theory or validate critical claims.

This chapter has four main objectives. First, we discuss principles of collaboration, specifically focusing on how the structure and process of collaboration differs from more traditional decision-making processes. In this section, we review the literature on collaboration particularly with respect to key concepts and organizing principles characteristic of a multi-stakeholder, consensus-driven approach.

Second, we review the rhetorical and theoretical debate on collaboration to answer questions such as: What are the driving forces behind this movement toward collaboration and partnerships? What are some of the benefits and pitfalls of using a collaborative approach? In our discussion of the theoretical underpinnings of collaboration, we address both the expected outcomes and critical concerns of collaboration as they relate to conservation planning at any number of spatial scales and geographic regions.

Third, we assess the limitations of past and current approaches used to measure the success of collaboration in practice. We follow this assessment with a practical application of a performance evaluation framework to investigate and link the process and outputs of a multi-stakeholder, collaborative planning process in Maine to social and environmental outcomes. In our analysis, we focus on a 10-year collaborative communication process – the Vernal Pool Working Group, a state-led collaborative planning initiative. While our chapter focuses on a case of vernal pool conservation planning, problems associated with natural resource conservation on private land transcends vernal pools and also relates to conservation planning at any number of spatial scales and geographic regions. Given the nature of vernal pool habitat (small and difficult to map, ephemeral, dependent on wetland and upland components, and widely distributed), we believe it is an important focal topic because conservation of this resource will be as challenging as any, and the results will be widely applicable to other natural resource protection issues (Hunter 2008).

We then conclude the chapter with lessons learned on the barriers and opportunities for using collaboration as a planning tool for protecting natural resources on private lands. Our goal in offering an example of progressive collaborative effort at conservation planning for vernal pools is to illustrate the central role collaborative communication can play in shaping the character of local-level planning efforts and, by extension, planning at larger spatial scales.

10.2 Traditional and Collaborative Planning in the United States

Traditional models of environmental governance (now commonly referred to as 'command-and-control') are characterized by a 'top-down' hierarchical structure, emphasizing rules and regulations promulgated and enforced from above. Authority is centralized with the federal government delegating responsibility to specialized agencies, states, and local governments. Within this fragmented system of government, resource management agencies (at least prior to the 1990s) rarely cooperated with one another or with other agencies (Thomas 2003). Each agency carried out public functions following different missions, cultures, and 'standard operating procedures.'

Traditional governance systems tend to be reactive, often evolving in response to public outcry and concern. They focus on remedial rather than preventive actions (Meiners and Yandle 1993). Environmental laws, policies, and programs are compartmentalized to address a specific medium – air, land, or water. Decision-making is technocratic or expert-driven and public involvement is encouraged or allowed only at certain entry points in the policy process as permitted by formal administrative procedures. For example, public laws including the National Environmental Policy Act (NEPA; 1969), the National Forest Management Act (1976), the Federal Advisory Committee Act (FACA), PL 92463 (1972), the Freedom of Information Act, 5 U.S.C.A. §552 (1966), and the Administrative Procedures Act (APA), 5 U.S.C.A. §501 et seq. (1946) ensure public access to agency records and decision-making processes for public land management.

In contrast to the technocratic model of environmental governance, the collaborative partnership model emphasizes a consensus-based decision-making process. Authority and responsibility is decentralized and shared horizontally among agencies, organizations, and individuals with a direct stake in the outcome. Collaboration infers shared power, and ideally all participants in a collaborative partnership have a high degree of freedom over the process and influence over decision-making. Collaborative partnerships encourage voluntary, face-to-face information exchange and problem solving in which multiple stakeholders can voice opinions in a consensus-driven decision-making process (Conley and Moote 2003). Some degree of public interaction is encouraged from the onset and not necessarily restricted to certain entry points as defined by formal administrative procedures. Rather than pursue narrow objectives such as water quality or habitat restoration, partnership objectives tend to be more broad-based, and collaborative initiatives often pursue more than one resource-related issue at a time. Collaborative partnerships are often formed proactively, organizing before an issue reaches a critical turning point.

Ranging along a continuum of formality, collaborative partnerships and planning processes vary considerably along several dimensions distinguished by the legal framework or form of agreement, by the specific issues they face, and by the character of its membership. The varying role of government in partnerships (e.g., leader, facilitator [through grants or non-regulatory incentives], or follower) may influence the structure and process of collaborative partnerships (Koontz et al. 2004). For example, the government's role may affect the way issues are defined, the resources available for collaboration, and the organizational processes that are established (Koontz et al. 2004). Thus, collaboration can be either 'top-down' – and often initiated in response to impending legislation – or 'bottom-up' partnerships originating and sustained at the grassroots or community level.

10.3 Rhetorical and Theoretical Benefits and Limits of Collaboration

In theory, a collaborative approach is a logical framework for decision-making and action, and the benefits of this inclusive approach are touted in the academic literature and popular press. For the most part, those who write about collaboration tend

10.3.1 Expected Outcomes

Much of the impetus for a collaborative approach is attributed to perceived shortcomings of traditional models of environmental governance. Collaboration is offered as a better way to address issues of diffuse pollution sources and overlapping jurisdictions and to resolve environmental disputes on private lands.

Diffuse Pollution Sources The traditional regulatory model of environmental governance is credited with many successes. By setting tough regulatory standards and procedures, federal statutes including the Clean Air Act, 42 U.S.C. § 7401 et seq. (1970) and the Clean Water Act, 33 U.S.C. § et seq. (1972) significantly curbed the emission of hazardous substances into the environment. As a result, surface waters are cleaner today than at the onset of the modern environmental movement (Council on Environmental Quality 1997; Mazmanian and Kraft 1999). While technocratic, regulatory fixes worked well for point-source pollution, non-point source pollution (e.g., agriculture runoff) proved more challenging to control under a regulatory approach. Despite recent improvements in environmental quality over the past 3 decades, reliance on traditional 'command-and-control' regulation is not sufficient to achieve government-mandated environmental objectives (Chertow and Esty 1997; John 1994; Mazmanian and Kraft 1999), particularly where private lands are concerned.

Overlapping Jurisdictions Overlaying the ecological landscape is a political, legal, and administrative landscape. Natural resources do not conform to these arbitrary political boundaries (Thomas 2003). Wildlife species often use multiple habitats to meet their life-history needs, and wetlands and other ecological systems are rarely confined within the boundaries of a single jurisdiction or ownership. In the U.S., the landscape is further complicated by a system of government that is fragmented among specialized agencies with different missions, culture, and methods of operation and by a series of environmental laws that tend to be limited in purpose, focusing on a single species, patch of habitat, or medium (air, land, or water). Current policies and programs are often criticized for being costly to administer and enforce (Meiners and Yandle 1993), and in many instances, regulations are inconsistent and difficult to enforce across administrative boundaries.

Because species and ecosystems transcend human-imposed boundaries, jurisdictional and habitat fragmentation necessitates both interagency cooperation (Thomas 2003) and the involvement of private interests in conservation planning decisions. When management units are defined ecologically rather than politically, greater coordination among local landowners and between private landowners and natural resource management agencies is required (Cortner and Moote 1999). This partnership idea is a cornerstone principle of 'ecosystem management' (Cortner and Moote 1999; Grumbine 1994; Kernohan and Haufler 1999; Noss and Cooperrider 1994; Norse 1993). Under the rubric of ecosystem management, collaborative partnerships grow from the involvement of all those affected in the decision-making process. In theory then, by partnering with various levels of government and the private sector, collaboration can facilitate greater coordination among stakeholders, offering a diversity of expertise and financial assistance not available in a single agency or organization (Endicott 1993; Chap. 4).

Conflict Resolution Participatory strategies are expected to temper the confrontational politics that typify environmental policy decisions (Beierle and Cayford 2002; Busenberg 1999; Kemmis 1990). Government regulation of private property for environmental purposes is politically unpopular, and emotionally charged debates between conservation and development interests have been common when wildlife and wetlands are involved (Bean and Rowland 1997; Freyfogle 2003; Meltz et al. 1999; Noss et al. 1997; Shogren 1998). Often developers and landowners find traditional regulatory models intrusive, cumbersome, adversarial, and in some instances, insufficient to address economic concerns (Ceplo 1995). They argue that environmental laws create uncertainty in planning, imposing costly delays on development projects (e.g., Marceau 2009; Pierce Atwood LLP 2006). They are also concerned that layers of regulation will lower the value of their land, raise the costs of operation, or impose restrictions on the use of their land.

Manifestations of property rights interests have a long history in the U.S. reaching back first to the Sagebrush Rebellion of the late 1970s and later the Wise Use Movement and the County Rights Movement of the 1980s and 1990s, respectively. These movements took place in the western states and were based largely on claims that federal resource management agencies were applying rules and regulations to landowners' operations in ways that made their properties less profitable (Wise 2004). Over the last 2 decades, heightened tensions between a landowner's perceived property rights (especially in terms of potential economic gains) and the legislative mandates of federal, state, and local agencies has galvanized the property-rights movement (Jansujwicz 1999), and property rights claims are increasingly being played out on a case-by-case basis in federal and state courts across the nation. The standard objection raised by property-rights advocates is that regulation 'takes' private land without compensation in violation of the Fifth Amendment to the U.S. Constitution. In the mid-1990s, as these interests began to question or resist land-use regulations, a reinvigorated property-rights movement gained increased momentum and visibility. After the 1994 congressional elections, a surging wave of anti-government, pro-property rights rhetoric swept the nation and dozens of grassroots groups became organized in opposition to the power of government to regulate private property for environmental or other purposes without compensation (Jansujwicz 1999).

Today, property-rights advocates continue to exert considerable political pressure, resulting in a regulatory climate where government often lacks the political will to impose strict regulations. Thus, while private lands harbor valuable habitat for flora and fauna and perform numerous environmental services, access, data collection, and relationships with landowners impede the protection of significant natural resources on private lands (Hilty and Merenlender 2003). To reconcile the increasing number of conservation-development conflicts on private lands, government agencies responsible for managing natural resources are embracing collaborative communication processes.

In theory, by involving the affected community throughout the planning process, adversarial decision-making is avoided, local citizens become invested in the process, and better environmental outcomes result (Sabel et al. 2000; Susskind and Cruikshank 1987; Wakeman 1997; Wondolleck and Yaffee 2000). Theory suggests that collaborative approaches are more likely to achieve program objectives because participants work together to identify mutually acceptable goals (Susskind and Cruikshank 1987). Collaboration is perceived as 'a process through which parties that see different aspects of a problem can constructively explore their differences and search for solutions that go beyond their own limited vision of what is possible' (Gray 1989). Wondolleck and Yaffee (2000), for example, cite a case in California (Quincy Library Group) where environmentalists and loggers were able to draw on their common interests, fears, and perceptions to craft a joint vision statement in a process that encouraged communication between disparate interests. In this case, theory holds that participants were more likely to accept the outcomes of a process that they perceived as fair and legitimate. Moreover, as Innes and Booher (1999) found based on their empirical research and practice in a wide range of consensus building cases, social learning during a consensus building process changes a participant's understanding of their own interests, leading them to conclude that consensus building can work more effectively than confrontational tactics.

In an idealized narrative, collaboration with stakeholders builds trust, support, and local capacity by fostering a sense of place, responsibility, and commitment (Brick et al. 2001; Wondolleck and Yaffee 2000). By involving the affected community throughout the planning process, adversarial decision-making is avoided, local interests become invested in the process and better environmental outcomes result (Sabel et al. 2000; Susskind and Cruikshank 1987; Wakeman 1997; Wondolleck and Yaffee 2000; Chap. 3).

Stakeholder participation provides a foundation for the development of social capital (that is, social networks and the associated norms of reciprocity) (Coleman 1988; Pretty and Smith 2004; Putnam 1995, 2001; Putnam et al. 1993), leading to more resilient decisions (Sabatier et al. 2005; Salamon et al. 1998). For example, in a review of international agriculture and rural conservation programs, Pretty and Smith (2004) found that stronger bonds within and between groups lead to more positive outcomes for both biological diversity and human livelihoods. In this example, bringing together farmers to deliberate on how to make changes to food production systems fostered new social relations and created new stores of social capital, which in turn helped sustain change. Not surprisingly, Pretty and Smith (2004) found that where social capital was high, new ideas spread more rapidly. Locally led cooperative planning also creates new social capital that supports further planning (Salamon et al. 1998). A study of local advisory groups (or task forces) participating in the Ohio Farmland Preservation Planning Program suggests that collaborative communication processes provide a useful first step in building community capacity to address future land-use issues (Koontz 2005). By engaging local communities, collaborative processes can generate innovative solutions tailored to local conditions (Landy et al. 1999).

Although many studies point to the benefits of collaboration, such a strategy is not always appropriate, and critics have raised important concerns of accountability, legitimacy, representation, and scientific credibility.

10.3.2 Critical Concerns

Accountability and Legitimacy Many fear that an arrangement involving multiple stakeholders in an open collaborative process slows decision-making (Coglianese 1999) and reduces accountability (Wondolleck and Yaffee 2000). Because management is horizontal under the collaborative paradigm, unclear lines of authority and responsibility result, and critics are particularly concerned about this devolution of agency power. They argue that it is not fair, legitimate, or wise to devolve the authority invested in federal agencies by Congress to implement laws and regulations to an unelected and perhaps unrepresentative collaborative group (McCloskey 1996; Wondolleck and Yaffee 2000). Moreover, the structure of collaboration often makes it difficult to determine whether partnerships remain accountable to the interests they serve (McCloskey 1996; Moote 2008; Weber 2003) or whether policy outcomes of collaboration serve few at the expense of many (Weber 2003). For example, agencies and interest groups that delegate decision-making authority to stakeholder partnerships need to know whether priorities established at the national or regional level are upheld locally. In many instances, collaborative exercises are designed to address local concerns and not the interests of the broader public.

Representation Critics also argue that collaborative initiatives lack adequate representation (Weber 2003). They suggest that stakeholders with the best access to current information tend to dominate collaborative exercises, and often few participants are members of the general public and unaffiliated, undermining any claim that these forums have some larger civic importance (McCloskey 2004–2005). Concerns over inequities in power and resources between members of a consensus group align with the principles of communication theory. This theory recognizes that communication practices are infused with power (Martin 2007), and these existing power relations may undermine meaningful citizen participation in collaborative efforts (e.g., Moote 2008). Recent communications research has questioned whether collaborative communication processes privilege the objectives of entities that already hold the decision-making power or serve the interests of dominant actors in the larger socio-political context in which they are embedded (Martin 2007). This raises concerns about whether a collaborative process is easily captured by interest groups with economic and political power (Katz and Miller 1996).

Scientific Credibility Opponents of collaboration stress that the outcomes of collaboration may lack scientific credibility (Coglianese 1999; Weber 2003). They argue that because consensus is the primary mechanism for reaching decisions, any agreements, plans, or policies chosen risk representing the decision causing the least controversy, and this may not necessarily be the one that is best for the resource (Coglianese 1999). Critics argue that the most intractable disputes are 'sidestepped' and others 'glossed over' with 'broad language acceptable to all sides' (Coglianese 1999). In an effort to attain consensus, 'extreme' views may be excluded or marginalized, more contentious issues ignored or avoided, and solutions imposing costs on participating stakeholders with the most power may not be considered (Beierle and Cayford 2002; Coglianese 1999; Peterson et al. 2002).

These critical concerns highlight the growing importance of empirical analysis. Because collaborative planning processes represent a new management tool with uncertain success (and because defining and measuring 'success' is difficult and often problematic), it is important to proceed with caution. Empirically derived evidence must be generated to support, refute, or elaborate on critic's claims. Such evaluation is necessary both to guide future efforts and policies and to identify variables associated with success.

10.4 Evaluating Collaboration in Practice

In practice, success is frequently assessed using two criteria: (1) evaluation of process and (2) a measure of outcome. For the first criterion, researchers identify the factors that contribute to or impede the success of collaborative partnerships. This assumes that the quality of a process influences the effectiveness of collaborative planning (Margerum 2002) and that several process factors can positively influence the chances of success (Gray 1989; Wondolleck and Yaffee 2000). In general, a 'quality' process meets certain criteria, including sufficient representation, effective leadership and facilitation, an efficient organizational structure (e.g., well-managed meetings), committed, knowledgeable participants, and the use of the best science available. In addition to these criteria, a quality process is also measured by determining whether the effort builds future capacity.

The second measure of success is based on outcomes: Do collaborative efforts achieve on-the-ground objectives? Do they result in a measurable improvement of the resource? This criterion is measured by a number of outcomes including the adoption and implementation of plans, projects, or policies, a measurable change in the resource (e.g., restored wetlands, improved water quality), or a change in land use or in local-level planning processes.

In theory, where process criteria are met and where the process is perceived as fair, legitimate, and transparent, better outputs and outcomes result. Outcomes of collaborative planning are directly related to the strength or weakness of the process, which affect long-term implementation (Margerum 2002). While it may not be possible for a process to fully meet all the criteria, failure to meet any one of them hinders the effectiveness of the process and the quality of its outcomes (Innes and Booher 1999).

10.4.1 Process Evaluation

Since the late 1980s, collaborative scholars have developed a set of principles and criteria against which collaborative efforts can be evaluated (e.g., Born and Genskow 1999; Coughlin et al. 1999; Gray 1989; Innes and Booher 1999; Kenney et al. 2000; Leach et al. 2002; Margerum and Born 1995; Moote et al. 1997; Susskind and Cruikshank 1987; Yaffee et al. 1996). From these studies we now know a great deal about the process of collaboration and can readily refer to a long list of ingredients, including both member factors and organizational factors that are recommended for success. However, while these studies offer important insight on the collaborative process, they offer comparatively little about whether a representative and well-structured process leads to better policy decisions and social and environmental outcomes.

10.4.2 Outcome Effects

Defining outcomes is often problematic. For one thing, the literature on collaboration does not clearly distinguish between outputs and outcomes. Thomas (2008), for example, finds that in some instances, studies that claim to measure environmental outcomes actually use outputs as proxies for outcomes. Without a clear definition of outputs and outcomes, the line between them is blurred. Outputs are the plans, projects, and other tangible items generated by collaborative planning efforts (Koontz and Thomas 2006). These are products that can be easily pointed to and recognized, including a set of agreements generated by the collaborative planning process (Margerum 2002). Agreements may be formal (e.g., final plans, policy statements, legislation, and new regulations) or informal proposals for voters or public officials to consider. Outcomes are defined as 'the effects of outputs on environmental and social conditions' (Koontz and Thomas 2006). Innes and Booher (1999) identify both tangible and intangible products as outcomes of collaboration. In their definition, tangible products include formal agreements such as plans, policies, legislation, and new regulations. Aligning with Margerum (2002), however, we consider agreements as outputs and choose to look beyond the plans to determine outcomes. To define outcomes, we use Innes and Booher's (1999) definition of 'second and third order effects' or 'activities triggered by the consensus building process,' including 'spin-off' partnerships (consensus building groups set up to work on implementation), collaborative projects, and innovations (e.g., strategies, actions, and new ideas). Environmental outcomes can be described as tangible outcomes (e.g., improved water quality, changed land management practices), and social outcomes are best described as intangible outcomes (e.g., increased trust, new relationships, or knowledge gained by participants). Intangible outcomes are often thought of as 'social, intellectual, and political capital' (Gruber 1994). Again, social capital refers to the social networks and the associated norms of reciprocity (Coleman 1988; Putnam 1995, 2001; Putnam et al. 1993). Intellectual capital includes mutual understanding of each others' shared interests, shared definitions of the problem, and agreement on data, models, projections, or other quantitative or scientific descriptions of the issue (Innes and Booher 1999). Political capital is defined as the ability to work together outside the consensus-building process to influence public action in ways they were unable to when acting individually (Innes and Booher 1999).

Once defined, significant methodological constraints also impede evaluation of environmental outcomes. Evaluations require assessments over a long time frame, and sampling methods amenable to statistical evaluations require large sample sizes of comparable entities. Identification of causal links between management activities and ecological trends are often difficult to make (Conley and Moote 2003; Thomas 2008). Moreover, because in many cases the only readily accessible data regarding partnership initiatives are provided by the members through newsletters, websites, videos, and presentations or through surveys completed by the very same participants, an underlying bias may result in an overly optimistic assessment of the effort's progress (Kenney 2000). Collaborative partnerships also compete for grant funds and other sources of financial support, and this provides an incentive to exaggerate the positive attributes of the effort, while downplaying the negative. While consideration of active participants is valid and even necessary, the research challenge is to balance insights of that population with other sources of information and analysis (Kenney 2000).

Given the significant methodological constraints, it is not surprising that most of the literature on collaboration has focused on process (e.g., organizational and membership factors). With the exception of social outcomes (e.g., Sabatier et al. 2005), little empirical research links collaborative outputs with environmental outcomes (Koontz and Thomas 2006), and few studies assess the long-term effects of collaboration on the development and implementation of natural resources policy. To fill this gap in knowledge, researchers are slowly shifting their focus, moving beyond a process-oriented approach to include in their analyses consideration of outcomes, including environmental outcomes (Layzer 2008). Mandarano (2008), for example, evaluates the process, outputs, and long-term effects of a specific collaborative planning effort, the Habitat Workgroup of the New York-New Jersey Harbor Estuary Program. Using a set of performance criteria, Mandarano (2008) described observed changes in social and environmental conditions and the apparent linkages between the Habitat Workgroup's process and outputs. In another study, Koontz (2005) used a multiple-case analysis of county-level, community-based task forces working on farmland preservation in Ohio to examine the impact of stakeholder participation on policymaking at the local level. While the quality of the process remains important, these studies go a step further to link the quality of the process with the quality of outputs and social and environmental outcomes. In the next section, we follow the lead of these investigators and use a performance evaluation framework to assess the process, outputs, and outcomes of a collaborative vernal pool conservation planning process in Maine.

10.5 Collaborative Management in Practice: The Vernal Pool Working Group

Using a case study of vernal pool conservation planning in Maine, we examine the role of collaboration and evaluate whether consensus-based decision making was a more efficient and effective way to meet regulatory objectives and ensure the long-term viability of the State's vernal pool resources. In the following sections we also discuss how proactive, multi-stakeholder decision-making processes can be integrated with traditional planning strategies. For example, we investigate whether engaging stakeholders in an open dialogue about vernal pool conservation and management ultimately led to better policy outcomes and greater 'buy-in' than a sole reliance on traditional forms of environmental governance and formal administrative procedures. By linking theory to empirical data, we also hope to identify the barriers and opportunities for using collaboration as a planning tool to manage natural resources on private lands.

In the following sections, we review the ecology and regulatory context for vernal pool conservation planning at the state and local level in Maine. These sections provide an overview of the origin and organization of the Vernal Pool Working Group (VPWG) and then apply criteria integrated from the various published performance evaluation frameworks to evaluate the process, outputs, and social and environmental outcomes of VPWG deliberations. The process and outcomes described below can serve as a template for approaching any conservation issue that requires management of resources on private lands. The framework offered here may be applied to the management of any natural resources on private lands that, due to their transboundary nature, require action by multiple stakeholders at the local and higher level.

10.5.1 *Ecology and Management of Vernal Pools in Maine*

Vernal pools in Northeastern North America are ephemeral to semi-permanent wetlands that obtain maximum depths in spring or fall and lack permanent surface water connections with other wetlands or water bodies. Pools typically fill with snowmelt or runoff in the spring, although some may be fed primarily by groundwater sources and may begin to refill in the fall. Pools are generally less than 0.4 ha, with the extent and type of vegetation varying widely. They provide optimal breeding habitat for animals adapted to temporary, fishless waters including, but not limited to mole salamanders (*Ambystoma* spp.), wood frog (*Rana sylvatica*),

Eastern spadefoot toad (*Scaphiopus holbrookii*), and fairy shrimp (*Eubranchipus* spp.) (Calhoun and deMaynadier 2008; Colburn 2004; Semlitsch and Skelly 2008). In addition, vernal pools provide foraging and resting habitat for many state-listed species in the Northeastern U.S. In Maine, these include spotted turtle (*Clemmys guttata*), wood turtle (*C. insculpta*), Blanding's turtle (*Emydoidea blandingii*), and ringed boghaunter dragonfly (*Williamsonia lintneri*).

While vernal pools are unique ecosystems that perform important functions at the landscape scale (Hunter 2008), protecting pools is a challenge for natural resource managers because they are small, ephemeral wetlands that are difficult to remotely identify. Furthermore, animals that breed in vernal pools require additional, adjacent terrestrial habitat for migrating, dispersing, foraging, and hibernation (Faccio 2003; Semlitsch 2002; Semlitsch and Skelly 2008). At the state level in the U.S., a number of approaches protect wetland resources (ELI 2005) and currently 15 states have their own comprehensive wetland regulatory programs (Mahaney and Klemens 2008). Within the Northeastern United States, Maine currently has the strongest vernal pool protections, designating a subset of ecologically outstanding vernal pools as 'significant wildlife habitat' under the Natural Resources Protection Act (NRPA), which provides for the regulation of wetlands and other important natural resources (38 M. R. S. A. §§ 480-A to 480-Z).

Although a subset of exemplary pools were designated as 'significant wildlife habitat' by the State in 1995, the requirement that these Significant Vernal Pools (SVP's) be defined and mapped by the Maine Department of Inland Fisheries and Wildlife (MDIFW) before they could be regulated was never acted on due to lack of agency resources. After 10 years of work by stakeholders, in April 2006, Maine adopted a definition for identifying SVP's (Significant Wildlife Habitat Rules, Chapter 335, Section 9 under NRPA) based on the abundance and presence of vernal pool indicator species – fairy shrimp, wood frogs, and blue-spotted (*Ambystoma laterale*) and spotted salamanders (*A. maculatum*) – or use by state-listed threatened or endangered species. An SVP includes the adjacent terrestrial habitat within a 76-m radius around the pool from the high-water mark. New regulatory protections became effective on September 1, 2007. While still short of the 159–290-m conservation zone recommended as essential for the long-term survival of pool-breeding amphibian populations in human-dominated landscapes (Calhoun and Klemens 2002; Semlitsch 1998), the enactment of this legislation marked a positive step toward protecting vernal pool resources. By extending the area of terrestrial habitat that is regulated around SVP's and by removing the requirement that vernal pools needed to be 'mapped' to be 'identified,' Maine established the most comprehensive and stringent measures for protecting vernal pools in Northeastern North America (Mahaney and Klemens 2008).

Maine's role in proactive management of vernal pools evolved slowly, taking more than 10 years to address the regulatory gaps for their protection. This protracted decision-making process highlights the confusing array of factors that can influence the pace at which institutional change occurs. In the following sections we discuss the evolving process of vernal pool conservation planning in Maine, specifically focusing on the origin and activities of the VPWG.

10.5.2 Origins of the Vernal Pool Working Group

Historically vernal pools did not receive much attention except on a case-by-case basis by government agencies charged with protecting wetland resources. This often resulted in conflicting signals from regulatory agencies weighing in on the same proposed project. For example, in the mid-1990s, a number of projects in the mid-coast area of Maine passed through the Maine Department of Environmental Protection (DEP) screening and were significantly delayed by review at the federal level (Army Corps of Engineers, ACOE). Whatever the reason (e.g., concerns from EPA or U.S. Fish and Wildlife Service, [USFWS], or even neighbors), incidents such as these highlighted the overlapping and often confusing regulatory process governing activities affecting vernal pools.

Perhaps as a response to these or similar incidents, Maine legislators began hearing much discontent from their constituents about the lack of coordination between federal and state wetland regulations. Prior to the revised 1995 state legislation that streamlined the permitting process, applicants would have to apply for permits from both federal and state agencies, each with differing requirements. In response to this, the state legislature passed a Legislative Resolve in 1993 that set up a Wetlands Task Force to recommend changes to the state wetland program and charged the DEP and the Maine State Planning Office (SPO) to oversee this process. The SPO also received EPA funding to produce a Wetland Conservation Plan for the State (Maine State Planning Office 2001). The Wetlands Task Force set up a number of working groups to address wetland conservation issues, including regulation, assessment, inventory, and mitigation. The VPWG had many of the same members as the Assessment Work Group under the broader Wetlands Task Force but was specifically formed to address the vernal pool issues that were never adequately addressed in the 1995 legislation. Under the 1995 NRPA, Significant Vernal Pool rules were added as a placeholder, and the VPWG was charged with implementing the changes to the legislation.

10.5.3 Process

Chronologically, the VPWG can be divided into two different processes. An earlier process (1995–2003) convened by SPO shortly after adoption of the 1995 legislation and a later process (2004–2006) reconvened by DEP. In the earlier process, VPWG participants met regularly at the SPO in Augusta, Maine. Employees of SPO facilitated the meetings, took and distributed minutes, and coordinated and disseminated materials for review prior to meetings. SPO's role in facilitation ended in 2002 when the lead facilitator left public service. A vacuum in leadership followed the departure of SPO as facilitator, and momentum was lost. The VPWG remained without direction until a representative of DEP reconvened the group in 2004. Despite changes in leadership, however, membership and process elements remained fairly consistent over time.

The VPWG included key stakeholders from federal, state, private, academic, and non-profit NGO's each contributing expertise in science, forestry, outreach, natural resource planning, and regulation. Stakeholders, many of whom were also members of the larger Wetlands Task Force, included MDIFW, Maine Forest Service (MFS), DEP, SPO, Maine Audubon Society, Maine Natural Areas Program (MNAP), University of Maine, and private environmental consultants. Although primarily a state-driven work group, federal agency representatives also attended meetings. The U.S. Army Corps of Engineers attended meetings as regulator, and the U.S. Fish and Wildlife Service, while not a formal partner, attended occasionally meetings to share their perspective on vernal pool issues.

Typically, the VPWG met several times a year, but the frequency of meetings varied from year to year depending on the issues and tasks at hand. Not every meeting was fully attended (and even when the table was full, not all members contributed to the discussion). Those absent had the opportunity to contribute through electronic mail. Membership of key interests remained consistent over time, although the group expanded as participating agencies brought in additional representatives with specific expertise to address emerging issues. While most decisions were made by the larger policy group, an ad-hoc technical group met to address issues, concerns, and topics identified by the broader group. In addition, while division leaders did not always 'sit' at the table, they remained actively involved in the policy decisions of the group.

The process was largely a state-driven interagency committee charged with a specific objective and was not a stakeholder process or broad collaboration. The process consisted largely of internal meetings of biologically based and oriented stakeholders and did not explicitly include public participation 'at the table.' Efforts were made to represent these interests by proxy of the invited stakeholders, and each stakeholder had input from his or her constituents throughout the process.

The VPWG had no formal mission statement. All members, however, had a basic understanding of their objective: to come to terms on the science of vernal pools and to discuss mechanisms to fulfill the legislative mandate designed to protect them. An agenda was loosely followed and decisions were made by an informal consensus rather than formal voting procedures. All members of the VPWG were considered equals and opportunities to contribute were given to all stakeholders at the table.

10.5.4 Outputs

Outputs can be divided into two general categories: (1) principal outputs that emerged as a result of face-to-face deliberations between VPWG members ('at the table') and directly addressed the mission to implement the NRPA and (2) ancillary outputs that were accomplished in tandem with these efforts but addressed non-regulatory concerns (e.g., public education, outreach, and local stewardship). We use the term ancillary to describe activities occurring outside of the VPWG's stated mission to fulfill the legislative mandate to define vernal pools and determine significance. While we distinguish between these outputs, the two approaches were not mutually exclusive. We acknowledge that non-regulatory approaches fostered public acceptance of vernal pool resources and protection mechanisms, thereby adding significant support to the mission of the VPWG.

Principle Outputs First, a scientific foundation was laid for developing a conservation policy based on the best available science. Research gaps noted by the VPWG developed into research projects for University of Maine graduate students, often partially funded and overseen by MDIFW and University of Maine faculty. During this time, five master's and five doctoral students produced data on life history needs of pool-breeding amphibians, two state-listed species of turtles dependent upon pools, and on amphibian responses to forestry practices (e.g., Baldwin et al. 2006a, b; Joyal et al. 2001; Lichko and Calhoun 2003; Oscarson and Calhoun 2007; Patrick et al. 2007; Vasconcelos and Calhoun 2004, 2006).

Second, definitions of vernal pools and Significant Vernal Pools were developed. The VPWG worked for 10 years to develop a definition of a vernal pool and the criteria for designating a subset of ecologically outstanding SVP's. The following definition was finally accepted by the State of Maine in April 2006, with new regulatory protections becoming affective on September 1, 2007:

> *A vernal pool, also referred to as a seasonal forest pool, is a natural, temporary to semi-permanent body of water occurring in a shallow depression that typically fills during the spring or fall and may dry during the summer. Vernal pools have no permanent inlet or outlet and no viable populations of predatory fish. A vernal pool may provide the primary breeding habitat for wood frogs* (Rana sylvatica), *spotted salamanders* (Ambystoma maculatum), *blue-spotted salamanders* (Ambystoma laterale), *and fairy shrimp* (Eubranchipus spp.), *as well as valuable habitat for other plants and wildlife, including several rare, threatened, and endangered species. A vernal pool intentionally created for the purposes of compensatory mitigation is included in this definition (Significant Wildlife Habitat Rules, Chapter 335 Section 9 under NRPA).*

SVP's were defined based on research results that described the range of egg mass numbers found in Maine vernal pools from a citizen-science program (VIP program discussed below). Ranges of egg mass numbers for each breeding amphibian were calculated, and the definition of SVP's was based on the intent of DEP that no more than half of the identified pools would potentially be regulated in the future (representing a political and biological compromise). Hence, significance was based on egg mass abundances to meet this criterion and the presence of state-listed threatened and endangered species.

Ancillary Outputs These were seen primarily in the numerous documents and citizen-science programs that were developed. Three representatives of the VPWG – Maine Audubon Society, MDIFW, and the University of Maine – designed and implemented projects to address the education, public outreach, and research gaps identified by the VPWG. Using the best available information on vernal pool ecology, including vernal pool manuals produced by other New England states, Maine Audubon Society produced *The Maine Citizen's Guide to Identifying and Documenting Vernal Pools* in 1999, with a second edition in 2003 (Calhoun 2003). Two more manuals, *Best Development Practices: Conserving Pool-Breeding Amphibians in*

Residential and Commercial Developments (Calhoun and Klemens 2002) and *Forestry Habitat Management Guidelines for Conserving Vernal Pool Wildlife* (Calhoun and deMaynadier 2004) were developed through a multi-year stakeholder process and published to promote voluntary protections. These documents targeted two practices likely to directly impact vernal pools and the adjacent terrestrial habitat: forestry and development. Dozens of workshops were given around the State to introduce the concepts developed in these voluntary guidelines to key stakeholders, including the industrial and small-woodlot forest communities and private landowners.

Maine Audubon Society developed a citizen volunteer program, the Very Important Pool (VIP) program, to inventory vernal pools statewide using the previously mentioned *The Maine Citizen's Guide to Identifying and Documenting Vernal Pools* as a training tool. This outreach program was initiated in 1999 and ran for 5 years to collect data on pool-breeding amphibians and their reproductive behavior in pools in Southern, Central, and Northern Maine (see Calhoun et al. 2003 for a summary). The goal of the VIP program was to raise the profile of vernal pools through statewide citizen participation, to engage the news media to help introduce vernal pool ecology and the importance of these small wetlands to the public, and to gather baseline inventory and assessment data on vernal pools in Maine that could help the VPWG understand the resource statewide and craft a definition of vernal pools and SVP's.

10.5.5 Environmental and Social Outcomes

While even a process without any agreement may be a success if participants have learned about the problem, about each other's interests, and about what may be possible (Innes and Booher 1999), an emphasis on both environmental and social outcomes requires looking beyond the process to assess the implementation of VPWG outputs. Several specific principal and ancillary outcomes can be identified as having emerged from the VPWG process:

Principal Outcomes First, deliberations surrounding the new legislation raised the visibility of vernal pools, creating increased interest in federal agencies, the State legislature, and the general public. Regulatory agencies (ACOE and DEP) requested training workshops for upper-level enforcement personnel on vernal pool identification and ecology. Personnel were requested to be enthusiastic when relaying information about vernal pool values and services to the public. Even though the regulation represents a political compromise (and hence not completely grounded in the best-available science), it has fostered discussions on vernal pool conservation at all political levels, most markedly, at the local level where science-based policies have greater potential to be implemented (Klemens 2000; Preisser et al. 2000).

While it may be difficult to precisely measure how the implementation of the new vernal pool rules affects habitat conditions and, by extension, populations of pool-breeding amphibians, the VPWG has stimulated an interest in these ecosystems. Growing interest, knowledge, and concern for vernal pools continues to motivate academic research and to build new partnerships. As a result, support for graduate student research has continued at the University of Maine, and a new multidisciplinary team is currently designing a research program focused on the social, ecological, and economic aspects of vernal pools (www.umaine.edu/sustainability solutions).

Second, as mentioned above, the VPWG directly or indirectly contributed to an evolving literature on vernal pools. In addition as a result of the VIP program, 120 trained citizen scientists collected amphibian breeding data on 97 'adopted' pools over 5 years. A description and the results of this initial study, as well as recommendations for advancing vernal pool conservation in New England, are described by Calhoun et al. (2003).

Third, relationships among stakeholders were strengthened. Perhaps one of the best illustrations of the benefits of personal, long-term relationships cultivated by the VPWG is the Significant Vernal Pool legislation drafted by the MDEP, MDIFW, Maine Audubon Society, the University of Maine and others. Agreement on the substance and wording of vernal pool and significant vernal pool definitions was not easy. The definition of vernal pools required compromises from both biologists and regulators, reflecting science tempered by political and practical exigencies. It had to incorporate language that was clear to lay people, supported previous legislative efforts, addressed stakeholder concerns, and practical for enforcement. For example, in the regulatory definition of vernal pool, anthropogenic breeding habitats (e.g., gravel pits, roadside ditches, and farm ponds) were excluded in the definition to avoid public concern that 'every mud puddle' would be regulated. Also, the wording (emphasis added) that a vernal pool '...*typically* fills during the spring or fall and *may* dry during the summer...' provided for a more flexible hydrologic regime. And lastly, vernal pools 'intentionally created for the purposes of compensatory mitigation' were added to the definition so as not to undermine past mitigation practices.

Clearly, the eventual adoption of these definitions would not have been possible without the significant stock of social capital (trust, relationships), intellectual capital (mutual understanding, agreed upon data), and political capital (ability to work together for agreed ends) created by the deliberative planning process. In the process, stakeholders had to appreciate varying views and learn to consider the potential impacts of the proposed legislation on interests other than their own. For example, scientists had to consider the increased burden of the proposed regulations on regulators (e.g., increased workforce), while regulators needed to appreciate the ecological implications of weakening the definition.

In the case of the VPWG, relationships among federal, state, local, and private interests were strengthened, and collaborations created during the process persist today. For example, relationships forged among the University of Maine and environmental consultants during the early stages of the VPWG process resulted in later collaborations such as the vernal pool town mapping projects that shared funds, technology, and expertise. Because of the strength of this partnering, the University gained access to mapping technology that would not otherwise have been available. In investing in new technology, the consulting firm expanded their business, created productive relationships with the University and local towns, and improved the accuracy of potential vernal pool maps. Towns have begun directly contacting the consulting firm for help in custom designing projects to meet their town's needs.

Finally, the VPWG accomplished an incredible 'coup.' While most participants acknowledge that the new vernal pools rules fall short of adequately protecting vernal pool resources (in terms of species requirements), 10 years of deliberation resulted in the strongest vernal pool mechanisms in the country.

Ancillary Outcomes First, the process led to the adoption of Best Development Practices (BDP's) (Calhoun and Klemens 2002) and Forestry Habitat Management Guidelines (Calhoun and deMaynadier 2004) by key resource managers. The New England District of the ACOE issues State Programmatic General Permits (PGP's) that expedite review of minimal impact work in wetlands within each New England state. To date, the Vermont and New Hampshire PGP's use the standards set forth in Calhoun and Klemens (2002) for evaluating impacts to vernal pools. ACOE also expects to incorporate language from the BDP's into their permit review process. Similarly, the USFWS in New England uses the BDP's as a standard when reviewing impacts to ecologically significant vernal pools that may not be regulated by the State. The Forestry Habitat Management Guidelines were embraced by the Maine Forest Service and Bureau of Public Lands and by a number of private commercial forestry companies. These guidelines must be followed in order to receive 'green certification' from the Sustainable Forestry Initiative or the Forest Stewardship Council.

Second, initiatives for mapping vernal pools by towns were accelerated. Fourteen towns in Maine have or are in the process of doing town-wide mapping and assessment projects in collaboration with Maine Audubon Society and the University of Maine. One town is considering an ordinance that provides stricter regulations for vernal pools than the State model. Justifications for this are based on the Town's mission to base town natural resources policy on the best-available science. Research based on gaps identified by the VPWG provided this scientific foundation. In 2008 and 2009, Maine Audubon Society received grants to provide seed money for seven towns to use the most advanced technology to map and assess vernal pools in collaboration with the University of Maine and a private environmental consulting firm. SPO has also contributed funds to a University of Maine project to assess the economic cost of conservation on public lands using five of the 15 towns engaged in the vernal pool project.

Fourteen Maine towns are at some stage of completing vernal pool mapping projects. Many more 'potential vernal pools' have been mapped but assessments are limited by a typically less than 50% rate of permission for access by private landowners. However, the towns still retain a map of potential vernal pools in their databases to help in permit review and natural resource planning exercises.

Finally, public attitudes, values, and behavior related to vernal pools have changed. In the case of the VPWG, these changes were an outcome of the development of personal, hands-on experience with vernal pools by local citizens.

10 Protecting Natural Resources on Private Lands

In evaluating vernal pool citizen-scientist programs, for example, Oscarson and Calhoun (2007) found that as a result of volunteering as citizen scientists, 40% of 30 survey respondents had become more active by attending conservation commission meetings, joining committees, and bringing more knowledge to commissions and land trusts. The majority of respondents indicated that they had increased awareness and concern for the impacts of development in their town. Ninety-four percent of the volunteers shared knowledge about the importance of conserving vernal pools with friends, family, and co-workers.

10.6 Lessons Learned

Collaboration represents a promising communication tool for managing transboundary natural resources in a way that links actions at the local level to landscape-scale conservation goals. But collaboration should not be considered a panacea (Koontz and Thomas 2006). Caution should be used in accepting overly optimistic views of partnership accomplishments advocated in the academic literature and popular press. By using the performance evaluation framework presented here, both the theoretical principles (or assumptions) supporting collaboration as well as the critical concerns can be evaluated and the ways in which multi-stakeholder collaborative communication processes can work alongside traditional forms of environmental governance can be better understood.

While we focus on collaborative conservation of vernal pools primarily at the state level, this case study illuminates the barriers and opportunities of using a collaborative strategy for other natural resources such as a listed species or timber management at various spatial scales. In our case, the substance and process of the VPWG offers noteworthy lessons about the possibilities and limits of collaborative communication processes.

First, collaboration coordinates activities, promoting more efficient use of limited human and financial resources. The VPWG brought together the capabilities and expertise of multiple stakeholders (and their associations) that otherwise may not have been united to work on issues of common concern. As our example of a statewide, vernal pool conservation initiative suggests, collaboration can support the sharing of financial and technical resources, stretching already tight agency and municipal budgets. Collaboration among agencies, private companies, municipalities, and academia can produce a prolific amount of research to support conservation strategies, including the new legislation, and improve access by town planners to state-of-the-art technology.

Collaborative vernal pool conservation planning has had other effects as well. Because vernal pools are difficult to remotely identify and are ubiquitous across the landscape, agencies with regulatory authority over vernal pools simply cannot be aware of every vernal pool and every project potentially affecting them. Federal agencies and state agencies often regulate the entire state from one (ACOE) or three (DEP) regional field offices. The ACOE has a Maine Project Office in Manchester and often weighs in on projects they consider important, yet a large area of the State is perhaps not regulated as thoroughly as it should be. Many pools go undetected and enforcement remains limited by personnel and financial constraints. In such areas, a municipal role will be critical. Mapping and assessment efforts at the town-level have the potential to proactively protect vernal pools that may have otherwise gone undetected by regulatory agencies. Thus, an important product of VPWG deliberation has been an increase in municipal awareness of vernal pools that has motivated a greater participation by local interests.

Second, collaboration is promoted by a shared sense of place or community, a focus on local problems and a common concern. While motivations (and willingness) varied, federal, state, local agencies, non-profit organizations, and consultants agreed to 'come to the table' to discuss their ideas and concerns regarding vernal pool protections. All participants were committed to using the best science available, and collaboration by VPWG members was motivated by a common concern: meet the State's mandate to protect vernal pools. Federal and state agencies participated to fulfill their mandate. MDIFW, for example, participated to ensure their input on policy decisions, as efforts of the VPWG would ultimately lead to guidelines for land-use regulation. SPO played a major role in shepherding the 1995 revisions to NRPA through the legislature, and they participated in the VPWG to develop a way forward and fulfill the legislative mandate to protect vernal pools.

Interests without legal requirements chose to participate for other reasons. Maine Audubon Society (2008), for example, participated to 'help put a little-known but all-important wildlife resource on the map,' and to protect essential breeding, feeding, and resting areas for a large number of species in Maine, such as blue-spotted salamander, Blanding's turtle, and eastern ribbon snake (*Thamnophis s. sauritus*). Consultants participated to find answers to questions they were confronted with in the field.

Third, collaboration allows for the representation of individuals and groups affected by the decision-making process. One of the tenants of collaboration is that individuals come to the table with varying levels of knowledge, skills, levels of power, and resources. While at times a power differential was present between members, all received equal representation at the table. However, while individuals with diverse backgrounds were represented at the table, the VPWG was, in fact, homogeneous in terms of interest. All participants expressed an interest in finding a way forward to protecting vernal pools, albeit at a different pace. In terms of inclusiveness and representation, certain interests were underrepresented by the VPWG. Most obviously, landowners, realtors, and developers were not directly involved. Their absence may be a legitimate concern because the revised SVP rules could potentially alter development plans on private lands. Private interests fear that the new rules will 'increase the time, expense, and uncertainty of all types of development projects that impact significant vernal pool habitat – from residential subdivisions to shopping centers to landfill expansions,' place the burden of identification and delineation of SVP habitat on the developer (a task previously assigned to MDIFW), and delay many projects until spring when developers could conclusively determine what permits would be required (Pierce Atwood LLP 2006).

However, the absence of landowners 'at the table' may have been appropriate in this case. While the absence of the regulated community was notable, the VPWG process was not designed to include public involvement. Rather, it was largely at the level of deliberation where professional and technical representatives used the best available science to define vernal pools and identify 'significance' criteria that would pass political and public scrutiny. In many ways, the VPWG needed to determine what would pass the 'straight face test' before getting input from the public. The regulatory mandate rested with the state, and it was not viewed as an appropriate venue to have individual landowners at the table, yet any decisions made would have to be palatable to state legislatures and their constituents.

Eventually, the process did proceed through a formal public review process (e.g., agency rule-making), allowing for citizen input. Thus, the VPWG provides an excellent example of the way in which multi-stakeholder collaboration can complement traditional administrative procedures. In this case, collaboration served as a mechanism to ensure 'buy in' by VPWG members and stakeholders with diverse interests and backgrounds. Because the VPWG was able to come to consensus on the definition of vernal pools as well as the criteria to determine a SVP, they represented a united front as the proposed rule went before the Maine state legislature.

Fourth, intangible outcomes such as building relationships, establishing trust, and sharing information are some of the most beneficial aspects of collaborative planning. Participants of the VPWG attribute successful outcomes to strong, personal relationships that developed during the process. Certainly, consensus on the language of the new vernal pool legislation and 'spin off partnerships' such as the vernal pool mapping and assessment project would not have been possible without the stock of social, political, and intellectual capital developed during deliberation. As a result of improved communication among stakeholders, vernal pools are now on the radar of regulatory agencies and the general public, more stringent regulations are in place, and towns are taking steps to identify and map their pools to allow for streamlined and proactive management. In addressing potential opportunities and barriers to collaborative planning, however, future empirical analysis must determine whether similar relationships develop between decision-makers and the public or whether decision-makers missed an opportunity to engage meaningfully with landowners and community members. Understanding how to secure landowner cooperation is particularly important because natural resource conservation increasingly depends on securing the cooperation of private-property owners in local communities (Peterson and Horton 1995; Peterson et al. 2002). By paying closer attention to the dynamics of stakeholder involvement and to issues of communication (e.g., Depoe et al. 2004), future research can assess how strategies used to engage landowners such as public workshops on the new Significant Wildlife Habitat rules, fact sheets from DEP, MDIFW, and Maine Audubon Society, and web-based resource materials (e.g., www. umaine. edu/vernalpools) affect information transfer, trust, and relationship building.

Fifth, diverse perspectives encourage a more broad-based understanding of the issues at stake, allowing for the design of more innovative solutions. The VPWG supports this conclusion. By supporting and disseminating ecological research and

Best Management Practices, the 10-year process encouraged social learning. Both the process and outputs of the collaborative effort improved stakeholders' understanding of vernal pool ecology and the challenges associated with conservation on private lands. The collaborative process created a feedback loop whereby knowledge gained through education and outreach programs influenced the ultimate vernal pool conservation strategy crafted by the VPWG. With regards to the design of innovative solutions, as of this writing a growing number of towns are involved with the vernal pool mapping and assessment project and citizen science program. Their involvement and commitment suggests that these communities are beginning to embrace alternative actions to protect natural resources, and particularly vernal pools on private property. Moreover, as more towns become engaged with the project, the original MDIFW goal of mapping and assessing pools can eventually be met through town initiatives founded upon town consensus. Conservation of pool-breeding amphibian habitat, like many conservation goals, is often most effective at the local level where neighbors, planners, and other concerned citizens play an active stewardship role (Klemens 2000; Preisser et al. 2000), and our example illustrated how local community engagement in collaborative processes can generate innovative solutions tailored to local conditions (Landy et al. 1999).

Sixth, it is clear that collaboration slows decision-making. Collaborative planning is oftentimes slow, difficult work, and the nearly 10-year process of the VPWG is no exception. For some participants, the collaborative process was painstakingly slow and frustrating with uncertain benefits. There are, however, plausible explanations to support the length of time required for a group such as VPWG to reach consensus: (1) the length of time required to design and implement natural resource management strategies is influenced by both the level of knowledge of the resource and by how controversial the regulation may be; (2) the possibility of more regulation on a seemingly ubiquitous and misunderstood resource was controversial; and (3) the many voices 'at the table' slowed decision-making, and interpersonal dynamics caused temporary stalemates.

Finally, one of the arguments against collaboration is that it results in the 'lowest common denominator solution' or the alternative supported by the most participants. The definition of vernal pools and the criteria for determining significance, although driven by science, were indeed a political compromise and did not completely reflect the best-available science. At the same time, however, VPWG members acknowledge that if the rule had been based on criteria better supported by science (e.g., minimum number of egg masses and width of buffer zones), more pools would have been captured, and the rules may not have passed muster with the state legislature and, as a consequence, vernal pools would have ended up with less regulation.

Collaborative planning is a slow, laborious process. It is often difficult, complicated, and challenging, and in general success requires time, patience, and perseverance (Diamant et al. 2003). Yet collaborative processes have the potential to achieve conservation goals associated with vernal pools and other ubiquitous natural resources at any number of spatial scales, from local to ecoregional. Thus, the challenges are well worth confronting.

10 Protecting Natural Resources on Private Lands

Acknowledgements We would like to thank members and stakeholders of the Vernal Pool Working Group for generously sharing their time and perspective on the process and outcomes of multi-stakeholder decision-making. We also thank Rob Lilieholm for his comments on earlier drafts. J. S. Jansujwicz gratefully acknowledges her advisors, Aram Calhoun and Rob Lilieholm; without their encouragement and continued support her research would not be possible. The School of Forest Resources, Department of Wildlife Ecology, and the Sustainability Solutions Initiative at the University of Maine also provided invaluable resources. This research was supported, in part, by National Science Foundation award EPS-0904155 to Maine EPSCoR at the University of Maine.

References

Baldwin, R. F., Calhoun, A. J. K., & deMaynadier, P. G. (2006a). The significance of hydroperiod and stand maturity for pool-breeding amphibians in forested landscapes. *Canadian Journal of Zoology, 84*, 1604–1615.

Baldwin, R. F., Calhoun, A. J. K., & deMaynadier, P. G. (2006b). Conservation planning for amphibian species with complex habitat requirements: A case study using movements and habitat selection of the wood frog (*Rana sylvatica*). *Journal of Herpetology, 40*, 442–454.

Bean, M. J., & Rowland, M. J. (1997). *The evolution of national wildlife law* (3rd ed.). Westport, CT: Praeger.

Beierle, T. C., & Cayford, J. (2002). *Democracy in practice: Public participation in environmental decisions*. Washington, DC: Resources for the Future.

Born, S. M., & Genskow, K. D. (1999). *Exploring the watershed approach: Critical dimensions of state-local partnerships. (Final report: the four corners watershed innovators initiative)*. Portland, OR: River Network.

Brick, P., Snow, D., & Van De Wetering, S. (Eds.). (2001). *Across the great divide: Explorations in collaborative conservation and the American West*. Washington, DC: Island Press.

Busenberg, G. J. (1999). Collaborative and adversarial analysis in environmental policy. *Policy Sciences, 32*, 1–11.

Calhoun, A. J. K. (2003). *Maine citizen's guide to locating and documenting vernal pools*. Falmouth, ME: Maine Audubon Society.

Calhoun, A. J. K., & deMaynadier, P. G. (2004). *Forestry habitat management guidelines for vernal pool wildlife (MCA Technical Paper No. 6)*. Bronx, NY: Metropolitan Conservation Alliance, Wildlife Conservation Society.

Calhoun, A. J. K., & deMaynadier, P. G. (2008). *Science and conservation of vernal pools in Northeastern North America*. Boca Raton, FL: CRC.

Calhoun, A. J. K., & Klemens, M. W. (2002). *Best development practices: Conserving pool-breeding amphibians in residential and commercial developments in the Northeastern United States (MCA Technical Paper No. 5)*. Bronx, NY: Metropolitan Conservation Alliance, Wildlife Conservation Society.

Calhoun, A. J. K., & Reilly, P. (2008). Conserving vernal pool habitat through community-based conservation. In A. J. K. Calhoun & P. G. de Maynadier (Eds.), *Science and conservation of vernal pools in the Northeastern United States* (pp. 319–336). Boca Raton, FL: CRC.

Calhoun, A. J. K., Walls, T. E., McCollough, M., & Stockwell, S. S. (2003). Developing conservation strategies for vernal pools: a Maine case study. *Wetlands, 23*, 70–81.

Carr, D. S., Selin, S. W., & Schuett, M. A. (1998). Managing public forests: Understanding the role of collaborative planning. *Environmental Management, 22*, 767–776.

Ceplo, K. J. (1995). Land-rights conflict in the regulation of wetlands. In B. Yandle (Ed.), *Land rights: The 1990s property rights rebellion* (pp. 104–149). Lanham, MD: Rowman & Littlefield.

Chertow, M. R., & Esty, D. C. (Eds.). (1997). *Thinking ecologically: The next generation of environmental policy*. New Haven, CT: Yale University Press.

Coglianese, C. (1999). The limits of consensus. *Environment, 41*, 28–33.

Colburn, E. A. (2004). *Vernal pools: Natural history and conservation*. Granville, OH: McDonald & Woodward.

Coleman, J. S. (1988). Social capital in the creation of human capital. *American Journal of Sociology, 94*, S95–S120.

Conley, A., & Moote, M. A. (2003). Evaluating collaborative natural resource management. *Society and Natural Resources, 16*, 371–386.

Cortner, H. J., & Moote, M. A. (1999). *The politics of ecosystem management*. Washington, DC: Island Press.

Coughlin, C. W., Hoben, M. L., Manskopf, D. W., Quesada, S. W. (1999). *A systematic assessment of collaborative resource management partnerships*. Ann Arbor, MI Master's project: University of Michigan.

Council on Environmental Quality. (1997). *Environmental quality: 25th anniversary report of the Council on Environmental Quality*. Washington, DC: Government Printing Office.

Danks, C. (2008). Institutional arrangements in community-based forestry. In E. M. Donoghue & V. E. Sturtevant (Eds.), *Forest community connections: Implications for research, management, and governance* (pp. 185–204). Washington, DC: Resources for the Future.

DePoe, S. P., Delicath, J. W., & Elsenbeer, M. A. (Eds.). (2004). *Communication and public participation in environmental decision making*. Albany, NY: State University of New York Press.

Diamant, R., Eugster, J. G., & Mitchell, N. J. (2003). Reinventing conservation: A practitioner's view. In B. A. Minteer & R. E. Manning (Eds.), *Reconstructing conservation: Finding common ground* (pp. 313–326). Washington, DC: Island Press.

Dwyer, L. E., Murphy, D. D., & Ehrlich, P. R. (1995). Property rights, case law, and the challenge to the Endangered Species Act. *Conservation Biology, 9*, 725–741.

ELI [Environmental Law Institute]. (2005). *State wetland program evaluation phase I*. Washington, DC: Environmental Law Institute.

Endicott, E. (1993). *Land conservation through public/private partnerships*. Washington, DC: Island Press.

EPA. (1996). *Watershed approach framework*. Retrieved February 2, 2010, from EPA Web site: http://www.epa.gov/owow/watershed/framework

EPA. (1998). *Clean water action plan: Restoring and protecting America's waters*. Washington, DC: Office of Wetlands, Oceans, and Watersheds, EPA.

Faccio, S. D. (2003). Postbreeding emigration and habitat use by Jefferson and spotted salamanders in Vermont. *Journal of Herpetology, 37*, 479–489.

Freyfogle, E. T. (2003). *The land we share: Private property and the common good*. Washington, DC: Island Press.

Gray, B. (1989). *Collaborating: Finding common ground for multiparty problems*. San Francisco, CA: Jossey-Bass.

Gruber, J. (1994). *Coordinating growth management through consensus-building: Incentives and the generation of social, intellectual, and political capital* (Working Paper, No. 617). Berkeley, CA: Institute of Urban and Regional Development, University of California, Berkeley.

Grumbine, R. E. (1994). What is ecosystem management? *Conservation Biology, 8*, 27–38.

Hilty, J. A., & Merenlender, A. M. (2003). Studying biodiversity on private lands. *Conservation Biology, 17*, 132–137.

Hunter, M. L. (2008). Valuing and conserving vernal pools as small scale ecosystems. In A. J. K. Calhoun & P. G. de Maynadier (Eds.), *Science and conservation of vernal pools in Northeastern North America* (pp. 1–10). Boca Raton, FL: CRC.

Innes, J. E., & Booher, D. E. (1999). Consensus building and complex adaptive systems. *Journal of the American Planning Association, 65*, 412–423.

10 Protecting Natural Resources on Private Lands

Jansujwicz, J. S. (1999). Property rights organizations: Backlash against regulation. In R. H. Platt (Ed.), *Disasters and democracy: The politics of extreme natural events* (pp. 111–130). Washington, DC: Island Press.

John, D. (1994). *Civic environmentalism: Alternatives to regulation in states and communities.* Washington, DC: CQ Press.

Joyal, L. A., McCollough, M., & Hunter, M. L., Jr. (2001). Landscape ecology approaches to wetland species conservation: A case study of two turtle species in southern Maine. *Conservation Biology, 15*, 1755–1762.

Katz, S. B., & Miller, C. R. (1996). The low-level radioactive waste citing controversy in North Carolina: Toward a rhetorical model of risk communication. In C. G. Herndl & S. C. Brown (Eds.), *Green culture: Environmental rhetoric in contemporary America* (pp. 111–140). Madison, WI: University of Wisconsin Press.

Kemmis, D. (1990). *Community and the politics of place.* Norman, OK: University of Oklahoma Press.

Kenney, D. S. (2000). *Arguing about consensus.* Boulder, CO: University of Colorado School of Law. Retrieved February 2, 2010, from http://www.cde.state.co.us/artemis/ucb6/UCB6582C762000INTERNET.pdf

Kenney, D. S., McAllister, S., Caile, W., & Peckham, J. (2000). *The new watershed source book: a directory and review of watershed initiatives in the western United States.* Boulder, CO: Natural Resources Law Center, University of Colorado School of Law.

Kernohan, B. J., & Haufler, J. B. (1999). Implementation of an effective process for the conservation of biological diversity. In R. K. Baydack, H. Campa, III, & J. B. Haufler (Eds.), *Practical approaches to the conservation of biological diversity* (pp. 233–249). Washington, DC: Island Press.

Klemens, M. W. (2000). *Amphibians and reptiles in Connecticut: A checklist with notes on conservation status, identification, and distribution* (DEP Bulletin No. 32). Hartford, CT: Connecticut Department of Environmental Protection.

Koontz, T. M. (2005). We finished the plan, so now what? Impacts of collaborative stakeholder participation on land use policy. *Policy Studies Journal, 33*, 459–481.

Koontz, T. M., & Thomas, C. W. (2006). What do we know and need to know about the environmental outcomes of collaborative management? *Public Administration Review* (December, Special Issue), 111–121.

Koontz, T. M., Steelman, T. A., Carmin, J., Korfmacher, K. S., Moseley, C., & Thomas, C. W. (2004). *Collaborative environmental management: What roles for government?* Washington, DC: Resources for the Future.

Landy, M. K., Susman, M. M., & Knopman, D. S. (1999). *Civic environmentalism in action: A field guide to regional and local initiatives.* Washington, DC: Progressive Policy Institute.

Layzer, J. A. (2008). *Natural experiments: Ecosystem-based management and the environment.* Cambridge, MA: MIT Press.

Leach, W. D., & Pelkey, N. W. (2001). Making watershed partnerships work: a review of the empirical literature. *Journal of Water Resources Planning and Management, 127*, 378–385.

Leach, W. D., Pelkey, N. W., & Sabatier, P. A. (2002). Stakeholder partnerships as collaborative policymaking: evaluation criteria applied to watershed management in California and Washington. *Journal of Policy Analysis and Management, 21*, 645–670.

Lichko, L. E., & Calhoun, A. J. K. (2003). An evaluation of vernal pool creation projects in New England: project documentation from 1991–2000. *Environmental Management, 32*, 141–151.

Mahaney, W. S., & Klemens, M. W. (2008). Vernal pool conservation policy: The federal, state, and local context. In A. J. K. Calhoun & P. G. de Maynadier (Eds.), *Science and conservation of vernal pools in Northeastern North America* (pp. 193–212). Boca Raton, FL: CRC.

Maine Audubon Society (2008). Vernal pools Retrieved September 1, 2009, from http://www.maineaudubon.org/conserve/citsci/vip.shtml

Maine State Planning Office. (2001). *Maine State Wetlands Conservation Plan.* Augusta, ME: Maine State Planning Office.

Mandarano, L. A. (2008). Evaluating collaborative environmental planning outputs and outcomes: Restoring and protecting habitat and the New York-New Jersey harbor estuary program. *Journal of Planning Education and Research, 27*, 456–468.

Marceau, D. (2009). Vernal pools: One consultant's perspective. Maine Association of Site Evaluators (MASE) Newsletter. Retrieved February 2, 2010, from http://www.mainese.com/documents/VernalPoolArticle.pdf

Margerum, R. D. (2002). Evaluating collaborative planning: Implications from an empirical analysis of growth management. *Journal of the American Planning Association, 68*, 179–193.

Margerum, R. D., & Born, S. M. (1995). Integrated environmental management: The foundations for successful practice. *Journal of Environmental Planning and Management, 38*, 371–391.

Martin, T. (2007). Muting the voice of the local in the age of the global: How communication practices compromised public participation in India's Allain Dunhangan environmental impact assessment. *Environmental Communication, 1*, 171–193.

Mazmanian, D. A., & Kraft, M. E. (1999). *Toward sustainable communities: Transitions and transformations in environmental policy*. Cambridge, MA: MIT Press.

McCloskey, M. (1996). The skeptic: Collaboration has its limits. *High Country News, 59* (May 13). Retrieved 2 February 2, 2010, from http://www.hcn.org/issues/59/1839

McCloskey, M. (2004–2005). What we have learned from William Leach's study of 76 collaborative watershed partnerships in California and Washington. *The Collaborative Edge*. Retrieved July 15, 2008, from http://www.csus.edu/ccp/newsletter/2005/winter

Meiners, R. E., & Yandle, B. (Eds.). (1993). *Taking the environment seriously*. Lanham, MD: Rowan & Littlefield.

Meltz, R., Merriam, D. H., & Frank, R. M. (1999). *The takings issue: Constitutional limits on land use control and environmental regulation*. Washington, DC: Island Press.

Moote, M. A. (2008). Collaborative forest management. In E. M. Donoghue & V. E. Sturtevant (Eds.), *Forest community connections: Implications for research, management, and governance* (pp. 243–260). Washington, DC: Resources for the Future.

Moote, M. A., McClaran, M. P., & Chickering, D. K. (1997). Theory in practice: Applying participatory democracy theory to public land planning. *Environmental Management, 21*, 877–889.

Norse, E. A. (1993). *Global marine biological diversity: a strategy for building conservation into decision making*. Washington, DC: Island Press.

Noss, R. F., & Cooperrider, A. Y. (1994). *Saving nature's legacy: Protecting and restoring biodiversity*. Washington, DC: Island Press.

Noss, R. F., O'Connell, M. A., & Murphy, D. D. (1997). *The science of conservation planning: Habitat conservation under the Endangered Species Act*. Washington, DC: Island Press.

Oscarson, D. B., & Calhoun, A. J. K. (2007). Developing vernal pool conservation plans at the local level using citizen-scientists. *Wetlands, 27*, 80–95.

Patrick, D., Calhoun, A. J. K., & Hunter, M. L., Jr. (2007). The orientation of juvenile wood frogs, *Rana sylvatica*, leaving experimental ponds. *Journal of Herpetology, 41*, 158–163.

Peterson, T. R., & Horton, C. C. (1995). Rooted in the soil: How understanding the perspectives of landowners can enhance the management of environmental disputes. *The Quarterly Journal of Speech, 81*, 139–166.

Peterson, M. N., Peterson, T. R., Peterson, M. J., Lopez, R. R., & Silvy, N. J. (2002). Cultural conflict and the endangered Florida Key Deer. *The Journal of Wildlife Management, 66*, 947–968.

Pierce Atwood LLP. (2006). Maine legislature sets stringent rules for vernal pools. Retrieved February 2, 2010, from http://www.pierceatwood.com/files/69_w0474905.pdf

Preisser, E. L., Kefer, J. Y., Lawrence, J. D., & Clark, T. W. (2000). Vernal pool conservation in Connecticut: an assessment and recommendations. *Environmental Management, 26*, 503–513.

Pretty, J., & Smith, D. (2004). Social capital in biodiversity conservation and management. *Conservation Biology, 18*, 631–638.

Putnam, R. D. (1995). Bowling alone: America's declining social capital. *Journal of Democracy, 6*, 65–78.

Putnam, R. D. (2001). *Bowling alone: The collapse and revival of American community*. New York: Simon & Schuster.

10 Protecting Natural Resources on Private Lands

Putnam, R. D., Leonardi, R., & Nanetti, R. (1993). *Making democracy work: Civic traditions in modern Italy*. Princeton, NJ: Princeton University Press.

Sabatier, P. A., Focht, W., Lubell, M., Trachtenberg, Z., Vedlitz, A., & Matlock, M. (2005). *Swimming upstream: Collaborative approaches to watershed management*. Cambridge, MA: MIT Press.

Sabel, C., Fung, A., & Karkkainen, B. (2000). *Beyond backyard environmentalism*. Boston, MA: Beacon.

Salamon, S., Farnsworth, R. L., & Rendziak, J. A. (1998). Is locally led conservation working? A farm town case study. *Rural Sociology, 63*, 214–234.

Semlitsch, R. D. (1998). Biological delineation of terrestrial buffer zones for pond-breeding salamanders. *Conservation Biology, 12*, 1113–1119.

Semlitsch, R. D. (2002). Critical elements for biologically based recovery plans of aquatic-breeding amphibians. *Conservation Biology, 16*, 619–629.

Semlitsch, R. D., & Skelly, D. K. (2008). Ecology and conservation of pool-breeding amphibians. In A. J. K. Calhoun & P. G. de Maynadier (Eds.), *Science and conservation of vernal pools in Northeastern North America* (pp. 127–148). Boca Raton, FL: CRC.

Shogren, J. F. (Ed.). (1998). *Private property and the Endangered Species Act: saving habitats, protecting homes*. Austin, TX: University of Texas Press.

Susskind, L., & Cruikshank, J. (1987). *Breaking the impasse: Consensual approaches to resolving public disputes*. New York: Basic Books.

Thomas, C. W. (2001). Habitat conservation planning: Certainly empowered, somewhat deliberative, questionably democratic. *Politics and Society, 29*, 105–130.

Thomas, C. W. (2003). *Bureaucratic landscapes: Interagency cooperation and the preservation of biodiversity*. Cambridge, MA: MIT Press.

Thomas, C. W. (2008, April). Evaluating the performance of collaborative environmental governance. Paper presented at the Consortium on Collaborative Governance Mini-Conference, Santa Monica, CA.

U. S. Government Accountability Office. (2008). *Natural resource management: Opportunities exist to enhance federal participation in collaborative efforts to reduce conflicts and improve natural resource conditions (GAO-08-262)*. Washington, DC: US Government Accountability Office.

USDA. (2002). *Major uses of land in the United States, 2002 (Economic Information Bulletin Number 14)*. Washington, DC: USDA Economic Research Service.

USDA and U.S. DOC (Department of Commerce). (2000). Unified federal policy for a watershed approach to federal land and resource management. *Federal Register, 65*(2002), 62566–62572.

Vasconcelos, D., & Calhoun, A. J. K. (2004). Movement patterns of adult and juvenile wood frogs (*Rana sylvatica*) and spotted salamanders (*Ambystoma maculatum*) in three restored vernal pools. *Journal of Herpetology, 38*, 551–561.

Vasconcelos, D., & Calhoun, A. J. K. (2006). Monitoring created seasonal pools for functional success: a six-year case study of amphibian responses, Sears Island, Maine, USA. *Wetlands, 26*, 992–1003.

Wakeman, T. H., III. (1997). Building sustainable public policy decisions through partnerships. *Journal of Management in Engineering, 13*(3), 40–48.

Weber, E. P. (2003). *Bringing society back in: Grassroots ecosystem management, accountability, and sustainable communities*. Cambridge, MA: MIT Press.

Wise, C. R. (2004). Property rights and regulatory takings. In R. F. Durant, D. J. Fiorino, & R. O'Leary (Eds.), *Environmental governance reconsidered: Challenges, choices, and opportunities* (pp. 289–321). Cambridge, MA: MIT Press.

Wondolleck, J. M., & Yaffee, S. L. (2000). *Making collaboration work: Lessons from innovation in natural resource management*. Washington, DC: Island Press.

Yaffee, S. L., Phillips, A. F., Frentz, I. C., Hardy, P. W., Maleki, S. M., & Thorpe, B. E. (1996). *Ecosystem management in the United States: An assessment of current experience*. Washington, DC: Island Press.

Chapter 11
Integrating Expert Judgment into Systematic Ecoregional Conservation Planning

Karen F. Beazley, Elizabeth Dennis Baldwin, and Conrad Reining

Abstract This chapter offers insights on integrating expert judgment into ecoregional conservation planning. We describe three examples that focus on benefits and challenges of (1) delivering GIS-based expert systems in stakeholder-based contexts, (2) integrating expert judgment and computer-based site selection scenarios, and (3) reaching expert consensus on delineating conservation planning areas across a diverse ecoregion. The examples highlight several important lessons. First, engagement of experts should not be simply about gaining approval. To maximize the extent of buy-in by experts, they need to be legitimately involved in the creation of methodology and results. Second, experts need to be distinguished in the planning process from stakeholders and local residents. While precise definitions are elusive and likely to vary from one region to another, a transparent methodology for assessing and weighting each group's input is important. Finally, the methods used for engaging expert participation need to match the experts' technological capabilities and conceptual understandings. While a lack of familiarity with certain aspects should not disqualify an expert from participation, it does highlight the importance of advanced preparation on the part of those facilitating the process. Beyond these more technical issues are those related to the social sciences of expert engagement. Social and qualitative forms of data are needed to build this understanding.

Keywords Collaboration • consensus • expert • ecoregional conservation planning • participation

---

K.F. Beazley (✉)
School for Resource and Environmental Studies, Dalhousie University, 1459 Oxford Street, Halifax NS B3H 4R2, Canada
e-mail: karen.beazley@dal.ca

E.D. Baldwin
Parks, Recreation, and Tourism Department, Clemson University, 271A Lehotsky Hall, Clemson, SC 29634
e-mail: ebaldwn@clemson.edu

C. Reining
Eastern Program Director, Wildlands Network, P. O. Box 225, East Thetford, VT 05043
e-mail: conrad@wildlandsnetwork.org

S.C. Trombulak and R.F. Baldwin (eds.), *Landscape-scale Conservation Planning*,
DOI 10.1007/978-90-481-9575-6_11, © Springer Science+Business Media B.V. 2010

11.1 Introduction

Despite its ecological cohesion, the Northern Appalachian/Acadian ecoregion is culturally complex, characterized by two dominant languages (French and English); multiple overlapping political systems; a mix of large cities, small towns, and villages; and land-use patterns shaped by rural forestry, agricultural areas, remnant large tracts of wilderness, both public and private land holdings, and many extractive industries. Numerous collaborative initiatives have been developed over many years to promote conservation throughout this ecoregion, initiatives that at their core involve careful planning that considers the views of experts, stakeholders, local residents, spatial diversity, temporal changes, and multiple scales of perspective.

Our intention in this chapter is to offer insights on various approaches to and challenges of integrating expert judgment into ecoregional conservation planning. We draw on examples from the Northern Appalachian/Acadian ecoregion to focus on specific issues, methodological approaches, participatory processes, and tools associated with such engagement. We also provide recommendations that should be of value to those outside of this ecoregion who themselves wish to integrate expert judgment into ecoregional conservation planning and are looking for lessons and strategies on how this might be done effectively.

Expert input is often sought in conservation planning, in part to compensate for deficiencies in data and to benefit from their tacit knowledge gained from experience in conservation-related initiatives and landscapes. Experts can lend greater legitimacy and robustness to the planning process and broader buy-in to its results. At the same time, conservation practitioners are increasingly using computer-based expert systems, particularly geographic information systems (GIS; Chap. 12) and site-selection software (Chap. 14), while recognizing that these are decision support tools rather than decision makers themselves. Although such tools can generate a range of planning scenarios and efficient solutions, they themselves do not indicate which solution would work best in a particular context or location. Given the multiple factors at play in broad regional or landscape-scale planning, a broader expert engagement in decision making is often warranted.

Given the benefits of expert engagement, many conservation planning efforts aim to incorporate it; however, it is not without its challenges and some applications are more successful than others. Since conservation planning is a relatively new venue for expert engagement, particularly in the integration of expert judgment and expert systems, effective methods of engaging and integrating experts are generally not well developed. Expert engagement is often presented as a 'black box,' with little or no methodological description or analytical reflection on its successes and limitations. To advance the discussion, we describe three examples of expert engagement in conservation planning that focus explicitly on benefits and challenges associated with (1) delivering GIS-based expert systems in stakeholder-based contexts, (2) integrating expert judgment and computer-based site selection scenarios, and (3) reaching expert consensus on delineating conservation planning areas across a diverse ecoregion.

11.2 Methods

We examine here three examples of expert engagement in systematic regional conservation planning across the Northern Appalachian/Acadian ecoregion, in which one or more of us had been involved: (1) assessing protected-area potential in the Nova Forest Alliance landbase in Nova Scotia (Anderson et al. 2009); (2) a Wildlands Network Design for the Greater Northern Appalachians (Beazley et al. 2010; Reining et al. 2006); and (3) delineating planning areas for ecoregional conservation planning in the Northern Appalachian/Acadian ecoregion. Each of these cases aimed to integrate expert systems, such as site selection software (e.g., MARXAN [Ball and Possingham 2000; Possingham et al. 2000]), and expert judgment into decision making for systematic conservation planning. We briefly describe the intent and methodological approaches of each, assess each in terms of the pros, cons, and lessons learned, and identify the broader issues that emerge when considering all three together, along with other examples and findings reported in the literature.

For the purpose of this research, we use the term 'integrating expert judgment,' and we differentiate experts from both stakeholders and local residents. Each of the following examples uses a different configuration of inclusion, from all three groups of potential participants to only one. We define experts as those who can integrate multiple elements of a whole system in their decision making and have detailed expertise related to topics of interest, and therefore require little education from planners to engage with decision making (Tynjälä 1999). Stakeholders are those that own affected businesses or land, or who otherwise self-define as having a 'stake' in the outcome of a conservation plan. They and local residents in an area offer more detailed input at a smaller scale on specific elements of the system that may otherwise go undetected by larger scale inclusion of experts (Hmelo-Silver and Pfeffer 2004). All three groups offer valuable information and knowledge; however, costs are associated with the inclusion of each. We address here primarily the inclusion of experts with an assessment of the pros and cons of that inclusion, referencing other groups to the extent that the consequences of their inclusion intersect with that of experts.

11.3 Integrating Expert Judgment in the Northern Appalachian/Acadian Ecoregion

11.3.1 Assessing Protected Area Potential in the Nova Forest Alliance Landbase in Nova Scotia

The Nova Forest Alliance (NFA) is part of the Canadian Model Forest Network and is located in Central Nova Scotia. The NFA community (including private woodland owners, government agencies, forest companies, non-government organizations, and academics) initiated a process to assess the protected-area potential of land parcels within the NFA landbase, with potential application within other model forests in Canada. Accordingly, we (Anderson et al. 2009) sought to create an integrated GIS-based decision-support tool for conservation assessment within the NFA. The use of decision-support tools such as GIS for conservation planning can be an effective means of engaging experts and other communities (Jordan 1998; Sieber 2000). Our objective was to develop a GIS application to compile relevant data, create a spatially explicit database, display potential values for protected areas across the study region, and provide a preliminary ranking of sites based on biological diversity. Such applications can provide decision support for community-focused decision making and provide spatially explicit representations of conservation values and land-ownership patterns. Our intent was that the resulting maps effectively communicate conservation knowledge to individual property owners, potentially motivating their engagement in conservation, to bring credence to the decision-making process by generating and visualizing conservation alternatives, and thus facilitate a collaborative approach for integrating expert and community judgment in the NFA.

From a participatory perspective, however, our GIS-based application in the NFA faced challenges that influenced the effectiveness of its implementation (described in Anderson et al. 2009). Despite the opportunities we initiated for participation by the NFA partnership, these proved insufficient to achieve adequate engagement and support for our process and its outcomes. This and other challenges associated with data sharing and lack of local GIS expertise limited the ongoing use of this tool in the NFA and its potential usefulness in other model forests. While not overly successful at integrating expert judgment, the experiential knowledge we gained through this application can be used to address questions about the effectiveness of participatory GIS applications for ecoregional conservation planning. Accordingly, we now strongly believe that expert as well as stakeholder engagement in the initial project design, goal-setting, and all subsequent stages is critical, even if this requires extensions to preferred or anticipated timelines.

In the NFA, specific methodological processes for engaging experts were not integrated into the study design. Study methods included our participation in regularly scheduled meetings and workshops delivered by the NFA, as well as informal telephone, electronic, and in-person communications with specific NFA partners and data providers. Through these communications, we interacted with knowledgeable NFA experts and stakeholders and attempted to solicit their input concerning conceptual and methodological approaches, the selection of conservation features, and definition of targets, data types, and data availability. While these methods provided opportunities for participation by the NFA partners, we received little input from them apart from the identification of potential data sources and provision of datasets. We used a combination of these recommended and available localized data sources and methodological literature on ecological reserve design (e.g., Beazley et al. 2005; Groves 2003; Margules et al. 1988, 2002; Noss and Cooperrider 1994) to define the relevant data for our conservation assessment. We subsequently developed a spatially explicit decision-support tool (based on both MARXAN

[Ball and Possingham 2000] and C-Plan [The University of Queensland n.d.]) and conducted various analyses to demonstrate the tool and the assessment process, as well as to generate visual map outputs. Written progress and final reports were delivered to the NFA, followed by presentations to experts and stakeholders at the 2004 and 2005 Annual General Meetings, where we sought additional input and feedback, but to little avail. We provided a final NFA in-house demonstration of the GIS-based decision-support tool in June 2005. Through these processes, we delivered datasets, maps, and tools to the NFA, along with key findings relevant to protected-area potential within the landbase. To date, resource limitations at the NFA have limited the adoption of these support tools and materials.

We subsequently identified several issues that contributed to the lack of participation, along with suggestions for overcoming them in future applications (Anderson et al. 2009). Other studies have found that community-based GIS can enhance participation and encourage open communication among participants (Jankowski and Nyerges 2001; Weiner et al. 2001), and thereby increase effective decision making and buy-in from interested groups (Carver 2003; Kyem 2000; Schlossberg and Shuford 2005; Sieber 2000). While this may indeed be the case in some applications, we encountered a significant challenge in stimulating sufficient direct participation among experts and stakeholders, although we observed that experts seemed more willing to offer suggestions than were stakeholders. This translated into a lack of buy-in and inadequate understanding among key participants as to the usefulness, benefits, and operation of the decision-support tool. As a consequence, neither the mapped results showing protected-area potential nor the decision-support tool itself are currently being used by NFA.

We concluded that integrating direct participatory processes earlier may have produced more satisfactory results. 'Involving community members in the selection of conservation features and targets in the early phases of the project might have stimulated more interest in the assessment stage' (Anderson et al. 2009). Subsequent workshops could have garnered more participation, in which experts and stakeholders could witness, discuss, and revise different map-based scenarios and other decision rules that would shape the results. This engagement could have helped to ensure that the decision tool adequately represented the conservation goals of the community and that its potential flexibility and utility were more widely understood, lending greater legitimacy to the process and ownership of the results. This observation is consistent with that of other conservation theorists, who suggest that involving concerned and informed individuals creates more realistic and sound conservation goals when targets are "clear, explicit and defensible" (Groves 2003).

With respect to the utility of the GIS technology, we naively assumed that by de-emphasizing the reliance on GIS expertise – by providing organized data in a readily usable form and by choosing relatively user-friendly programs – the complexity of using the tool would be reduced (Anderson et al. 2009). The lack of adoption of this tool by the NFA, however, forces us to recognize that even relatively simple GIS-based programs are still expert-grounded tools that require a greater degree of technical skill and appreciation than may be p resent among experts or in many communities (Chap. 12), thus limiting its use after project delivery.

The NFA is an appropriate community and geographic region for a collaborative or participatory conservation initiative since it includes a community of experts and stakeholders interested in sustainable development, all of whom have informed views on how land resources can be managed. Paradoxically, this study (Anderson et al. 2009) illustrates the significant challenges to successful implementation of GIS-based community research that include complexities of data acquisition and sharing policies, challenges in initiating and sustaining meaningful engagement by experts and other participants, longevity of resources, and use of the decision-support tool at the local community level.

To overcome these obstacles, active and early participation is critical to community buy-in; this can occur using a variety of methods such as interactive workshops and web forums to share and create knowledge, develop specific conservation objectives, explore alternative scenarios using visual outputs such as maps, and work towards mutually agreeable decisions. Also, data sharing networks among universities, governments, and community organizations, such as the NFA, could greatly enhance local capacity to acquire the geospatial information necessary for effective decision support (Chap. 12). Such policies would reduce the data procurement efforts needed and allow for more concentration on community engagement. Finally, effective participatory technical systems for community-based decision support must be developed in ways that meet the needs of participants without being so complex as to create barriers to their use. 'Creating decision-support tools, information, and knowledge that integrate community objectives and reflect community capacities are critical elements in both the development and on-going application stages. As such, considerable attention should be given to whether or not GIS technology is appropriate and sustainable as part of a decision-support tool in a community-based, public-participatory context' (Anderson et al. 2009).

11.3.2 A Wildlands Network Design for the Greater Northern Appalachians

As part of its work on the design and implementation of a continental-scale network of protected area, the Wildlands Network initiated a conservation planning process in the Greater Northern Appalachians (GNA) of Eastern Canada and the U.S. Accordingly, we (Reining et al. 2006) conducted a map-based methodology designed to systematically identify a network of areas of high conservation priority within the Northern Appalachian/Acadian and St. Lawrence/Champlain Valley ecoregions. We used a three-track approach (Noss 2003) intended to (1) represent environmental variation across these ecoregions, (2) protect special elements such as occurrences of rare species or communities and other sites with high ecological value, and (3) conserve sufficient habitat to support viable populations of focal species (Lambeck 1997; Miller et al. 1999; Noss and Cooperrider 1994; Noss et al. 1999).

11 Integrating Expert Judgment into Systematic Ecoregional Conservation Planning 241

To establish the location and extent of the network design elements, we used three major sources of information: (1) the results of site selection analyses; (2) The Nature Conservancy's (TNC) and Nature Conservancy of Canada's (NCC) Tier 1 matrix forest blocks in the Northern Appalachian/Acadian ecoregion (Anderson et al. 2006); and (3) input from experts from within environmental non-governmental organizations, government wildlife agencies, and academia (Beazley et al. 2010). The site selection analyses were based on MARXAN-generated solutions for representation of ecological land units, special elements, and source and threatened source habitat for three focal species (Canada lynx [*Lynx canadensis*], American marten [*Martes pennanti*], and wolf [*Canis lupus* or *lycaon*]). These were overlaid with Tier 1 and Tier 2 matrix forest blocks identified by TNC/NCC, and assessed and refined through expert input. To obtain expert input, we conducted a series of day-long workshops in Nova Scotia, New Brunswick, Québec, Vermont, New York, and Maine from January through May 2006. Some experts were also consulted by telephone (Reining et al. 2006).

The intent of the expert consultation was twofold: (1) to integrate expert systems (e.g., MARXAN) with expert judgment, and (2) to move from decision support (e.g., outputs of MARXAN) to decision making (e.g., proposed network design) (Beazley et al. 2010). Twelve different scenarios were run in MARXAN, defined by four different target levels (low, medium, high-low, and high) and three different values for a boundary length modifier variable, which influences the degree of fragmentation allowed in a solution (Chap. 14). MARXAN generates many kinds of outputs, including best runs and summed runs for each scenario. The best run is a near-optimal network solution that meets the goals with the least amount of land area. The summed run shows the number of times a planning unit was selected over several separate runs based on a given scenario. The more often a planning unit is selected, the more important it is to meeting the goals for that scenario. In addition, summed-summed runs show how often planning units are selected across separate runs and different planning scenarios. Those areas that are selected repeatedly across scenarios can be interpreted as having a high ecological irreplaceability (Chap. 14). Those selected infrequently are considered more 'replaceable' – that is, their relative conservation importance is lower than areas selected more frequently because their contribution to achieving conservation goals can more easily be replaced by other locations.

These outputs from the MARXAN analyses provide useful information to support decisions about the extent and the elements of a network design. However, since several potential solutions are equally valid, additional steps need to be taken to incorporate expert and local knowledge and other data that have not been captured by the site selection algorithm in order to make defensible decisions regarding the network design. Consequently, the best-run, summed-run, and summed-summed-run outputs were combined with the TNC/NCC Tier 1 matrix blocks and we consulted with experts to establish the location and extent of network design elements.

Through such consultations, we overlaid the matrix blocks with the results of the site selection analyses and used them to refine the preferred scenarios. In addition to refinements made to the network design based on TNC's matrix forest blocks, we also made refinements on the basis of expert input received during the workshops and telephone calls with local experts in each state or province. In those meetings, the results of the site selection analysis were presented and then the meeting facilitators sought to achieve the following six goals:

1. Determine the preferred scenario, or combination of scenarios, for the state or province based on local conservation knowledge.
2. Determine overlap with known areas of conservation value.
3. Identify areas of known conservation value that were not captured.
4. Discuss deficiencies in the analysis.
5. Delineate potential boundary revisions that could readdress these deficiencies.
6. Discuss how this study should be communicated to other audiences.

The meeting process unfolded quite differently in each state and province, as described in Reining et al. (2006) and Beazley et al. (2010), but all helped to refine the final network design and ensure greater accuracy and relevance in the local context. In all Canadian provinces (Nova Scotia, New Brunswick, and Québec), experts chose to emphasize a combination of scenarios that set medium, high-low, and high goals and the lowest degree of allowable fragmentation. Experts there also recommended higher goals in areas with greater extent of natural cover or wildness, and lower goal scenarios in areas with more human development (e.g., agricultural regions). In Québec, experts explicitly divided the region into three sections and chose different scenarios for each. In contrast, experts in New Brunswick recommended a 5-km buffer around key network elements, such as linkages, to provide flexibility both in ensuring that a portion of the area will always be managed in support of the larger network and in determining how and where resources such as timber will be managed within the area.

Conversely, in the U.S. (Maine, Vermont, and New Hampshire), experts chose scenarios that allowed for moderate fragmentation. Similar to experts in Canada, however, those in the U.S., particularly in Maine, recommended different approaches or changes to the scenario outputs in regions that are more developed (e.g., Downeast Maine) versus those that are more wild (e.g., Northern Maine). The expert input also served to incorporate important linkage areas for focal species that do not necessarily emerge from site selection analyses, thus capturing additional information that would have been missing without expert engagement.

As a consequence of subregional differences in the selection of scenarios and outputs and the inclusion of other features introduced by the experts, the resulting network incorporates subregional goals. This result may thus be more consistent with those of other studies (e.g., Carroll et al. 2003) in which subregional goals were explicitly set. We recommend that future site selection analyses at ecoregional scales stratify the study area into smaller subregions and establish goals for those regions, while perhaps trying to maintain overall goals for the entire ecoregion. This should avoid the concentration of network elements within discrete areas of the ecoregion and result in a more distributive network.

11.3.3 Delineating Conservation Planning Areas in the Northern Appalachian/Acadian Ecoregion

Beginning in 2004, as part of a consortium of scientists from throughout the Northern Appalachian/Acadian ecoregion, we (Trombulak et al. 2008) began collaborating on a systematic conservation plan that transcends Canadian-U.S. political boundaries and encapsulates a range of subregions, including the Acadian Forest, the Northern Appalachian Mountains, and the Adirondack Mountains (Chap. 1). Our overarching goal, developed under the umbrella of Two Countries, One Forest (2C1Forest), was to produce a scientifically valid conservation plan for the Northern Appalachian/Acadian ecoregion. Our approach incorporated a three-track strategy – ecological representation, habitat for focal species, and rare species – to assess site-specific ecological irreplaceability (Chap. 14). We also assessed current (Woolmer et al. 2008) and future (Baldwin et al. 2007) forecasts of threats from human activity (Chap. 13) to incorporate site-specific vulnerability and urgency of threat. These we combined to assess conservation priorities, based on both irreplaceability and threat (Trombulak et al. 2008).

To make it easier to understand and visualize how areas that are threatened or important for achieving regional conservation goals are distributed across the landscape, we first sought to subdivide the ecoregion into a set of conservation planning areas, modeled on the approach used by Noss et al. (2002).

In brief, current technologies for remote sensing and GIS allow for spatial resolutions far greater than what is usable for conservation planning at the scale of the ecoregion. For example, our threats analysis for the Northern Appalachian/Acadian ecoregion was based on 90-m resolution, which results in over 43 million separate planning units, and our irreplaceability analysis was based on 10-km^2 hexagonal resolution, resulting in over 63,000 planning units. Clearly, these are too numerous to be the basis for identifying on-the-ground conservation priorities or for graphing variation in threat or importance at an ecoregional scale.

As a consequence, 2C1Forest sought to aggregate planning units together into a smaller number of conservation planning areas. Each conservation planning area would ideally represent a collection of contiguous planning units with similar measured levels of irreplaceability and have boundaries based upon easily understood geographic or cultural features (e.g., rivers, ridgelines, or major highways). We did not have any predetermined target number of conservation planning areas, although we thought that between 100 and 200 of these areas might provide an optimal balance between fine-scale resolution of regional variation and a manageable number of planning areas.

In January 2007, 2C1Forest identified an initial set of 76 conservation planning areas that encompassed the entire ecoregion, based on similarities in importance scores and other biophysical characteristics such as rivers, ridgelines, and major highways (Fig. 11.1). We then set out to validate and measure resonance of the boundaries of the proposed conservation planning areas with conservation experts throughout the ecoregion, solicit feedback, and integrate this knowledge into a refined set of conservation planning areas. The use of experts as opposed to local residents and stakeholders was deliberate and beneficial for many reasons (Hmelo-Silver and Pfeffer 2004). If there can only be a few meetings, and the task requires a base level of knowledge, experts are usually ready to perform such tasks with less preparation and explanation. Additionally, if experts have a long tenure in a region, they are

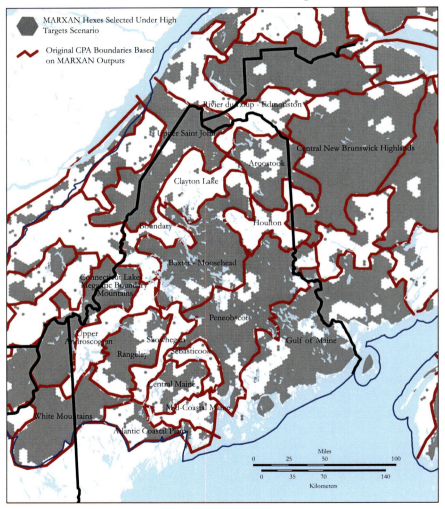

Fig. 11.1 Map composition showing evolution of conservation planning units in Maine. Step 1: initial set of conservation planning areas delineated by 2C1Forest. Step 2: revised set of conservation planning areas based on irreplaceability values and geographic or cultural features. Step 3: final set of conservation planning areas after review by experts

Step 2: Conservation Planning Areas Based on Geographical Features

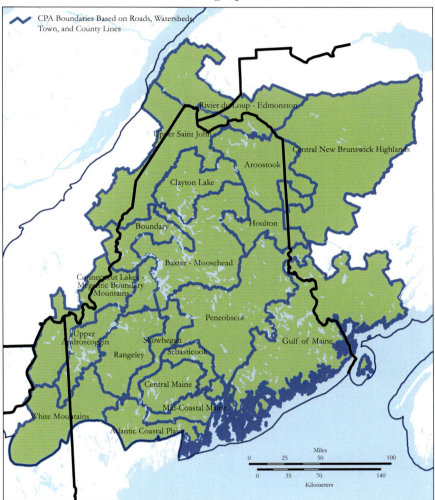

Fig. 11.1 (continued)

often sensitive to issues and needs of stakeholders not represented in meetings. Accordingly, we arranged for a series of meetings and workshops with experts from non-governmental organizations, government agencies, and academia, as well as unaffiliated local experts engaged in conservation activities across the ecoregion.

Going into this exercise, we were aware that expert judgment is frequently included in conservation planning and is often presented without a clear methodological framework. As a consequence, we developed methods to investigate the

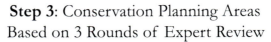

Step 3: Conservation Planning Areas
Based on 3 Rounds of Expert Review

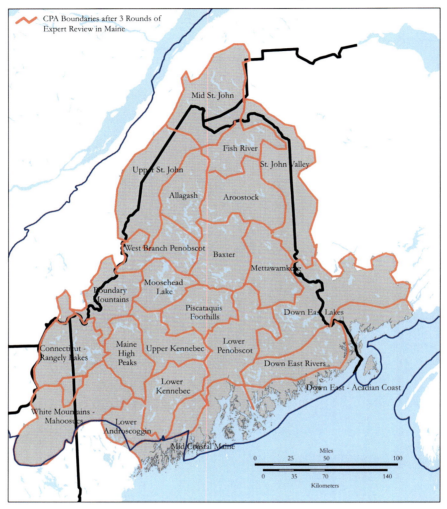

Fig. 11.1 (continued)

use of expert judgment as a conservation planning tool in a systematic transparent manner while conducting our expert engagement processes. We documented in detail four planning meetings in Maine and Québec, using the services of a professional notetaker during three of these meetings, allowing us to concentrate on facilitating the meetings and workshops. Verification of planning unit boundaries was conducted in real time with maps. Reasons for suggested changes, motivations, and the thoughts and feelings presented were documented for later analysis.

Using qualitative methodology, we subsequently analyzed the text from four meetings, in which 28 conservation planning experts participated.

Experts from natural resource agencies, conservation organizations, and academia were invited to these meetings. Attendees were given an overview presentation of the scientific work of 2C1Forest and the approach the consortium was taking toward conservation planning in the ecoregion. The concept of conservation planning units was introduced, together with the draft maps of conservation planning areas that had been developed by the group in January 2007. Large format paper maps of Maine and Québec, and surrounding states and provinces, were provided. GIS capacity was also available at each meeting. Workshop participants were asked to review, in plenary, the boundaries of each conservation planning area, the ecological and cultural features contained therein, and names that had been assigned to the conservation planning area. We asked the group to propose changes that they thought were needed to the boundaries and names.

Themes emerged from these meetings and workshops that transcend the particular case of the Northern Appalachian/Acadian ecoregion and relate to overall issues of including expert judgment in conservation planning. Experts tended to converge on a set of recommendations, suggesting that a 'consensus atmosphere' may lead a group of experts to the same conclusion. We also found that the expert recommendations that emerge from any particular meeting related strongly to the missions and goals of the organizations represented at the meeting. For example, in Maine, the set of conservation planning areas agreed upon by the experts blended boundaries based on scientific rationales as well as local knowledge and local needs, such as area-specific initiatives or campaigns (Fig. 11.1).

We experienced challenges in coming to consensus in the meeting held in Québec, due primarily to two factors. First, communication challenges emerged from language-related differences between primarily English-speaking facilitators and primarily French-speaking or bilingual experts. Second, among the experts in Québec, understanding of the ecoregional planning process was more limited than in Maine, possibly because several of the experts in Maine had previously been engaged as experts in the Wildlands Network Design process (Sect. 11.3.2). The difficulties experienced in the Québec meeting changed our approach to delineating conservation planning areas to one that avoided the use of expert engagement and instead used pre-defined spatial units, such as watersheds and biophysical regions. Although the solicitation of further diverse expert opinion was abandoned, the planning process benefited from a more thorough understanding of the local and regional conservation landscape.

11.4 Discussion

Several issues emerge from our consideration of these examples of conservation initiatives that attempted to integrate expert judgment with other methods and sources of data. One key issue is the importance of early involvement by experts to increase legitimacy and promote buy-in to both the process and its results. Experts (and by extension, stakeholders and local residents as well) should be engaged in defining the problems and objectives, delineating planning units, setting goals and targets, selecting preferred scenarios and solutions, and other key aspects of decision making associated with large-scale conservation initiatives. This engagement in fundamental decisions and rule-setting creates enthusiasm for the process and its implementation, without which a plan is unlikely to be realized on the ground.

If the notion that legitimacy and buy-in from a broad base is accepted as being important for successful implementation of a conservation plan, then the question arises as to who should be involved. Are experts alone enough, or should the process include a cross-section of experts, stakeholders, and local residents? Each of these groups has relevant and potentially non-overlapping interests, knowledge, and perspectives. Stakeholder and local views warrant integration as well, but require separate or different processes than those designed for experts because of the different levels and types of information needed to facilitate informed participation by these groups. However, many of the same points made with respect to expert involvement may well apply to these other groups, such as early engagement and the importance of gaining their buy-in and support for the process and its outcome.

Beyond the question of engaging experts are the issues of defining expertise and distinguishing experts from other participants (Tynjälä 1999). Some experts are particularly good at integrating information across fields, and others have detailed knowledge in a specialized or localized field. Both forms of expertise are important (Doswald et al. 2007). Many factors come into play in ecoregional conservation planning. Who then is considered an expert, in which fields, and who is a stakeholder or interested party? If each of these groups are to be involved in the process, then it can be a complex task to determine who makes the decisions and on what grounds, whose expertise or interest or knowledge counts, and who is to be invited to participate. For example, how do the traditional ecological knowledges of indigenous First Nations become incorporated into the planning process and how is it weighted relative that of other local people who have lived in and worked with the land and water for many generations? Such questions need to be addressed explicitly and transparently.

If one wants to include experts (or other stakeholders and local residents) that do not have technological or conceptual familiarity with GIS-based or expert systems or tools, then the issue of technological capacity needs to be addressed. If the experts, stakeholders, or residents are not familiar with these tools, then the efficacy of participatory GIS applications is limited. GIS-based mapping tools can integrate a wealth of relevant information and generate several alternative scenarios relatively quickly, but as demonstrated by the Nova Forest Alliance example, participatory GIS methods that aim to provide tools for use by experts and stakeholders will not work unless the intended users understand how to integrate these tools into their decision making on an on-going basis.

A tension also exists between the accuracy of expert judgment regarding conservation values and features and the results of empirical and expert literature-based models.

As Clevenger et al. (2002) note, the widespread availability of geographic information systems allows for a 'more explicitly reasoned environmental decision-making process based on qualitative or expert-judgment data in multi-criteria evaluations.' But expert-judgment data are not guaranteed to be as accurate as data derived by more empirical means, as Clevenger et al. (2002) found when they compared three spatially explicit habitat models for back bear (*Ursus americanus*), one based on empirical data, one based on a review of the literature on black bear habitat requirements, and one based solely on expert judgment. The expert literature-model provided results that were a good approximation of those obtained by the empirical model, and both performed much better than the expert-judgment model.

Empirical data may not always be available, however, and are often time-consuming and expensive to produce. Doswald et al. (2007), in studies of lynx in Switzerland, found that 'expert knowledge, and especially local knowledge, can be employed to create a good habitat suitability model.' The researchers concluded that 'this has implications for conservation and science because it shows not only that expert knowledge may be used when no other data exist, but also that local 'ground workers' should be employed more often in the development of habitat suitability models or conservation plans.'

Yet, even when empirical data exist, expert judgment can lead to information not available through other scientific methods. Experts in a local area are more likely to understand cultural and political barriers to conservation, which may mitigate or avoid conflict inherent in conservation planning (Dorussen et al. 2005; Hmelo-Silver and Pfeffer 2004). In contrast, stakeholders, while more focused on specific elements of a system, as opposed to integrated information, can shed light on 'hotspots of value' in a given landscape, which may help guide planning and communication in conservation planning (Brown et al. 2004; Raymond and Brown 2006).

The issue of scale is important when integrating expert judgment. Local experts tend to think locally and have detailed knowledge and deep concern about localized areas, whereas ecoregional planners tend to think at larger spatial scales. This difference in perspective was illustrated in both the Wildlands Network Design exercise in the Greater Northern Appalachians (Sect. 11.3.2) and 2C1Forest's efforts to delineate conservation planning units for the ecoregion (Sect. 11.3.3). In both cases, local experts defined locally-relevant approaches for their provinces or states, or for subregions within them. In the Wildlands Network Design, local experts selected a mix of solutions so as to combine those that they thought were better tailored to natural versus culturally transformed subregions and that delineated areas that were more consistent with those previously defined as locally important. While such inputs may inject an important aspect of subregional stratification in ecoregional conservation planning, they may also detract from the systematic character of the planning exercise. Subregional or post-expert-integration analyses may serve to determine the influence of input by local experts on the degree to which a conservation plan achieves ecoregional goals.

In 2C1Forest's attempts to delineate consistent conservation planning areas for the Northern Appalachian/Acadian ecoregion, local experts sought consensus, with mixed success, around planning area that best matched pre-existing or emerging ecological classifications in their own state or province. This points to at least two issues important to conservation planners when engaging and integrating expert input. First, it is unlikely that ecoregional consensus will be reached through separate expert processes conducted in separate subregions (such as provinces and states). The degree to which incorporating conservation planning areas delineated on different bases would affect the results is unknown. Thus, it would be useful to (1) create processes that combine experts from various subregions to attempt to define a consistent ecoregion-wide conservation planning area scheme and/or (2) conduct analyses to determine the effects of using subregionally-defined conservation planning areas that differ across an ecoregion.

The second issue is ensuring that key concepts are clearly understood. In the 2C1Forest exercise, the concept and analytical purpose of conservation planning areas were not well understood by many of the experts participating in the meetings and workshops. The interplay and distinctions in roles between planning areas and ecological classifications is subtle and perhaps not adequately understood even among some experts. Further, using ecological classes as conservation planning areas would have implications for assessments of ecological irreplaceability that incorporate representation as a conservation goal, since every planning unit/ecological class would need to be represented. The analytical effect is that the comparative component around representation values would essentially be eliminated from the assessment of irreplaceability of conservation planning areas. Thus, expert involvement in delineating conservation planning areas requires clear understanding and communication of sometimes subtle, fundamental concepts and methodologies and their analytical purposes.

Regardless of the need to systematically integrate expert input, we stress the importance of flexibility. Flexibility in approach to conservation planning initiatives is necessary given cultural diversity across broad ecoregions, diversity both in scales of perspective among experts and in types of experts. While every ecoregion may not include two official languages and two nations, cultural diversity will inevitably exist nonetheless. Even within one country with one language, significant cultural differences often exist within large landscapes. Often these differences derive from complex and interrelated factors such as long histories of diverse livelihoods and land and resource use, population densities, the degree of cultural transformation of the landscape, and the history of settlement and occupancy by different cultures, including First Nations. Across large ecoregions, it is inevitable that cultural, social, and economic diversity, as well as biogeographic diversity, will occur, and the processes for integrating expert input into ecoregional planning will need to respond accordingly.

Both consensus and conflict are part of decision making processes among experts. It has been demonstrated that groups may tend to come to a consensus-based decision irrespective of what experts might decide were they to provide their input outside of the group (Bojórquez-Tapia et al. 2003). Did this occur in the three processes examined here? We believe that in the 2C1Forest exercise, consensus was achieved from each individual meeting in Maine, consistent with the findings of others (e.g., Bojórquez-Tapia et al. 2003) that experts tend to reach consensus on 'salient' issues. If a planning group does a good job of communicating to the experts the importance of their participation, then experts are more likely to engage in problem-solving directed toward the desired goal. In this scenario, the experts had a task that was likely to engage the 'consensus phenomenon,' as well as mitigate potential conflict over issues of 'turf.' However, the experts did not come to consensus in Québec, which may in part have been due to inadequate communication of the importance of their contribution to the overall planning initiative. As mentioned earlier, the experts in Maine had a clearer understanding of the ecoregional planning process based on a longer history of exposure in Maine to this perspective on conservation planning (Baldwin et al. 2007; Beazley et al. 2010; Reining et al. 2006). The participants in Québec may have benefited from additional preparatory materials and processes, including explicit assurances about the importance and value of their input.

Regardless, ecoregion-wide consensus was not reached and would be unlikely to be reached through separate independent meetings in various provinces and states, as consensus emerges within but not between such meeting groups. This reinforces the notion that some common meetings or cross-participation of experts among various groups across the ecoregion might be necessary for ecoregion-wide consensus to emerge, should this be determined to be necessary or desirable.

Ecoregional conservation planning is another example of a broader movement in resource management in which support is shifting to approaches that include multiple centers of interaction and away from those in which a single agency holds power (Conca 2005; Hajer and Wagenaar 2003; Plummer et al. 2005). In such contexts, integration of expert and other judgments is a fundamental component. In interactive approaches, questions arise around how much authority to make decisions should be devolved to each group (experts, conservation organizations, stakeholders, or local residents) and how such groups should coordinate or collaborate to make decisions. The potential for conflict is great, as various experts and stakeholders have different interests and will therefore support varying views and approaches. Conflict is an inevitable part of participatory and collaborative processes and should be embraced. Conflict resolution techniques are thus important to participatory processes, including those involving experts, particularly if experts represent a variety of fields or come at their task with localized versus integrative skills.

Working through conflicting views, however, leads to creation of new knowledge, as participants confront the views of others and strive toward resolution. Lee (1994) suggested that 'political conflict can provide ways to recognize errors, completing and reinforcing the self-conscious learning of adaptive management.' Acknowledging and addressing conflicts can serve to generate new information, strengthen outcomes, and reinforce ownership by experts (Conca 2005). Conflict and its resolution are important components of ecoregional conservation planning. The tensions between consensus and conflict are key to legitimacy and consensus. Only when multiple views are considered and genuinely incorporated will a strong network of supporters emerge to enable implementation of the conservation plan.

11.5 Lessons Learned

The documented benefits of expert engagement in ecoregional planning include increased support and legitimacy for the process and its results, and increased buy-in to its implementation. Experts bring a source of data, tacit knowledge, and local context and perspectives to the conservation planning process, and often infuse ecoregional planning with subregional diversity reflective of the ecoregion. They can provide guidance in key aspects, such as setting goals and targets, and assessing and combining various scenarios and solutions generated by expert systems. Together, a group of experts can provide both ecoregional overviews and local perspectives. As conservation planning becomes decentralized and multi-centered interactions and initiatives are more broadly supported, the engagement of experts will become more important to successful conservation initiatives.

As important as these benefits and imperatives are, challenges inherent in integrating the views of a wide range of experts remain, both among the experts and with expert systems. The tensions and synergies between conflicting views and processes towards consensus require negotiation processes and skills. With these, potentially disruptive differences can be embraced and channelled into creative new solutions and new knowledge. Questions remain as to how much consensus is necessary or desirable, and how much divergence or diversity can be incorporated into conservation planning across areas as large as ecoregions. Tensions exist between the need for consistency across the ecoregion and the imperative of flexibility deriving from diversity.

Studies call for the incorporation of expert judgment to be more systematic, which we support. Close attention to methods of expert engagement, the results accruing from such engagement, and the implications of expert influence on ecoregional planning is important to better understand how to best engage experts and integrate their input with other ecoregional planning methodologies, data sources, and tools, such as GIS-based expert systems. Various fundamental planning components that experts may influence, such as planning units, goals, targets, scenario development, and selection of solutions, should be analysed and compared to understand the implications of such decisions, such as their sensitivity to subregional variation. Opportunities to use nested hierarchies or hybridized approaches that combine high-level ecoregion-wide classifications and rules with localized subregional ones (e.g., MARXAN with Zones [http://www.uq.edu.au/marxan/]) should be explored and tested for their effectiveness. Such approaches could allow for systematic applications of consistent protocols across ecoregions, as well as accommodate diverse subregions. In the meantime, subsequent analyses of expert-driven results should determine the extent to which they retain spatial cohesion and achieve conservation goals at the ecoregional scale.

However, the examples described here of integrating expert judgment into systematic ecoregional conservation planning highlight several important lessons. First, engagement of experts should not be simply about gaining after-the-fact opinions or approval. To maximize the extent of buy-in by experts, many of whom would be responsible for ultimately implementing any resulting conservation plan, they need to be legitimately involved in the creation of methodology and results. Second, experts need to be distinguished in the planning process from stakeholders and local residents. While precise definitions of these different groups of participants are elusive and are likely to vary from one region to another, a transparent methodology for assessing and weighting each group's input is important. Finally, the methods used for engaging expert participation need to match the experts' technological capabilities and conceptual understandings. Planners should not assume that all participants have the same level of experience with methodologies or computer-based decision-support tools. While a lack of familiarity with such methodologies should not disqualify an expert from participation, it does highlight the importance of advanced preparation and planning on the part of those facilitating the process. In other words, simply inviting a group of experts to show up at a meeting and expecting significant results to emerge is unlikely to be successful.

Beyond these more technical issues are those related to the social sciences of expert engagement. How might we take what we have learned here to lead to more efficient and successful participatory conservation initiatives in the future? Are there ways to harness the consensus atmosphere in an expert meeting to help lead to more resilient conservation strategies? We have seen that local experts can drive results to match local priorities, and large-scale planning efforts thus risk becoming less spatially coherent if separate meetings are held strictly within subregions. Are there ways to build consensus across broad ecoregions, or is subregional consensus adequate, or alternatively, desirable? These questions can only be answered if the social and qualitative forms of data to build this understanding are tracked. Taken together with scientific measures of successful conservation, a more integrated picture of expert engagement will begin to emerge that will help conservation planners globally.

References

Anderson, C., Beazley, K., & Boxall, J. (2009). Lessons for PPGIS from the application of a decision-support tool in the Nova Forest Alliance of Nova Scotia, Canada. *Journal of Environmental Management, 90*, 2081–2089.

Anderson, M. G., Vickery, B., Gorman, M., Gratton, L., Morrison, M., Maillet, J., et al. (2006). The northern Appalachian/Acadian ecoregion: Ecoregional assessment, conservation status and resource CD. The nature conservancy, eastern conservation science and the nature conservancy of Canada: Atlantic and Quebec regions. Retrieved June 10, 2009, from http://conserveonline.org/workspaces/ecs/napaj/nap

Baldwin, R. F., Trombulak, S. C., Beazley, K., Reining, C., Woolmer, G., Nordgren, J. R., et al. (2007). The importance of Maine for ecoregional conservation planning. *The Maine Policy Review, 16*, 66–77.

Ball, I. R., & Possingham, H. P. (2000). *MARXAN (V1.8.2): Marine reserve design using spatially explicit annealing, a manual*. Retrieved November 10, 2005, from University of Queensland, Marxan Web site: http://www.uq.edu.au/marxan/docs/marxan_manual_1_8_2.pdf

Beazley, K., Smandych, L., Snaith, T., MacKinnon, F., Austen-Smith, P., Jr., & Duinker, P. (2005). Biodiversity considerations in conservation system planning: Map based approach for Nova Scotia, Canada. *Ecological Applications, 15*, 2192–2208.

Beazley, K., Reining, C., Doran, P., & Bettigole, C. (2010). Integrating site-selection tools and expert judgment for conservation system design in the Greater Northern Appalachians of Canada and the United States. In S. Bondrup-Nielsen, K. Beazley, G. Bissix, D. Colville, S. Flemming, T. Herman, et al. (Eds.), *Ecosystem based management: Beyond boundaries (Proceedings of the sixth international conference of science and management of protected areas, 21–26 May, 2007, Wolfville, NS)*. Retrieved February 10, 2010, from http://www.sampaa.org/publications.htm

Bojórquez-Tapia, L. A., Brower, L. P., Castilleja, G., Sánchez-Colón, S., Hernández, M., Calvert, W., et al. (2003). Mapping expert knowledge: Redesigning the monarch butterfly biosphere reserve. *Conservation Biology, 17*, 367–369.

Brown, G., Smith, C., Alessa, L., & Kliskey, A. (2004). A comparison of perceptions of biological value with scientific assessment of biological importance. *Applied Geography, 24*, 161–180.

Carroll, C., Noss, R. F., Paquet, P. C., & Schumaker, N. H. (2003). Use of population viability analysis and reserve selection algorithms in regional conservation plans. *Ecological Applications, 13*, 1773–1789.

Carver, S. (2003). The future of participatory approaches using geographic information: Developing a research agenda for the 21st century. *Urban and Regional Information Systems Association Journal, 15*, 61–71.

Clevenger, A. P., Wierzchowski, J., Chruszcz, B., & Gunson, K. (2002). GIS-generated, expert-based models for identifying wildlife habitat linkages and planning mitigation passages. *Conservation Biology, 16*, 503–514.

Conca, K. (2005). *Governing water: Contentious transnational politics and global institution building*. Cambridge, MA: MIT Press.

Dorussen, H., Lenz, H., & Blavoukos, S. (2005). Assessing the reliability and validity of expert interviews. *European Union Politics, 6*, 315–337.

Doswald, N., Zimmermann, F., & Breitenmoser, U. (2007). Testing expert groups for a habitat suitability model for the lynx (*Lynx lynx*) in the Swiss Alps. *Wildlife Biology, 13*, 430–446.

Groves, C. R. (2003). *Drafting a conservation blueprint: A practitioner's guide to planning for biodiversity*. Washington, DC: Island Press.

Hajer, M. A., & Wagenaar, H. (2003). *Deliberative policy analysis: Understanding governance in the network society*. Cambridge, UK: Cambridge University Press.

Hmelo-Silver, C. E., & Pfeffer, M. G. (2004). Comparing expert and novice understanding of a complex system from the perspective of structures, behaviors, and functions. *Cognitive Science, 28*, 127–138.

Jankowski, P., & Nyerges, T. L. (2001). *Geographic information systems for group decision making: Towards a participatory, geographic information science*. London: Taylor and Francis.

Jordan, G. (1998, October). *A public participation GIS for community forestry user groups in Nepal: putting people before the technology.* (Paper presented at the International Conference on Empowerment, Marginalization and Public Participation GIS, Santa Barbara, California).

Kyem, P. A. K. (2000). Embedding GIS applications into resource management and planning activities of local and indigenous communities: A desirable innovation or a destabilizing enterprise? *Journal of Planning Education and Research, 20*, 176–186.

Lambeck, R. J. (1997). Focal species: A multi-species umbrella for nature conservation. *Conservation Biology, 11*, 849–856.

Lee, K. N. (1994). *Compass and gyroscope: Integrating science and politics for the environment*. Washington, DC: Island Press.

Margules, C. R., Nicholls, A. O., & Pressey, R. L. (1988). Selecting networks of reserves to maximise biological diversity. *Biological Conservation, 43*, 63–76.

Margules, C. R., Pressey, R. L., & Williams, P. H. (2002). Representing biodiversity: Data and procedures for identifying priority areas for conservation. *Journal of Biosciences, 27*, 309–326.

Miller, B., Reading, R., Stritholt, J., Carroll, C., Noss, R., Soulé, M., et al. (1999). Using focal species in the design of nature reserve networks. *Wild Earth, 8*, 81–92.

Noss, R. (2003). A checklist for wildlands network designs. *Conservation Biology, 17*, 1270–1275.

Noss, R., & Cooperrider, A. (1994). *Saving nature's legacy: Protecting and restoring biodiversity.* Washington, DC: Island Press.

Noss, R. F., Dinerstein, E., Gilbert, B., Gilpin, M., Miller, B., Terborgh, J., et al. (1999). Core areas: Where nature reigns. In M. Soulé & J. Terborgh (Eds.), *Continental conservation: Scientific foundations of regional reserve networks* (pp. 99–128). Washington, DC: Island Press.

Noss, R. F., Carroll, C., Vance-Borland, K., & Wuerthner, G. (2002). A multicriteria assessment of the irreplaceability and vulnerability of sites in the Greater Yellowstone Ecosystem. *Conservation Biology, 16*, 895–908.

Plummer, R., Spiers, A., FitzGibbon, J., & Imhof, J. (2005). The expanding institutional context for water resources management: The case of the Grand River Watershed. *Canadian Water Resources Journal, 30*, 227–244.

Possingham, H., Ball, I., & Andelman, S. (2000). Mathematical methods for identifying representative reserve networks. In S. Ferson & M. Burgman (Eds.), *Quantitative methods for conservation biology* (pp. 291–306). New York: Springer.

Raymond, C., & Brown, G. (2006). A method for assessing protected area allocations using a typology of landscape values. *Journal of Environmental Planning and Management, 49*, 797–812.

Reining, C., Beazley, K., Doran, P., & Bettigole, C. (2006). *From the Adirondacks to Acadia: A wildlands network design for the Greater Northern Appalachians (Wildlands Project Special Paper No. 7).* Richmond, VT: Wildlands Project.

Schlossberg, M., & Shuford, E. (2005). Delineating "public" and "participation" in PPGIS. *Urban and Regional Information Systems Association Journal, 16*, 15–26.

Sieber, R. E. (2000). GIS implementation in the grassroots. *Urban and Regional Information Systems Association Journal, 12*, 15–29.

Trombulak, S. C., Anderson, M. G., Baldwin, R. F., Beazley, K., Ray, J. C., Reining, C., et al. (2008). *The northern Appalachian/Acadian ecoregion: Priority locations for conservation action (Special Report No. 1).* Warner, NH: Two Countries, One Forest.

Tynjälä, P. (1999). Towards expert knowledge? A comparison between a constructivist and a traditional learning environment in the university. *International Journal of Educational Research, 31*, 357–442.

The University of Queensland (n.d.). The C-plan conservation planning system. Retrieved February 1, 2010, from The Ecology Centre Web site: http://www.uq.edu.au/ecology/index. html?page=101951

Weiner, D., Harris, T. M., & Craig, W.J. (2001, December). *Community participation and geographic information systems.* (Paper presented at the Workshop on Access and Participatory Approaches in Using Geographic Information, Spoleto, Italy).

Woolmer, G., Trombulak, S. C., Ray, J. C., Doran, P. J., Anderson, M. G., Baldwin, R. F., et al. (2008). Rescaling the human footprint: A tool for conservation planning at an ecoregional scale. *Landscape and Urban Planning, 87*, 42–53.

Chapter 12
The GIS Challenges of Ecoregional Conservation Planning

Gillian Woolmer

Abstract The tools of Geographic Information Systems (GIS) are well suited to the application of conservation planning, a pursuit that requires the overlay and analysis of often large volumes of geographic information, including the locations and distribution of multiple conservation targets and threats. During any conservation planning process, challenges related to the use of GIS can be expected, particularly for large planning areas that span multiple administrative jurisdictions. Challenges likely to be encountered relate to (1) the complex nature of spatial data, including data sources, access, licensing, quality, and compatibility, (2) the need to develop adequate capacity for GIS for the duration of the planning process, and (3) making spatial information generated by the GIS based planning process available to partners and stakeholders. By understanding the nature of the GIS challenges to be expected, conservation managers and GIS professionals can plan for the resources necessary to successfully achieve the goals of the planning process. In this chapter, I share the GIS experiences, challenges, and lessons learned from a multi-year, multiple-partner conservation planning effort for the transboundary Northern Appalachian/Acadian Ecoregion of North America.

Keywords Geographic Information Systems • GIS • Mapping • Spatial analysis • Spatial data

12.1 Introduction

With the analysis of information on the locations and distribution of multiple conservation targets and threats a central feature of conservation planning (Margules and Pressey 2000; O'Neil et al. 2005), a system to manage and analyze digital spatial data is critical. A geographic information system (GIS) is designed to collect, store,

---

G. Woolmer (✉)
WCS Canada, Suite 600, 720 Spadina Ave., Toronto, ON M5S 2T9, Canada
e-mail: gwoolmer@wcs.org

S.C. Trombulak and R.F. Baldwin (eds.), *Landscape-scale Conservation Planning*,   257
DOI 10.1007/978-90-481-9575-6_12, © Springer Science+Business Media B.V. 2010

transform, analyze, and display such spatial information (Burrough 1986; Burrough and McDonnell 1998; ESRI 1997), which make the tools of a GIS ideally suited to conservation planning.

The use of GIS technologies by the conservation community has increased significantly since the mid-1990s. Today, GIS is a ubiquitous tool for conservation practitioners, a development that can be attributed to several factors: (1) the emergence of accessible desktop GIS software, (2) increased affordability of GIS-capable computers, (3) increased availability of spatial data and information that are published in GIS-compatible formats by both the public and private sectors at relatively low or no cost, and (4) increased availability of specialized GIS training at colleges and universities.

Conservation planning is a particularly challenging task when undertaken at the scale of ecoregions that transcend state, provincial, and/or country boundaries. Such projects require the collation, processing, and analysis of large quantities of complex spatial information of different types and quality, collected by various parties for different purposes, at varying levels of detail and published from multiple sources. Effective conservation planning at such scales, therefore, requires the development of a correspondingly complex information database that has been organized into a coherent, transparent system that can be readily accessed and used by multiple collaborators. Such database management tasks are not trivial and yet are frequently underestimated and not adequately planned for at the outset of many conservation planning exercises.

Conservation GIS dates back to the use of paper maps, printed aerial photographs, satellite imagery, and mylar overlays. The first national-level computerized GIS, known as the 'Canada Geographic Information System' (CGIS), was released in the early 1960s in Canada by the federal Department of Forestry and Rural Development (Tomlinson 1984; Wing and Bettinger 2008) to manage spatial information about soils, agriculture, recreation, wildlife, waterfowl, forestry, and land use collected for the Canadian Land Inventory (CLI). An early example of GIS relevant for conservation planning was the USGS Gap Analysis Program (GAP), launched in 1989 (Scott et al. 1993). The simple but powerful concept of overlaying information on the distributions of native species and natural communities with that of land protection status was aimed at assessing the degree to which important areas for biological diversity were represented by the network of conservation lands, and correspondingly to identify 'gaps' to guide future conservation action. These early approaches to GIS-based land use and conservation planning evolved rapidly into computer-driven models designed to assess threats due to land-use change (Chap. 2), optimize representation (Chap. 14), analyze connectivity (Chap. 16), and predict climate influences (Chap. 15).

When embarking on a conservation planning initiative, it is important that project leaders be well-prepared for the GIS-related challenges that may arise during the planning process to avoid any potential delays or even derailment of the process. The goal of this chapter is to review these challenges and offer guidance based on the lessons I have learned from participation in Two Countries, One Forest (2C1Forest), a bi-national, multi-stakeholder conservation planning collaborative (Bateson 2005) for the Northern Appalachians/Acadian ecoregion of North America (Chap. 1). I will cover several important issues related to the use

12 The GIS Challenges of Ecoregional Conservation Planning

of GIS in conservation planning: (1) the complex nature of spatial data, including data sources, access, licensing, quality, and compatibility, (2) the need to develop adequate capacity for GIS for the duration of the planning process, and (3) the importance of making spatial information generated by GIS for ecoregional conservation planning available to partners and stakeholders – a critical outreach component required to develop support for plan implementation.

The Northern Appalachian/Acadian ecoregion encompasses portions of four U.S. states (New York, Vermont, New Hampshire, and Maine) and all or part of four Canadian provinces (Québec, New Brunswick, Nova Scotia, and Prince Edward Island). This geographic context presented numerous challenges to developing a coherent plan derived from spatial data, primarily because these data were published by numerous NGOs and government agencies for multiple jurisdictions, levels of governance (two countries, multiple states and provinces), and distinct cultural and linguistic traditions.

12.2 Building a GIS for Ecoregional Conservation Planning

A major component of any conservation planning process is a GIS that is tailored to the goals of the planning effort and the extent of the planning region. The building of a GIS is often described as simply the compilation and overlaying of digital spatial data; however, a more complete definition is that a GIS is actually an organized system of interrelated geographic data, computer hardware, software, and personnel (ESRI 1997). A comprehensive view of GIS as both an information-management system and decision-support tool is essential for undertaking a successful ecoregional conservation planning project. This section will review and provide guidance on building such a GIS with a focus on the aspects related to data and personnel, which comprise the most challenging and complex aspects of the overall system. A discussion of hardware and software options is beyond the scope of this chapter, and these facets of GIS systems have seen significant and rapid improvement in recent years. Specifically, GIS-capable desktop computers tend to be affordable and accessible in most areas of the world, and a variety of GIS software application options are now available, including no-cost open source and freeware (e.g., GRASS and Quantum GIS) and proprietary software (e.g., ArcGIS by ESRI, MapInfo by Rockware), some of which can be acquired at significantly reduced cost for conservation applications through grant programs, such as the ESRI Conservation Program (www.conservationgis.org).

12.2.1 Data

Spatially explicit data provide the foundation for any ecoregional conservation planning process, and governments – nations, states and provinces, and municipalities – increasingly publish spatial information suitable for ecoregional-scale planning in

GIS formats. However, because ecoregions are largely defined by their climate and biogeography (Bailey 2004), their boundaries rarely align with or fall entirely within the boundaries of a single administrative jurisdiction. More often, ecoregions are 'transboundary' in that they span multiple political and administrative jurisdictions, and therefore ecoregional conservation planning typically requires building transboundary spatial data layers.

GIS data represent real-world features (e.g., towns, roads, land use, elevation) in digital formats as one of two abstractions (Burrough and McDonnell 1998): discrete objects (e.g., a house or recorded species location) or continuous fields (e.g., rainfall amount, elevation, or land cover). The latter are either quantitative in nature, like millimeters of rainfall or qualitative like land-cover classes. For conservation planning, many different types of spatial information will be required, and although these data can be categorized in many different ways, common types of information will likely be required. 'Base data' refers to map layers that serve as components of maps over which other spatially explicit information are placed for reference. These provide the underlying context for the planning exercise upon which other data can be positioned. Common base data themes include administrative and jurisdictional boundaries (e.g., municipalities, state/provincial/national borders, protected areas), elevation, hydrology (e.g., lakes, rivers, watersheds), populated places (e.g., cities, towns, villages, urban areas), and transportation networks (e.g., roads and railways).

Other types of spatial information that are often required for conservation planning include data on human population (e.g., census), human land uses (e.g., agriculture, mining, residential and industrial development), natural land cover (e.g., vegetation classes), and geology. Desired biological information includes the locations and distributions of rare and endangered species, other focal species, communities, ecosystems, and processes that may have been selected as important conservation features.

Five steps are essential to building a transboundary GIS database for a conservation planning project: (1) determine the required map scale and data resolution, (2) find and access the datasets, (3) assess the quality of the datasets, (4) assess the compatibility of the datasets, and (5) combine datasets to create single transboundary GIS data layers. These steps are described below using examples from the Northern Appalachian/Acadian ecoregion conservation planning initiative (Trombulak et al. 2008).

Data Scale and Resolution For any project that involves spatial analysis, it is critical to use GIS data that represent features at an appropriate scale or resolution for the task. For conservation planning across large landscapes, such as ecoregions, the most appropriate spatial data are those that represent geographic features either as vector data ranging between the map scales of 1:25,000 and 1:100,000 or as raster data with cell sizes ranging between 25 and 100 meters. Choosing the appropriate scale and resolution of GIS data for conservation planning involves a balance between meeting broad-scale and local-level conservation objectives. For example, while it may be quite useful to discern individual trees or houses for planning at the scale of a municipality,

12 The GIS Challenges of Ecoregional Conservation Planning   261

working with data at a higher resolution than forest blocks and residential areas is seldom necessary from a regional perspective. In fact, dealing with data that are too fine-scale for ecoregional planning can restrict the GIS process unnecessarily through the need to append thousands of data tiles and handle enormous volumes of data. Spatial data need only possess the spatial accuracy and information accuracy required to capture the degree of local variation in the geographic distribution and types of features on the landscape that is necessary for the results of a conservation planning process to be believable, relevant, and implementable at the scales at which local decision makers and land-use planners operate (Rejeski 1993).

The optimum scenario for transboundary ecoregional planning is to obtain existing published GIS data layers that contain features mapped at the appropriate scale and cover the full extent of the planning region. While digital spatial data for such large regions can be found as part of freely available continental or global datasets, most such datasets contain geographic features represented at relatively coarse scales. Examples include the North American Environmental Atlas (1:10,000,000; National Atlas 2009) and the global Vector Map Level 0 database (1:1,000,000; NIMA National Imagery and Mapping Agency 2000). This trade-off between spatial extent and resolution is not always the case; the Protected Areas Database of the U.S. (PAD-US Protected Areas Database of the United States 2010), for example, is far more comprehensive than most state databases. Yet, in general, using already-amalgamated data from multiple jurisdictions that have large geographic extents often comes at the expense of the grain of such data. Even then, regional or global datasets are not often available at a suitable scale, and the only option in that case is to 'stitch together' the more local datasets of the appropriate map scale that have been published by the individual jurisdictions that comprise the planning region (Chap. 18).

Published spatial data between the scales of 1:25,000 and 1:100,000 exist for most jurisdictions in North America, Europe, and Australasia. However, for other regions of the world, spatial data at these scales may not be freely available. Under such circumstances, it is necessary either to use GIS data from less detailed continental or global datasets or, if possible, obtain data from other sources such as NGOs, natural resource extraction companies working in the region, or for-profit digital data producers.

Searching for, Accessing, and Using Data Although any good web search engine can be used to search for spatial data, the best tools for locating high-quality spatial data are web-based GIS clearinghouses and portals. These provide a variety of ways to search for data, using a combination of keyword, interactive maps, and thematic data listings, and they promise the highest likelihood of identifying base and environmental data layers. The clearinghouses and portals that were most useful for identifying GIS data for the Northern Appalachians/Acadian ecoregion are listed in Table 12.1.

In the U.S., any spatial data generated through tax monies is generally distributed freely, unless security or other sensitivity concerns (e.g., rights of private landowners) exist. In Canada, however, the tradition of cost recovery for the distribution of government-published spatial data makes the situation markedly different.

Table 12.1 Government and private GIS clearinghouses and portals used to find published GIS data for the Northern Appalachian/Acadian ecoregion

Government	
Canada	
National	
GeoConnections Discovery Portal	http://geodiscover.cgdi.ca
GeoBase – Canadian Council of Geomatics	http://www.geobase.ca
GeoGratis – National Resources Canada	http://geogratis.cgdi.gc.ca
Provincial	
Nova Scotia 'GeoNOVA'	http://www.geonova.ca
Prince Edward Island	http://www.gov.pe.ca/gis
New Brunswick – Service New Brunswick	http://www.snb.ca/gdam-igec/e/2900e.asp
USA	
National	
GeoData.Gov	http://gos2.geodata.gov/wps/portal/gos
State	
New York State GIS Clearing house	http://www.nysgis.state.ny.us
Vermont Center for Geographic Information	http://www.vcgi.org/
New Hampshire 'GRANIT'	http://www.granit.unh.edu
Maine Geographic Information Systems	http://megis.maine.gov/
Private	
ESRI	
Geography Network	http://www.geographynetwork.com
Geography Network Canada	http://www.geographynetwork.ca

For example, digital census data in the U.S. are distributed freely online by the U.S. Census Bureau, whereas in Canada, GIS and statistical data for the finest level of census units can be acquired only at a price. If these data are required for large areas, like the Northern Appalachian/Acadian ecoregion, it can cost thousands of dollars. However, a trend is emerging in Canada toward making many types of government published spatial data freely available through federal and provincial portals; examples include the GeoNova portal for Nova Scotia and the Land Information Ontario (LIO) portal.

If a required data layer does not exist, transboundary digital datasets need to be created from scratch, which may involve digitizing hardcopy maps, interpreting and georeferencing orthophotos, or classifying satellite images. For example, a road data layer can be created by digitizing road features from aerial photos, slope and aspect can be derived from a digital elevation model, and species distributions can be derived from predictive habitat suitability models (Guisan and Zimmermann 2000). In any case, the potential need to purchase good-quality data must be factored into the costs of any conservation planning project, whether a needed layer involves a simple fee purchase from a vendor or paying someone to create, reclassify, or model it.

Regardless of the costs involved, data access and use may require a data license with a publisher to protect both intellectual property and remove the publisher of any potential liabilities that result from data use (Longhorn et al. 2002). Such data licenses are likely to restrict data use and re-distribution, requiring the user to specify exactly how the data will be used and for what purpose. The location where the data will reside and the names, or positions, of staff that will have access to the data may also be required. This type of data licensing helps the publisher, often a government agency, ensure that data are not used for inappropriate applications and reduce risk of data being re-distributed, intentionally or unintentionally, to other parties without permission.

Data licenses are legal documents and, like any other, should be managed and stored securely with minimal risk of loss so they can be accessed and referenced as needed. Conservation planning is often a multi-year process that involves collaborations among multiple organizations and leads to new analyses years after a data license was acquired. Therefore, data licenses need to be revisited frequently to ensure compliance and to renegotiate licenses to permit additional uses.

Data Quality Quite apart from the challenges of obtaining the variety of desirable GIS data for conservation planning, a further complexity is introduced once the data are in hand and interpretation is required. Practitioners will need to understand how a dataset was created and evaluate the quality of the data it contains. For this reason, it has become standard for data publishers to produce metadata – 'data about data' – to accompany a dataset. Metadata come most commonly as an additional file to the digital GIS layer, for example as a text file, an HTML files, or in the case of ESRI format GIS files, as an XML file. GIS metadata allow the user to determine if a dataset is suitable for their specific needs and allows for GIS data to be searchable through GIS clearinghouses and portals. The two most widely used GIS metadata standards – the ones by the U.S. Federal Geographic Data Committee (FGDC Federal Geographic Data Committee 2000) and the International Organization of Standards (ISO International Organization of Standards 2010) – stipulate what types of information about a dataset must be included in GIS metadata, including information about the data publisher, data quality, spatial reference, data attribute fields, and attribute codes.

Because conservation planning involves decisions about particular pieces of geography, it is important to understand the nature and quality of the underlying data at any given location. In this vein, GIS data may represent discrete features or may represent derived information. For example, spot height elevation data on a topographic map represent actual on-the-ground field observations, whereas a continuous elevation surface dataset in the form of a digital elevation model (DEM) is not created from a continuum of direct observations. Rather, elevation values for locations between discrete elevation observations are modeled using interpolation techniques. The quality of an interpolated surface depends on the accuracy, number, and distribution of known data points and the suitability of the interpolation method (Aronoff 1995). Continuous thematic data, like land use, are frequently derived from the interpretation and classification of satellite images. The accuracy of such classifications can be evaluated (Stehman and Czaplewski 1998) and can be surprisingly low. The 1992 U.S. National Land Cover Dataset (NLCD) derived from the Landsat TM images, has only 80.5% and 59.7% accuracy rates for the general and detailed land cover classification levels (Yang et al. 2001), meaning that even at the coarsest grain of classification, land cover was incorrectly classified 19.5% of the time.

Likewise, habitat suitability models meant to predict the distribution and status of a species are generally derived from characteristics associated with point locations of animal occurrences and are not always assessed for their accuracy (Groves 2003). This underscores the importance of not accepting all data at face value, and instead highlights the need to examine the manner in which data are collected and understand the quality of information associated with modeling efforts prior to using them. Knowing how a dataset was created or derived and the accuracy of the derivation method will allow the user make choices about which datasets are best to use.

Additional complexities related to data accuracy are illustrated by two map layers that are widely used for conservation purposes – protected areas and natural heritage databases. Having accurate and up-to-date information on the protected status of a given land base is key to understanding the degree to which conservation targets are protected in the planning area and where additional conservation measures are needed. Acquiring this information for an ecoregion spanning multiple jurisdictions is complicated by the fact that accurate and complete spatial data on the locations of protected areas for a nation, state, or province are often not managed by a single agency because multiple agencies have responsibility for managing these lands. For example, Canadian national parks are managed by the federal agency Parks Canada, while provincial parks are managed by provincial natural resource agencies (with a similar hierarchy for different protected area designations in the U.S.). Some provinces and states have an agency that maintains GIS data but others do not, sometimes making it necessary to acquire data from multiple agencies even within a single jurisdiction. In recent years, global and regional protected areas datasets have been published to combat this challenge, by compiling data from multiple jurisdictions into a single database. For example, the World Database on Protected Areas (WDPA World Database on Protected Areas 2009), the Protected Areas Database of the United States (PAD-US Protected Areas Database of the United States 2010), and the North American Environmental Atlas (National Atlas 2009) are assembled from multiple governmental (e.g., provinces, states, countries) and private (e.g., non-governmental organizations) sources.

However, some databases are more comprehensive than others in the sense of what gets included as a protected area. For example, while the most recent PAD-US contains over 700,000 terrestrial protected area polygons assembled from states and participating agencies, the 2009 WDPA, which incorporates an earlier version of PAD-US, only has 6,770 for the United States and 112,725 for the entire globe. This discrepancy derives from the stricter definition for protected areas used by the WDPA. Continental efforts such as the North American Environmental Atlas seek to harmonize data across neighboring country boundaries for the very purpose of aiding conservation planning at multiple scales. These efforts are more inclusive than the WDPA and rely on how each country defines their protected areas and how frequently they update their GIS data. Conservation lands are not limited, however, to those under government jurisdiction. Privately owned lands in North America,

Europe, and elsewhere with conservation easements or 'servitudes' need to be assessed for their conservation value (Jenkins 2008). As with protected areas, these lands can be classified according to established protected area classification schemes, such as those defined by the International Union for Conservation of Nature (Dudley 2008) or the U.S. GAP Analysis Program (Crist 2000). Information about the location of these privately owned lands has historically been difficult to gather and tends to change rapidly because new easements are constantly being purchased. Fortunately, in the U.S. a multi-partner initiative is underway to track and map lands protected by conservation easements through the National Conservation Easement Database (NatureServe 2009a).

All of this illustrates the complexities involved in the interpretation and use of seemingly straightforward data on the locations of protected areas. Every conservation planning project should consider the source of these data in the context of the scale of the study (e.g., local-global), the definitions of protected area status employed (e.g., IUCN or GAP codes), and the age and accuracy of the dataset.

Obtaining spatial data on rare species, focal species, and ecosystems provides an additional example of challenges inherent in interpreting and applying GIS data for conservation planning. First, the availability of suitable data depicting occurrence, distribution, and/or abundances of such conservation features varies from place to place around the world. In many parts of the world, databases of the known locations of rare, threatened, and endangered species are maintained. For example, in the U.S. and Canada information from Natural Heritage Programs within the majority of states and provinces can be accessed centrally through NatureServe (NatureServe 2009b; NatureServe Canada 2009). However, because such data are not always collected systematically, the strength of a given dataset depends on collection effort; hence, the most robust datasets of this nature tend to be spatially biased in relation to access (e.g., roads, rivers, populated areas; Pressey 2004). Moreover, data gaps that result from unsurveyed areas can lead to false interpretation of species absences. An additional consideration is that the distribution of rare species, focal species, and/or ecosystems may not be entirely based on actual field observations but derived from computer-based models. When modeling species occurrences and suitable habitat, it is important for conservation planners to evaluate sources of species distribution data, including potential biases of locality information, and to determine the need to improve the underlying data (Akcakaya 2004; Mackenzie et al. 2006). Where efforts to collect such information have not been consistent across a landscape, it is important for conservation planners to acknowledge and account for the inherent limitations of this information (Pressey 2004).

Data Compatibility A significant challenge for building transboundary GIS data layers is indentifying GIS datasets from adjacent administrative jurisdictions and multiple agencies that are compatible and can be appended together to create GIS layers for the full extent of the planning region. The most compatible datasets are those that contain geographic features represented in the same data model (vector or raster), at the same scale or resolution with similar data accuracy, have comparable dates of data collection and/or publication, and have been created for the same purpose.

The compatibility of feature attributes and attribute codes must also be considered for datasets to be combined across boundaries. If two roads datasets have been created by different jurisdictions, for example, it is likely that different road classification schemes (based on road width, number of lanes, and surface type) and sets of attribute codes will have been used, requiring a complex cross-walk of the road class codes of the two datasets.

When harmonizing the attributes and attribute codes of datasets so they can be combined, it is important to decide how much attribute information to transfer from the input datasets to the resulting output transboundary data layer, decisions that are typically based on the intended uses for the transboundary data. For thematic data-sets (e.g., land use/land cover, roads), cross-walking the attribute codes of multiple input datasets most often requires reclassifying the feature types of each input data-sets to a reduced set. Such an approach will result in a loss of information, but this is often the only option in order to build a data layer from multiple sources.

One example of the process involved in ensuring that spatial datasets are comparable were the U.S. and Canadian census datasets used for the Northern Appalachian/Acadian ecoregional planning effort. The U.S. census is executed fully every 10 years, with an intermediate partial census survey from which census statistics are estimated every fifth year after a full census. With every census, a GIS dataset called the TIGER/Line Files is updated from the previous census. This dataset has a spatial accuracy and map scale of 1:100,000 and contains GIS data layers that include census mapping units (irregular polygons called census blocks), administrative boundaries, roads, rail lines, and many other geographic features (U.S. Census Bureau 2000). The Canadian census is likewise fully executed every 10 years, but 1 year after the U.S. Census, and like in the U.S., a partial intermedi-ate census is conducted 5 years after a full census. An accompanying GIS dataset is also updated with every census, including census mapping units, administrative boundaries, and roads data. These data have a spatial accuracy of between 1:50,000 and 1:250,000 (Statistics Canada 2002a, b). The smallest census mapping unit in Canada is the Dissemination Area (DA), which is also an irregular shape and size.

Achieving a single transboundary human population data layer that covered the whole Northern Appalachian/Acadian ecoregion required the amalgamation of the smallest census mapping units with attribute fields for total population and total number of dwellings (or housing units) per census unit – statistics collected by both the U.S. Census Bureau and Statistics Canada. This data layer was created by appending the 2000 U.S. census blocks data layer with the 2001 Canadian dissemination areas data layer (Fig. 12.1). These census datasets were deemed compatible because (1) they were created for the same purpose, (2) they are mapped at comparable (although not exactly the same) map scales, (3) the dates of data collection and publication represent the most recent versions of census data available at the time our study, and (4) the key attributes – total population and number of dwellings – were present in both datasets.

Geoprocessing to Create Transboundary Datasets Appending adjacent datasets into a single transboundary GIS data layer requires that both the spatial features and associated attributes of the input datasets be combined. Conservation managers

12 The GIS Challenges of Ecoregional Conservation Planning

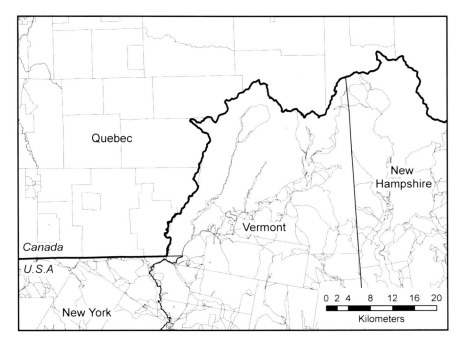

Fig. 12.1 A map of the smallest census mapping units for the Northern Appalachian/Acadian ecoregion – the 2000 U.S. census blocks and the 2001 Canadian dissemination areas – for a region at the U.S./Canadian border

need to be aware that this can entail considerable time and effort depending on input data format (raster or vector), number of input datasets, the magnitude of feature misalignments across dataset boundaries, and the congruity of feature attribute codes for cross-walking.

To illustrate, creating a seamless ecoregional data layer of census blocks for the Northern Appalachian/Acadian ecoregion required edge matching misaligned polygons because when the U.S. and Canadian census datasets were appended, topological errors occurred along the data boundary (Fig. 12.2a). The resulting data gaps and overlaps were removed through a combination of automated and manual GIS procedures (Fig. 12.2b).

When creating a transboundary roads data layers from U.S. and Canadian datasets (Fig. 12.3), on the other hand, the misalignments between road features at the border were less severe. We chose not to perform any edge matching in this instance because (1) the output roads dataset was not going to be used for any GIS applications that would require line features to be connected, such as networking applications, and (2) the misalignments were generally less than 30 m, or less than half the cell size (90 m) of the common raster analysis grid for the ecoregion. If misalignments of vector features across a data boundary are greater than half the intended raster analysis cell size, edge matching is recommended so that misalignments are not carried forward when the transboundary vector layer is converted to raster format for analytical purposes.

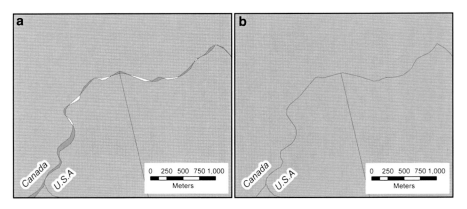

Fig. 12.2 Edge matching to align census mapping units at the boundaries of the datasets to a common edge: (**a**) polygon features before edge matching, showing data gaps in white and overlaps in darker grey, and (**b**) polygon features after edge matching

Cross-walking feature attribute codes across multiple datasets generally requires reducing the codes that represent feature classes in the input datasets to a smaller set of classes in the output dataset. The output set of feature classes may be derived from one of the input datasets or may be a new, simplified set of feature classes. For example, when we created the transboundary roads GIS layer, reconciling the 45 road categories in the U.S. TIGER/Line Files roads dataset with the Canadian roads dataset (DMTI Spatial 2009) that contained only 6 categories necessitated the reduction of the input road classes to four output roads classes (Table 12.2).

All conservation planning ultimately requires characterization of land cover and land use (LULC). Creating the transboundary LULC data layer in the Northern Appalachian/Acadian ecoregion was not straightforward. We had to combine five input datasets from different sources, all with different sets of LULC classes, two of which were in raster format and three in vector format. For the U.S. portion of the ecoregion, we used the 1992 National Land Cover Dataset (NLCD), a 30-m raster dataset published by the USGS. Because no equivalent to this dataset exists nationally for Canada, LULC datasets had to be sourced for each of the four provinces within the ecoregion. To combine LULC datasets, we cross-walked the input LULC attribute codes of the five input datasets by reducing them to a set of 15 LULC categories, which were derived from the 24 classes of the U.S. NLCD dataset. This required that we significantly condense the LULC classes of the input datasets, some of which contained up to 64 LULC categories (Table 12.3).

12.2.2 GIS Capacity

Any conservation planning initiative requires a team that includes personnel who are skilled in GIS and who can dedicate time towards the management and analysis of large volumes of spatial data. Additional GIS-related tasks necessary to the planning process

12 The GIS Challenges of Ecoregional Conservation Planning

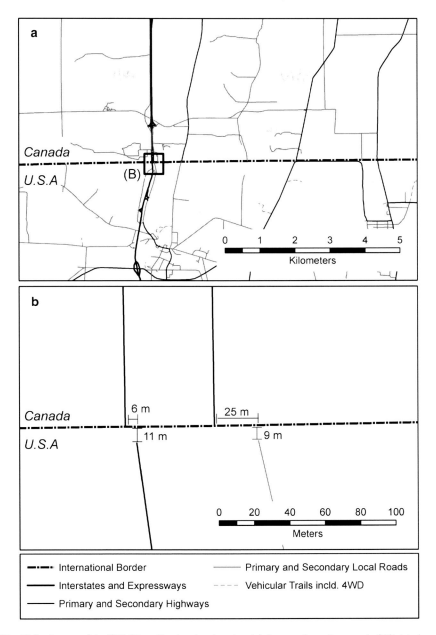

Fig. 12.3 A map of the U.S./Canadian border showing (**a**) the transboundary roads GIS data layer comprised of roads data from the U.S. and Canada, and (**b**) the misalignment distances between features across the data boundary

270 G. Woolmer

Table 12.2 The cross-walk table for the road-class attribute codes of two GIS roads data layers from the U.S. and Canada

U.S. input	Transboundary output	Canada input
TIGER/line files road classes	Transboundary roads data layer road classes	DMTI spatial roads road classes
Primary highway with limited access A11, A12, A13, A14, A15, A16, A17, A18 Road with special characteristics A63	1 – expressways and interstates	1 – expressway
Primary road without limited access A21, A22, A23, A24, A25, A26, A27, A28 Secondary and connecting road A31, A32 A33, A34, A35, A36	2 – principle and secondary highways	2 – principle highway 3 – secondary highway
Secondary and connecting road A37 and A38 Local, neighborhood, and rural road A41, A42, A43, A44, A45, A46, A47, A48 Road as other thoroughfare A70, A73 Road with special characteristics A60, A61, A62, A64	3 – major and local roads	4 – major road 5 – local road
Vehicular trail (4WD) A51, A52, A53 Road as other thoroughfare A71, A72, A74	4 – vehicular trails incld. 4WD	6 – trail

include cartographic support for the production of maps for meetings, presentations, and reports, and interactive mapping support to facilitate active map-based collaborations during the workshops necessary to engage participants (Chaps. 4 and 11).

Ideally, a conservation collaborative involving multiple organizations, agencies, and/or research institutions working at an ecoregional scale will include a number of professionals who share data and collaborate on GIS-related analyses. Individuals contributing GIS expertise may include full-time paid technicians, analysts, and researchers, with additional mapping, analyses, and models produced by consultants as needed. Indeed, it is becoming increasingly common for graduate students to emerge from M.Sc. and Ph.D. degrees with technical GIS expertise, which has translated into a general increase in GIS-related capacity for conservation planning projects.

Our experience in the Northern Appalachian/Acadian ecoregional planning initiative underscored the importance of GIS capacity across multiple organizations, because one person alone was not able to support the full GIS needs of the

12 The GIS Challenges of Ecoregional Conservation Planning

Table 12.3 The land use/land cover (LULC) classes in the LULC transboundary GIS data layer created for the Northern Appalachian/Acadian ecoregion and the number of LULC classes from each of the five U.S. and Canadian datasets that were combined to create each output LULC category

LULC classes in output transboundary dataset	Number of LULC classes in input datasets				
	U.S.	Canada			
	NLCD	Quebec	New Brunswick	Nova Scotia	PEI
Open water	1	4	5	4	1
Ocean	1	–	–	–	–
Residential	3	1	1	3	2
Commercial or indust. or trans.	1	1	7	6	2
Bare rock/sand/clay	1	0	2	12	1
Quarries, mines, gravel pits, peat bogs	1	1	4	3	1
Regenerating forest	1	6	7	1	7
Deciduous forest	1	1	5	1	1
Conifer forest	1	3	8	1	1
Mixed forest	1	1	8	1	2
Shrubland	1	1	2	13	1
Agriculture/plantations/ cultivated	5	6	6	3	1
Forested or shrub wetland	1	1	4	3	2
Emergent herbaceous wetlands, marsh or open bogs	1	1	1	11	1
No data/unclassified	1	1	3	0	0
Total number of classes per dataset	21	28	63	62	23

collaborative. Having at least some continuity of core GIS personnel for the duration of a conservation planning process is a real advantage, in that it helps maintain and retain a GIS history. GIS personnel transitions can result in loss of project knowledge related to data acquisition, treatments, compilation, and analyses. Such knowledge loss can be reduced by keeping good metadata, documenting data processing and analytical procedures, maintaining well-organized databases, storing data licenses securely, and managing staff transitions so as to overlap outgoing and incoming staff to facilitate knowledge exchange.

12.3 Distributing Outputs of Conservation Plans

The goal of an ecoregional conservation plan is to provide a scientifically sound conservation vision for an ecoregion. For such a vision to have any chance of being realized, it must be embraced and implemented by multiple individuals, organizations, and agencies across the region (Chap. 4). GIS plays a critical communication role because of the ease of interpreting GIS-generated maps as compared to raw data (Theobald 2009). To increase the probability of implementation, the conservation plan must be communicated clearly to those responsible for conservation management in the planning region, including government, industry, NGOs, and private landowners. One way to facilitate such communication is to ensure that the results of a conservation plan, accompanied with associated reports, maps, and GIS data, are easily accessible to these audiences.

12.3.1 Making GIS-Based Information Accessible

While papers, books, and reports are useful for communicating in-depth information regarding the methods, results, and implications of conservation planning exercises, the map-related information associated with such publications is unavoidably static and limited in the amount of information it contains. Distribution of GIS-based map layers generated by conservation planning efforts, on the other hand, provides a more direct means of enabling users to create customized products that meet their specific needs.

When detailed, GIS-based spatial information has been generated from a planning process, it is important to find a means to make that information available in such a way that it can be explored by different users, all of whom may be operating at different scales. Until recently, agencies and organizations that publish GIS-based spatial information most commonly distributed their data products in GIS-compatible digital file formats. While appropriate for the community of GIS professionals, such a distribution model unfortunately means that GIS data and the spatial information they contain are inaccessible to those without access to GIS resources. Increasingly, agencies and organizations that publish spatial information are communicating to broader audiences via web-based portals that include interactive map viewers. As our own experience in the Northern Appalachian/Acadian ecoregion demonstrated, presenting analyses at meetings stimulated interest by participants in obtaining results in map format customized for their local geographies to support their specific conservation activities. Given the dispersed nature of the collaborative network of 2C1Forest and the limited resources of the affiliated organizations that have strong GIS and mapping capacity, a web-based mapping solution appeared to be the most effective way to communicate and disseminate the data that had been generated.

12.3.2 Developing a Web-Based Mapping Tool

The solution my colleagues and I identified for distributing the results of GIS-based information that had been created for the ecoregion was to create the Northern Appalachian/Acadian Ecoregion Conservation Planning Atlas, or the '2C1Forest

Atlas' (Two Countries, One Forest 2009), an online mapping tool that would allow users to interact with GIS data and create custom maps to meet their specific needs.

The earliest web-based mapping tools available to the public were those that provided users with driving directions. These kinds of tools have become more numerous and sophisticated, and by 1998 almost every GIS company had a web-based mapping product (Li 2008). However, early applications were relatively slow and had limited functionality. Major advances in mapping technologies over the last 10 years, along with increased internet access speeds, have made it possible to develop more complex web-based mapping tools with dramatically improved functionality.

To develop and sustain a web-based mapping application, it is necessary to (1) define the purpose of the application and needs of the users, (2) develop content, (3) design and develop the application, and (4) host and maintain the application throughout its intended lifespan. To develop our atlas, we formed a seven-person Atlas Project Team made up of GIS professionals from the various organizations that published the related data, a 2C1Forest staff person with communications skills and authority over the 2C1Forest web site where the project was to be hosted, and a potential atlas user with no mapping or GIS skills.

Application Purpose and User Needs Because the development of an internet mapping application represents a significant investment of resources, it is vital to clearly define its purpose prior to the commencement of design or development. For example, the purpose of the 2C1Forest Atlas was to communicate the GIS-based results of ecoregional-scale conservation planning analyses conducted by 2C1Forest and partners for the Northern Appalachian/Acadian ecoregion to conservation practitioners in the region.

The Atlas Project Team was aware that the GIS and mapping skills of the potential users of the Atlas varied greatly, ranging from users who were merely comfortable surfing the internet to those who were experienced GIS professionals. However, to more accurately assess the GIS and technical capacity of our intended audience, we conducted a user needs assessment using a questionnaire sent to members of the 2C1Forest community by email. The questionnaire asked about their organization's size, scope of work, GIS capacity, types of spatial data used and needed, and speed of internet connection.

As expected, we discovered that our users possessed a wide range of GIS skills. We found that the majority of organizations that responded (57 of 63, or 90.2%) had GIS software in their organizations and used it regularly (74.6%). A small proportion either didn't use their software (3.2%) or used it only occasionally (12.7%). Moreover, 34.7% of the organizations were without internal GIS expertise, indicating that at least one-third of organizations concerned with conservation and land-use planning in the Northern Appalachian/Acadian ecoregion needed to outsource their GIS and mapping needs. We also learned that the biggest barriers to organizations using GIS for conservation planning were limited staff time, a lack of funding for GIS products and services, the high cost of GIS software, difficulty in obtaining data, and low data quality. On the other hand, 68% of respondents had experience using online mapping tools, with only one indicating that they had never even used Google Earth, and 77% of respondents most often worked with a high-speed internet connection.

Atlas Content It is important when developing a web-based mapping tool that the intended content is well-planned, including not only the GIS data to be provided but also the associated contextual and supporting materials to help the user interpret the data. The Atlas Project Team learned that the initial GIS data to be provided must be complete and ready to be added before development of an atlas begins. Development of web-based mapping tools can be expensive; because GIS data can take considerable time to create, if they are not completed at the start of atlas development, significant delays and costs will likely be incurred.

Purchased data or data governed by licensing agreements cannot be added as content to an internet mapping application without the consent of the publishers. For example, we were unable to add data on protected areas in Canada due to licensing restrictions that existed at the time. By contrast, we were able to add the LULC data because we revisited the data licenses with the publishing agencies and they gave permission to include the transboundary LULC data layer we created. The rationale for this decision was that it would be impossible for users to derive the original licensed data from the derived transboundary dataset. To meet the atlas content needs of GIS professionals with access to GIS software and hardware, the GIS data files from the conservation planning process were posted on the 2C1Forest website in compressed downloadable file formats, accompanied by full Federal Geographic Data Committee (FGDC Federal Geographic Data Committee 2000) standard metadata.

Design and Development Even conservation organizations that have good scientific expertise and technical capabilities for GIS do not generally have the capacity to design and build an efficient online mapping application. For this reason, we chose to contract the development of the 2C1Forest Atlas to a private firm specializing in internet mapping. We investigated partnerships with a number of universities and consulting groups but ultimately chose to work with DM Solutions Group (DMSG 2009), an Ottawa-based consulting firm specializing in internet mapping using the free open-source software program MapServer (Kropla 2005; Ojeda-Zapata 2005). Additionally, DMSG had experience working with the Canadian Federal Agency, GeoConnections (GeoConnections 2008), which provided funding for the Atlas under their 2006 Regional Thematic Atlas program.

As a result of a planning session guided by DMSG, we determined that the Atlas should (1) be user friendly for non-GIS users, (2) adopt a simple 'stress free' look and integrate well with the 2C1Forest website, (3) feel like a hardcopy atlas in that it should contain multiple maps accessed through an index page, with each map accompanied by contextual materials to help with interpretation, and (4) allow users to easily customize, save, and share maps. From start to finish, the design and development of the Atlas took 9 months. This relatively rapid pace was facilitated by the fact that the Atlas had a well-defined user audience and publication-ready data content prior to the start of development.

Hosting and Maintenance For an internet mapping application to remain live and functional, hardware maintenance, regular backups, and the regular implementation of software updates are essential. Otherwise, if a web-based tool becomes unstable, it will be deemed unreliable and ultimately abandoned by the user audience. For the 2C1Forest Atlas, we chose to use the same consulting group that developed the Atlas to provide hosting and maintenance for an annual fee. This was the most cost-effective option compared to developing sufficient in-house capacity for all of these tasks. The application was designed with an initial 5-year lifespan because (1) internet mapping technologies are developing rapidly, and new technologies would likely be available after a 5-year period, (2) the base data layers from which many of our conservation data were derived (e.g., census data, roads data, and LULC data) would be updated and published by the end of 2012, and (3) this corresponded with a planned evaluation of the content and design of the Atlas to determine its ongoing usefulness to the user community.

Regional GIS-based atlases also need to be considered in the context of rapidly emerging web-based mapping projects that collate and provide data at greater and greater spatial scales to support larger conservation communities. In 2009, three new web-based mapping tools designed for this purpose emerged in North America. Two of these were specifically designed to provide web-based mapping tools for users to add their own spatial conservation information and data, inspired by the recognition that most conservation organizations do not have the capacity to develop their own web-based mapping applications. The first of these is the Conservation Registry (The Conservation Registry 2010), developed by Defenders of Wildlife, which enables users to register their conservation projects, with the idea that the more conservation groups that use the tool, the better understanding the conservation community will have about where conservation activities are occurring. Thus, the Conservation Registry essentially maps conservation capacity.

The second tool is Data Basin (Data Basin 2009), developed by the Conservation Biology Institute. This is essentially a GIS data warehouse and viewing tool. It allows users to post their GIS-based conservation data with accompanying contextual information, such as images and reports, while also allowing users to search for data and create custom maps that combine any of the datasets posted to the site. It also has the capacity to support online collaborative mapping workspaces, a functionality that can potentially support planning collaborations without the need for GIS analysts, hardware, and software.

12.3.3 Outreach and Training

Once an online mapping tool is completed, targeted outreach activities are necessary to inform the intended users about the new tool and its associated benefits. Outreach strategies will vary depending on intended audience and available resources. For the 2C1Forest Atlas, we (1) launched the Atlas with a live demonstration at an annual 2C1Forest conference, (2) delivered seven 1-day workshops in the Northern Appalachian/Acadian ecoregion in Spring 2008, training 114 individuals from conservation NGOs, land trusts, foundations, colleges, universities, and government agencies, (3) delivered presentations and demonstrations at professional conferences, and (4) informed potential users about the Atlas through emails and newsletters.

12.4 Lessons Learned

The collaborative mapping and analysis efforts of 2C1Forest and partners for the Northern Appalachian/Acadian ecoregion offer several lessons about GIS-related challenges that are likely to be encountered during any effort at landscape-scale conservation planning.

First, conservation managers must be prepared to dedicate the necessary resources to support the GIS needs of the planning process for the duration of the planning project with respect to GIS software, hardware, data, personnel, and expert modeling.

Second, published transboundary GIS data layers of the appropriate scale and resolution for ecoregional planning that seamlessly cover the geographic extent of the planning region are rare. Therefore, transboundary GIS data often need to be created by combining multiple datasets published for adjacent administrative jurisdictions. This has potential consequences for quality and resolution of the resulting transboundary data layer.

Third, when a transboundary GIS data layer is created from multiple input datasets, it is important that the input datasets be maximally compatible. The most compatible datasets are those that contain geographic features represented in the same data model (vector or raster), at the same scale or resolution with similar data accuracy, have comparable dates of data collection and/or publication, have been created for the same purpose, and use the same or similar attributes and attribute codes.

Fourth, information about the locations of ecological features that are the focus of conservation interest (e.g., species, habitats, and ecological processes) can be hard to find and are generally incomplete. Thus, conservation planners must anticipate and plan for the resources required to create complete datasets either through field observations or the development of predictive distribution models.

Fifth, data access and use can be restricted by data licenses. Data licenses are legal documents and should be managed and stored securely. It is likely that licenses will need to be accessed for years after they are acquired, and if they are lost or misplaced, considerable staff resources may be required to recover or renegotiate a license.

Sixth, for a collaborative conservation planning initiative to be successful, multiple organizations within the collaborative should collectively bring GIS resources to the project and not rely on any single organization to support the full GIS needs of the planning process.

Seventh, in the event a turnover in GIS staff occurs, the risk of loss of GIS knowledge regarding the conservation planning process can be minimized by adhering to the practice of documenting how data are processed and analyzed, creating standard metadata for the final versions of all datasets created, and maintaining an organized GIS database. Facilitating an overlap between outgoing and incoming personnel is also recommended to permit project knowledge transfer.

Finally, for the results of a conservation plan to be used and its recommendations implemented, it is important to make the map-based information generated during the process accessible to stakeholders and decision makers not only as GIS data files, but through the use of an interactive web-based mapping interface whereby access to the information is not reliant on in-depth GIS training.

Acknowledgments I want to recognize the GIS efforts of Greg Kehm and GIS colleagues at The Nature Conservancy and Nature Conservancy Canada who created a number of the transboundary datasets in the Northern Appalachian/Acadian ecoregion. I would like acknowledge Patrick Doran and Charlie Bettigole for their GIS contributions to their Northern Appalachian conservation vision, and Justina Ray, Steve Trombulak, and Rob Baldwin for their support and guidance over the years I have worked on conservation in the 2C1Forest region.

References

Akcakaya, H. R. (2004). Using models for species conservation and management. In H. R. Akcakaya, M. Burgman, O. Kindvall, C. C. Wood, P. Sjögren-Gulve, J. S. Hatfield, et al. (Eds.), *Species conservation and management: Case studies* (pp. 3–14). New York: Oxford University Press.

Aronoff, S. (1995). *Geographic information systems: A management perspective* (4th ed.). Ottawa: WDL Publications.

Bailey, R. G. (2004). Identifying ecoregion boundaries. *Environmental Management, 34*(Suppl 1), S14–S26.

Bateson, E. M. (2005). Two countries, one forest – deux pays, une forêt: Launching a landscape-scale conservation collaborative in the Northern Appalachian region of the United States and Canada. *George Wright Forum, 22,* 35–45.

Burrough, P. A. (1986). *Principles of geographical information systems and land resource assessment.* New York: Oxford University Press.

Burrough, P. A., & McDonnell, R. A. (1998). *Principles of geographical information systems.* New York: Oxford University Press.

Crist, P. J. (2000). Mapping and categorizing land stewardship. In J. M. Scott (Ed.), *A handbook for gap analysis (Version 2.1.0)* (pp. 119–136). Retrieved February 16, 2010, from http://www.gap.uidaho.edu/handbook/CompleteHandbook.pdf

Data Basin. (2009). *Data basin.* Retrieved February 20, 2010, from www.databasin.org

DMTI Spatial. (2009). *DMTI spatial.* Retrieved February 20, 2010, from http://www.dmtispatial.com/

DMSG [DM Solutions Group]. (2009). *Maps for MapServer data products.* Retrieved February 20, 2010, from DMSG Web site: www.dmsolutions.ca

Dudley, N. (Ed.) (2008). *Guidelines for applying protected area management categories.* Gland, Switzerland: IUCN. Retrieved February 16, 2010, from http://www.gap.uidaho.edu/Portal/Stewardship/IUCN_cat_guidelines_final_2008.pdf

ESRI (Environmental Science Research Institute). (1997). *Understanding GIS: The arc/info method* (4th ed.). Redlands, CA: ESRI Press.

278 G. Woolmer

FGDC (Federal Geographic Data Committee). (2000). *Content standard for digital geospatial metadata workbook (Version 2.0)*. Reston, VA: US Geological Survey, Federal Geographic Data Committee. Retrieved February 16, 2010, from http://www.fgdc.gov/metadata/documents/workbook_0501_bmk.pdf

GeoConnections (2008). *GeoConnections annual report 2006–2007*. Retrieved January 16, 2010, from http://www.cgdi.gc.ca/publications/reports/ar/200607_Annual_Report_DeskTopped.pdf

Groves, C. R. (2003). *Drafting a conservation blueprint: A practitioner's guide to planning for biodiversity*. Washington, DC: Island Press.

Guisan, A., & Zimmermann, N. E. (2000). Predictive habitat distribution models in ecology. *Ecological Modelling, 135*, 147–186.

ISO (International Organization of Standards). (2010). *ISO 19115: 2003–Geographic information – metadata*. Retrieved February 20, 2010, from ISO Web site: http://www.iso.org/iso/iso_catalogue/catalogue_tc/catalogue_detail.htm?csnumber=26020

Jenkins, J. (2008). *Conservation easements and biodiversity in the Northern Forest Region*. New York: Open Space Institute and Wildlife Conservation Society.

Kropla, B. (2005). *Beginning MapServer: Open source GIS development*. New York: Apress.

Li, S. (2008). Web mapping/GIS services and applications. In Z. Li, J. Chen, & E. Baltsavias (Eds.), *Advances in photogrammetry, remote sensing and spatial information sciences* (pp. 335–353). London: CRC Press.

Longhorn, R. A., Henson-Apollonio, V., & White, J. W. (2002). *Legal issues in the use of geospatial data and tools for agriculture and natural resource management: A primer*. Mexico, D.F.: International Maize and Wheat Improvement Center (CIMMYT). Retrieved February 12, 2010, from http://csi.cgiar.org/download/IPR_Primer.pdf

Mackenzie, D. I., Nichols, J. D., Royle, J. A., Pollock, K. H., Bailey, L. L., & Hines, J. E. (2006). *Occupancy estimation and modeling: Inferring patterns and dynamics of species occurrences*. London: Academic.

Margules, C. R., & Pressey, R. L. (2000). Systematic conservation planning. *Nature, 405*, 243–253.

National Atlas. (2009). *Raw data download – North American atlas*. Retrieved February 20, 2010, from NationalAtlas.gov Web site: http://www.nationalatlas.gov/atlasftp-na.html

NatureServe. (2009a). *National conservation easement database: A message for easement data holders*. Retrieved February 3, 2010, from http://www.natureserve.org/projects/pdfs/nced-Flyer.pdf

NatureServe. (2009b). *NatureServe: A network connecting science with conservation*. Retrieved February 20, 2010, from NatureServe Web site: http://www.natureserve.org/

NatureServe Canada. (2009). *NatureServe Canada: A network connecting science with conservation*. Retrieved February 20, 2010, from NatureServe Canada Web site: http://www.natureserve.ca

NIMA (National Imagery and Mapping Agency). (2000). *Vector map level 0 (digital chart of the world)* (5th ed.). Washington, DC: National Imagery and Mapping Agency.

Ojeda-Zapata, J. (2005). *Minnesota's MapServer flourishes in hot web-based mapping sector*. St. Paul, MN: St. Paul Pioneer Press.

O'Neil, T. A., Bettinger, P., Marcot, B. G., Luscombe, B. W., Koeln, G. T., Bruner, G. H. J., et al. (2005). Application of spatial technologies in wildlife biology. In C. E. Braun (Ed.), *Wildlife techniques manual* (6th ed., pp. 418–447). Washington, DC: The Wildlife Society.

PAD-US (Protected Areas Database of the United States). (2010). *Protected areas database f the United States: Improving the nation's data on natural resources and park lands*. Retrieved February 20, 2010, from PAD-US Web site: http://www.protectedlands.net

Pressey, R. L. (2004). Conservation planning and biodiversity: Assembling the best data for the job. *Conservation Biology, 18*, 1677–1681.

Rejeski, D. (1993). GIS and risk: A three-culture problem. In M. F. Goodchild, B. O. Parks, & L. T. Steyaert (Eds.), *Environmental modeling with GIS* (pp. 318–331). New York: Oxford University Press.

Scott, J. M., Davis, F., Csuti, B., Noss, R., Butterfield, B., Groves, C., et al. (1993). Gap analysis: A geographic approach to protection of biological diversity. *Wildlife Monographs, 57*, 5–41.

12 The GIS Challenges of Ecoregional Conservation Planning

Statistics Canada. (2002a). *Cartographic boundary files 2001 census second edition reference guide (No. 92F0171GIE)*. Ottawa, ON: Statistics Canada.

Statistics Canada. (2002b). *Road network files 2001 census reference guide (No. 92F0157GIE)*. Ottawa, ON: Statistics Canada.

Stehman, S. V., & Czaplewski, R. L. (1998). Design and analysis for thematic map accuracy assessment: Fundamental principles. *Remote Sensing of Environment, 64*, 331–344.

The Conservation Registry. (2010). *The conservation registry*. Retrieved February 20, 2010, from http://www.conservationregistry.org/

Theobald, D. M. (2009). *GIS concepts and ArcGIS methods* (4th ed.). Ft. Collins, CO: Conservation Planning Technologies.

Tomlinson, R. F. (1984). Geographic information systems – A new frontier. *The Operational Geographer, 5*, 31–35.

Trombulak, S. C., Anderson, M. G., Baldwin, R. F., Beazley, K., Ray, J. C., Reining, C., et al. (2008). *The Northern Appalachian/Acadian ecoregion: Priority locations for conservation action (Special Report No. 1)*. Warner, NH: Two Countries, One Forest. Retrieved February 16, 2010, from http://www.2c1forest.org/en/resources/Special_Report_1.pdf

Two Countries, One Forest. (2009). *Northern Appalachian/Acadian ecoregion conservation planning atlas*. Retrieved February 20, 2010, from 2C1Forest Web site: www.2c1forest.org/atlas

U.S. Census Bureau. (2000). *Census 2000 TIGER/line files technical documentation*. Washington, DC: US Census Bureau.

WDPA (World Database on Protected Areas). (2009). *2009 WDPA released*. Retrieved February 20, 2010, from WDPA Web site: www.wdpa.org

Wing, M. G., & Bettinger, P. (2008). *Geographic information systems: Applications in natural resource management* (2nd ed.). New York: Oxford University Press.

Yang, L., Stehman, S. V., Smith, J. H., & Wickham, J. D. (2001). Thematic accuracy of MRLC land cover for the eastern United States. *Remote Sensing of Environment, 76*, 418–422.

Chapter 13
The Human Footprint as a Conservation Planning Tool

Stephen C. Trombulak, Robert F. Baldwin, and Gillian Woolmer

Abstract Conservation planning is aided by an ability to view spatially explicit patterns of landscape transformation that are both multivariate and mapped with a fine-scale resolution. The Human Footprint is one such measure of transformation, integrating information on human access, settlement, transformation of land use/land cover, and development of energy infrastructure. We used this methodology to develop a fine-scale (90-m resolution) map of the degree of human transformation of the Northern Appalachian/Acadian ecoregion as well as develop models to project changes in key dynamic aspects of this map – roads, human population density, and land cover change due to amenities development – to identify in a comprehensive and systematic fashion locations that are currently highly transformed or vulnerable to transformation in the future. Although more than 90% of this ecoregion exhibits less than half of the maximum amount of transformation seen anywhere here, several regions, particular around urban areas and within major valleys, are already highly transformed. In addition, under reasonable scenarios of future population growth and development, threat levels for several areas currently with low levels of transformation are projected to increase, providing conservation planners a way to prioritize current conservation action to proactively achieve conservation goals for the future.

Keywords GIS • Human Footprint • Human influence • Northern Appalachian/Acadian ecoregion • Transformation

---

S.C. Trombulak (✉)
Department of Biology, Middlebury College, Middlebury VT 05753
e-mail: trombulak@middlebury.edu

R.F. Baldwin
Department of Forestry and Natural Resources, Clemson University, 261 Lehotsky Hall, Clemson, SC 29634-0317
e-mail: baldwi6@clemson.edu

G. Woolmer
WCS Canada Suite 600, 720 Spadina Ave., Toronto, ON M5S 2T9, Canada
e-mail: gwoolmer@wcs.org

S.C. Trombulak and R.F. Baldwin (eds.), *Landscape-scale Conservation Planning*, 281
DOI 10.1007/978-90-481-9575-6_13, © Springer Science+Business Media B.V. 2010

13.1 Introduction

That conservation planning needs to occur at all is a reflection of the fact that humans have influenced the global landscape in ways that threaten biological diversity. Proper planning therefore requires an understanding of these influences in terms of where they currently occur, their relative magnitude, and how they might change in the future. Without this systematic understanding, effective mitigation of current threats and proactive avoidance of future threats to biological diversity cannot occur.

Numerous measures have been advanced as metrics of human influence on the landscape, including human population and housing density (Parks and Harcourt 2002; Theobald 2003), city lights at night (WRI 2000), road density (Carroll 2005; Saunders et al. 2002), density of endangered species (Dobson et al. 1997), appropriation of net primary productivity (Haberl et al. 2007), and deposition of pollutants (Driscoll et al. 2001, 2007). All of these metrics are valid because all of them focus on anthropogenic influences that are either indicators or direct causes of degradation and loss of biological diversity.

As with any assessment, however, the more variables used to evaluate the magnitude and distribution of influences on the landscape, the more likely that the results of the assessment will not be biased toward any single variable. The importance of a multivariate approach to mapping human influences – and by extension, threats to biological diversity – is made clear through analogy with human health; because the human body is a complex system, a medical check-up (i.e., a human health threat assessment) involves attention to numerous variables (e.g., blood pressure, reflexes, and lymph nodes) to accurately determine one's condition. So too with natural systems, and because different threats have different effects on biological diversity (Chap. 2), integrating measures of threat – for example, housing density and roads – provides a more accurate depiction of the magnitude of human activity than when either of these threats are examined on their own (Woolmer et al. 2008). Further, integrating multiple variables that may have spatially disjunct effects (e.g., amenity housing and urban sprawl) better reveals the site-specific potential for cumulative transformative influences. Ideally, therefore, conservation planning is best informed by a spatially explicit, multivariate assessment of how the natural landscape has been transformed, an assessment that shows the relative magnitude and distribution of threats in a way that is easily interpreted.

13.2 The Human Footprint Methodology

A spatially explicit, multivariate threat assessment methodology was developed by Sanderson et al. (2002) to identify the least human-transformed landscapes – the 'Last of the Wild' – in each of the world's major biomes. Dubbed the Global Human Footprint, the methodology was straightforward: (1) compile spatially explicit data on anthropogenic influences to natural landscapes (e.g., populations density, roads, anthropogenic land cover); (2) create a scoring system so that higher scores are given for influences that alter natural conditions to a greater degree; (3) weight the scoring system to reflect a priori decisions on the relative importance of each influence in the final footprint score; (4) calculate the scores for each influence at each location based on a heuristic model to combine the scores in a manner that avoids redundancy; and (5) normalize the scores between 0 and 100 to bound the range of scores between the minimum and maximum amount of influence seen anywhere within the landscape.

As developed by Sanderson et al. (2002), sources of human influence fell into one of four categories: human settlement, human access, human land use, and electrical power infrastructure. Each human influence source was scored on a scale from 0 to 10 as to degree of human transformation and ecological impact (0 being no or minimal impact, 10 being maximum impact, generally reflecting complete and permanent conversion to development). The scores were then combined to produce a single index that was then normalized within ecological subregions (to reflect regional differences in ecological resiliency to different magnitudes of influence) and produce a map of relative human influence – or impact – on a scale from 0 to 100. What emerges is a map of Human Footprint (HF) that reflects the cumulative amount of transformation at each location relative to the least and greatest amount of transformation found anywhere in the target region. For example, a location with an HF score of 10 exhibits a level of transformation that is 10% greater than locations with the least amount of transformation.

The Human Footprint has many useful characteristics as an assessment tool for threats to biological diversity arising from anthropogenic landscape transformation, and it may be used in a conservation planning context. Conservation planning is interdisciplinary, involving collaborations among multiple professions, and a challenge of conservation science is to provide tools that are information-rich yet transparent. The Human Footprint methodology accomplishes this in that it is multivariate, integrative, and scaled within an intuitively logical range (0–100). Further, it can be applied on any spatial scale for which a relatively uniform set of data is available. Sanderson et al. (2002) applied their methodology to map the Human Footprint across the entire planet at a resolution of 1 km^2.

The map of the Global Human Footprint has been applied to a number of conservation planning initiatives, including the following:

1. A collaboration of the World Wildlife Fund, Save the Tiger Fund, Wildlife Conservation Society, and the Smithsonian's National Zoological Park, in which the Global Human Footprint was used to evaluate and prioritize the 76 identified Tiger Conservation Landscapes across Asia (Sanderson et al. 2006).
2. An initiative led by The Nature Conservancy, in which the Global Human Footprint was used to evaluate threat abatement as part of an assessment of 'effective conservation' (Boucher et al. 2006).
3. A National Geographic Society project called the 'MegaFlyover' conducted by Dr. Mike Fay of the Megatransect Project, in which the Global Human Footprint was used as the basis on which to assess the extent of human influence in Africa

by flying over the full gradient of human transformation in every biome on the continent (National Geographic Society n. d.).

4. The Wildlife Conservation Society and the Center for International Earth Science Information Network recently updated the Global Human Footprint analysis, demonstrating their long-term commitment to revise and improve this tool for global and regional conservation planning (CIESIN n. d.).

Woolmer et al. (2008), using a Human Footprint map they developed for the Northern Appalachian/Acadian ecoregion at a 90-m resolution (Sect. 13.3) showed that the increased resolution of the analysis revealed more detailed information about the distribution of transformation across that ecoregion compared to what was revealed for the same region in the global analysis. This demonstrated the importance of developing fine-scale, landscape-specific Human Footprint maps rather than relying on a subset of the coarse-scale map developed by Sanderson et al. (2002).

Since then, the methodology has been used to map smaller regions with finer resolution. For example, Leu et al. (2008) developed a Human Footprint for the Western U.S. Although their methodology varied from that of Sanderson et al. (2002), the goal was the same: to map the extent and intensity of human influence on the landscape. In their study, Leu et al. (2008) combined seven input models: three models that quantified top-down anthropogenic influences of synanthropic predators (avian, dog, and cat) and four models that quantified bottom-up anthropogenic influences on habitat (invasion of exotic plants, human-caused fires, energy extraction, and fragmentation).

More recently, the Model Forest of Newfoundland and Labrador program is currently mapping the Human Footprint of those provinces based on the approach taken to map the regional Human Footprint of the Northern Appalachians (Model Forest of Newfoundland and Labrador n. d.).

The Human Footprint methodology is not without its limitations. For one, it is dependent on the data that are available. As a geographic database, the Human Footprint layers are necessarily limited to those features that have been mapped and digitized at the scale at which the analysis is being conducted. Features such as nitrogen deposition, species extirpation, species invasions, surface water withdrawals, and climate change can be real threats to biological diversity (Chap. 2), but if spatially explicit data are not available for the entire planning region at a meaningful resolution, then they cannot be effectively incorporated into the analysis, thus making a map of the Human Footprint actually of the 'Human Footprint based on mapped influences.'

Further, the currency of available datasets generally lags behind their availability and, typically, differs among datasets. For example, the Human Footprint map for the Northern Appalachian/Acadian ecoregion (Woolmer et al. 2008) used population and housing density data from the 2000 U.S. census and 2001 Canadian census; land cover maps from various times for different states and provinces, ranging from 1992 to more recent times; information on the locations of large dams from 2003 in Canada and 2005 in the U.S.; and electrical utility corridors from 2000 in both countries (Table 13.1). Thus, the HF score for a location is neither current nor attributable to a single point in time. These limitations are not unique to the Human Footprint, however; any threat assessment using these component datasets will likewise have these temporal restrictions.

Finally, the influence scores assigned to different levels of transformation are ordinal in nature and relative with respect to each other, and ultimately the effects are assumed to be additive. While little information is available to suggest any other kind of mathematical relationship among variables, and human land-use transformation is known to be cumulative, further research using the Human Footprint should investigate its mathematical structure. Scaling continuous data on dwelling density into an ordinal scale between 0 and 10, for example, requires decisions on the nature of the relationship between changes in dwelling density and influence on biological diversity (e.g., choosing among linear, exponential, and logistic relationships) and the threshold density at which the maximum influence is reached, decisions that are rarely directly addressed in the literature.

Further, the scores themselves are based on expert opinion and literature review and are thus to a large degree subjective. Expert opinion is commonly used in conservation planning and is increasingly incorporated into quantitative indices such as the Human Footprint, habitat suitability, and other metrics (Noss et al. 2002). Nonetheless, a systematic, transparent approach to integrating expert opinion is desirable (Chap. 11). Despite these limitations, however, the Human Footprint methodology provides the most complete means yet devised for assessing the degree to which a landscape has been transformed by human action. The resulting map can serve as the inverse of habitat suitability for sensitive species and provide a cost surface for modeling connectivity (Chap. 16). It provides an integrated view of transformation, yielding interpretations that are counter to those derived from measures based on single influences. For example, maps of human population density fail to reveal the magnitude of landscape transformation in lightly settled regions, such as agricultural areas or lands that are the focus of nature-based amenities development (e.g., ski areas, remote lakeshore developments; Baldwin et al. 2009). By assuming that the lowest areas of human impact are the most suitable for those species most sensitive to human activities and are areas where natural rather than anthropogenic disturbances dominate, the Human Footprint can clearly delineate areas that may serve in a general sense as wildlife corridors or ecological reserves. Although it does not directly measure impact to biological diversity, it can be used to measure ecological threat because it focuses on those anthropogenic features that are the most completely documented as having negative impacts on numerous aspects of biological systems.

Importantly, the Human Footprint may also be the most effective tool yet devised for communicating complex information about landscape transformation to the public. Evidence for this is in its widespread adoption by non-governmental organizations, including National Geographic, Wildlife Conservation Society, and World Wildlife Fund. Such a map is also intuitively obvious; in many public meetings, we have found that its scientific basis and conservation implications are quickly understood even by people who have never seen a Human Footprint map

286 S.C. Trombulak et al.

Table 13.1 Source and resolution for the ten data sets used to map the regional human footprint of the Northern Appalachian/Acadian ecoregion

Feature	Source
Population density; dwelling density	USA: Census 2000 Tiger/Line Files – census blocks. 1:100,000
	Canada: Cartographic Boundary Files 2001 Census, Statistics Canada – dissemination areas. 1:50,000
Urban areas	USA: Census 2000 Tiger/Line Files – Urbanized Areas. 1:100,000
	Canada: Cartographic Boundary Files 2001 Census, Statistics Canada – Urban Areas. 1:50,000
Roads	USA: Census 2000 Tiger/Line Files. 1:100,000
	Canada: CanMap Route Logistics V8.2, DMTI Spatial 1:50,000
Rail	USA: Bureau of Transportation Statistics (BTS), National Rail Network 1:100,000
	Canada: CanMap Rail V8.2, DMTI Spatial, 1:50,000
Land use/land cover	New York, Vermont, New Hampshire, Massachusetts: USGS, National Land Cover Dataset (NLDC). 1992. 30 m resolution
	Maine: USGS GAP Analysis Program. 1993. 30 m resolution
	Quebec: Canadian Wildlife Service, Environment Canada. 30 m resolution
	New Brunswick: Department of Natural Resources & Environment. 1:10,000
	Nova Scotia: Department of Natural Resources, Ecosystem Management Group. 1:10,000
	PEI: Department of Agriculture, Fisheries, Aquaculture and Forestry. 1:10,000
Large dams	USA: U.S. Army Corps of Engineers National inventory of dams (NID), 2005 (scale unknown)
	Canada: Canadian Dam Association, 2003. Locations digitized using 1:50,000 topographic maps (www.etopo.ca)
Watersheds	USA: USGS, 1:250,000 scale Hydrologic Units of the United States (HUC8), 1994
	Canada: Atlas of Canada National Frameworks – Drainage Areas (2003). National Resources Canada 1:1,000,000
Mine sites	USA: USGS Mineral and Metal Operations, 1998 (scale unknown)
	Canada: Principal Mineral Areas of Canada – Map 900A, Natural Resources Canada. 2003. 1:6,000,000
Utility corridors	USA and Canada: NIMA Vector Map Level 0 Edition 5, 2000. 1:1,000,000

Land use/land cover data were compiled by The Eastern Resource Office of The Nature Conservancy (TNC). NIMA, National Imagery and Mapping Agency; USGS, United States Geological Survey; WWF, World Wildlife Fund before. A drawback to such tools, of course, is that they can be oversimplified, and their implications can be taken too literally; however, the task of the communicator using the Human Footprint is to make its underlying assumptions and inferential limitations clear.

13.3 The Current Human Footprint in the Northern Appalachian/Acadian Ecoregion

Threats to the Northern Appalachian/Acadian ecoregion from human activity are so pervasive as to affect almost every aquatic, terrestrial, and marine ecosystem (Chap. 2). Airborne pollutants from the Midwest of both the U.S. and Canada settle over vast areas of this ecoregion, contaminating rivers, lakes, ponds, and marine ecosystems as well as influencing the biogeochemistry of surrounding forests. Acid rain, mercury and other heavy metals, particulates, and ground level ozone penetrate even the most pristine areas and affect functioning of ecosystems (Driscoll et al. 2001, 2007; Evers et al. 2007). Meanwhile, industrial effluent enters food webs, and many compounds (e.g., chlorinated hydrocarbons) bioaccumulate in marine and terrestrial predators, affecting both reproduction and survival.

The very conditions for life in this ecoregion are also changing, as human-induced climate change threatens to affect the ranges of plants and animals here where many exist at the southern or northern limits of their physiological capacities (Carroll 2007; Frumhoff et al. 2007; Chap. 15).

While these threats are pervasive, no single factor affects biological diversity more than physical habitat destruction (Hunter and Gibbs 2007; Vitousek 1994; Wilcove et al. 1998). Although many species were driven to extirpation or extinction by overexploitation since the Pleistocene (Alroy 2001) and more recently following European colonization, intensive land use often results in permanent changes to habitats with lasting effects on wildlife populations. When humans need land for agriculture, mining, timber harvesting, housing, or transportation, natural landscapes are transformed to human landscapes. Often this process of habitat conversion introduces additional threats such as pollution and invasive species, and natural processes such as fire and water flow are altered. While not all human activities are detrimental to biological diversity, the cumulative effect of human activities on the land surface is the dominant force shaping ecosystems today (Haberl et al. 2007; Vitousek et al. 1997).

The global map of the Human Footprint (Sanderson et al. 2002) estimates that 83% of the Earth's land surface is measurably impacted by human activities. Transformation of natural land cover contributes to detrimental changes in the global carbon and other nutrient cycles, increases in soil erosion, degradation of freshwater ecosystems, and changes in climate, and is the single most important cause of the loss of biological diversity (Chap. 2). For example, in North America more than one-third of carnivore and ungulate species have experienced a range contraction of at least 20% due to human settlement patterns (Laliberte and Ripple 2004; Chap. 9). Geographic isolation of national parks – due to intensification of land use beyond park boundaries – has resulted in loss of native mammal species (Newmark 1995; Parks and Harcourt 2002) and the development of protocols for monitoring land-use change by the U.S National Park Service (Jones et al. 2009). In the Northern Appalachian/Acadian ecoregion, anthropogenic changes in land cover in recent centuries have predominated and underlie pronounced changes in both ecosystem structure and function (Foster and Aber 2004). Changes on the landscape include measurable shifts in plant and animal distributions in terrestrial and aquatic systems due to changes in the intensity, frequency, and duration of disturbances, overexploitation, and climate (Davis et al. 1980; Foster et al. 2002; Chaps. 2, 6, 9, and 15).

Because of the ecological significance of human transformation of the land's surface, a consortium of conservation organizations in the Northern Appalachian/ Acadian ecoregion, under the umbrella of the organization Two Countries, One Forest (2C1Forest), quantified the Human Footprint as a basis for assessing threats for setting conservation priorities in the ecoregion. Due to its complex settlement history (e.g., intensive Native American occupancy followed by European settlement as early as the mid-1600s), relative degree of geographic isolation, and elevational, ecological (e.g., marine-terrestrial), and latitudinal gradients, the Northern Appalachian/Acadian ecoregion is heterogeneous with regard to land use, land ownership, habitats, and degrees of transformation. Consequently, the multivariate approach of the Human Footprint is well-suited to modeling this inherent landscape-scale complexity. We applied assembled transboundary databases and established methods to map human impacts with the greatest accuracy possible ($90\ m^2$), and developed simple, repeatable models to project selected, salient aspects of those threats into the future (Future Human Footprints; Sect. 13.4). Our goal was to provide a time-sensitive picture of how threats are distributed on the landscape now and how they may be distributed in the future (ca. 2040).

To map the Current Human Footprint in the Northern Appalachian/Acadian ecoregion, we compiled spatial data layers comparable to those used to map the Global Human Footprint (Woolmer et al. 2008), and followed its general methodology by (1) selecting a spatial resolution of analysis based on the scale of the best available data, (2) selecting datasets representing the different sources of landscape transformation and then assigning aggregate Human Influence (HI) scores, (3) combining HI scores across datasets to quantify direct human influence, which results in a map of the Human Influence Index (HII), and (4) normalizing the HII scores across ecological subregions to calculate relative human influence within each subregion, resulting in an ecoregional map of the Human Footprint.

To fully capture the human influences on the periphery of the ecoregional boundary, we buffered our analytical boundary to 40 km and mapped the Human Footprint to a 20-km buffer around the ecoregion. We only assessed human influence

13 The Human Footprint as a Conservation Planning Tool

on terrestrial ecosystems and did not attempt to assess human influences on freshwater or coastal systems.

We used 10 datasets to represent the four categories of human influence used in the Global Human Footprint: (1) human settlement (population density, dwelling density, and urban areas); (2) human access (roads and rail lines); (3) human land use (land use/land cover, large dams, watersheds, and mines); and (4) energy infrastructure (utility corridors) (Table 13.1).

We chose datasets to capture those human activities and trends relevant to human influence in this ecoregion in the present time. For example, we included dwelling density to capture the influence of second homes related to amenity developments and decreasing household size, but we did not use navigable rivers as a source of human access (as was done by Sanderson et al. 2002) because they do not presently serve as significant transportation corridors in the ecoregion separate from the existing roads network. We assigned HI scores to each data layer to reflect their relative contribution to human influence on the land on a scale from 0 (low) to 10 (high). Scores were assigned based on published studies relevant to this ecoregion and on expert opinion.

The Current Human Footprint (CHF; Fig. 13.1) reveals where similar human influence scores are accumulated and land transformation to human uses is most intense. Three main patterns stand out. First, large areas with low Human Footprint scores still remain within this ecoregion – and only a portion of these (62% of all locations with an HF score ≤10) are found on GAP status 1–3 protected areas, which are lands that are permanently secured against conversion to development

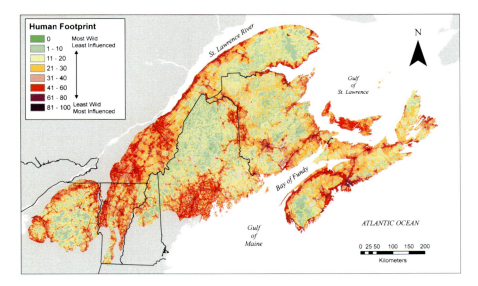

Fig. 13.1 The Current Human Footprint of the Northern Appalachian/Acadian ecoregion

(Fig. 13.2). Second, separating these areas are areas with high levels of human activity. These appear to fragment the region into large blocks of less-transformed land – the Adirondack Mountains, Northern New England, Gaspé Peninsula, New Brunswick, and parts of Nova Scotia. Third, even within these large blocks of land with low Human Footprint scores, human impacts are still present, suggesting that human land use is widespread even outside of the heavily settled valleys and coastlines.

On average, the region is still only moderately transformed by human impacts relative to the maximum amount present anywhere in the ecoregion. The distribution of HF scores peaks in the 11–20 range and declines steadily with greater scores (Fig. 13.3). Greater than 90% of the ecoregion has an HF ≤50. However, the vast majority of the area experiences some human influence; only 0.2% of the ecoregion has a score of HF = 0 (indicating no human transformation of the landscape given the measures we incorporated in our analysis).

Although 53,790 km^2 (16%) in the ecoregion has an HF score ≤10, these locations are distributed in 17,813 blocks ranging in size from <1 km^2 to 1,930 km^2. Most of these blocks are small: 14,368 (80.7%) are ≤1 km^2 in size and only 79 (0.004%) are >1,000 km^2. Thus, despite the appearance of large areas of land with low HF scores, most such areas in the ecoregion are quite small and fragmented.

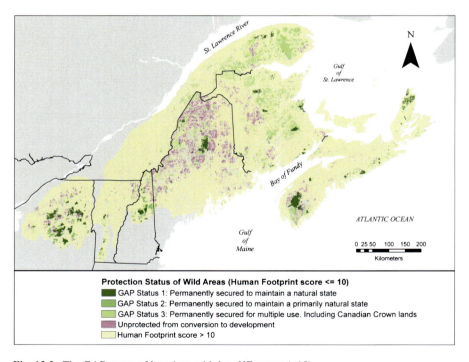

Fig. 13.2 The GAP status of locations with low HF scores (≤10)

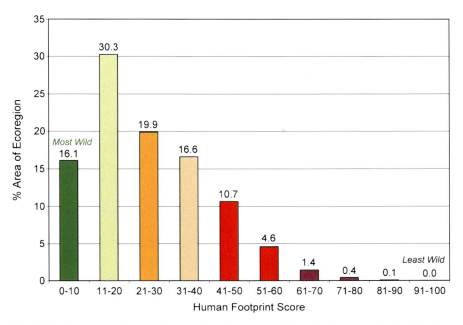

Fig. 13.3 Histogram of current human footprint scores of the Northern Appalachian/Acadian ecoregion

13.4 Future Human Footprints: Projecting Future Threats

The Future Human Footprints attempt to project a set of Human Footprint input parameters into the future. As much as possible, the goal of this effort was to project those parameters that were most likely to have an impact on biological diversity ('keystone threats'; Chap. 2) and could reasonably be modeled based on available information. Of course, predicting the behavior of natural systems over time can be uncertain. Thus, predictions must be based on fairly simple parameters and cover a range of scenarios so that decision makers can choose from among several plausible 'futures'; the purpose of the Future Human Footprints is not to say how the future will be but rather how it *might* be given a set of plausible conditions.

To map Future Human Footprint (FHF) scenarios, we chose features of the CHF that had 'keystone threat' characteristics (Chap. 2), were likely to change during the time scales we were interested in (several decades), and for which we had available data and/or modeling approaches. We adapted existing forecast models for land-use change to project them into the future under scenarios derived from examination of historical processes in the Northern Appalachian/Acadian ecoregion and other regions with analogous social, ecological, and economic conditions (i.e., the Pacific Northwest and Upper Midwest of the continental U.S.). After projecting these features, they were combined with the features of the CHF that were not forecast so as to provide a comparable surface to the CHF. In other words, the Current

Human Footprint was considered the baseline, and whatever features could be modeled that would make a difference were then added back into the CHF, replacing those original layers. Three salient features were chosen for modeling: (1) human settlement (the maximum of projected population density or current housing density); (2) residential, public roads; and (3) locations with high 'amenity-development' potential, likely to be the focus of new developments outside of existing settled areas.

As when we chose spatial resolution and data layers for the CHF, we applied the concept of parsimony (e.g., fewest parameters to explain the most variation) and the idea that it is more feasible to forecast scenarios than make predictions (Carpenter 2002). We chose to model the future based on best available data and simplest available models, and over time scales for which we felt confident, knowing that the further one projects into the future, the greater the uncertainty encountered. Human settlement was projected by taking the county-level 1990s growth rate in population density from the U.S. and Canadian census, and multiplying it by the year 2000 (U.S.) or 2001 (Canadian) census block (U.S.) or dissemination area (Canada) population densities, compounded by decade, over four decades. This approach conforms to the 'neighborhood' philosophy of modeling change, in which the conditions of a geographical neighborhood (being the county growth rate in this analysis) affects the smaller scale densities within it (Theobald 2003). However, this is cruder than hedonic modeling approaches, which attempt to predict transition probabilities of individual land parcels (Bell and Irwin 2002). It could also be advantageous to treat population or housing density as a dependent variable in a linear regression and use a number of geographic proxies (such as land use/land cover) as independent variables in order to select a model that could then be projected into the future. Both of these alternative approaches require data that are seldom available for the geographic extent, multiple countries (i.e., tax parcel data, economic data), and time series required for this kind of transboundary ecoregional analysis.

While modeling housing density is clearly advantageous for accounting for places with dwellings but few permanent residents (Theobald 2003), we chose to model human population density into the future and modify the final scenario to reflect disjunct housing influences in two ways. First, the final scenario represented the maximum of either the projected human population density or the current housing density. Second, the final scenario incorporated projected land-use transitions to amenity development in remote, lakeshore environments.

Roads are salient ecological features to model because roads have localized acute effects as well as far reaching and more chronic ecological effects (Trombulak and Frissell 2000). We chose to model the probability of occurrence (using a similar logit link approach as habitat occupancy modeling) of regular, public roads (i.e., primary roads, secondary roads, and highways) in the future because this class of roads is highly dynamic in our ecoregion due to low-density residential development, and because their expansion is directly related to human settlement (Baldwin et al. 2007a, b). We based this analysis on 17 years of historical data on road growth in Maine, one of the largest states in the region. Because this analysis is based on roughly two decades of change, we suggest that this projection points to areas of higher and lower risk for receiving new residential, public roads somewhere within a 10- to 25-year horizon.

Finally, we chose to model risk from amenity development (e.g., second homes, marinas) to undeveloped, lightly settled land. The Northern Appalachian/Acadian ecoregion includes much amenity-rich, undeveloped private land. As a result of recent dramatic changes in the economics of landownership in this ecoregion (Chap. 5), millions of hectares of private forestland with thousands of lakes, mountains, and other amenities have recently been transferred to real estate investment trust status (REIT) (Hagan et al. 2005). Amenity development represents new growth nodes disjunct from expanding urban areas (Bartlett et al. 2000). In many forested, remote regions in the Northern Hemisphere, these new growth nodes occur around ski areas, undeveloped lake and river shorelines, and coastlines. In this ecoregion, the most vulnerable of these amenities may be the thousands of lakes that occur on private forestlands where lakeshore development may be poorly regulated. In the Upper Midwest of the U.S., these kinds of lakes have attracted a boom in development in recent decades (Gustafson et al. 2005; Woodford and Meyer 2003), a boom modeled by one of our forecast scenarios for the Northern Appalachian/Acadian ecoregion. Our model thus included lakeshores on large, developable lakes that occur on lands owned by companies with a predisposition to sell for real estate (e.g., listed as REITs) and within a day's drive of the region's 16 major urban centers. We used these factors to select land in lightly settled landscapes likely to experience conversion to development in the near future, and the selected areas were incorporated into FHF scenarios.

Because of the importance of understanding the natural and socioeconomic processes by which threats arise (Chap. 2), we talked to regional experts in forestry, land-use economics, wildlife, and development in order to form a complete picture of the threats to the Northern Appalachian/Acadian ecoregion. Together, the projections comprising the Future Human Footprint represent both aggregate and disaggregate processes of development (Chap. 2). First, we modeled disaggregate process of incremental expansion in existing settled landscapes by decadal, exponential population expansion and residential road expansion (Baldwin et al. 2007b). Second, aggregate decisions of land-use change – risk to lakeshores of conversion to development model (Baldwin et al. 2009) – modeled the instantaneous establishment of new nodes of development.

The outputs of the projections were assigned corresponding impact (HI) scores, combined with the existing CHF layers that were not deemed salient, and normalized in the same way as the CHF to produce a FHF for three scenarios of future change. These scenarios were developed based on either of two assumptions: (1) that the region will continue to grow and change as it has in the recent past, or (2) that the region will grow and change in a manner analogous to similar regions of North America (the Pacific Northwest and the Upper Midwest of the continental U.S.). Each scenario incorporates both aggregate and disaggregate processes.

Specifically, the three FHF scenarios are as follows:

Scenario 1: Current Trends Under the Current Trends scenario, the rates of change in human settlement experienced during the 1990s continue to drive new settlement patterns into the future. Coupled with this is a modest rise in amenities development around heretofore undeveloped lakeshores – 'instantaneous transition' of forested landscapes to developed ones through aggregated decisions of large landowners. Incremental, disaggregated, cumulative change is modeled by (1) current trends of population growth projected 40 years, and (2) projected 80% probability surface for regular, public roads. Instantaneous, aggregated change is modeled by ownership-weighted risk to wilderness lakeshores within 100 km from major urban areas.

The second and third scenarios forecast what might happen in this ecoregion if the rates of change accelerate due to increased immigration of people into the region, such as might happen if new employment opportunities emerge in the region or living conditions became less desirable elsewhere (e.g., economic downturns or water shortages). We coupled this incremental process with a heavy rise in amenity development, such as might happen if there was a rise in wealth in the region due to new industries, as occurred in the Greater Seattle area of the Pacific Northwest during the 1990s.

Scenario 2: Rapid Influx – Pacific Northwest Model (High Urban Growth and Low Amenity Development) Incremental, disaggregated, cumulative change is modeled by 1990s population growth from Pacific Northwest counties, weighted as urban or non-urban, projected over 40 years. The projected population densities are used to produce new road projections for regular public roads. Instantaneous, aggregated change is modeled by risk to lakeshores within 100 km of major urban areas.

Scenario 3: Rapid Influx B – North Central Lakes Model (High Urban Growth and High Amenity Development) Incremental, disaggregated, cumulative change is modeled by 1990s population growth from North Central Lakes region counties of the Upper Midwest U.S., projected 40 years. The projected population densities are used to produce new road projections for regular, public roads. Instantaneous, aggregated change is modeled by risk to lakeshores within 200 km of major urban areas.

As an example of how one scenario forecasts the FHF, Fig. 13.4 shows the FHF based on growth patterns in the North Central Lakes region (Scenario 3: Rapid Influx B). With any FHF, a 'difference map' can be produced that shows the degree of difference, negative or positive, with the CHF; Fig. 13.5 shows such a difference map, illustrating where, compared to the present, impacts may accumulate (pink and red) and where they may abate (blue).

Regardless of scenario, the FHF analysis shows two important trends: (1) intensification and spreading outwards of human impact around settled areas, and (2) spreading of human impact throughout areas with low Human Footprint scores under the CHF. Both of these trends pose significant risks to biological diversity. Intensifying settlement (e.g., in the greater Montreal metropolitan area in Québec, or along the Green Mountains in Vermont) threatens wildlife that depend on local-scale habitat. For example, conditions for pool-breeding amphibians are projected to worsen under all future scenarios (Baldwin et al. 2007c). Likewise, intensification of settlement will cause greater landscape fragmentation throughout the ecoregion, threatening wildlife dependent on connectivity among and within large forest blocks.

13 The Human Footprint as a Conservation Planning Tool

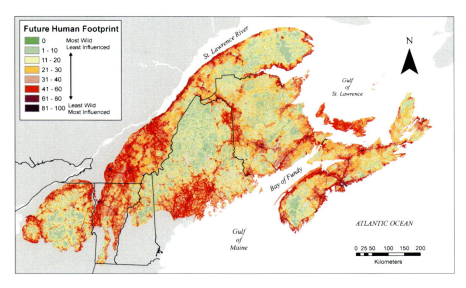

Fig. 13.4 The Future Human Footprint in the Northern Appalachian/Acadian ecoregion in Scenario 3: Rapid Influx B

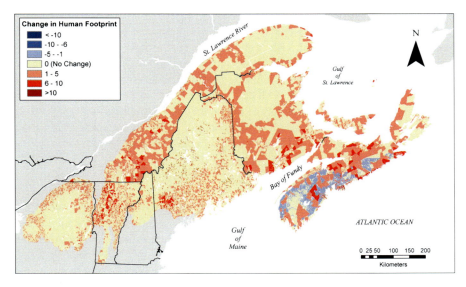

Fig. 13.5 The difference between the current human footprint and the Future Human Footprint (Scenario 3: Rapid Influx B) for the Northern Appalachian/Acadian ecoregion. Areas colored pink and red are projected to experience increased transformation – or threat – in future years. Areas in blue are projected to experience reduced threat

At the same time, spreading human impact through lightly settled areas introduces two new significant threats. First, it introduces and 'hardens' human infrastructure – including housing development, resorts, and paved roads – in areas previously dominated by timber harvesting. Second, these isolated resort developments can become new development nodes, leading to future incremental growth typical of settled landscapes.

A drawback of the FHF is that it does not model changes in forest cover and composition. In other words, it looks at human settlement proxies for land use/land cover change and not the change itself, such as might be derived from satellite imagery. We decided that such projections are dependent on management plans of individual landowners to such an extent that those kinds of projections are much more applicable to single large ownerships or collections of ownerships at the sub-regional scale. Similarly, the FHF does not model changes in the spatial distribution of logging roads (roads built specifically to access timber and often not maintained between harvests). While ecological impacts of logging roads are significant, in large part because they provide access to remote areas (Forman and Alexander 1998), we lacked the data to model such a dynamic phenomenon. Specifically, while we had good data on forest roads derived from various sources, it was impossible to predict the management plans of the various timber companies that would cause them to clear or abandon roads. Nonetheless, logging roads were represented in each scenario because they remained constant from the CHF. Finally, we suggest that future attempts to build Future Human Footprints take into account climate change, distribution of environmental toxins, and other threats that are difficult to map at these scales.

To the best of our knowledge, the FHF is the first such forecasting attempt to take a multivariate approach and explicitly incorporate two distinct processes of land-use change – one that is incremental expansion of settled areas, and one that represents the risk posed when undeveloped lands, far from towns and cities, instantaneously transition from existing land use/land cover to amenity development. As an indication of the power of incremental expansion to transform the landscape at the ecoregion scale, the accumulation of new, residential roads over a 20-year horizon will likely double the area susceptible to those roads, adding another 500,000 km to the existing network.

Likewise, instantaneous transition of currently little transformed areas poses a significant risk to landscape connectivity in the future. It is not the total amount of land near amenities that is vulnerable to transition that is noteworthy; rather, it is the dispersion of these lands over vast forested areas. Lakeshores we modeled as being vulnerable to development within 200 km of major urban centers represent only 1,118 km^2 (0.3% of the ecoregion), and those within 100 km represent only 625 km^2 (0.2% of ecoregion). At the same time, these areas are scattered throughout hundreds of thousands of hectares of the most wild and remote portions of the ecoregion (Baldwin et al. 2009). Changes such as these may transform what is now forest (albeit managed and often measurably transformed) to a landscape that has a new kind of human infrastructure: vacation homes, resorts, and roads to service them, further spreading the influence of humans outside of settled areas shown in the CHF.

The Current Human Footprint suggests that the settlement patterns of the region may still represent the 'primary productivity' and 'industrial' phases of settlement,

13 The Human Footprint as a Conservation Planning Tool 297

phases in which the most severe land transformation happens (that from systems dominated by natural disturbance to those dominated by anthropogenic disturbance) (Huston 2005). What the Future Human Footprint attempt to capture is a third, perhaps final phase of human settlement, the 'information/communication' phase where people settle and work from virtually anywhere. Areas at risk during this phase typically have high aesthetic values and reasonable access to urban areas and other service centers (Bartlett et al. 2000; Huston 2005). While we only provide forecasts based on analogous regions where such change has demonstrably taken place, significant evidence already suggests that this underway in the Northern Appalachian/Acadian ecoregion: parcelization of large farms and woodlots, development of shorelines and ridgetops, and increasing road infrastructure.

Based on our assessment of threats, we conclude that, given the trends forecast by our scenarios, adding more land to the protected areas network in a strategic fashion is the only way to mitigate permanent land conversion. In the Northern Appalachian/Acadian ecoregion, while slightly more than one-third (35%) of the land area is under some form of protection that prevents it from being converted to development (Anderson et al. 2006), only 7% of the landscape is designated as highly protected for conservation of biological diversity (GAP status 1), indicating that 93% is not managed primarily to protect ecosystems, ecosystem processes, populations of individual wildlife species, or other components biological diversity.

Certainly, the land-use changes forecast by Scenarios 2 and 3 (Rapid Influx) may not occur, and growth may continue only as in recent decades – rapid in some areas and slow in others (Scenario 1). The recent downturn in the global economy and the housing market in particular may change the factors that drove patterns of growth in the Pacific Northwest and Upper Midwest U.S. during the 1990s. Concern about climate change may lead to innovative new partnerships with landowners to sequester carbon that will compensate them equally as conversion to housing thereby satisfying the hedonistic principle (Chap. 2) for land conversion. In the true spirit of scenario modeling, we suggest that the Future Human Footprint methodology be viewed as a way to continually update forecasts for the future based on new and changing information. The only way to respond to uncertainty is to continually observe, document, monitor, and anticipate new changes.

13.5 Lessons Learned

While the Northern Appalachian/Acadian ecoregion is still one of the most forested and 'wild' ecoregions in Eastern North America, it may be one of the most vulnerable simply because so much undeveloped land is unprotected and within reach of densely populated areas. As is likely to be true in many other regions in North America, threats to the Northern Appalachian/Acadian ecoregion are currently concentrated in settled landscapes but may rapidly expand outwards given changes in social or ecological conditions that would encourage migration (e.g., climate, location of large industries, and availability of land with high amenity value).

Conservation planning needs to recognize the potential for human geography to rapidly change. In particular, conservation initiatives should not rely solely on the matrix forest being maintained primarily as managed forest; large tracts could and currently are being transformed to multiple uses including large-scale development for recreational housing and services. Conservation planners should seek partnerships with private landowners and government agencies to ensure that (1) large-scale fragmentation of existing forest blocks does not occur, and (2) new nodes of development inside large forest blocks are clustered and kept to a minimum, and that infrastructure to service them (roads, in particular) is built and maintained to minimize fragmentation and other adverse impacts (e.g., salt spray, collisions with wildlife, alterations in hydrology of wetlands and other water bodies).

As noted above, the basis for a map of the Human Footprint is intuitively obvious: large influences lead to large scores, and large scores imply large influences. However, our experience with presenting the Human Footprint maps (current and future under different scenarios) at public meetings throughout the region taught us that stakeholders need guidance in interpreting the implications of the map. First, people readily assume that areas with high HF scores do not have conservation value and should be ignored as priorities for conservation action. This is incorrect; a high score simply indicates a high degree of human transformation and is more properly interpreted as a potential measure of the threat and perhaps an additional index of priority for conservation action (Noss et al. 2002). It also gives some insight as to what conservation tools may be appropriate in any particular location; conservation actions that can be applied where human influence is low may be completely impossible where influence is high.

Second, people tend to focus on large blocks of land with low HF scores, assuming that all components of biological diversity require large, undisturbed areas for their conservation. This is also incorrect (Shafer 1995). Some species only require small areas to maintain viable populations, and as long as their diverse habitat requirements are met, small areas are sufficient for their conservation within a landscape. Abandoning conservation efforts in regions with overall high HF scores is precisely the wrong response for achieving comprehensive conservation goals.

Thus, when introducing maps of the Human Footprint to the public, considerable attention must be given to guiding their attention toward what the maps actually show: the distribution of and magnitude of human influences, knowledge of which can help guide effective and regionally comprehensive conservation action.

References

Alroy, J. (2001). A multispecies overkill simulation of the end-Pleistocene megafaunal mass extinction. *Science, 292*, 1893–1896.

Anderson, M. G., Vickery, B., Gorman, M., Gratton, L., Morrison, M., Maillet, J., et al. (2006). *The Northern Appalachian/Acadian ecoregion: Ecoregional assessment, conservation status and resource CD. The nature conservancy, eastern conservation science and the nature conservancy of Canada: Atlantic and Quebec regions.* Retrieved January 31, 2010, from http://conserveonline.org/workspaces/ecs/napaj/nap

Baldwin, R. F., Ray, J. C., Trombulak, S. C., & Woolmer, G. (2007a). Relationship between spatial distribution of urban sprawl and species imperilment: Response to Brown and Leband. *Conservation Biology, 21,* 546–548.

Baldwin, R. F., Trombulak, S. C., Anderson, M. G., & Woolmer, G. (2007b). Projecting transition probabilities for regular public roads at the ecoregion scale: A Northern Appalachian/Acadian case study. *Landscape and Urban Planning, 80,* 404–411.

Baldwin, R. F., Bell, K. P., & Sanderson, E. W. (2007c). Spatial tools for conserving pool-breeding amphibians: An application of the landscape species approach. In A. J. K. Calhoun & P. G. deMaynadier (Eds.), *Science and conservation of vernal pools in Northeastern North America* (pp. 281–297). Boca Raton, FL: CRC Press.

Baldwin, R. F., Trombulak, S. C., & Baldwin, E. D. (2009). Assessing risk of large-scale habitat conversion in lightly settled landscapes. *Landscape and Urban Planning, 91,* 219–225.

Bartlett, J. G., Mageean, D. M., & O'Connor, R. J. (2000). Residential expansion as a continental threat to U.S. coastal ecosystems. *Population and Environment, 21,* 429–446.

Bell, K. P., & Irwin, E. G. (2002). Spatially explicit micro-level modelling of land use change at the rural-urban interface. *Agricultural Economics, 27,* 217–232.

Boucher, T., Hoekstra, J., Jennings, M. & Ervin, J. (2006). *Global estimates of effective conservation.* (Paper presented at the 9th annual conference of the Society for Conservation GIS, San Jose, CA).

Carpenter, S. R. (2002). Ecological futures: Building and ecology of the long now. *Ecology, 83,* 2069–2083.

Carroll, C. (2005). *Carnivore restoration in the northeastern U.S. and southeastern Canada: A regional-scale analysis of habitat and population viability for wolf, lynx, and marten – Report 2: lynx and marten viability analysis (Wildlands Project Special Paper No. 6).* Richmond, VT: Wildlands Project. Retrieved December 15, 2009, from Klamath Center for Conservation Research Web site: http://www.klamathconservation.org/docs/Carroll_LynxMarten_hi.pdf

Carroll, C. (2007). Interacting effects of climate change, landscape conversion, and harvest on carnivore populations at the range margin: Marten and lynx in the Northern Appalachians. *Conservation Biology, 21,* 1092–1104.

CIESIN (Center for International Earth Science Information Network). (n. d.). *The last of the wild.* Retrieved February 3, 2010, from CIESIN Web site: http://www.ciesin.columbia.edu/wild_areas/

Davis, M. B., Spear, R. W., & Shane, L. C. K. (1980). Holocene climate of New England. *Quaternary Research, 14,* 240–250.

Dobson, A. P., Rodriguez, J. P., Roberts, W. M., & Wilcove, D. S. (1997). Geographic distribution of endangered species in the United States. *Science, 275,* 550–553.

Driscoll, C. T., Lawrence, G. B., Bulger, A. J., Butler, T. J., Cronan, C. S., Eagar, C., et al. (2001). Acidic deposition in the Northeastern United States: Sources and inputs, ecosystem effects, and management strategies. *BioScience, 51,* 180–198.

Driscoll, C. T., Han, Y. -J., Chen, C. Y., Evers, D. C., Lambert, K. F., Holsen, T. M., et al. (2007). Mercury contamination in forest and freshwater ecosystems in the northeastern United States. *BioScience, 57,* 17–28.

Evers, D. C., Han, Y. -J., Driscoll, C. T., Kamman, N. C., Goodale, M. W., Lambert, K. F., et al. (2007). Biological mercury hotspots in the Northeastern United States and Southeastern Canada. *BioScience, 57,* 29–43.

Forman, R. T. T., & Alexander, L. E. (1998). Roads and their major ecological effects. *Annual Review of Ecology and Systematics, 29,* 207–231.

Foster, D. R., & Aber, J. D. (2004). *Forests in time: The environmental consequences of 1,000 years of change in New England.* New Haven, CT: Yale University Press.

Foster, D. R., Motzkin, G., Bernardos, D., & Cardoza, J. (2002). Wildlife dynamics in the changing New England landscape. *Journal of Biogeography, 29,* 1337–1357.

Frumhoff, P. C., McCarthy, J. J., Melillo, J. M., Moser, S. C., & Wuebbles, D. J. (2007). *Confronting climate change in the U.S. Northeast: Science, impacts, and solutions.* Cambridge, MA: Union of Concerned Scientists. Retrieved February 3, 2010, from http://www.climatechoices.org/assets/documents/climatechoices/confronting-climate-change-in-the-u-s-northeast.pdf

Gustafson, E. J., Hammer, R. B., Radeloff, V. C., & Potts, R. S. (2005). The relationship between environmental amenities and changing human settlement patterns between 1980 and 2000 in the Midwestern USA. *Landscape Ecology, 20*, 773–789.

Haberl, H., Erb, K. H., Krausmann, F., Gaube, V., Bondeau, A., Plutzar, C., et al. (2007). Quantifying and mapping the human appropriation of net primary production in earth's terrestrial ecosystems. *Proceedings of the National Academy of Sciences, 104*, 12942–12947.

Hagan, J. M., Irland, L. C., & Whitman, A. A. (2005). *Changing timberland ownership in the Northern Forest and implications for biodiversity (No. MCCS-FCP-2005-1)*. Brunswick, ME: Manomet Center for Conservation Sciences.

Hunter, M. L., & Gibbs, J. P. (2007). *Fundamentals of conservation biology* (3rd ed.). Oxford, UK: Blackwell.

Huston, M. A. (2005). The three phases of land-use change: Implications for biodiversity. *Ecological Applications, 15*, 1864–1878.

Jones, D. A., Hansen, A. J., Bly, Doherty, K., Verschuyl, J. P., Paugh, J. I., et al. (2009). Monitoring land use and cover around parks: A conceptual approach. *Remote Sensing of Environment, 113*, 1346–1356.

Laliberte, A. S., & Ripple, W. J. (2004). Range contractions of North American carnivores and ungulates. *BioScience, 54*, 123–138.

Leu, M., Hanser, S. E., & Knick, S. T. (2008). The human footprint in the West: A large-scale analysis of anthropogenic impacts. *Ecological Applications, 18*, 1119–1139.

Model Forest of Newfoundland and Labrador. (n. d.) Model Forest of Newfoundland and Labrador. Retrieved February 3, 2010, from http://www.wnmf.com/partnerInititives

National Geographic Society. (n. d.). Africa megaflyover: Charting the last wild places on Earth. Retrieved February 3, 2010, from http://ngm.nationalgeographic.com/ngm/megaflyover/

Newmark, W. D. (1995). Extinction of mammal populations in Western North American National Parks. *Conservation Biology, 9*, 512–526.

Noss, R. F., Carroll, C., Vance-Borland, K., & Wuerthner, G. (2002). A multicriteria assessment of the irreplaceability and vulnerability of sites in the Greater Yellowstone Ecosystem. *Conservation Biology, 16*, 895–908.

Parks, S. A., & Harcourt, A. H. (2002). Reserve size, local human density, and mammalian extinctions in U.S. protected areas. *Conservation Biology, 16*, 800–808.

Sanderson, E. W., Jaiteh, M., Levy, M. A., Redford, K. H., Wannebo, A. V., & Woolmer, G. (2002). The human footprint and the last of the wild. *BioScience, 52*, 891–904.

Sanderson, E., Forrest, J., Loucks, C., Ginsberg, J., Dinerstein, E., Seidensticker, J., et al. (2006). *Setting priorities for the conservation and recovery of wild tigers: 2005–2015. The technical assessment.* New York and Washington, DC: Wildlife Conservation Society, World Wildlife Fund, and Save the Tiger Fund. Retrieved February 3, 2010, from http://www.savethetigerfund.org/AM/Template.cfm?Section=Full_Reports&Template=/CM/ContentDisplay.cfm&ContentID=2714

Saunders, S. C., Mislivets, M. R., Chen, J., & Cleland, D. T. (2002). Effects of roads on landscape structure within nested ecological units of the Northern Great Lakes Region, USA. *Biological Conservation, 103*, 209–225.

Shafer, C. L. (1995). Values and shortcomings of small reserves. *BioScience, 45*, 80–88.

Theobald, D. M. (2003). Targeting conservation action through assessment of protection and exurban threats. *Conservation Biology, 17*, 1624–1637.

Trombulak, S. C., & Frissell, C. A. (2000). Review of ecological effects of roads on terrestrial and aquatic communities. *Conservation Biology, 14*, 18–30.

Vitousek, P. M. (1994). Beyond global warming: Ecology and global change. *Ecology, 75*, 1861–1876.

Vitousek, P. M., Mooney, H. A., Lubchenco, J., & Melillo, J. M. (1997). Human domination of earth's ecosystems. *Science, 277*, 494–499.

Wilcove, D. S., Rothstein, D., Dubow, J., Phillips, A., & Losos, E. (1998). Quantifying threats to imperiled species in the United States. *Bioscience, 48*, 607–615.

Woodford, J. E., & Meyer, M. W. (2003). Impact of lakeshore development on green frog abundance. *Biological Conservation, 110*, 277–284.

Woolmer, G., Trombulak, S. C., Ray, J. C., Doran, P. J., Anderson, M. G., Baldwin, R. F., et al. (2008). Rescaling the human footprint: A tool for conservation planning at an ecoregional scale. *Landscape and Urban Planning, 87*, 42–53.

WRI (World Resources Institute) (2000). *City lights of the world: Based on NOAA-National Geophysical Data Center, Stable Lights and Radiance Calibrated Lights of the World (1998)*. Accessed on January 3, 2010, from http://earthtrends.wri.org/pdf_library/maps/4_m_citylights.pdf

Chapter 14
Assessing Irreplaceability for Systematic Conservation Planning

Stephen C. Trombulak

Abstract Systematic conservation planning requires that locations targeted for conservation action be prioritized, which can be difficult when planning across large landscapes because the possible sets of locations and conservation goals are all so large. These difficulties can be overcome with the use of computer programs that can handle large volumes of data and can identify sets of locations (called 'solutions') that achieve specified conservation goals. I describe the efforts of the conservation organization Two Countries, One Forest to identify priority locations in the Northern Appalachian/Acadian ecoregion using MARXAN to classify locations based on the number of times they are included in a solution. Priority scores range from highly irreplaceable (almost always required) to highly replaceable (almost never required). Conservation goals encompassed ecosystems, threatened and endangered species, geophysical landscape features, and focal carnivores. The amount of land at any particular level of priority varied depending on the target level set for each goal (ranging from low to high); however, target levels had only a small effect on the amount of highly irreplaceable lands (10.4–13.5% of the ecoregion), which were largely associated with existing conservation lands. Other lands also contribute to achieving regional conservation goals, but are generally interchangeable, providing flexibility for integrating conservation planning with broad public engagement.

Keywords Irreplaceability • MARXAN • Planning units • Priority • Systematic conservation planning

---

S.C. Trombulak (✉)
Department of Biology, Middlebury College, Middlebury, VT 05753, USA
e-mail: trombulak@middlebury.edu

S.C. Trombulak and R.F. Baldwin (eds.), *Landscape-scale Conservation Planning*, 303
DOI 10.1007/978-90-481-9575-6_14, © Springer Science+Business Media B.V. 2010

14.1 Introduction

Conservation planning is, ultimately, an exercise in setting priorities. Because no conservation initiative – whether it involves purchasing land, acquiring easements, or providing financial incentives – can be implemented everywhere simultaneously, conservation planners and practitioners must inevitably make choices about which locations are priorities for achieving their goals, locations that serve either as first steps in a series of actions taken over time or as standalone projects (Bottrill et al. 2009; Wilson et al. 2009).

Setting priorities can, on occasion, be straightforward because sometimes a planner simply has few choices to make. This may be particularly true when a planning effort takes place on a local scale. Determining which riparian corridor to protect, for example, is easy if only one river flows within the project region. Setting priorities can also be straightforward if conservation goals are defined narrowly, such as the protection of a single rare species or community; the mere presence of the desired organisms makes the location where they are found a priority by default even if the overall conservation goal is regional in scope. Places can also become priorities if they possess some unique, intrinsic characteristic, such as historical, scenic, or cultural importance. A famous river, a popular beach, or an historic town forest can become priorities for conservation simply because of what they are and not for their contribution or relationship to other locations or goals.

However, as the conservation goals become increasingly comprehensive and more ecological features are considered important (e.g., multiple species and communities, diverse geophysical characteristics, natural disturbance regimes, general responsiveness to environmental change) and the area over which those goals need to be achieved expands from local to regional scales, the number of possible locations that could help achieve the goals goes up dramatically, and the difficulty of setting priorities in a defensible and repeatable way becomes immense. The importance of one location, therefore, is based not only on what is present there but (1) on the distribution of the important features across the region, (2) the current level of conservation protection of the features at other locations, and (3) the spatial relationships among locations.

One way to set priorities is simply to assess locations based on their own intrinsic values or characteristics, independent of the values or characteristics of other locations or how complementary (e.g., completely redundant, completely different) they are with other locations. Such ad hoc conservation planning has historically been the norm. Parks, refuges, wild and scenic rivers, and wilderness areas – all have been created, and thus by default identified as priorities, simply because they were scenic, popular, available, uncontested, or contained habitat for a desirable species. In the Northern Appalachian/Acadian ecoregion, for example, ad hoc conservation planning has been the norm in both current and historic times. In recent years, such initiatives have included the International Appalachian Trail (Council of International Appalachian Trails. n.d.), Northern Forest Canoe Trail. (n.d.), Maine Forest Biodiversity Project (McMahon 1998), and the Wildland Area Project of the Northern Forest Alliance.

14 Assessing Irreplaceability for Systematic Conservation Planning

While ad hoc conservation planning can be – and has been – effective in achieving narrow conservation goals (e.g., protecting a scenic river, conserving an available parcel of land, protecting one representative of each ecosystem), it is at best inefficient and at worst counterproductive for achieving more comprehensive conservation goals. The second approach to setting priorities, systematic conservation planning, specifically avoids identifying priorities for locations without considering their relationship to other locations, in terms of both the ecological features they jointly contain and their geographical position with respect to one another (Groves et al. 2002; Margules and Pressey 2000; Margules and Sarkar 2007). Rather, the extent to which a location becomes a conservation priority for some type of action is contextual: the location's priority is based not only its own characteristics but on those of all other locations as well. Systematic conservation planning approaches the setting of priorities through the assessment of each location's irreplaceability in a greater context and, therefore, ideally results in a set of identified priority locations that are maximally efficient in achieving comprehensive goals, minimally redundant in applying limited time or money to conservation initiatives, and broadly defensible to stakeholders both in terms of why a location has been identified as a priority and how much of a priority it might actually be.

In this chapter, I describe an initiative to assess irreplaceability and identify priority locations for conservation action in the Northern Appalachian/Acadian ecoregion carried out by Two Countries, One Forest (2C1Forest), a confederation of conservation organizations that seeks to identify and pursue priorities for achieving landscape-scale conservation goals (Chapter 1). My particular emphasis is on describing the process we used for identifying priorities so that our approach could more easily be carried out in any ecoregion. Such a goal is not unreasonable since the methodology and analytical tools for doing so are readily available and reasonably straightforward to use (Smith et al. 2006). Yet any effort at systematic conservation planning, including the one described here, requires numerous subjective decisions, including which ecological features to include in the planning process, what conservation targets to set for each feature, what kind of spatial configuration is required for sets of locations, and how to treat locations based on existing patterns of ownership or occupancy. Therefore, I also describe the bases for the decisions we made in this process so that others can make the most informed decisions possible for planning in other regions.

14.2 The Theory of Systematic Conservation Planning

The first full conceptual model for systematic conservation planning was formulated by Margules and Pressey (2000) and most recently refined by Margules and Sarkar (2007) and Moilanen (2008). This framework takes the form of a set of sequential steps (adapted here from Margules and Sarkar 2007):

1. Identify stakeholders for the planning region.
2. Compile, assess, and refine relevant biological and socio-economic data.
3. Select the ecological features that will be used as surrogates for all biological diversity.
4. Determine how much of each feature is minimally desirable to conserve, referred to in systematic conservation planning as a 'target' for each feature. (It should be noted that the term 'target' takes on different meanings in other conservation contexts; for example, in The Nature Conservancy's 5-S Framework for Conservation Project Management (The Nature Conservancy 2009), 'target' is used to refer to an ecological feature; cf., Chapter 6.)
5. Identify the locations that are already being managed for conservation at some minimally acceptable level.
6. Identify new locations for conservation so that all selected ecological features (Step 3) are included at the desired target levels (Step 4) on existing (Step 5) and new locations combined.
7. Assess the likelihood that ecological features can persist over time at the new locations.
8. Delete from further consideration new locations where the likelihood of a feature's persistence is low, and then repeat Step 6–8 until all new locations contain features that are likely to persist over time.
9. Identify the best suite of new locations that together achieve the targets for all features while also considering other values (e.g., potential value of a location for extractive industry) and costs.
10. Implement a plan to conserve the features at each location selected.
11. Monitor and reassess progress and success.

The majority of the steps they outline are not unique to the systematic conservation planning framework; to some extent, identification and engagement of stakeholders, acquisition and analysis of data, and monitoring of success are (or should be) part of all conservation and natural resource management efforts, even those that are ad hoc.

What makes the approach advocated by Margules and Pressey (2000) and subsequent authors (reviewed in Margules and Sarkar 2007) a significant advance in conservation planning was their emphasis on complementarity (Pressey et al. 1993; Vane-Wright et al. 1991), the identification of new locations for conservation so that all identified ecological features are included at the desired target levels on existing and new locations combined (Step 6). The contribution of any single location, and hence its conservation value, is assessed in the context of both what is present at that location and elsewhere so that the ultimate suite of locations identified as important for conservation complement one another in terms of achieving conservation goals – used here to refer collectively to the specified features and their targets – throughout the region.

Much of the early literature on systematic conservation planning explicitly referred to priority locations as potential 'reserves' (Margules and Pressey 2000), which implicitly suggested a direct correspondence between the conservation value of a location and the management tool that should be used to conserve those values. Importantly, the language used to describe priority locations has subsequently shifted from 'reserve' to 'conservation area' (Margules and Sarkar 2007; Sarkar 2003), indicating that once a location has been identified as a priority, any tool in the conservation toolbox can be used to conserve the features there, including easements on private land, incentives, and regulations, in addition to public ownership and management as a traditional ecological reserve (e.g., park, wildlife refuge, wilderness area). Ultimately, what matters for conservation is not what tool is used but rather that it is effective.

Decoupling the method of identifying priorities from the methods of protection becomes especially important in systematic conservation planning within large landscapes. The larger the planning region, the more likely it is that socio-political attitudes regarding any single conservation tool will vary. Only in exceptional situations could all of the priority locations identified through systematic conservation planning within a large landscape be managed together as a single comprehensive system of publicly-owned ecological reserves.

When few locations need to be considered and few conservation goals achieved, assessing the complementarity of locations is relatively easy to do through visual inspection of data. For example, if the complete list of mammal species is known for each of five locations, it is straightforward to identify which suite of locations is needed in order to include each of the species at least once. But as the number of locations increases and the conservation goals become more complex (e.g., protect at least 5% of the total area of all common species, 50% of the total area for all rare species, and all populations of endemic species), efficient solutions rapidly become impossible to identify without the aid of computers.

Several different computer software packages are available for assessing complementarity, including MARXAN, C-Plan, and CLUZ (Ball and Possingham 2000; Ferrier et al. 2000; Smith 2004), which all give broadly similar results (Carwardine et al. 2007). One of the most widely used is MARXAN (Ball 2000; Ball and Possingham 2000; Ball et al. 2009). This is with good reason – MARXAN is well documented and supported by its developers, continuously improved for greater flexibility (Watts et al. 2009) and integration with geographic information systems (cf., The Nature Conservancy 2008), capable of simultaneously incorporating many conservation features and targets, and freely available in a format that runs on personal computers. Importantly, MARXAN is flexible to the needs of planners and provides realistic answers in the sense that the only realistic answer to the question, 'Where are the priority locations for achieving conservation goals in a region?' is, 'It depends.'

What ecological features are to be conserved? MARXAN allows planners to specify any number of features simultaneously, including populations, species, ecosystems, and non-living components of the environment (e.g., American marten [*Martes americana*], oak-hickory forests, granitic summits).

What are the targets that need to be achieved? For each ecological feature that is considered, planners can specify separate targets, characterized by areal extent, proportional representation, or absolute number (e.g., all granitic summits, 1,000 ha of oak-hickory forest, 20 known locations of marten). A suite of locations that achieves all of the specified targets is called a solution; locations that are part of a solution could thus be considered priorities.

Are some features more important than others? Features can be given relative weights, whereby when trade-offs need to be made, solutions can be biased toward achieving the targets for some features rather than others (e.g., meeting targets for endangered species might be considered more important than for soil types).

What locations can be considered? Locations can be specified in advance as 'required' (i.e., must always be included in a solution; for example, an existing national park), 'forbidden' (i.e., must never be included in a solution regardless of what ecological features are present there), or 'potential' (i.e., could or could not be included in a solution, depending on what ecological features are found there, its complementarity with respect to all other locations, and the specified targets).

How much spatial cohesion is desirable among priority locations? The extent of fragmentation allowed in a solution can be specified through a variable that gives greater or lesser weight to locations based on their proximity to other selected locations. The greater the value set for this variable, the greater the weight given to a location if it borders another one included in a solution, thus minimizing fragmentation. If the variable is set to zero, the planner is specifying that landscape fragmentation is not an issue in assessing a location's importance.

MARXAN identifies efficient suites of locations with an algorithm called simulated annealing, which maximally achieves the specified conservation goals while minimizing the amount of the landscape, either in terms of area or cost, identified as important (Ball 2000; Ball and Possingham 2000; Possingham et al. 2000). In brief, the steps in simulated annealing are (1) selection of an initial random suite of locations, (2) calculation of how well that suite satisfies the conservation goals, (3) addition or removal of locations to create a new suite, (4) comparison of how well the new and previous suites satisfy the conservation goals, (5) keeping the best these two suites, and (6) repeating, or iterating, Steps 3–5 a large number of times (e.g., 1 million). One complete cycle of these steps – from the random selection of an initial suite of locations to the final suite identified after a large number of iterations – is called a simulation, and as a simulation progresses, the suites of locations included tend toward being able to achieve more of the conservation goals with less total area and/or cost. The suite of locations identified at the end of a simulation is called the solution for that simulation.

It is important to understand that, with current computer technology, finding optimal or 'unambiguously best' solutions to complex planning problems, such as those with numerous conservation goals and potential locations, is quite difficult (Pressey et al. 1996). Thus, landscape-scale conservation planning, which almost by definition involves complex planning problems, requires choosing between identifying what is unambiguously best and achieving what is doable.

MARXAN offers a compromise between these two choices: It can be run on personal computers, it can find solutions to complex problems in short periods of time, and its solutions have been shown to be close to, even if not always identical to, the true optimal solution (McDonnell et al. 2002). Thus, the final solution may change slightly if the simulation is repeated. As a result, a better measure of a location's priority derives not from its inclusion in the final solution for one simulation but in the number of final solutions in which it is included. The more frequently a location is included in a solution, the more irreplaceable it is and therefore the greater of a priority it is for achieving the conservation goals.

Thus, each location falls into one of three broad categories:

1. Never included in a solution and therefore unimportant for achieving the conservation goals.
2. Always included in solutions and therefore considered to be completely irreplaceable (i.e., the conservation goals can never be met without those locations).
3. Included in some but not all solutions and therefore considered to have some level of replaceability (i.e., it can contribute to meeting the conservation goals, but what it contributes can be provided by other locations as well).

Thus, a location's irreplaceability can be scaled between 0 (never included) and 100 (always included); the higher a location's irreplaceability score, the greater of a priority for conservation action it is.

Assessing a location's importance for conservation through consideration of its irreplaceability has additional benefits for conservation planning. First, locations with intermediate irreplaceability scores are, by definition, interchangeable with one or more other locations for achieving the stated goals. In other words, locations A and B may each make the same contribution to achieving the goals, and if A is conserved, then B need not be. This helps identify both constraints and opportunities for compromise with regard to locations that need to be conserved.

Second, the implementation of conservation plans is more successful when the public is included in helping to craft them (Anderson et al. 2009; Chapters 10 and 17). By identifying locations that are interchangeable, the public can potentially use the results of an irreplaceability analysis as a basis for making further decisions about priorities, decisions where the trade-offs and consequences for achieving overall goals are made explicit. Thus, MARXAN and all other such computer programs are best thought of as decision support tools than as the source of a single best decision.

In summary, the analytical phase of systematic conservation planning when done at a landscape scale requires the use of computer software to find efficient solutions to what are exceedingly complex problems. MARXAN, which is widely used for conducting such analyses, makes generating potential solutions straightforward. Yet the ultimate value of the solutions depends almost entirely on decisions made before the analyses are run, decisions which fundamentally determine what problems they are solutions to. The planner needs to identify (1) the important ecological features to be considered, (2) the conservation targets for each feature, (3) the relative importance among the features, (4) the a priori status of each location with respect to its potential to be included in a solution, and (5) the importance of spatial cohesion among locations identified as priorities. None of these questions have absolute answers; thus, systematic conservation planning requires making subjective choices, comparing solutions across a range of choices to determine which identified priorities are robust, and accepting that the measure of a location's priority is relative, conditional, and unavoidably uncertain.

14.3 Irreplaceability in the Northern Appalachian/Acadian Ecoregion: Setting the Initial Parameters

As part of a larger planning initiative in the Northern Appalachian/Acadian ecoregion led by 2C1Forest, my colleagues and I used MARXAN to conduct an irreplaceability analysis to identify priority locations for conservation action in this region. As noted above, a series of questions needed to be answered in order to establish the input parameters for the analyses.

14.3.1 Selecting Ecological Features

We focused on 178 ecological features organized into four categories: special ecosystems; rare, threatened, or endangered species; focal carnivores; and ecological land units. These features were chosen for a number of reasons. First, they span a wide range of ecological characteristics – including both living and non-living features, organisms and communities at risk, and wide-ranging megafauna – all of which could potential act as surrogates whose protection would serve to protect a myriad of other features as well. Second, they were ecological features of interest to the conservation organizations that participate in 2C1Forest, particularly The Nature Conservancy/Nature Conservancy Canada (TNC/NCC) and Wildlands Network (formerly known as the Wildlands Project). Third, spatially explicit, high-resolution data were available for each of them.

Data on special ecosystems, rare, threatened, and endangered (RTE) species, and ecological land units (ELU's) were obtained from TNC/NCC's work on the ecoregional portfolio in the Northern Appalachian/Acadian ecoregion (Anderson et al. 2006). From 1999 to 2006, TNC/NCC prepared a series of ecoregional assessments for the Northern Appalachian/Acadian ecoregion. One of the outcomes of these assessments was the identification of locations of several rare, small-scale ecosystems. We chose seven of these as ecological features for our analysis: wetland basins, mountain summits, steep slopes, ravines, floodplains, coastal wetlands, and streams flowing through high-quality blocks of forest (Tier 1 matrix blocks, see Section 14.3.4). These seven include a range of ecosystems that are non-forested terrestrial or aquatic, thus expanding the breadth of the ecosystems used in identifying priorities in this largely forested ecoregion.

Another product of TNC/NCC's ecoregional portfolio was the development of a classification system for ELU's, non-living aspects of the environment that are likely to serve both as surrogates for biological diversity and as enduring features

14 Assessing Irreplaceability for Systematic Conservation Planning

of the landscape less likely to be altered in response to long-term environmental change like global warming. ELU's are based on three geographical/geological features: elevation (6 ranges), topography (14 categories), and bedrock geology (9 categories). Of the 756 possible combinations of these 29 categories, 164 distinct ELU's are found in the Northern Appalachian/Acadian ecoregion, all of which were included as ecological features.

TNC/NCC has also historically been instrumental in the documentation of locations of RTE species. Based on these data, we included all species that were globally ranked as critically imperiled, imperiled, vulnerable, or apparently secure or uncertain (G1, G2, G3, and G4–G?) as ecological features in our analysis.

The Wildlands Network provided information on source habitat for three focal carnivores – Canada lynx (*Lynx canadensis*), American marten, and Eastern gray wolf (*Canis lupus* or *lycaon*) – under different scenarios of landscape condition. Carroll (2003, 2005) identified areas of source habitat for these three species, all of which are native to the ecoregion but are considered threatened or extirpated in all (wolf) or part (lynx and American marten) of it. The Wildlands Network selected these as focal species for several reasons, including (1) their contribution to top-down ecological regulation within the region, (2) their sensitivity to human activities and human-induced landscape change (Carroll et al. 2003), and (3) for lynx and marten, their populations in the Northern Appalachian/Acadian ecoregion are peninsular extensions of broader boreal ranges (Carroll 2005). From Carroll's (2003, 2005) analyses, we selected three scenarios for inclusion in our analysis, one for each of the focal species: wolf source habitat under current landscape conditions, marten source habitat with continued trapping, and lynx source habitat.

14.3.2 Selecting Conservation Targets

For each ecological feature, we needed to select a target – or minimum amount that needs to be achieved in order for a solution to be considered successful. Targets were defined individually for each feature and were all expressed as a percent of all locations where the feature is present. For example, a 30% target for a feature present at 100 locations means that a solution must include at least 30 locations where the feature is present. It can include more than 30 locations with that feature, but it cannot include fewer. In practical terms, targets influence the number of locations included in solutions and the level of ecological redundancy obtained; the higher the target levels, the greater the redundancy and the greater the number of locations included in a solution.

Setting targets for ecological features is one of the most challenging aspects of systematic conservation planning (Pressey et al. 2003; Warman et al. 2004). The more redundancy specified as a target, the more locations identified as priorities. While objectively evaluating the optimal level of redundancy is possible when planning for single species (McCarthy et al. 2005), it is exceedingly difficult to do in complex systems, thus leading to potential conflicts among stakeholders.

Table 14.1 The percentage targets for each of the four categories of ecological features under the three target scenarios

	Target Scenario		
Feature Type	Low	Medium	High
Special ecosystems	50	65	80
RTE species	50	65	80
Focal carnivores	30	45	60
ELU's	5–20	25–40	45–60

Table 14.2 The percentage targets for ELU's as a function of target scenario and commonality in the ecoregion

Proportional Representation	Low	Medium	High
>1%	5	25	45
0.1–1%	10	30	50
0.01–0.1%	15	35	55
<0.01%	20	40	60

To minimize the polarization that can occur among stakeholders if only one set of targets is made, and thus risking the withdrawal from the planning process of participants who feel the targets were set too high or low, we developed a series of target scenarios to explore what would emerge as priorities under a range of targets. The three scenarios were labeled (1) low, (2) medium, and (3) high, and were defined as low, medium, and high percentages of the occurrences of the ecological features that must be included for a solution to be considered successful (Table 14.1).

For special ecosystems, focal carnivores, and RTE species, target levels were the same for all features within a feature type. For ELU's, the target percentage varied according to how common an ELU is in the ecoregion, with greater percentages set for rare ELU's than for common ones (Table 14.2). For example, under the low target scenario, a solution is only successful if it includes 50% of the occurrences of each special ecosystem and RTE species, 30% of the critical habitat for each of the focal carnivores, and 5–20% of the occurrences of each of the ELU's (5% of common ELU's, 20% of rare ELU's, and 10–15% of ELU's in between).

14.3.3 The Relative Importance Among Features

MARXAN allows ecological features to be weighted with respect to each other through a penalty or cost imposed on a solution that fails to meet specified targets. This is an inherent aspect of the simulated annealing algorithm; the efficiency of a solution is a function of both the solution's benefit (the extent to which targets for each feature are met) and cost (amount of area or monetary cost, plus the penalty for failing to meet a target). If the penalty is the same for all features, then they are considered to be equally important. However, penalties can be set differently for each feature. Penalties are assessed in relative terms; a feature with an assigned penalty of 4 is considered to be four times more important than a feature with an assigned penalty of 1.

In our analyses, we assigned penalties of 4, 2, 1, and 1 for special ecosystems, ELU's, RTE species, and focal carnivores, respectively, based on the belief that ecosystems and ELU's have broad utility as conservation umbrellas for many different species, including those that may still be undiscovered in the ecoregion or that may migrate into the region in the future.

14.3.4 Locations, the Units of Planning

Conservation planners must consider three different aspects of the locations, or planning units in the language of MARXAN, that will be assessed for their priority: the geographic basis for delineating locations (their size and shape), their a priori status with respect to inclusion in a solution, and the weighing given to potential locations based on some aspect of their quality or condition.

Planning units can be delineated in any manner by which a polygon can be drawn, such as watersheds, soil types, townships, a grid of equal-area hexagons, a grid of equal-area squares, or a set of land parcels available for purchase. Each has an advantage, and planners should select a manner of delineation that best complements the purpose of the planning effort (e.g., watersheds for aquatic conservation planning [Chapter 6]).

Similarly, the area of the planning units must be determined. This choice should be based on the resolution and accuracy of the data associated with each ecological feature as well as the scale on which conservation action resulting from irreplaceability analysis will take place. For example, terrestrial conservation initiatives might consider planning units with areas that compare to the areas of ecological reserves or other conservation lands that may eventually be created.

We chose to use a grid of 65,378 separate 10-km^2 hexagons that was spatially homogenous across the entire ecoregion. This allowed us to (1) assess the irreplaceability of all possible locations in the region, not just ones that were preselected, (2) factor area out of our analysis, treating all planning units the same with respect to their potential to contribute to achieving the conservation goals, and (3) strike a balance between the resolution of the data and the local spatial scale at which many of the subsequent conservation actions will ultimately need to be carried out (Chapter 10).

Each hexagon was a priori classified as one of three types: forbidden (or 'locked out'), required (or 'locked in'), and potential. All hexagons that were 50% or more classified by the U.S. or Canadian census bureaus as 'urban areas' were locked out of all solutions. Although this resulted in the inability to meet all conservation goals (because the targets for some features could only be achieved by including urban areas), it provided a way to avoid having solutions biased toward the inclusion of urban, and therefore generally more expensive, areas. Although this goal could have been achieved by including property values in the analysis, we found these data impossible to obtain over an area as large as the Northern Appalachian/Acadian ecoregion.

Two types of lands were locked into all solutions. The first was the set of existing GAP status 1 and 2 protected areas. The locations of these protected areas were previously identified and classified by TNC/NCC during their ecoregional planning process (Anderson et al. 2006; Section 14.3.1). TNC/NCC employed a crosswalk analysis to transform the IUCN protected area categories used in Canada to equivalencies in the GAP status system used in the U.S. Locking these protected areas into solutions was based on pragmatism; lands that are already publicly owned and managed primarily for conservation need to be taken into account when looking for new locations to include as conservation areas (Step 5 of the systematic conservation planning framework, Section 14.2).

The second type of land that was locked into all solutions was a set of high-quality matrix forest blocks, also identified previously by TNC/NCC. Because forests are the dominant ecosystem type in much of Eastern North America, TNC/NCC were interested in identifying large (greater than 10,000 ha), unfragmented blocks of matrix forest in good ecological condition as priorities for conservation. These blocks would also ideally contain at least some mature forests, have outstanding features like high-quality headwaters, and be surrounded primarily by natural land cover. Additionally, taken as a whole, blocks were selected from across the ecoregion and the full range of ELU's in order to maximize their collective contribution to including the full scope of biological diversity found in the ecoregion. Ultimately, 174 'Tier 1' matrix blocks were identified and in the irreplaceability analysis described here were locked into solutions.

All locations that were not locked into (GAP status 1–2 protected areas and Tier 1 matrix blocks) or out of (urban areas) solutions were available for inclusion in solutions. As with ecological features, locations can be weighted by their characteristics. This weighting was applied through a penalty factor: locations assigned a high-penalty factor could be included in solutions only if the benefits they provided to achieving the conservation targets outweighed the penalty of including them. As with the weighting of ecological features, penalty factors are relative. A landscape condition with a penalty factor of 2 is twice as 'costly' to include in a solution as a condition with a penalty factor of 1. We decided to weight locations with respect to two criteria: level of protection and quality of land cover (Table 14.3). As the level of protection and the quality of the land cover declined, the greater the penalty we imposed for including a location into a solution.

Table 14.3 Cost incurred for planning units based on their status as protected areas and land cover. All planning units identified as GAP status 1–2 or Tier 1 forest blocks are already locked into the solutions

Protection status	Land cover			
	Tier 1	Tier 2	Not Tier 1–2/ Natural	Not Tier 1–2/ Unnatural
Gap 1–2	1	1	1	1
Gap 3	1	2	3	5
Not Gap 1–3	1	3	4	6

14 Assessing Irreplaceability for Systematic Conservation Planning   315

As noted above, all locations identified as GAP status 1 and 2 protected areas or Tier 1 matrix blocks were locked into solutions. Locations with lower levels of protection (GAP status 3 or no protection) or a lower quality of land cover had increasingly more severe penalties imposed for inclusion in a solution.

14.3.5 Spatial Cohesion Among Planning Units

An additional characteristic that can be used to weight the priority of a location is its proximity to other priority locations. Valuing spatial proximity among locations is justified if some conservation goals cannot be achieved by a single location but instead require multiple locations taken together. This is likely to be the case if planning units are small (such as in this analysis) and ecological features include species that require large areas in order to maintain viable populations.

One measure of cohesiveness (and its inverse, fragmentation) is the length of the boundary of a group of planning units relative to the area of those units. For a given total area of a set of planning units, a longer total boundary length would be characteristic of low cohesion. In MARXAN, the weighting given to cohesion is controlled by a 'boundary length modifier' (BLM). The greater this value, the greater the cost imposed on a planning unit that is not adjacent to another planning unit already in a solution.

Weighting solutions for spatial cohesion may come at a cost, however, in terms of requiring more locations to be identified as priorities in order to achieve the conservation goals. Thus, selecting the optimal level of spatial cohesion requires finding a balance between the total amount of area included in a solution and the total length of the boundary. The optimal value is the one where decreasing the BLM does not meaningfully decrease the number of locations in a solution, and increasing the BLM does not meaningfully decrease the total length of boundary (Stewart and Possingham 2005). This value is easy to determine empirically for any given set of ecological features and targets. For our analysis in the Northern Appalachian/Acadian ecoregion, we empirically determined the optimal BLM to be equal to 0.00035 (Fig. 14.1); BLM's less than 0.00035 dramatically increased fragmentation with no decrease in the total area required to meet conservation goals and BLM's greater than 0.00035 dramatically increased the area required with little decrease in fragmentation.

14.3.6 Calculating Irreplaceability

As noted in Section 14.2, one MARXAN simulation compares a vast number of different solutions to identify an efficient solution. Yet because (1) the simulated annealing algorithm identifies solutions that are close to but not necessarily the absolute most efficient in terms of monetary cost or area, and (2) more than one

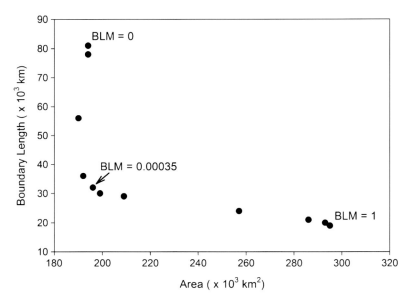

Fig. 14.1 The relationship between fragmentation (as measured by total boundary length) and the amount of land in a best solution as a function of the boundary length modifier in the Northern Appalachian/Acadian ecoregion

suite of locations may equally achieve the conservation goals, no single solution can be called the absolute best. Therefore, determining the importance of a location requires that numerous simulations be run and their results combined. Each location can then be assessed in terms of what percentage of the simulations it is included in a solution (Noss et al. 2002).

For our analyses, we ran 100 separate simulations to create 100 solutions. A planning unit's irreplaceability is thus a score between 0 (never present in a solution and therefore never required to achieve the specified conservation goals) and 100 (always present in a solution and always required to achieve the goals). The higher a location's score, the more irreplaceable it is for conservation in this ecoregion.

It needs to be remembered that irreplaceability scores do not define solutions. Solutions are suites of locations that collectively achieve the conservation goals, and they will include locations with intermediate irreplaceability scores (i.e., locations that contribute to achieving the goals but could be replaced by other locations). The final solution from among all of the simulations that is the most efficient at achieving the stated conservation goals could be considered the 'best solution,' although such a determination could really only occur after greater scrutiny of all the identified locations (Step 7 of the systematic conservation planning framework, Section 14.2).

Calculating irreplaceability, however, permits the identification of priority locations for conservation action. The more irreplaceable a location, the more critical it is for achieving conservation goals. Irreplaceable locations become priorities in

14 Assessing Irreplaceability for Systematic Conservation Planning

terms of how much and how quickly time and money ought to be invested to achieving conservation there. Highly irreplaceable locations can also be thought of as 'no-regrets' locations: places where additional information, changes in targets, and changes in conditions elsewhere within the region are unlikely to change their importance for conservation, so actions taken now are less likely to be regretted later. This is especially true when conservation actions – such as land acquisition and purchase of conservation easements – cannot be implemented simultaneously at all locations identified in a best solution. If conservation actions need to be implemented in stages over time, then the most effective strategy appears to be to begin with highly irreplaceable locations (Meir et al. 2004) and then reassess priorities for subsequent action repeatedly over time.

This approach to implementing a comprehensive conservation area network is relevant to the Northern Appalachian/Acadian ecoregion. Such a network will not be implemented all at once, but only incrementally as a consequence of numerous separate actions by each of the conservation organizations and government agencies participating in 2C1Forest. What is needed at the present time is an assessment of where the highly irreplaceable locations are and how those locations relate to the project areas for each of the participating groups to help them set priorities for their on-the-ground actions. This approach does not replace or negate the eventual need to identify a suite of locations that comprehensively achieves the conservation goals; it is merely the critical first step in that process.

14.4 Irreplaceability in the Northern Appalachian/Acadian Ecoregion: the Results

As might be expected with low targets, measures of irreplaceability are strongly influenced by the planning units that are locked into and out of solutions (Fig. 14.2, Table 14.4). Under the low targets scenario, 141,250 km^2 (27.6% of the ecoregion) have an irreplaceability score of 100 (planning units are always included in solutions). Conversely, 260,770 km^2 (51.0%) are unimportant or unavailable for achieving conservation goals (irreplaceability scores of 0).

The remaining 109,370 km^2 (21.4%) are neither locked into nor out of solutions, yet have intermediate irreplaceability scores ranging between 1 and 99, and are strongly skewed toward low values (1–20, Table 14.4), indicating that most of the locations in the ecoregion that are neither highly irreplaceable (100) nor unimportant (0) are highly replaceable, being included in at most 20% of the solutions.

This pattern suggests that under the low target scenario (a) roughly one-quarter (27.6%) of the ecoregion is highly irreplaceable for achieving the conservation goals under the constraints we set, and (b) the conservation goals that cannot be met on the highly irreplaceable lands can be met by a wide variety of other locations. Furthermore, a large amount of ecoregion (51.0%) is never included in a solution, indicating that given the availability of other locations for achieving the conservation goals, they are not needed under the low target scenario.

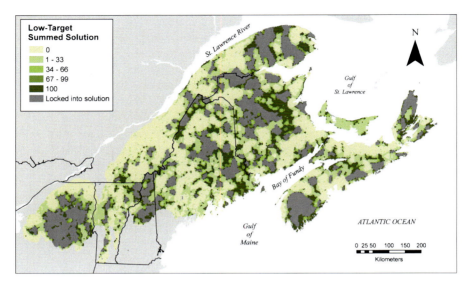

Fig. 14.2 Irreplaceability of planning units under the low target scenario with scores shown only for planning units that were not locked into solutions (Tier 1 matrix blocks and GAP status 1–2 protected areas)

Almost two-thirds (92,960 km^2) of the area that scores as highly irreplaceable does so because it is locked into solutions by virtue of being in a Tier 1 matrix block already prioritized by The Nature Conservancy or in an existing GAP status 1 or 2 protected areas. However, another one-third (48,920 km^2) is highly irreplaceable even though it is not locked into a solution (Table 14.4). These lands tend to be adjacent to lands that are locked into solutions (Fig. 14.2), indicating the tendency for solutions to prioritize locations that will minimize fragmentation of priority lands throughout the ecoregion. In contrast, only about 4% (9,530 km^2) of the unimportant lands are deemed so because they have been locked out of the solutions (i.e., urban areas).

As noted above (Section 14.3.6), the highly irreplaceable locations by themselves do not constitute a solution to achieving the conservation goals. Figure 14.3 shows the best solution from among the 100 simulations run under the low targets scenario. Of course, much work remains to be done before it could be argued that this is indeed the 'best' solution for implementation; particularly important would be on-the-ground assessment of the likelihood that the ecological features are viable at each location where they are found, reassessment of solutions once locations without viable representatives of ecological features are eliminated, and consideration of other values present at identified locations (Steps 7–9 of the systematic conservation planning framework, Section 14.2), all through participatory processes that include the regional stakeholders (Chapters 3 and 10).

Locations that are locked in have a similar influence under the medium targets scenario (Table 14.4). The primary changes observed relative to the low targets scenario are (a) a decrease in the amount of land that is never required to achieve

Table 14.4 The amount of land in square kilometers (with percentage of total area in parentheses) in different ranges of irreplaceability (percent inclusion in 100 solutions) under different target scenarios. Irreplaceability scores are shown in the second line. Unimportant lands (Irreplaceability score = 0) are subdivided into locations that were locked out of all solutions and locations that were not locked out but never appeared in any solution. Highly irreplaceable lands (Irreplaceability score = 100) are subdivided into locations that were locked into solutions and locations that were not locked in but that appear in all solutions anyway

Target Scenario	Locked out	Neither locked in nor locked out							Locked in	Total
	0	*0*	*1–20*	*21–40*	*41–60*	*61–80*	*81–99*	*100*	*100*	
Low	9,530 (1.9)	251,240 (49.1)	81,390 (15.9)	10,620 (2.1)	7160 (1.4)	4860 (1.0)	5340 (1.0)	48,290 (9.4)	92,960 (18.2)	511,390
Medium	9,530 (1.9)	174,670 (34.1)	130,020 (25.4)	28,410 (5.6)	12,800 (2.5)	7450 (1.5)	6390 (1.2)	49,160 (9.6)	92,960 (18.2)	511,390
High	9,530 (1.9)	118,310 (23.1)	106,520 (20.8)	54,940 (10.7)	34,300 (6.7)	25,900 (5.1)	17,470 (3.4)	51,460 (10.1)	92,960 (18.2)	511,390

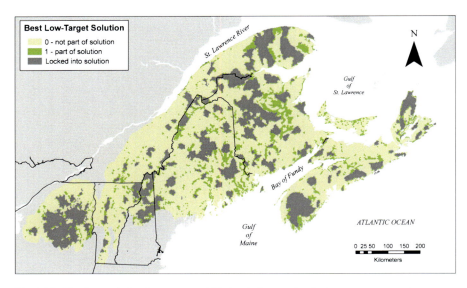

Fig. 14.3 The best solution from among 100 simulations under the low targets scenario

conservation goals (from 51.0% to 36.0%), (b) a negligible increase in the amount of highly irreplaceable land (27.6–27.8%), and (c) a slight shift among intermediate irreplaceability lands to be included in more solutions. Thus, the highly irreplaceable lands largely remain the same, less land never contributes to achieving conservation solutions, and the increased target levels result in the inclusion of a larger range of the lands that remain.

Under the high targets scenario, roughly the same amount of area is highly irreplaceable (28.3%) as under both low and medium target levels (Table 14.4). However, only 25.0% of the land is deemed unimportant (compared to 51.0% and 36.0% under the low and medium target levels, respectively). Intermediate irreplaceability lands are further skewed toward higher values, indicating that under high target levels, specific locations become increasingly irreplaceable for achieving conservation goals.

From the analyses of these three scenarios, four key messages emerge. First, solutions are greatly influenced by lands that are locked in: a large fraction of the specified conservation goals, even under high target levels, can be achieved by the Tier 1 matrix blocks and the existing public lands managed primarily for ecological values (GAP status 1–2 protected areas). Including these lands as required parts of conservation solutions results in a relatively small amount of additional lands to capture all highly irreplaceable areas (Table 14.4), and these lands are largely located adjacent to or as connections between Tier 1 matrix blocks and GAP status 1–2 protected areas (Fig. 14.2).

Second, and unsurprisingly, as target levels increase, the amount of land needed to meet overall conservation goals likewise increases. Yet despite the greater levels of selection, those additional lands show a great deal of replaceability; achieving higher target levels requires greater replication of protected lands for ecological features, but this replication can be achieved with many different configurations of lands apart from the limited amount of land identified as completely irreplaceable.

Third, when target levels are low, broad areas of the ecoregion never contribute to achieving the specified conservation goals. However, as targets increase, the potential contribution of much of these areas also increases, indicating that a majority of locations in the ecoregion have the capacity to contribute to achieving conservation goals if the desired level of ecological replication is high enough.

Finally, with all three target scenarios, highly irreplaceable lands that are not Tier 1 or GAP status 1–2 protected areas are found throughout the ecoregion and thus represent important priorities for conservation action. The highly irreplaceable lands by themselves would be insufficient to conserve the ecological features at the specified targets; a complete system of conservation lands would require inclusion of many locations that are considered replaceable. However, the highly irreplaceable lands are priorities for conservation action (Meir et al. 2004) – such as public acquisition and management as GAP status 1–2 protected areas, purchase of conservation easements, or education and incentives for conservation by private landowners – and provide conservation organizations and government agencies good, no-regrets places to focus their attention, especially when time, money, and information may be limiting (McDonald-Madden et al. 2008).

14.5 Lessons Learned

The work of 2C1Forest to identify priority locations for conservation action in the Northern Appalachian/Acadian ecoregion highlights several important lessons relevant to landscape-scale conservation initiatives anywhere. First, no one conservation organization or government agency is likely to have all of the data that are important for conducting comprehensive irreplaceability analyses. Because acquiring such data can be expensive and time consuming, coalitions of conservation groups are critical for this kind of analysis (Chapters 4 and 10).

While much has been written about the necessary conditions for effectively including the public in conservation planning, much less has been said about how to manage coalitions of organizations. I found that the most important trait that allowed 2C1Forest to complete its analyses was compromise. Every organization and agency has its own history, mission, internal management structure, priorities, competing projects, and need to produce products that can be branded for fundraising. Compromise on all of these constraints as well as on the selection of conservation features and targets, planning units, and other modeling decisions is critical to success.

One characteristic of an irreplaceability analysis that makes compromise easier is inherent in the fact that the selection of input parameters in conservation planning software, such as MARXAN, is subjective. Thus, their selection needs to be transparent and the solutions tested for robustness under a range of choices. If opinions differ among participants in what values to set for any initial parameter, an assessment of the extent to which different choices change the results (i.e., sensitivity analysis) will indicate whether or not it even matters (Chapter 11). Thus, sensitivity analyses are important as tools both for the analysis as well as management of group dynamics because group members can see quantitatively the relative influence on final modeling outputs of parameters that may be important to them (Butler et al. 1997).

Second, landscape-scale planning is important for achieving broad conservation goals, but ultimately the implementation of any action whose value is identified through such an analysis has to take place on a local level (Chapter 10). Thus, landscape-scale planning has to be undertaken with the participation of people and organizations working on the local level. 2C1Forest worked toward this by inviting frequent evaluation of our analyses by the conservation groups (including NGO's, government agencies, and private companies) that participate in the 2C1Forest confederation. That way, each group, even if their work focused on only a small area, could see the larger landscape as a context for their local efforts, and 2C1Forest could see what was needed in its analyses – both their formulation and communication – to be useful at the local level (Pierce et al. 2005).

Finally, working across multiple political boundaries (e.g., two countries, four U.S. states, and four Canadian provinces) makes data acquisition challenging both because of inherent differences in data structures (e.g., population census data) and availability. These challenges can be overcome with time, money, and patience, and they need to be adequately accounted for in both work plans and budgets.

Acknowledgments I would like to thank my colleagues in the Science Working Group of Two Countries, One Forest – M. Anderson, R. Baldwin, K. Beazley, P. Doran, G. Forbes, L. Gratton, J. Ray, C. Reining, and G. Woolmer – for everything they contributed to our work together.

References

Anderson, M. G., Vickery, B., Gorman, M., Gratton, L., Morrison, M., Maillet, J., et al. (2006). The Northern Appalachian/Acadian ecoregion: Ecoregional assessment, conservation status and resource CD. The nature conservancy, eastern conservation science and the nature Conservancy of Canada: Atlantic and Quebec regions. Retrieved January 31, 2010, from http://conserveonline.org/workspaces/ecs/napaj/nap

Anderson, C., Beazley, K., & Boxall, J. (2009). Lessons for PPGIS from the application of a decision-support tool in the Nova Forest Alliance of Nova Scotia, Canada. *Journal of Environmental Management, 90,* 2081–2089.

Ball, I. R. (2000). Mathematical applications for conservation ecology: The dynamics of tree hollows and the design of nature reserves. Dissertation, University of Adelaide, Adelaide, Australia.

Ball, I. R., & Possingham, H. P. (2000). *MARXAN (V1.8.2): Marine reserve design using spatially explicit annealing, a manual.* University of Queensland, Marxan Web site: Retrieved December 15, 2009, from http://www.uq.edu.au/marxan/docs/marxan_manual_1_8_2.pdf

Ball, I. R., Possingham, H. P., & Watts, M. E. (2009). Marxan and relatives: Software for spatial conservation prioritisation. In A. Moilanen, K. A. Wilson, & H. P. Possingham (Eds.), *Spatial conservation prioritization: Quantitative methods and computational tools* (pp. 185–195). Oxford: Oxford University Press.

14 Assessing Irreplaceability for Systematic Conservation Planning

Bottrill, M. C., Joseph, L. N., Carwardine, J., Bode, M., Cook, C., Game, E. T., et al. (2009). Finite conservation funds mean triage is unavoidable. *Trends in Ecology and Evolution, 24*, 183–184.

Butler, J., Jia, J., & Dyer, J. S. (1997). Simulation techniques for the sensitivity analysis of multi-criteria decision models. *European Journal of Operations Research, 103*, 531–546.

Carroll, C. (2003). *Impacts of landscape change on wolf viability in the northeastern U.S. and southeastern Canada: Implications for wolf recovery (Wildlands Project Special Paper No. 5).* Richmond, VT: Wildlands Project. Retrieved January 31, 2010, from The Wildlands Network Web site: http://www.wildlandsproject.org/ files/pdf/carroll_wolf_lo.pdf

Carroll, C. (2005). *Carnivore restoration in the northeastern U.S. and southeastern Canada: A regional-scale analysis of habitat and population viability for wolf, lynx, and marten – Report 2: lynx and marten viability analysis (Wildlands Project Special Paper No. 6).* Richmond, VT: Wildlands Project. Klamath Center for Conservation Research. Retrieved December 15, 2009, from http://www.klamathconservation.org/docs/Carroll_LynxMarten_hi.pdf

Carroll, C., Noss, R. F., Paquet, P. C., & Schumaker, N. H. (2003). Use of population viability analysis and reserve selection algorithms in regional conservation plans. *Ecological Applications, 13*, 1773–1789.

Carwardine, J., Rochester, W. A., Richardson, K. S., Williams, K. J., Pressey, R. L., & Possingham, H. P. (2007). Conservation planning with irreplaceability: does the method matter? *Biodiversity and Conservation, 16*, 245–258.

Council of International Appalachian Trails. (n.d.). Retrieved January 28, 2010, from http://www.internationalat.org/Pages/index

Ferrier, S., Pressey, R. L., & Barrett, T. W. (2000). A new predictor of the irreplaceability of areas for achieving a conservation goal, its application to real-world planning, and a research agenda for further refinement. *Biological Conservation, 93*, 303–325.

Groves, C. R., Jensen, D. B., Valutis, L. L., Redford, K. H., Shaffer, M. L., Scott, J. M., et al. (2002). Planning for biodiversity conservation: putting conservation science into practice. *BioScience, 52*, 499–512.

Margules, C. R., & Pressey, R. L. (2000). Systematic conservation planning. *Nature, 405*, 243–253.

Margules, C. R., & Sarkar, S. (2007). *Systematic conservation planning.* New York NY: Cambridge University Press.

McCarthy, M. A., Thompson, C. J., & Possingham, H. P. (2005). Theory for designing nature reserves for single species. *American Naturalist, 165*, 250–257.

McDonald-Madden, E., Baxter, P. W. J., & Possingham, H. P. (2008). Making robust decisions for conservation with restricted money and knowledge. *Journal of Applied Ecology, 45*, 1630–1638.

McDonnell, M. D., Possingham, H. P., Ball, I. R., & Cousins, E. A. (2002). Mathematical methods for spatially cohesive reserve design. *Environmental Modeling and Assessment, 7*, 107–114.

McMahon, J. (1998). *An ecological reserves system inventory: Potential ecological reserves on Maine's existing public and private conservation lands.* Augusta, ME: Maine State Planning Office.

Meir, E., Andelman, S., & Possingham, H. P. (2004). Does conservation planning matter in a dynamic and uncertain world? *Ecology Letters, 7*, 615–622.

Moilanen, A. (2008). Generalized complementarity and mapping of the concepts of systematic conservation planning. *Conservation Biology, 22*, 1655–1658.

Northern Forest Canoe Trail. (n.d.). Retrieved January 28, 2010, from http://www.northernforest-canoetrail.org/

Noss, R. F., Carroll, C., Vance-Borland, K., & Wuerthner, G. (2002). A multicriteria assessment of the irreplaceability and vulnerability of sites in the Greater Yellowstone Ecosystem. *Conservation Biology, 16*, 895–908.

Pierce, S. M., Cowling, R. M., Knight, A. T., Lombard, A. T., Rouget, M., & Wolf, T. (2005). Systematic conservation planning products for land-use planning: interpretation for implementation. *Biological Conservation, 125*, 441–458.

Possingham, H., Ball, I., & Andelman, S. (2000). Mathematical methods for identifying representative reserve networks. In S. Ferson & M. Burgman (Eds.), *Quantitative methods for conservation biology* (pp. 291–306). New York: Springer-Verlag.

Pressey, R. L., Humphries, C. J., Margules, C. R., Vane-Wright, R. I., & Williams, P. H. (1993). Beyond opportunism: Key principles for systematic reserve selection. *Trends in Ecology and Systematics, 8*, 124–128.

Pressey, R. L., Possingham, H. P., & Margules, C. R. (1996). Optimality in reserve selection algorithms: when does it matter and how much? *Biological Conservation, 76*, 259–267.

Pressey, R. L., Cowling, R. M., & Rouget, M. (2003). Formulating conservation targets for biodiversity pattern and process in the Cape Floristic Region, South Africa. *Biological Conservation, 112*, 99–127.

Sarkar, S. (2003). Conservation area networks. *Conservation and Society, 1*, v–vii.

Smith, R. J. (2004). Conservation Land Use Zoning (CLUZ) Software. Durrell Institute of Conservation and Ecology, Canterbury, UK. Retrieved January 28, 2010, from http://www.mosaic-conservation.org/cluz

Smith, R. J., Goodman, P. S., & Matthews, W. S. (2006). Systematic conservation planning: a review of perceived limitations and an illustration of the benefits, using a case study from Maputaland, South Africa. *Oryx, 40*, 400–410.

Stewart, R. R., & Possingham, H. P. (2005). Efficiency, costs and trade-offs in marine reserve system design. *Environmental Modeling and Assessment, 10*, 203–213.

The Nature Conservancy. (2008). Protected Area Tools for ArcGIS 9.3 Version 3.0. Retrieved January 28, 2010, from http://gg.usm.edu/pat/index.htm

The Nature Conservancy. (2009). How we work: Conservation by design. Retrieved January 28, 2010, from http://www.nature.org/aboutus/howwework/cbd/science/art14309.html

Vane-Wright, R. I., Humphries, C. J., & Williams, P. H. (1991). What to protect? Systematics and the agony of choice. *Biological Conservation, 55*, 235–254.

Warman, L. D., Sinclair, A. R. E., Scudder, G. G. E., Klinkenberg, B., & Pressey, R. L. (2004). Sensitivity of systematic reserve selection to decisions about scale, biological data, and targets: case study from Southern British Columbia. *Conservation Biology, 18*, 655–666.

Watts, M. E., Ball, I. R., Stewart, R. S., Klein, C. J., Wilson, K., Steinback, C., et al. (2009). Marxan with Zones: software for optimal conservation based land- and sea-use zoning. *Environmental Modeling and Software*. doi:10.1016/j.envsoft.2009.06.005.

Wilson, K. A., Carwardine, J., & Possingham, H. P. (2009). Setting conservation priorities. *Annals of the New York Academy of Sciences, 1162*, 237–264.

Chapter 15
Conservation Planning in a Changing Climate: Assessing the Impacts of Potential Range Shifts on a Reserve Network

Joshua J. Lawler and Jeffrey Hepinstall-Cymerman

Abstract As climates change over the coming century, many species will experience range shifts. Some species that currently inhabit protected areas will move out of those areas and others will move in. Drawing on model projections from previous studies, we assessed potential changes in the representation of trees, birds, mammals, and amphibians in the protected areas of the Northern Appalachian/Acadian ecoregion of North America. Six of 17 tree species were projected to experience a reduction in the areas suitable for growth in the region's protected areas and 11 of the 17 were projected to gain representation. Seven of 14 bird species were projected to experience losses in representation of their suitable habitat and the other seven were projected to experience gains. Range-shift projections for mammals and amphibians indicated that the protected areas would likely experience 13% and 21% turnover in these species, respectively with roughly half of the species experiencing losses of suitable habitat in the reserves and half experiencing gains. Despite these potential changes, protected areas are still likely to be one of the best tools for protecting biodiversity in a changing climate. One of the major challenges for the coming decades will be to provide the connectivity that will facilitate movement out of, and into, protected areas.

Keywords Climate change • Conservation planning • Protected areas • Range shifts • Species

---

J.J. Lawler (✉)
School of Forest Resources, University of Washington, Box 352100, Seattle, WA 98195-2100
e-mail: jlawler@u.washington.edu

J. Hepinstall-Cymerman
Warnell School of Forestry & Natural Resources, University of Georgia, 180 E. Green St., Athens, GA 30602-2152
e-mail: jhepinstall@warnell.uga.edu

S.C. Trombulak and R.F. Baldwin (eds.), *Landscape-scale Conservation Planning*, 325
DOI 10.1007/978-90-481-9575-6_15, © Springer Science+Business Media B.V. 2010

15.1 Introduction

Implicit in the notion of a reserve system designed to protect biodiversity is the assumption that the areas within a reserve network will provide for the persistence of species and ecosystem functions through time (Margules and Pressey 2000). This assumption is challenged by reserve size, the lack of redundancy in the reserve system, and threats from outside the reserves. In many cases, these challenges will pale in comparison to the challenges posed by climate change.

In the past, many species responded to major changes in the Earth's climate by moving to track suitable temperature and moisture regimes (Davis and Shaw 2001). Following major changes in climate, some species that were rare became quite common, while others that were common became rare or even went extinct. Many ecological communities were reshuffled and in some cases new 'no-analog' assemblages were formed (Brubaker 1988). The climatic changes projected for the twenty-first century have the potential to alter today's ecological systems to a similar, if not greater degree. Even the lower end of the range of projected warming for the middle of the century will likely result in major changes to some ecosystems and communities.

As climatic conditions change, some species will likely move out of current protected areas in search of more suitable climates. Other species may be able to move into protected areas as new climatic regimes and new habitats emerge. Thus, one obvious and critical question is: how well will current reserve systems protect biological diversity in a rapidly changing climate? A second, equally important question is: how can additional areas be added to reserve networks to increase protection of biological diversity as climate changes? Will adding areas to a reserve network be an effective solution? If not, the entire concept of conservation planning will need to be revisited.

Here, we explore some of these questions in a case study focused on the Northern Appalachian/Acadian ecoregion of North America, a nearly 330,000 km^2 area including portions of four U.S. states (New York, New Hampshire, Vermont, and Maine) and four Canadian Provinces (Québec, New Brunswick, Nova Scotia, and Prince Edward Island) (Fig. 15.1, Chap. 1). We begin by providing some background on recent and projected future climatic changes in general and in the region. We go on to discuss recent and potential future climate impacts, specifically describing projected climate-driven range shifts for trees, birds, and other vertebrates in the region. We then use these projected range shifts to evaluate the potential effect of climate-driven shifts in species distributions on the current reserve network of the Northern Appalachian/Acadian ecoregion. We conclude with a general discussion of some of the ways that we can begin to adapt conservation-planning approaches to address climate change.

15.2 Climate Change

Global average temperatures have risen by 0.7°C over the last 100 years (IPCC 2007a). The rate of warming has increased over the past century such that the rate over the last 50 years was greater than over the past 100, and the rate over the last

15 Conservation Planning in a Changing Climate   327

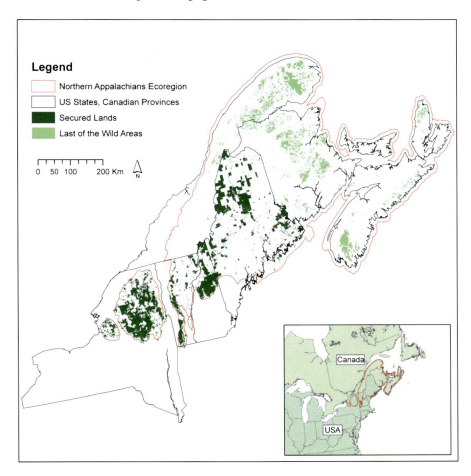

Fig. 15.1 Study area boundary encompassing the Northern Appalachians/Acadian ecoregion. Secured lands were available only for the United States and overlap substantially with the Last of the Wild Areas

20 years was greater still. Furthermore, the years from 1995 to 2007 included 11 of the highest global average temperatures on record. Temperatures have increased more dramatically at higher northern latitudes. For example, average annual temperatures in Alaska have risen 2–4°C over the last century (Houghton et al. 2001). Recent changes in precipitation have been less consistent. Some regions of the globe have experienced increases in precipitation whereas others have experienced decreases (IPCC 2007a).

Global average temperatures are projected to rise between 1.1°C and 6.4°C by 2100 (IPCC 2007a). More recent studies have concluded that this range is likely to be conservative and thus a larger range with a higher maximum temperature is likely (Oppenheimer et al. 2007). High northern latitudes are projected to experience some of the greatest increases in temperature – potentially over 7.5°C. Similar to patterns in recent changes in global precipitation, projected changes in precipitation are more inconsistent than projected changes in temperature. In general, some areas of the globe are projected to experience increases in precipitation and others are expected to experience decreases. In still other regions, little agreement exists with respect to even the direction of precipitation changes across the projections from different general circulation models (GCMs).

15.2.1 *Projected Climate Change for the Northern Appalachian/ Acadian Ecoregion*

As for the rest of the globe, temperatures are consistently projected to increase in the Northern Appalachian/Acadian ecoregion over the coming century, with observed increases of 0.86–1.86°C over the last century for New England and New York (Trombulak and Wolfson 2004). Based on simulations from three different GCM's run for a mid-high emissions scenario (SRES A2, Nakicenovic et al. 2000) average annual temperatures are projected to increase by 3.5–5.4°C by 2099 (Fig. 15.2a). Based on predictions from the same three models run for the same emission scenario, total annual precipitation was projected to increase slightly by 2 to 12 mm (Fig. 15.2b). These projections were based on differences in the average of mean annual temperatures and average total annual precipitation projections from a 10-year period at the end of the century (2090–2099) and the period from 2000 to 2009.

15.3 Climate Impacts

Recent climatic changes have already begun to alter ecosystems. Changes in hydrology, fire regimes, glaciers, and sea levels have all been linked to recent temperature trends. For example, increased temperatures have resulted in reduced snowpack in some regions (Mote 2003). Reduced snowpack and an increase in the amount of precipitation falling as rain in turn have the potential to alter the timing and quantity of stream flow, shifting peak flows earlier in the spring, and reducing summer flows. Changes in flow regimes have implications for stream fish (Battin et al. 2007) and changes in snowpack may directly affect some terrestrial species (Carroll 2007).

Other changes in hydrology will also have implications for many species. Changes in the timing and amount of precipitation and increased evaporation due to rising temperatures are likely to have the most profound effects on shallow rain-fed wetlands (Burkett and Keusler 2000; Winter 2000). Reductions in water levels and increases in water temperatures have the potential to increase turbidity and decrease dissolved oxygen concentrations (Poff et al. 2002). Furthermore, increases in temperature may lead to increased productivity resulting in more frequent algal blooms and anoxic conditions (Allan et al. 2005).

As a result of increasing temperatures, global average sea levels are projected to rise between 18 and 59 cm by the year 2100 (IPCC 2007a). Because this estimate

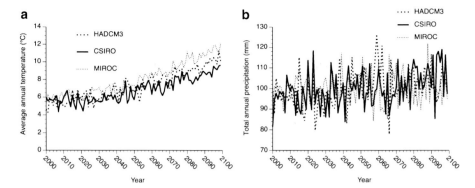

Fig. 15.2 Projected future average annual temperatures (**a**) and total annual precipitation (**b**) for the Northern Appalachians/Acadian ecoregion. Projections were based on simulations of the UKMO HADCM3.1, CSIRO MK3, and MIROC3.2 (medres) general circulation models run for a mid-high (SRES A2) emission scenario. The original climate projections were taken from the World Climate Research Programme's (WCRPs) Coupled Model Intercomparison Project phase 3 (CMIP3) multi-model dataset. These projections were then downscaled by the Lawrence Livermore National Laboratory (LLNL), Reclamation, and Santa Clara University (SCU) and are stored and served at the LLNL Green Data Oasis

is primarily based on the thermal expansion of the oceans, a gradual melting of the ice sheets, and the loss of non-polar glaciers and does not account for the potential for rapid changes in the Greenland and West Antarctic ice sheets, these are likely to be underestimates (Oppenheimer et al. 2007). Sea-level rise has begun to alter many low-lying coastal systems (IPCC 2007b) with estuaries and coastal wetlands and marshes among the most heavily affected. In many cases, habitats will need to be created upslope of these systems to allow species to move in response to the loss of these sensitive habitats (Pearsall 2005; Scott et al. 2008).

Fire plays a major role in the dynamics of many ecosystems. Increases in temperature and decreases in precipitation have the potential to produce drier fuels and thus more frequent or larger fires. In addition, changes in atmospheric carbon dioxide (CO_2) concentrations, precipitation patterns, and temperatures may alter vegetation dynamics resulting in changes in fuel loads. Thus, it is not surprising that recent increases in the frequency of large wildfires in the Western U.S. have been linked to changes in climate (Westerling et al. 2006). Projected future climatic trends will likely result in even more dramatic increases in the frequency of large fires across the Western U.S. (McKenzie et al. 2004). These increases will result in changes in landscape patterns, forest structure and composition, and habitat availability for many forest-dependent animal species.

Although all of the climate-driven changes discussed above can have indirect effects on biological diversity, climate change is also having direct effects on plants and animals. The majority of the recent ecological changes that have been attributed to climate change are changes in phenology or species distributions. Many spring events occur earlier, advancing at a rate of 2.3 days per decade (Parmesan and Yohe 2003). These events include breeding in birds (Crick et al. 1997; Dunn and Winkler 1999), emergence of insects (Roy and Sparks 2000; Stefanescu et al. 2003), and advanced calling and breeding in amphibians (Beebee 1995; Gibbs and Breisch 2001). Likewise, many shifts in species distributions have been recorded, including poleward and elevational shifts in birds (Root 1992, 1993; Thomas and Lennon 1999) and butterflies (Parmesan 1996; Parmesan et al. 1999). In a review of documented effects of climate change on a wide range of species, Parmesan and Yohe (2003) concluded that range shifts, for species in which they have been recorded, have been occurring, on average, at a rate of 6.1 km per decade towards the poles and 6.1 m per decade upward in elevation. These responses to recent changes beg the question of how species will modify their ranges in response to projected future climate changes. We focus most of the remainder of the chapter on these potential future range shifts and their implications for conservation planning.

15.3.1 *Projecting Potential Future Range Shifts*

Several studies have used the relationships between current species' distributions and current climate to project potential shifts in species ranges in response future climate-change projections (Thuiller et al. 2005). Most such studies use correlative bioclimatic models that define the current range of a species as a function of the current climate. These models describe the climatic space or climate envelope that a species currently occupies. Projected future climate data are then used as inputs to the models to project where the climate space of a species might be in the future. These models can provide a preliminary assessment of how climate change might affect a species' distribution or the flora or fauna of a region.

Although correlative bioclimatic models are arguably the best tools currently available for projecting the potential effects of climate change on the distributions of large numbers of species, they are not without their limitations (Pearson and Dawson 2003, 2004). Correlative bioclimatic models generally do not directly account for biotic interactions, evolution, land-use change, dispersal, or time lags in the response of vegetation to climate change. In addition, any number of factors may alter the mechanisms that determine how a species responds to climatic factors. For example, increases in atmospheric CO_2 concentrations have the potential to increase water-use efficiency in plants, allowing them to persist in drier environments. Thus, a correlative model that defines the current range of a plant species as a function of historic climatic conditions may fail to capture the future potential range of the species in an atmosphere with twice as much CO_2.

An additional limitation to the correlative bioclimatic modeling approach involves the uncertainty inherent in the models themselves. Several studies have demonstrated how different approaches based on statistical or machine-learning techniques can provide very different projected future ranges for the same species given the same projected future climate data (Pearson et al. 2006). These differences are due to errors in the bioclimatic models, uncertainties resulting from limited species distribution data, and differences in the approaches themselves. Fortunately,

15 Conservation Planning in a Changing Climate

several approaches have been shown to reduce the errors in bioclimatic models (Elith et al. 2006; Lawler et al. 2006; Prasad et al. 2006). Among these approaches are model-averaging techniques that base predictions on multiple models built with subsets of a given dataset.

Despite the limitations of correlative bioclimatic models, they provide a preliminary estimate of how large numbers of species may respond to climate change at coarse spatial scales. Although it is often difficult to find independent datasets to test bioclimatic models, evidence suggests that these models can capture recent shifts in the ranges of some species (Araújo et al. 2005). Furthermore, although many species will respond to climate change in complex and unpredictable ways, many of the recent observed range shifts have been similar to the types of shifts predicted by correlative models – species have generally moved upwards in elevation and poleward in latitude.

15.3.2 Projected Changes in Habitat Suitability for Tree Species in the Eastern United States

Iverson et al. (2008a, b) modeled potential changes in the suitability of habitat for 134 tree species in response to six potential future climate-change scenarios. Current species distributions were taken from Forest Inventory and Analysis (FIA) plots in the Eastern U.S. FIA plot data were used to calculate importance values based on the number of stems and basal area of each species in a plot following the methods of Iverson and Prasad (1998). In addition to current climate data, soils, topography, and land-cover data were used to build the models for the 134 species. All data were gridded with a 20-km resolution. Random forests, a machine-learning-based model-averaging approach (Breiman 2001; Cutler et al. 2007), was used to model the importance value (habitat suitability) as a function of current climate. Because Iverson et al. used FIA data, their modeling was limited to the U.S.

Future climate projections were derived from three different GCM's run for low (SRES B1) and high (SRES A1FI) emissions scenarios (Iverson et al. 2008b). Mean projections for high and low emissions were also derived from the three GCM's projections. These climate projections were generated for the IPCC fourth assessment report (IPCC 2007a) and downscaled by Hayhoe et al. (2006). Projected future climate data were then used to project shifts in habitat suitability (importance values for each 20×20-km cell) for the end of the century (Prasad et al. 2006). Area-weighted importance values (AWIV) were used to explore the changes expected in future suitable habitat for a species. Because both the area of suitable habitat and the importance value can change independently, a ratio of future AWIV to current AWIV is more informative in understanding future conditions (Iverson et al. 2008b).

Suitable climatic conditions for many tree species in the Eastern U.S. are projected to shift northward. For some species, the loss of climatic suitability at the southern end of their range may result in a severe contraction of the species' distribution in the

U.S. For example, balsam fir (*Abies balsamea*), paper birch (*Betula papyrifera*), and black spruce (*Picea mariana*) are all projected to experience decreases in climatic suitability over much of their U.S. ranges. Other species, whose northern range limits occur in the U.S., may experience range expansions as a result of an increase in the area of suitable climatic conditions. For example, suitable climatic conditions for black hickory (*Carya texana*), black oak (*Quercus velutina*), and longleaf pine (*Pinus palustris*) are all projected to expand in response to climate change.

Iverson et al. (2008b) also projected potential shifts in the climate space of ten different general forest types. Even the lower emissions scenario resulted in the projected near complete loss of the climate space for the spruce-fir and white-red-jack pine forest types in the Eastern U.S. In addition, the high emissions scenario resulted in a mass contraction in the Eastern U.S. of the climates suitable for the maple-beech-birch forest type and projected expansions of suitable areas for oak-hickory and oak-pine forest types. That is not to say, however, that these forest types or the species that make them up will necessarily disappear from the landscape. Some species will be able to move north into Canada and others will find climatic refugia at higher elevations.

15.3.3 Projected Bird Species Range Shifts in the Eastern United States

Predicted changes in climate such as warmer and wetter conditions as well as increased variability of weather and increased probability of extreme weather events all have the potential to affect bird populations (Saether et al. 2000). Furthermore, at broad spatial scales, bird species distributions are often associated with climatic factors. Rodenhouse et al. (2008) used bioclimatic models of 150 bird species to assess the potential effects of climate change on birds in the Northeastern United States. The bioclimatic models were based on Breeding Bird Survey data, climate, elevation, and tree abundance data (Matthews et al. 2004). Matthews et al. (2004) also used random forest predictors to build their bioclimatic models. Matthews et al. (2004) and Rodenhouse et al. (2008) used these bird bioclimatic models with future climate projections from Hayhoe et al. (2006) and projected future tree species abundance (i.e., importance values) from Iverson and Prasad (2001) and Iverson et al. (2008a, b) to predict changes in potential future bird species distribution (incidence) and abundance (area-weighted incidence).

Rodenhouse et al. (2008) concluded that climate change was likely to result in large changes in the distribution of potential habitat for many species. In general, many species were projected to experience increases in potential habitat in Maine and New Hampshire and conversely, many were projected to experience losses in potential habitat in Pennsylvania and Western New York (Rodenhouse et al. 2008). However, on a species-by-species basis, the projections were highly variable – some species were projected to gain habitat and others were expected to lose habitat.

15.3.4 Projected Mammal and Amphibian Range Shifts in North and South America

In an unrelated study, Lawler et al. (2009) projected climate-driven shifts in the potential ranges of 2,954 bird, mammal, and amphibian species in the Western Hemisphere. Like the projections for the trees and birds described above, these projections were made with correlative bioclimatic models. Lawler et al. (2009) built models with species distribution data derived from gridded range maps and current climate data modeled with a 50-km resolution. As in the tree and bird analyses described above, Lawler et al. (2009) also used random forest predictors to model species occurrences as a function of current climate. Models were built with a subset of the data and tested on a reserved, semi-independent dataset. All models described by Lawler et al. (2009), and all of those used in the analyses described in Sect. 15.4, correctly predicted at least 90% of the presences (grid cells within the current range of the species) and at least 80% of the absences (grid cells outside the current range of the species) in the test-datasets.

To project climate-driven shifts in the potential ranges of the 2,954 species, Lawler et al. (2009) used 30 different climate-change projections derived from 10 different GCM's run for three different greenhouse-gas emissions scenarios (a low [SRES B1], a mid [SRES A1B], and a mid-high [SRES A2] scenario). As with the projected future climate data used in the two studies described above, these future climate projections were produced for the IPCC fourth assessment report (IPCC 2007a). The climate projections were obtained from the World Climate Research Programme's Coupled Model Intercomparison Project phase 3 multi-model archive (http://www-pcmdi.llnl.gov/ipcc/about_ipcc.php), and downscaled to the 50-km grid using an approach described in Lawler et al. (2009).

Lawler et al. (2009) concluded that many species would likely experience large shifts in the distribution of their potential ranges. Although many potential ranges were projected to shift poleward and towards higher elevations, others were projected to shift in less predictable ways. The less predictable shifts in many of these potential ranges are likely to have impacts on the Northern Appalachian/Acadian ecoregion. For example, based on one particular climate-change projection, the potential ranges of the moose (*Alces alces*) and the eastern mole (*Scalopus aquaticus*) were projected to contract from and expand into, respectively, the ecoregion (Fig. 15.3).

15.4 Implication of Range Shifts for the Northern Appalachian/Acadian Ecoregion

We used projected shifts in potential ranges of trees, birds, mammals, and amphibians from the three studies described above to explore the potential effect of climate change on the protection afforded by the current reserves in the Northern Appalachian/Acadian ecoregion. Because the bioclimatic models from the three

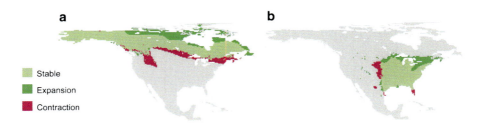

Fig. 15.3 Projected climate-induced shifts in the potential ranges of the moose (**a**) and the eastern mole (**b**). Projections are based on average climatic conditions simulated for 2071–2099 by the UKMO-HADCM3 general circulation model run for a mid-high (SRES A2) emissions scenario (Lawler et al. 2009)

studies differed in extent (Eastern U.S. only versus all of North America), our analyses for birds and trees are limited to the U.S. portion of the Northern Appalachian/Acadian ecoregion whereas our analyses for mammals and amphibians cover the entire ecoregion. To address the issue of having different study extents for these analyses, we defined our reserve network in two different ways. For analyzing the effects of climate-driven shifts in the potential ranges of trees and birds, we used the Permanently Secured Lands (i.e., protected parcels managed for natural resource conservation) developed as part of the ecoregional assessment conducted by The Nature Conservancy and available from the Two Countries, One Forest Conservation Atlas (www.2c1forest.org; Chap. 12). These areas covered 32,077 km^2 (25.4% of the ecoregion within the U.S. [126,077 km^2]). For analyzing the effects of climate-driven shifts in the potential ranges of mammals and amphibians, we used the Wild Areas of the Northern Appalachian/Acadian ecoregion that were assigned GAP status 1–3, derived as part the Human Footprint analysis conducted by the Wildlife Conservation Society and available from Two Countries, One Forest (Chap. 13). The 'Last of the Wild' map represents areas that are the '10% wildest' areas within the ecoregion (Woolmer et al. 2008), a much more conservative estimate of 'conserved areas' and covered only 33,590 km^2 of the ecoregion (10.2%).

We asked two related, yet different questions about the potential effects of climate-driven shifts in potential ranges. First, we asked how individual species would be represented in a reserve network in a changing climate. To answer this question, we chose 17 tree species and 14 bird species from the work of Iverson et al. (2008b) and Matthews et al. (2004). Eleven of the tree and seven of the bird species were projected to experience overall contractions (i.e., decreases in importance values and losses of suitable habitat) in their potential ranges within the Eastern U.S. and six and seven, respectively, were projected to experience overall expansions in their potential ranges in the U.S. Again, given the extent of the original range projections, our assessment for these trees and birds was limited to the portion of Northern Appalachian/Acadian ecoregion within the U.S. We overlaid the ecoregion as well as the reserve network boundaries with both the current modeled range and the projected future potential ranges of each of the species. We used two ensembles of three projected future potential ranges for each species.

15 Conservation Planning in a Changing Climate

These ensembles were averages of projections from three GCM's run for a low and a high emissions scenario, respectively (Sect. 15.3.2). We calculated the area-weighted sum of the importance values (for trees) or incidence values (for birds) for both the current day and for 2100. We then calculated the percent change in summed importance values (future/current *100) within the ecoregion as a whole and within the reserve network for each of the 17 tree and 14 bird species for low and high emissions climate-projection ensembles.

For our second question, we asked how the ecoregion, the reserve network, and one reserve in particular might change with respect to species composition as a result of projected range shifts. To answer this question, we used 26 amphibians and 73 mammals from the Lawler et al. (2009) study. This was the total number of amphibians and mammals in the region for which Lawler et al. (2009) were able to build relatively accurate models. We examined the projected shifts in the potential ranges of the 96 species with respect to (1) the ecoregion, (2) the reserve network, and (3) Adirondack Park. We overlaid the three areas of interest with 10 projected future potential ranges for each species. These projections were based on the simulated climates for the years 2071–2100 from 10 GCM's run for a mid-high (SRES A2) emissions scenario. We then calculated three measures of faunal change for the ecoregion, the network, and Adirondack Park. First, we calculated the number and percentage of species projected to completely lose all climatically suitable space within each area. This can be seen as a projection of the magnitude of potential loss in the current amphibian and mammal faunas for each region. Second, we calculated species turnover for each region. We summed the number of species projected to completely lose climatic space within the area and the number of species that could potentially move into the area due to the development new climatically suitable space. We then divided this number by the number of species currently in the area (and multiplied by 100) to provide a measure of the percentage of species turnover in the area. Our estimate of potential species turnover assumes species will be able to disperse into newly suitable areas. Finally, we calculated the percentage of species projected to experience contractions or expansions in their potential ranges within the ecoregion, reserve network, and Adirondack Park based on assumptions of full dispersal and no dispersal to newly generated potential ranges.

15.4.1 Trees and Birds of the Northern Appalachian/Acadian Ecoregion

Of the 17 tree species included in our analyses, six were projected to experience decreases in area-weighted importance values within the U.S. portion of the Northern Appalachian/Acadian ecoregion, eight were projected to experience increases in area-weighted importance values, and three (red maple [*Acer rubrum*], eastern hemlock [*Tsuga canadensis*], and eastern white pine [*Pinus strobus*]) had mixed results depending on the emissions scenario (Table 15.1). The area-weighted importance values for three species (balsam fir, yellow birch [*Betula alleghaniensis*],

Table 15.1 Current summed importance values and percent change in summed importance values for areas within the Northern Appalachians/Acadian ecoregion and the reserve network within that ecoregion for each of 17 tree species by the year 2100. The percent change values are based on the mean of three climate simulations from three different general circulation models run for the low (SRES B1) and high (SRES A1FI) emissions scenarios, respectively

	Ecoregion			Reserve network		
	Current	Low	High	Current	Low	High
Balsam fir, *Abies balsamea*	4,865	34	20	1,111	45	23
Yellow birch, *Betula alleghaniensis*	1,908	61	32	518	65	33
Red spruce, *Picea rubens*	2,480	40	34	675	40	33
Quaking aspen, *Populus tremuloides*	1,078	85	37	192	114	58
American beech, *Fagus grandifolia*	2,845	81	50	803	84	47
Sugar maple, *Acer saccharum*	3,544	85	68	898	81	64
Eastern hemlock, *Tsuga canadensis*	1,716	117	80	287	164	126
Red maple, *Acer rubrum*	4,192	114	90	862	122	106
Eastern white pine, *Pinus strobus*	1,661	100	91	242	131	159
White ash, *Fraxinus americana*	1,039	151	144	180	165	175
Black cherry, *Prunus serotina*	813	169	176	165	190	213
Red oak, *Quercus rubra*	592	283	293	71	495	579
Sweet birch, *Betula lenta*	149	377	423	25	525	692
Silver maple, *Acer saccharinum*	87	415	831	14	341	871
White oak, *Quercus alba*	110	775	1,579	8	1,953	4,469
Chestnut oak, *Quercus prinus*	47	632	1,651	10	549	2,032
Black oak, *Quercus velutina*	72	924	1,835	5	2,150	5,233

and red spruce [*Picea rubens*]) declined by 35–60%, with patterns generally similar for the ecoregion and the reserve network. These results indicate that balsam fir-red spruce, the dominant forest type in much of the Northeastern U.S., may occupy only 20–30% of its former range if the warmer climate predictions occur (Fig. 15.4).

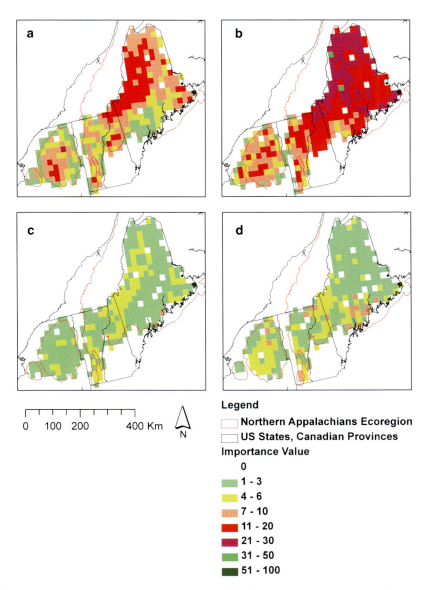

Fig. 15.4 Current species distribution by Importance Value for red spruce (**a**) and balsam fir (**b**) and future potential distribution based on mean predictions of three GCM models using a high emission scenario for red spruce (**c**) and balsam fir (**d**)

Sugar maple [*Acer saccharum*], which has large economic value in the region, is also predicted to decline by 15–36%.

All six tree species predicted by Iverson et al. (2008b) to have increased habitat in the Eastern U.S. had increased importance values within both the ecoregion and the reserve network. In addition, white ash (*Fraxinus americana*) and black cherry (*Prunus serotina*), species expected to decline across the Eastern U.S., were predicted to increase in importance within the ecoregion and reserve network. This may be due to a northward shift in the ranges of these species. Neither of the two species is currently prevalent in the ecoregion. The high elevations present in the ecoregion may also provide refugia for these species as the climate warms. Quaking aspen (*Populus tremuloides*), a species also expected to decline in the Eastern U.S., showed a slight increase in the reserve network under low climate-change predictions. The distribution of quaking aspen is projected to shift northward as temperatures increase. Thus, aspen may become more prevalent in the protected areas of the region.

Of the six bird species projected to experience general decreases in their potential ranges, the Black-throated Green Warbler (*Dendroica virens*), was projected to experience the greatest decreases in summed incidence values for both the low and high emissions scenarios across the ecoregion and the network (Table 15.2). By 2100, summed incidence values for the Black-throated Green Warbler were projected to be 38% and 43% of their current values given a high emissions scenario for the ecoregion and network, respectively. In contrast, the Summer Tanager (*Piranga rubra*) was projected to experience the greatest increases in summed importance values across the region and network (77- and 71-fold increases, respectively, given the high emissions scenario).

15.4.2 Amphibians and Mammals of the Northern Appalachian/Acadian Ecoregion

Projected shifts in the potential ranges of amphibians and mammals resulted in projected changes of 13–25% in the amphibian fauna and 7–21% in the mammal fauna depending on the spatial scale (ecoregion, network, or park) and whether unlimited or conversely no dispersal was assumed (Table 15.3). Turnover rates for both taxa were highest in Adirondack Park. This is generally expected, as isolated or smaller areas should experience a higher percentage of turnover in species as a result of shifting ranges than should larger areas. For example, in smaller areas, small range shifts can result in a species moving completely out of the area, but for larger areas, much larger shifts will be required to move a species completely out of the area.

Assuming unlimited dispersal, a little more than half of the amphibian and mammal species were projected to experience increases in their potential ranges within the ecoregion and the network. This pattern did not hold for Adirondack Park in which only 29% of the amphibian species and 17% of the mammal species were projected to experience expansions of their potential ranges within the park. As one would expect, under the no-dispersal scenario (i.e., when species were assumed to

Table 15.2 Current and percent change in summed habitat suitability values (incidence values) for areas within the Northern Appalachians/Acadian ecoregion and the reserve network within that ecoregion for each of 14 bird species by the year 2100. The percent change values are based on the mean of three climate simulations from three different general circulation models run for a low (SRES B1) and a high (SRES A1FI) emissions scenario, respectively. The 14 species are divided into a group of seven that were projected to experience overall contractions in potential habitat and a group projected to experience overall expansions in potential habitat

	Ecoregion			Reserve network		
	Current	Low	High	Current	Low	High
Overall projected habitat losses						
Black-throated green warbler, *Dendroica virens*	235	63	38	56	69	43
White-throated sparrow, *Zonotrichia albicollis*	322	68	41	72	77	45
Veery, *Catharus fuscescens*	335	83	47	74	90	51
Rose-breasted grosbeak, *Pheucticus ludovicianus*	312	96	66	68	99	76
Black-capped chickadee, *Poecile atricapilla*	333	96	66	73	100	74
Yellow warbler, *Dendroica petechia*	298	106	73	62	113	85
Song sparrow, *Melospiza melodia*	337	102	79	73	102	88
Overall projected habitat gains						
Yellow-billed cuckoo, *Coccyzus americanus*	15	511	1,437	3	502	1,641
Blue-grey gnatcatcher, *Polioptila caerulea*	7	626	2,038	2	371	1,505
Red-bellied woodpecker, *Melanerpes carolinus*	5	1,523	4,199	0.5	1,762	7,315
Yellow-breasted chat, *Icteria virens*	2	1,505	8,313	0.3	1,212	11,104
Orchard oriole, *Icterus spurius*	2	1,714	8,425	0.5	693	5,654
Blue grosbeak, *Passerina caerulea*	1	945	20,517	0.2	274	10,305
Summer tanager, *Piranga rubra*	0.1	4,715	77,226	0.02	2,251	70,915

J.J. Lawler and J. Hepinstall-Cymerman

Table 15.3 Projected climate-induced changes in amphibian and mammal faunas in the Northern Appalachians/Acadian ecoregion, the reserve network within that ecoregion, and Adirondack Park for the end of the century based on a mid-high (SRES A2) emissions scenario. The percent turnover was calculated as the sum of the number of species currently absent from the area with potential ranges that are projected to expand into the area and the number of species currently in the area whose projected future potential ranges do not include any part of the area divided by the total number of species currently in the area, multiplied by 100. The percent contraction, expansion, and equal are the percentages of the species currently in the area that are projected to experience contractions, expansions, or no change in potential range within the area. All values were calculated assuming unlimited dispersal into newly created potential range (left-hand columns) and no dispersal into newly created potential range (right-hand columns)

	Unlimitwed dispersal				No dispersal			
	Percent turnover	Percent contraction	Percent expansion	Percent equal	Percent turnover	Percent contraction	Percent expansion	Percent equal
Amphibians								
Ecoregion	13	46	54	0	17	96	0	1
Reserve network	21	46	54	0	17	96	0	1
Adirondack park	25	42	29	29	21	63	0	37
Mammals								
Ecoregion	13	33	66	1	10	79	0	21
Reserve network	1 3	41	57	2	7	75	0	25
Adirondack park	21	37	17	46	17	46	0	54

be restricted to the portion of their current potential range that overlapped their projected future potential range), many more species were projected to experience contractions in their potential ranges. A much smaller percentage of species was projected to experience contractions of their ranges within Adirondack Park than in the network or the ecoregion as a whole. For example, 96% of the amphibian species were projected to experience contractions in their potential ranges in the ecoregion and the network, but only 63% were projected to experience contractions within the Adirondack Park. Again, this is in part intuitive because range contractions will be more evident across larger spatial extents.

15.5   Conservation Planning in a Changing Climate

Although our analyses only investigated one aspect of climate change – climate-driven range shifts – and did not account for changes in hydrology, fire regimes, interspecific interactions, or changes in disease dynamics, they indicate that the composition of some regions and protected areas will be greatly altered in a changing climate. The ability of the reserve system to capture species as they move from protected areas will depend on species' dispersal abilities, the spatial distribution of protected areas, and the nature, location, and prevalence of barriers to dispersal.

The types of changes our analyses predict bring into question the efficacy of current approaches to reserve selection. Today's protected area networks are the result of a combination of different events, policies, and strategies. For much of the twentieth century, reserves were selected to protect scenic beauty, to provide habitat for specific groups of species, or they were added to the network opportunistically (Chap. 14). Even though more recently, conservation organizations have applied systematic approaches to reserve selection based on the current distribution of various elements of, or surrogates for, biological diversity (Margules and Pressey 2000; Chap. 14), the current reserve systems in most places were not specifically designed to protect biological diversity. Furthermore, none of these reserve systems were designed to address climate change. How then can these reserve networks be augmented to better protect flora and fauna in a changing climate?

Several suggestions have been made for how to augment reserve networks to address climate change. Some of the most basic of these involve adding more reserves to increase redundancy, designating buffers around reserves, increasing the size of existing reserves, and adding larger reserves to the network (Halpin 1997; Noss 2001; Shafer 1999). Slightly more nuanced suggestions include placing new reserves at the poleward edges of species ranges, between existing reserves to enhance connectivity, or in latitudinal arrays (Pearson and Dawson 2005; Shafer 1999). In the following three sections, we review some of the other approaches that have been proposed to improve the effectiveness of protected-area networks in a changing climate. These approaches vary in the degree to which they depend on projected changes in climate and projected biotic responses to climate change. Given the uncertainty in projected future climate projections and the even greater uncertainty in many of the projected biotic responses to climate change, the success of approaches that rely more heavily on projected climate impacts will likely be more uncertain (Lawler et al. 2010). We describe the various approaches that have been discussed in order, beginning with those that are less reliant on climate and climate-impact projections and concluding with those that are more reliant on such projections.

15.5.1 Planning for the Abiotic Template

Current systematic conservation planning efforts most often use maps of the current distribution of biota to select areas to maximally represent biological diversity (Margules and Pressey 2000). Given that the biota will likely move in response to climate change, one alternative would be to base reserve selection on the stable abiotic elements that in part determine species distributions. Landforms, soils, and elevation all play critical roles in determining the ecological community at a given site. Elevation, latitude, water-bodies, and local topography help to define local climatic conditions. Regardless of global climatic changes, many local and regional climatic gradients (but not actual conditions) will likely persist. Thus, it may be possible to use current climatic gradients, in conjunction with soils and landform data, to select areas that will provide the most diverse set of abiotic conditions under which new ecosystems and new communities can develop. Furthermore, it may be possible to combine species-based approaches and approaches designed to capture current climatic gradients. Pyke and Fischer (2005) suggest selecting multiple sites that capture the range of climatic conditions across a species' distribution.

15.5.2 Identifying Climate Refugia

In addition to using edaphic conditions and current climatic gradients to select conservation areas, projected future climatic conditions may be useful for guiding reserve selection in some cases. Saxon et al. (2005) mapped projected changes in environmental conditions defined by climate, soils, and topography. Areas in which environmental conditions are less likely to change may serve as climatic refugia. By protecting these areas, it may be possible to preserve some species that otherwise would not find suitable environments in a rapidly changing climate. On the other hand, climate-change projections can provide estimates of where changes are likely to be the most severe and thus where species movements and hence connectivity might be most crucial.

15.5.3 Selecting Sites Based on Range-Shift Projections

Finally, some have suggested that range-shift projections can be used to design reserve networks that will be resilient to climate change (Hannah et al. 2007; Williams et al. 2005). For example, Hannah et al. (2007) compared approaches for selecting reserve networks based on a combination of current species distributions and projected future distributions derived from climate-envelope models. At present, however, projected climate-induced range shifts are too uncertain for these projections to be the basis for reserve selection. Reserve networks selected on the basis of complementarity can be heavily influenced by the distribution of a relatively small number of rare species. Because rare species will be some of the most difficult to model, their projected future ranges will likely be highly uncertain. Thus, reserve networks based on these projections will likely fail to protect many of the species in most need of protection.

A more appropriate use of range-shift projections will be to test some of the more basic suggestions for expanding reserve networks in a changing climate. For example, will protecting areas based on their geology, soils, and topography capture species as they move in response to climate change? Will positioning new reserves along latitudinal or elevational gradients facilitate protection in a changing climate? These simple approaches can be tested using bioclimatic models built with a variety of different assumptions and with a range of different climate-change scenarios to assess their robustness to uncertainty in future range projections.

15.6 Conclusions

Although the approaches described above make intuitive sense, we must ask whether they can address the magnitude of change natural systems are likely to experience in coming centuries. Although some species will be able to disperse to new protected areas, many will not. In the past, species readily moved across continents in response to changing climates. As glaciers advanced, many plants and animals retreated southward across the Northern Hemisphere, in some cases taking refuge in isolated strongholds with suitable climates. As the glaciers retreated again, some of those species moved northward greatly expanding their ranges (Davis and Shaw 2001). The climatic changes projected for the coming century will likely necessitate such dramatic range shifts for many species.

Today's landscape, however, differs markedly from past landscapes over which many species could move largely unhindered. Today, roads, agricultural fields, settlements, and other highly altered landscapes cover much of the Earth's surface. For many species, adding reserves to a network, even strategically as stepping-stones, may not be enough to ensure the long-distance movements required by climate change. It may be possible to enhance landscape connectivity for some species by building corridors through or bridges over barriers to dispersal (Chap. 16). However, such tactics will often be species-specific and likely only address a selected group of species. More general approaches that involve managing the matrix lands between reserves to increase the availability of habitat and facilitate movement have the potential to increase the connectivity of the landscape for a wider range of species (Franklin and Lindenmayer 2009; Noss 2001). Nonetheless, for some species, more drastic measures will likely be needed. For these species it may be necessary to pick them up and move them to new areas. Such assisted colonization will require considering many factors, not the least of which include feasibility and potential detrimental effects (Hoegh-Guldberg et al. 2008).

Even if species are able to move to newly suitable climates, the other conditions that are necessary for survival and reproduction may not follow. Our analyses only considered climate-driven shifts in species' potential ranges. As discussed above, species and ecosystems will experience many different changes in addition to shifts in their potential ranges. Phenological changes (Beebee 1995), changes in disturbance regimes (Westerling et al. 2006), altered disease dynamics (Pounds et al. 2006), and changes in the distributions of predators, prey, and competitors all have the potential to have complex cascading effects.

Despite these many challenges, protected areas are still likely to be one of the best tools we have for protecting biological diversity in a changing climate. Thus, one critical task for conservation biologists in the coming decade will be to test the basic strategies that have been proposed for increasing the resilience of the reserve system. Such tests will likely involve a combination of simulation models and well-designed experiments. If future tests indicate that these approaches have a high probability of protecting species in future climates and are economically and socially feasible, it may be prudent to augment our current protected area networks with additional lands. If, however, the results of these tests prove to be less positive, it may be necessary to reconsider our primarily place-based approach to conservation.

15.7 Lessons Learned

The analyses described here reveal a number of important lessons for how landscape-scale conservation planning needs to consider the biological consequences of future climate change. First, climate change will likely result in large changes to some existing protected area networks. The ability of a network to protect biological diversity in a changing climate will depend in part on the spatial distribution of the individual protected areas and the ability of species to move across the landscape.

Second, predictive models can be used to assess the potential impacts of climate change on reserve systems. In addition to the correlative bioclimatic models used here, mechanistic simulation models can also be used to investigate potential shifts in species distributions or changes in ecosystem functioning in response to climate change. Other models can be used to project potential changes in fire regimes, hydrology, or pests and pathogens. However, it is critical to understand the assumptions and the limitations of these models.

Third, because of the discontinuous boundary of the Northern Appalachian/Acadian ecoregion (Fig. 15.4) and the coarse resolution of the species predictions, some of our results may be non-intuitive and may, in fact, be artifacts of the scale mismatch between our datasets. Any modeling effort that attempts to downscale climate predictions to specific regions will encounter similar scaling issues.

Fourth, all models and predictions are approximations of what may happen in the future. The limitations of each data source need to be considered when inter-

15 Conservation Planning in a Changing Climate

preting any result. All of the datasets used in our analyses contain their own sources of error, which are then compounded when combined. Our results should be viewed primarily as a potentially useful rubric for examining the location and connectivity of existing and future conservation lands with respect to potential changes in habitat for tree, mammal, bird, and amphibian species.

Finally, however, we can still conclude that current protected area networks can be augmented to increase their ability to protect biological diversity in a changing climate. New reserves can be strategically placed to protect the abiotic template on which ecological systems evolve, environmental gradients, potential climate refugia, and areas that are likely to be critical connectivity points for species moving in response to climate change.

Acknowledgements We thank Dr. Evan Girvetz for assistance with the climate change analyses. We acknowledge the Program for Climate Model Diagnosis and Intercomparison (PCMDI) and the WCRP's Working Group on Coupled Modelling (WGCM) for their roles in making available the WCRP CMIP3 multi-model dataset. Support of this dataset is provided by the Office of Science, U.S. Department of Energy. We thank L. Iverson, S. Matthews, and M. Peters of the U.S. Forest Service for access to their tree and bird spatial data.

References

Allan, J. D., Palmer, M., & Poff, N. L. (2005). Climate change and freshwater ecosystems. In T. E. Lovejoy & L. Hannah (Eds.), *Climate change and biodiversity*. New Haven, CT: Yale University Press.

Araújo, M. B., Pearson, R. G., Thuiller, W., & Erhard, M. (2005). Validation of species–climate impact models under climate change. *Global Change Biology, 11*, 1504–1513.

Battin, J., Wiley, M. W., Ruckelshaus, M. H., Palmer, R. N., Korb, E., Bartz, K. K., et al. (2007). Projected impacts of climate change on salmon habitat restoration. *Proceedings of the National Academy of Sciences, 104*, 6720–6725.

Beebee, T. J. C. (1995). Amphibian breeding and climate. *Nature, 374*, 219–220.

Breiman, L. (2001). Random forests. *Machine Learning, 45*, 5–32.

Brubaker, L. (1988). Vegetation history and anticipating future vegetation change. In J. K. Agee & D. R. Johnson (Eds.), *Ecosystem management for parks and wilderness* (pp. 41–61). Seattle, WA: University of Washington Press.

Burkett, V., & Keusler, J. (2000). Climate change: Potential impacts and interactions in wetlands of the United States. *Journal of the American Water Resources Association, 36*, 313–320.

Carroll, C. (2007). Interacting effects of climate change, landscape conversion, and harvest on carnivore populations at the range margin: Marten and lynx in the Northern Appalachians. *Conservation Biology, 21*, 1092–1104.

Crick, H. Q. P., Dudley, C., Glue, D. E., & Thomson, D. L. (1997). UK birds are laying eggs earlier. *Nature, 388*, 526–526.

Cutler, D. R., Edwards, T. C., Jr., Beard, K. H., Cutler, A., Hess, K. T., Gibson, J., et al. (2007). Random forests for classification in ecology. *Ecology, 88*, 2783–2792.

Davis, M. B., & Shaw, R. G. (2001). Range shifts and adaptive responses to quaternary climate change. *Science, 292*, 673–679.

Dunn, P. O., & Winkler, D. W. (1999). Climate change has affected the breeding date of tree swallows throughout North America. *Proceedings of the Royal Society, London, 266*, 2487–2490.

Elith, J., Graham, C. H., Anderson, R. P., Dudík, M., Ferrier, S., Guisan, A., et al. (2006). Novel methods improve prediction of species' distributions from occurrence data. *Ecography, 29*, 129–151.

Franklin, J. F., & Lindenmayer, D. B. (2009). Importance of matrix habitats in maintaining biological diversity. *Proceedings of the National Academy of Sciences, 106*, 349–350.

Gibbs, J. P., & Breisch, A. R. (2001). Climate warming and calling phenology of frogs near Ithaca, New York, 1900–1999. *Conservation Biology, 15*, 1175–1178.

Halpin, P. N. (1997). Global climate change and natural-area protection: Management responses and research directions. *Ecological Applications, 7*, 828–843.

Hannah, L., Midgley, G. F., Andelman, S., Araújo, M. B., Hughes, G., Martinez-Meyer, E., et al. (2007). Protected area needs in a changing climate. *Frontiers in Ecology and the Environment, 5*, 131–138.

Hayhoe, K., Wake, C. P., Huntington, T. G., Luo, L., Schwartz, M. D., Scheffield, J., et al. (2006). Past and future changes in climate and hydrological indicators in the U.S northeast. *Climate Dynamics, 28*, 381–407.

Hoegh-Guldberg, O., Hughes, L., McIntyre, S., Lindenmayer, D. B., Parmesan, C., Possingham, H. P., et al. (2008). Assisted colonization and rapid climate change. *Science, 321*, 345–346.

Houghton, J. T., Ding, Y., Griggs, D. J., Noguer, M., van der Linden, P. J., Dai, X., et al. (Eds.). (2001). *Climate change 2001: The scientific basis*. Cambridge, UK: Cambridge University Press.

IPCC. (2007a). *Climate change 2007: The physical science basis. Contribution of Working Group I to the Fourth Assessment Report of the Intergovernmental Panel on Climate Change*. Cambridge, UK: Cambridge University Press.

IPCC. (2007b). *Climate change 2007: Impacts, adaptation and vulnerability. Contribution of Working Group II to the Fourth Assessment Report of the Intergovernmental Panel on Climate Change*. Cambridge, UK: Cambridge University Press.

Iverson, L. R., & Prasad, A. M. (1998). Predicting abundance of 80 tree species following climate change in the eastern United States. *Ecological Monographs, 68*, 465–485.

Iverson, L. R., & Prasad, A. M. (2001). Potential changes in tree species richness and forest community types following climate change. *Ecosystems, 4*, 186–199.

Iverson, L., Prasad, A., & Matthews, S. (2008a). Modeling potential climate change impacts on the trees of the northeastern United States. *Mitigation and Adaptation Strategies for Global Change, 13*, 487–516.

Iverson, L. R., Prasad, A. M., Matthews, S. N., & Peters, M. (2008b). Estimating potential habitat for 134 eastern US tree species under six climate scenarios. *Forest Ecology and Management, 254*, 390–406.

Lawler, J. J., White, D., Neilson, R. P., & Blaustein, A. R. (2006). Predicting climate-induced range shifts: Model differences and model reliability. *Global Change Biology, 12*, 1568–1584.

Lawler, J. J., Shafer, S. L., White, D., Kareiva, P., Maurer, E. P., Blaustein, A. R., et al. (2009). Projected climate-induced faunal change in the western hemisphere. *Ecology, 90*, 588–597.

Lawler, J. J., Tear, T. H., Pyke, C., Shaw, M. R., Gonzalez, P., Kareiva, P., et al. (2010). Resource management in a changing and uncertain climate. *Frontiers in Ecology and the Environment, 8*, 35–43.

Margules, C. R., & Pressey, R. L. (2000). Systematic conservation planning. *Nature, 405*, 243–253.

Matthews, S. N., O'Connor, R. J., Iverson, L. R., & Prasad, A. M. (2004). *Atlas of climate change effects in 150 bird species of the eastern United States*. Newtown Square, PA: USDA Forest Service, Northeastern Research Station.

McKenzie, D., Gedalof, Z., Peterson, D. L., & Mote, P. (2004). Climatic change, wildfire, and conservation. *Conservation Biology, 18*, 890–902.

Mote, P. W. (2003). Trends in snow and water equivalent in the Pacific Northwest and their climatic causes. *Geophysical Research Letters, 30*, 1601–1604.

Nakicenovic, N., Alcamo, J., Davis, G., de Vries, B., Fenhann, J., Gaffin, S., et al. (2000). *Special report on emissions scenarios. A Special Report of Working Group III of the Intergovernmental Panel on Climate Change*. Cambridge, UK: Cambridge University Press.

Noss, R. F. (2001). Beyond Kyoto: Forest management in a time of rapid climate change. *Conservation Biology, 15*, 578–590.

15 Conservation Planning in a Changing Climate

Oppenheimer, M., O'Neill, B. C., Webster, M., & Agrawala, S. (2007). The limits of consensus. *Science, 317*, 1505–1506.

Parmesan, C. (1996). Climate and species' range. *Nature, 382*, 765–766.

Parmesan, C., & Yohe, G. (2003). A globally coherent fingerprint of climate change impacts across natural systems. *Nature, 421*, 37–42.

Parmesan, C., Ryrholm, N., Stefanescu, C., Hill, J. K., Thomas, C. D., Descimon, H., et al. (1999). Poleward shifts in geographical ranges of butterfly species associated with regional warming. *Nature, 399*, 579–583.

Pearsall, S. H., III. (2005). Managing for future change on the Albemarle Sound. In T. E. Lovejoy & L. Hannah (Eds.), *Climate change and biodiversity* (pp. 359–362). New Haven, CT: Yale University Press.

Pearson, R. G., & Dawson, T. P. (2003). Predicting the impacts of climate change on the distribution of species: Are bioclimate envelope models useful? *Global Ecology and Biogeography, 12*, 361–371.

Pearson, R. G., & Dawson, T. P. (2004). Bioclimate envelope models: What they detect and what they hide; response to Hampe (2004). *Global Ecology and Biogeography, 13*, 471–473.

Pearson, R. G., & Dawson, T. P. (2005). Long-distance plant dispersal and habitat fragmentation: Identifying conservation targets for spatial landscape planning under climate change. *Biological Conservation, 123*, 389–401.

Pearson, R. G., Thuiller, W., Araújo, M. B., Martinez-Meyer, E., Brotons, L., McClean, C., et al. (2006). Model-based uncertainty in species range prediction. *Journal of Biogeography, 33*, 1704–1711.

Poff, N. L., Brinson, M. M., & Day, J. W. J. (2002). *Aquatic ecosystems and global climate change: Potential impacts on inland freshwater and coastal wetland ecosystems in the United States*. Arlington, VA: Pew Center on Global Climate Change.

Pounds, J. A., Bustamante, M. R., Coloma, L. A., Consuegra, J. A., Fogden, M. P. L., Foster, P. N., et al. (2006). Widespread amphibian extinctions from epidemic disease driven by global warming. *Nature, 439*, 161–167.

Prasad, A. M., Iverson, L. R., & Liaw, A. (2006). Newer classification and regression tree techniques: Bagging and random forests for ecological prediction. *Ecosystems, 9*, 181–199.

Pyke, C. R., & Fischer, D. T. (2005). Selection of bioclimatically representative biological reserve systems under climate change. *Biological Conservation, 121*, 429–441.

Rodenhouse, N., Matthews, S., McFarland, K., Lambert, J., Iverson, L., Prasad, A., et al. (2008). Potential effects of climate change on birds of the Northeast. *Mitigation and Adaptation Strategies for Global Change, 13*, 517–540.

Root, T. L. (1992). Temperature mediated range changes in wintering passerine birds. *Bulletin of the Ecological Society of America, 73*, 327.

Root, T. L. (1993). Effects of global climate change on North American birds and their communities. In P. M. Kareiva, J. G. Kingsolver, & R. B. Huey (Eds.), *Biotic interactions and global change* (pp. 280–292). Sunderland, MA: Sinauer Associates.

Roy, D. B., & Sparks, T. H. (2000). Phenology of British butterflies and climate change. *Global Change Biology, 6*, 407–416.

Saether, B. E., Engen, S., Lande, R., Arcese, P., & Smith, J. N. M. (2000). Estimating the time to extinction in an island population of song sparrows. *Proceedings of the Royal Society of London – Series B: Biological Sciences, 267*, 621–626.

Saxon, E., Baker, B., Hargrove, W., Hoffman, F., & Zganjar, C. (2005). Mapping environments at risk under different global climate change scenarios. *Ecology Letters, 8*, 53–60.

Scott, J. M., Griffith, B., Adamcik, R. S., Ashe, D. M., Czech, B., Fischman, R. L., et al. (2008). National Wildlife Refuges. In S. H. Julius & J. M. West (Eds.), *Preliminary review of adaptation options for climate-sensitive ecosystems and resources. A Report by the U.S. Climate Change Science Program and the Subcommittee on Global Change Research* (pp. 8-1–8-95). Washington, DC: U.S. Environmental Protection Agency.

Shafer, C. L. (1999). National park and reserve planning to protect biological diversity: Some basic elements. *Landscape and Urban Planning, 44*, 123–153.

Stefanescu, C., Peñuelas, J., & Filella, I. (2003). Effects of climatic change on the phenology of butterflies in the northwest Mediterranean Basin. *Global Change Biology, 9,* 1494–1506.

Thomas, C. D., & Lennon, J. J. (1999). Birds extend their ranges northwards. *Nature, 399,* 213.

Thuiller, W., Lavorel, S., Araújo, M. B., Sykes, M. T., & Prentice, I. C. (2005). Climate change threats to plant diversity in Europe. *Proceedings of the National Academy of Sciences of the United States of America, 102,* 8245–8250.

Trombulak, S. C., & Wolfson, R. (2004). Twentieth-century climate change in New England and New York, USA. *Geophysical Research Letters, 31,* L19202. doi:10.1029/2004GL020574.

Westerling, A. L., Hidalgo, H. G., Cayan, D. R., & Swetnam, T. W. (2006). Warming and earlier spring increase western U.S. forest wildfire activity. *Science, 313,* 940–943.

Williams, P. H., Hannah, L., Andelman, S. J., Midgley, G. F., Araújo, M. B., Hughes, G., et al. (2005). Planning for climate change: Identifying minimum-dispersal corridors for the Cape Proteaceae. *Conservation Biology, 19,* 1063–1074.

Winter, T. C. (2000). The vulnerability of wetlands to climate change: A hydrologic landscape perspective. *Journal of the American Water Resources Association, 36,* 305–311.

Woolmer, G., Trombulak, S. C., Ray, J. C., Doran, P. J., Anderson, M. G., Baldwin, R. F., et al. (2008). Rescaling the human footprint: A tool for conservation planning at an ecoregional scale. *Landscape and Urban Planning, 87,* 42–53.

Chapter 16
Modeling Ecoregional Connectivity

Robert F. Baldwin, Ryan M. Perkl, Stephen C. Trombulak, and Walter B. Burwell III

Abstract Nature reserves increasingly function as islands in a human-dominated matrix. Habitat conservation initiatives that seek to reconnect patches using functional corridors are increasingly part and parcel of conservation planning projects. Numerous methods have evolved all based on similar of least cost paths, functional rather than structural connectivity, and landscape resistance. However significant differences in approach exist including whether a network of patches is considered simultaneously or as patch–patch pairs, and whether the goal is to model spatially explicit corridors or movement bottlenecks. We review these approaches and then describe ecological connectivity modeling for an ecoregion using the graph-theoretic approach considering two different patch-node scenarios, at the ecoregion scale and apply a more localized connectivity modeling exercise for a subregion and a single focal species the American Black bear (*Ursus americanus*). We discuss the difficulties of attempting to model functional corridors for focal species over heterogenous landscapes, and the potential benefits of using 'naturalness' or Human Footprint surrogates for connectivity.

R.F. Baldwin(✉)
Department of Forestry and Natural Resources, Clemson University, 261 Lehotsky Hall, Clemson, SC 29634-0317
e-mail: baldwi6@clemson.edu

R.M. Perkl
Department of Planning and Landscape Architecture
Clemson University, 121 Lee Hall, Clemson, SC 29634-0511
e-mail: rperkl@clemson.edu

S.C. Trombulak
Department of Biology, Middlebury College, Middlebury, VT 05753
e-mail: trombulak@middlebury.edu

W.B. Burwell
2115 Reaves Drive, Raleigh, NC 27608
e-mail: Tripp.burwell@gmail.com

S.C. Trombulak and R.F. Baldwin (eds.), *Landscape-scale Conservation Planning*, DOI 10.1007/978-90-481-9575-6_16, © Springer Science+Business Media B.V. 2010

Keywords Connectivity models • Focal species • Graph theory • Habitat connectivity • Human Footprint

16.1 Introduction

It became obvious to conservation biologists early in the discipline's history that even in large, well-protected natural areas such as U.S. National Parks species would disappear if connectivity was lost with other habitats (Newmark 1987, 1995; Wilson and Peter 1988). Protected areas alone are insufficient to achieve conservation goals across broad landscapes; they need to be ecologically connected to one another in order to allow gene flow, larger effective population sizes, and recolonization following local population extirpation. Aided by computer mapping technologies, conservation biologists have subsequently modeled population processes at scales greater than parks and reserves themselves (Noss 1983). Conservation planners increasingly recognize the importance of managing the matrix in which 'island' reserves exist. Consequently, focus has shifted to identifying and protecting critical habitat linkages, building reserve networks as opposed to independent sets of reserves, and viewing landscapes as shifting mosaics of disturbances (Clark 1991; Soulé and Terborgh 1999).

As with most topics in conservation biology, habitat connectivity is complex due to the inherent variability of natural systems. Divergent life histories of focal species, unpredictable behavioral decisions about movement pathways, temporal and spatial variation in habitat composition and structure, environmental change, and dynamic human social factors make the challenge of modeling the optimal placement of corridors as complex as any aspect of conservation planning.

Yet much progress has been made recently in meeting that challenge. The technical and scientific approaches to these problems have coalesced over the last 10 years in the emerging field of 'connectivity conservation' – a subfield of conservation biology with far-reaching implications for understanding basic ecological and evolutionary processes, as well as long-term maintenance of species and their habitats in the face of land-use and climate change (Beier et al. 2008; Crooks and Sanjayan 2006; McRae et al. 2008).

Before the advent of geographical information systems (GIS), the implementation of most aspects of conservation planning were predominantly intuitive, opportunistic, and biased. Connectivity conservation is no exception. The purpose of this chapter is to explore connectivity conservation in the context of ecoregional planning, using the Northern Appalachian/Acadian ecoregion as a case study. Ultimately, we hope that conservation planners will be better able to choose among approaches to connectivity modeling – selecting those that make the most sense for their particular applications and understanding well the strengths and limitations of the results their models provide.

16.2 What Are We Connecting and Why?

Population processes operate at all spatial and temporal scales: connectivity can be studied from local to global scales, for any time frame, and for any group of focal species of conservation concern. Effort can be made to maintain connectivity in widely divergent settings, such as for gene flow among populations of frogs that use isolated ponds for breeding (Berven and Grudzien 1990; Hitchings and Beebee 1998; Reh and Seitz 1990), for viability of wide-ranging predators facing interacting effects of land-use and climate change (Carroll 2007), and for responsiveness to climate change among all species present in shifting continental biomes (Williams et al. 2005).

Traditional conservation planning has emphasized a coarse-scale approach to connectivity. The foundation for this coarse-scale thinking is the regional reserve network (e.g., Yellowstone to Yukon Conservation Initiative 2010), conceptually justified in large part on the spatial requirements of large, wide-ranging umbrella species (Soulé et al. 2005) or coarse-filter biotic representation (Groves et al. 2002; Hunter 1991).

While a focus on large landscapes dominates landscape-scale planning, it is also important to understand connectivity at multiple scales. As the cumulative positive impacts of grassroots conservation efforts, such as land trusts, come to be better appreciated (Merenlender et al. 2004; Theobald et al. 2000), it is important to understand how local conservation efforts do and do not meet landscape-scale conservation goals. For example, if an effort to connect frog breeding ponds is made at the local scale, then the cumulative effect of such local connectivity projects could result in a positive gain for connectivity at a larger scale, depending on the degree of overlap of breeding ponds with the habitats needed by wide-ranging species. Conversely, coarse-scale connectivity and reserve selection may have an umbrella effect on local ecosystem processes. The large landscapes needed to sustain populations of large mammals will most likely include wetlands and neighboring uplands, habitats that are critical for supporting frogs. However, these relationships may break down on a case-by-case basis. For example, a linkage that is appropriate for wolf passage may include large clear-cuts that are less permeable for frogs (deMaynadier and Hunter 1998; Mech et al. 1995).

Thus, the first step in any connectivity modeling process is to ask and answer the questions, which habitats are being connected, and why? Is the focus on a single species of conservation concern or several? At which spatial and temporal scales is connectivity needed? What is needed for functional connectivity: gene flow over generations or the opportunity for seasonal migrations? What threats need to be mitigated: road networks, land-use change, plant community shifts in response to climate change? Model inputs, parameters, assumptions, and outputs will all vary depending on the answers to these questions.

A valid conservation goal is to connect habitat patches for focal species. Focal species are usually selected because they (1) represent specific ecological features or processes (e.g., riparian areas), (2) are sensitivity to landscape barriers (e.g., roads), and/or (3) serve as an umbrella for other species (e.g., planning for connectivity for one species results in meeting the needs of other species as well) (Beier et al. 2008). In this context, a 'corridor' is an area of the landscape (including aquatic systems) selected for its capacity to promote gene flow among neighboring populations or seasonal migratory movements. In order to define habitat patches and corridors for a given focal species, its habitat needs – including variability in these requirements over time and space – must be understood as much as possible. In many cases, habitat use by a species can differ from one location to another. For example, wood frog (*Rana sylvatica*) habitats in Maine differ markedly from those in Missouri (Baldwin et al. 2006; Rittenhouse and Semlitsch 2007), cougar (*Puma concolor*) diets and habitats vary substantially throughout their geographic range (Iriarte et al. 1990), and black bear (*Ursus americanus*) home ranges vary markedly even among forest types (Powell et al. 1997). Finally, field studies themselves are often biased; abiotic conditions and populations vary within and among years to such an extent that conclusions derived from studies of habitat selection occurring over only a short duration may not reveal a complete understanding of habitat use for any species (Morrison et al. 2006).

The availability of data to make these determinations varies widely among habitats within species, and among taxa. Even widespread species are much better studied in some parts of their range than others. For example, the wolf (*Canis lupus*) has been better studied in the field in the western than eastern parts of its range primarily because it is more abundant in the West; yet habitats and prey are different in Eastern North America, and if wolves expand eastward it should not be assumed that they will exhibit the same habitat relationships (Mladenoff and Sickley 1998). Thus, before beginning the process of modeling connectivity, an answer is needed to the question, how strong are the data and assumptions used as inputs into the model? As has been noted throughout the history of habitat modeling, even though a GIS program identifies a habitat patch on the computer screen, a high level of uncertainty remains as to the patch's real characteristics.

16.3 Ecoregional Functional Connectivity

Structural connectivity refers to an area that physically connects two habitats of interest (e.g., two patches, two reserves). Planning for structural connectivity typically involves models using a habitat surrogate (e.g., intact forest cover, riparian zone, or 'greenway'), often in the absence of any real information on how the movement behaviors of focal species respond to that surrogate (Kindlmann and Burel 2008). In contrast, functional connectivity implies that the area actually has been or is likely to be used by focal species in the future (Theobald 2006). In a sense, functional connectivity is the ultimate goal for connectivity modeling because it directly contributes in a measureable way to a primary conservation goal, the viability of the population (Carroll 2006); it is hoped that connectivity will mitigate habitat fragmentation by connecting smaller patches or reserves into a landscape network that functions more like a larger reserve.

However, despite strong theoretical support for its importance in population biology, few empirical studies demonstrate that connectivity provided by habitat conservation programs actually helps populations (Beier and Noss 1998; Haddad and Tewksbury 2006). Nevertheless, the performance of a landscape with respect to connectivity is often assumed based on the study of habitat selection and movements of focal species (Beier and Noss 1998). Assessment of functional connectivity requires knowledge of the habitat preferences and movements of the focal species in the study area, preferably obtained over multiple seasons. Increasingly, telemetry data are being used to construct and validate site-specific connectivity models, but much more often the model parameters are derived by review and/or meta-analysis of the scientific literature and by soliciting 'expert opinion' from field biologists (Beier et al. 2008; Chap. 11).

The connectivity modeling programs discussed below use geographic proxies for habitat quality (e.g., topography, elevation, land use/land cover, wetlands), landscape permeability, and spatial configuration, interpreted through statistical models and/or network analyses that combine proxies in ways that mimic how an organism evaluates a landscape with respect to movement. The process typically incorporates one or more of the following concepts: (1) habitat suitability, (2) least-cost paths, (3) hotspots of resistance, and (4) proximity and arrangement of habitat patches in landscape networks.

16.3.1 General Concepts

The likelihood of an organism moving from one place to another across the landscape is a function of three interrelated factors. First is the basic life history of the animal, especially its patterns of dispersal and migration. Dispersal is the movement of genetic material from one breeding population to another. Migration generally refers to seasonal movements of individuals from one habitat to another (for example, annual migrations from breeding to overwintering habitat). Dispersal and migration can occur over different spatial scales, which can greatly affect connectivity modeling. For example, dispersal in a migratory passerine bird can involve moving from one breeding population to another over several hundred meters. Conversely, migration in the same animals can occur over thousands of kilometers. On the other hand, the scale across which migration occurs is not necessarily always greater than dispersal. Juvenile amphibians, for example, generally move much further from their natal sites to adult home ranges or territories than do adults during their seasonal migration from breeding to overwintering sites (Semlitsch 2008).

Second, all movements are influenced in part by the quality of the habitat that lies between habitat patches, called the 'matrix' (Ricketts 2001). Landscape ecologists use two related terms to describe the degree to which the matrix allows passage of an organism: 'permeability' to refer to how easily an animal can pass through the matrix, and 'resistance' to measure how difficult such passage may be.

Third is the inherent variability in biological systems and how it relates to modeling connectivity. Animals within a population may vary individually in terms of behavioral responses to habitat characteristics, based on their genetic makeup and learning. Individuals may vary temporally (e.g., year to year, season to season) in how they respond to environmental cues (Gaines and McClenaghan 1980). Furthermore, compounding the effects of individual behavioral plasticity is the fact that the environment itself changes over time (Walther et al. 2002). The distributions of suitable habitats will change, and consequently the optimal locations for corridors will also change.

16.3.2 Specific Concepts of Connectivity Modeling

Selecting Focal Species Generally speaking, the goal of connectivity modeling is to increase the effective size of areas conserved for their conservation values to allow for individual movement among patches. As such, conservation planners often select focal species to serve as umbrellas such that meeting their needs for connectivity should provide connectivity for other species and ecological processes (Beier et al. 2008). Important processes in this regard include hydrology, nutrient flow, carbon source-sink dynamics, and natural disturbance (e.g., wind, fire, insects), among others.

Much care needs to be taken in making assumptions about how well a focal species serves as an umbrella (Chap. 17). Focal species often have widely divergent habitat needs and will not serve as umbrellas for each other. For example, based on the tutorial for the software program CorridorDesigner (CorridorDesign 2009), modeled corridors for the black bear and javelina (*Tayassu tajacu*) in Arizona follow different pathways because the javelina is a low-elevation, desert-dwelling species, and the black bear is a high-elevation, forest-dwelling species. Consequently, even between these two larger mammals from that region, one will not typically serve as a corridor umbrella for the other.

Large carnivores are often chosen as umbrella species for connectivity modeling (Seidensticker et al. 1999). They can be a good place to start the modeling process because of their extensive spatial needs, specific habitat needs for at least some aspects of their life history, and, generally speaking, rarity; they are often given a high conservation priority, and their needs for connectivity merit attention independent of the extent of their appropriateness as focal species. Care needs to be taken, however, in making assumptions about carnivores as umbrella species because many move through marginal and degraded habitats, especially as juveniles (Harrison 1992).

16 Modeling Ecoregional Connectivity

Such habitats – e.g., early successional forest in managed landscapes – may serve for carnivore dispersal but will not support populations of plant and animal species that depend on mature forests.

Ideally, for any given project a set of focal species would be strategically selected to model connectivity. When considering a particular species to be focal species for connectivity modeling, the following questions should be addressed:

1. Do its movements occur at the scale appropriate for the planning initiative?
2. How narrow or specific are its habitat requirements during its movements?
3. Is there enough known about its habitat requirements and preferences to answer this with confidence? For example, one of us (WBB) conducted a separate connectivity modeling study on black bear in Eastern New York to Western Vermont using the software program FunConn (Theobald et al. 2006). A sensitivity analysis on the overlap in modeled corridors varied dramatically in response to inputs describing a bear's perception of and relationship to its environment. Not surprisingly, modeled corridors show virtually no overlap when a bear's preferences are alternatively described as being for forests or open habitat. More interestingly, however, other aspects of the movement ecology of bears also lead to divergent results. Different grains of discrimination of habitat types by a bear (8 vs. 16 categories) results in less than 65% overlap between modeled corridors, and whether a large lake acts as a barrier to movement (yes vs. no) results in less than 80% overlap (Burwell 2009). All of these results indicate the critical importance of accurate information on habitat ecology in order for a focal species approach to be meaningful.
4. To what degree would it serve as an umbrella for other species and/or ecosystem processes?
5. Have permeability studies been conducted on this species, especially related to the habitat types found in the study area? For example, several studies have used taxa-specific land-use indices to estimate landscape resistance (Baldwin and deMaynadier 2009; Compton et al. 2007).
6. What degree does the selection improve or diminish the opportunities to engage the public in a connectivity planning effort? Stakeholders respond differently to conservation efforts for 'flagship species,' and responses to large carnivores are not universally favorable. For example, in some parts of North America talking about wolf conservation will elicit only blank stares or worse from landowners in the corridor matrix, while different species (e.g., mesocarnivores, such as marten [*Martes americana*], bobcats [*Lynx rufus*], or lynx [*Lynx canadensis*]) may be perceived as less threatening to their livelihoods while accomplishing umbrella goals for connectivity modeling.

Habitat Suitability Regardless of specific approach, all connectivity modeling involves assessments of habitat suitability. Habitat suitability is assumed to be indicated by the degree to which a location is used by a particular species. Generally, habitat suitability maps are modeled 'surfaces' in which each cell of a grid contains a value denoting the quality of the habitat within that cell for that species. In habitat occupancy models, which are one class of suitability models, values range between

0 and 1 because they are outputs of logistic regressions, while in more general suitability approaches an ordinal habitat quality ranking system may be used.

It is important to note that for many species – particularly habitat generalists – habitat suitability can be a poor predictor for specific habitat corridors because they are likely to be less selective during migration and dispersal than during other phases of their life history (Baldwin et al. 2006; Haddad and Tewksbury 2006). Regardless of whether the output is occupancy based on detection probabilities or a more generalized index of suitability, habitat models conducted in a GIS uses geographical proxies for actual habitat. These typically include topography, elevation, land use/land cover, human infrastructure (e.g., roads), and other relevant landscape features, e.g., rivers and wetlands. Input data for habitat models can vary widely in accuracy. For example, land use/land cover data that typically form the basis for habitat models are relatively coarse (e.g., 30-m resolution) relative to the spatial scale at which many organisms make choices about movement or occupancy. Such data are useful for modeling habitats for many wide-ranging species but are too coarse for species dependent upon localized habitats (e.g., small meadows, wetlands). Many types of errors can compromise connectivity analyses, including errors of classification (e.g., evergreen broadleafed or evergreen coniferous), position (bad editing, multiple joins and unions, or incorrect coordinate systems), and modeling (Scott et al. 2002). Modeling errors include errors of model specification (what parameters are included and in what combinations), model selection (interpretation of p-values, AIC values, or other model selection devices), and statistics.

Least-Cost Path Landscape permeability is a key component of every approach to connectivity (Compton et al. 2007; McRae and Beier 2007; Theobald 2006). Permeability and its inverse, resistance, are intuitive concepts: how difficult is it for an organism to move through the landscape?

Least-cost path is a common modeling application that calculates effective distances and optimal pathways based on resistance values that are assigned to cells in a grid. Resistance can be high and be based on obvious barriers (e.g., a busy 4-lane highway) or resistance can take the form of more subtle gradients. In a least-cost path analysis, it is assumed that an animal seeks paths with the least cumulative cost for traveling from one patch to another.

When modeling connectivity, roads need to be considered in greater depth than other landscape features because they provide different kinds of barriers for different species. Specifically, they vary widely in size, transportation function, and traffic volume, influencing their relative degree of landscape permeability (Forman et al. 2003). For some animals, a road functions as a behavioral barrier because visual or auditory stimuli cause road avoidance. For such species, a road increases resistance but doesn't cause direct mortality. Other organisms may not avoid roads but instead experience increased mortality when they try to cross. However, while roads are generally barriers for most wildlife species, they can also sometimes be an asset for others. For example, wolves and grizzly bears (*Ursus arctos*) use logging and other low-volume roads to facilitate their own movements through a landscape (Mace et al. 1996; Mladenoff et al. 1999).

Problems in predicting an animal's response to landscape resistance (e.g., roads and forest management activities) arise when their tolerance of habitat types differs depending on life-history stage. For many vertebrates, juvenile dispersal is a life-history phase that has very broad habitat parameters (Eisenberg 1981). For example, juvenile coyotes (*Canis latrans*) show little directionality to dispersal movements (Harrison 1992). Furthermore, migrating adults from various taxa will freely travel through habitats that are unsuitable for reproduction, foraging, or hibernation in order to access seasonally available resources (Berger 2004).

Variability in all relevant parameters is likely to influence resistance and should be considered in least-cost path modeling. Populations contain variability in how individual organisms respond to the landscape during movement. Also, seasonal changes will influence least-cost paths. A ridge may be passable in the summer but not the winter. Seasonal flooding of landscapes lowers resistance for aquatic and semiaquatic species, but increases it for those that are more fully terrestrial. Storms, rock slides, treefalls, and other stochastic, catastrophic events can change resistance. Management actions by people create disturbance in landscapes at various temporal and spatial scales, and patterns of landscape resistance will vary over time. For example, managed forest landscapes are often described as a 'shifting mosaic' of habitat types. Frequency, extent, and duration of timber harvests combined with pre- and post-harvest management techniques, including herbicide treatments, thinning, and cultivation, can vastly influence the paths of organisms (Chapin et al. 1998; deMaynadier and Houlahan 2008). More field study on migration and dispersal in relation to resistance (Gobeil and Villard 2002; Rothermel and Semlitsch 2002) will improve the accuracy of connectivity models.

Landscape Networks Modeling the landscape as a network of habitat nodes permits identification of essential nodes and alternative least-cost paths among them (Urban and Keitt 2001). A landscape network is a topologically related graph (Theobald 2006). Fundamental metrics are (1) patch size and configuration (nodes), and (2) corridor orientation, length, and width (edge angles and effective distance) (Table 16.1; Fig. 16.1). Nodes represent patches of habitat or ecological reserves, and attributes are assigned to nodes that reflect various metrics of the patch or reserve that they represent. Euclidean distances among patches are modified based on the modeled least-cost paths among them. This modification produces the effective distances, which represent more realistic distances along optimal pathways. Additional edges can be generated that represent multiple pathways or the next optimal connection among multiple nodes. Adding or deleting potential nodes and edges can allow for landscape-level assessments to prioritize nodes and/or edges based on their contributions to overall landscape connectivity.

Identifying multiple potential corridors allows for the retention of functional redundancy. This is an important aspect of landscape networks because it aids in addressing the often unrealistic assumption that a single least-cost path will be used by a species. A modeling run will identify an optimal least-cost path for each pair of patches and additional corridors corresponding to a specified threshold for the next most optimal routes (Theobald 2006).

Landscape networks may be used for the design of reserve networks at multiple scales. Below (Section 16.4), we model connectivity among multiple patches at the ecoregion scale. However, networks may also be effective at a very local scale

Table 16.1 Terms used in ecoregional connectivity modeling as adapted from the connectivity literature

Structural connectivity	Physically connected habitat for which its value for connecting ecological processes is either unknown or doesn't exist.
Functional connectivity	Connected habitat that provides for movement and gene flow.
Corridor	A design feature of a reserve network that allows movement and gene flow.
Permeability	Degree to which a landscape feature is a barrier to movement and gene flow.
Linkage	(1) Loosely defined area having corridor values at the ecoregion scale; or (2) polyline connecting nearest patch boundaries within the graph when modeling landscape networks.
Resistance	Inverse of permeability.
Least-cost path	Potential route offering the least landscape resistance for movement and gene flow.
Graph	Collection of habitat patches, blocks, or reserves found within the landscape being evaluated.
Node	Vector data point that represents the location of the habitat patch, block, or reserve.
Edge	Polyline that connects each connected node within the graph (not to be confused with other usages of the term edge as in 'edge habitat') or with 'linkage' above.
Path	Sequence of connected nodes that when taken together form a 'walk.'
Walk	Unique combination of nodes connected by their respective edges. The length of a walk is represented by the sum of each edge. Walks represents unique paths, and/or alternative, potential corridors.
Cycle	Walk that represents a closed loop of nodes, considered closed if the starting node and the ending node are the same.
Tree	Walk that does not represent a closed loop of nodes or cycles. This represents a path that is linear in nature similar to the configuration of a stream network.
Spanning tree	Tree that includes every node, patch, block, or reserve found within the graph or landscape. Each graph may have several spanning trees, each comprised of a unique combination of paths.
Minimum spanning tree	Spanning tree that includes the shortest total distance of edge length.
Connected graph	Graph where a path exists between every two nodes, thus allowing every node within the landscape to be reachable or connected in some sequence.
Subgraph	Subset of the total graph that is isolated from the rest. This would be likely in highly fragmented landscapes where edge distances are great or effective distances are too high.
Graph component	Connected subgraph where a path exists between any two nodes, thus allowing every node within the component to be reachable. A subgraph can be comprised of multiple isolated components.
Circuit	Network of nodes, connected by resistors, as in an electrical circuit and with electron 'resistance' analogous to landscape 'resistance.'

16 Modeling Ecoregional Connectivity

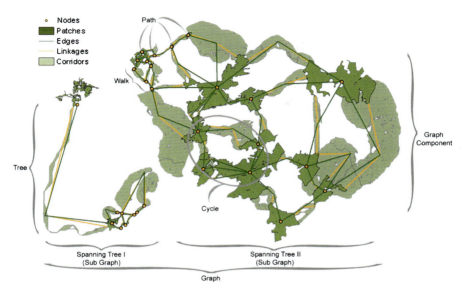

Fig. 16.1 A hypothetical landscape network, exemplifying the terms in Table 16.1

where they can establish functional habitat networks for pool-breeding amphibians (Pyke 2005).

Circuits A more recent evolution in connectivity research has involved electrical circuit theory. Application of electrical circuit theory is particularly valuable for identifying 'pinch points' in habitat networks – places where both current flow and resistance are very high (McRae and Beier 2007; McRae et al. 2008). Circuits are networks of habitat nodes connected by resistors, as in an electrical circuit connected by wires. McRae and Beier (2007) describe it as follows: 'as multiple or wider conductors connecting two electrical nodes allow greater current flow than would a single, narrow conductor, multiple or wider habitat swaths connecting populations allow greater gene flow.' More pathways, wider pathways, and greater density of patches provide greater opportunities for animal movement. Very few connectivity models of any kind have actually been tested against real-world animal movements. Therefore, the fact that McRae and Beier (2007) validated their connectivity modeling software Circuitscape (Circuitscape.org n.d.), using actual gene flow data from both plant and animal populations is a significant advance for the science of corridor ecology.

16.4 A Landscape Network for the Northern Appalachian/Acadian Ecoregion

Historically, ecoregional connectivity planning has been hindered by lack of suitable datasets, lack of software capable of handling multiple habitat patches, and inexact ecological information on focal species. Here we demonstrate an application of landscape network approach to modeling connectivity in the Northern Appalachian/Acadian ecoregion. We used FunConn (v1) (Theobald et al. 2006), a connectivity analyses program designed to operate in the ArcGIS 9.x environment (Theobald 2006). The sequence of steps required for the use of FunConn is representative of most approaches to connectivity modeling in that it involves (1) creating a habitat-quality surface, (2) using landscape resistance, and (3) defining and connecting functional patches; however, it diverges from other approaches by building and evaluating landscape networks.

Ecoregional planning can be viewed as being, in part, a coarse-filter process of identifying ecological reserves and linkages between them based on representation of the region's natural features (Chap. 14). It generally does not focus on a single focal species, or even a suite of focal species, unless those species are endangered or otherwise critically important. In a similar manner for ecoregional connectivity, we chose not to model habitat suitability for individual focal species, but rather adopted a coarse-scale methodology that adopts naturalness or the inverse, human impact, as a surrogate for suitability for any species that is sensitive to human-induced land transformation. While focal species connectivity planning is a good idea in many circumstances, we recommend our approach when (1) an effective focal species set is not well defined, (2) habitat parameters for focal species are not well understood, (3) the study area is broad and heterogenous, within which habitat patterns and behavior of focal species may vary substantially, (4) a fine-scale, generalized resistance surface is available, and (5) when the concern for connectivity primarily focuses on species sensitive to transformation of land cover (e.g., species with limited dispersal capabilities, a high degree of sensitivity to edge habitat and/or affinity for core habitat, vulnerability to land use/land cover change, or a known affinity for 'naturalness').

Methods We built landscape networks under two connectivity scenarios using the Human Footprint developed at a 90-m resolution for this ecoregion (Woolmer et al. 2008; Chap. 13) as the resistance layer. Under Scenario A, we used existing GAP Status 1 protected areas (highest protection levels) as habitat nodes. For this region, these include many areas of the Adirondack State Park, Baxter State Park, and national parks in both Canada and the U.S. (N = 95). Under this scenario, we assume that we are evaluating connectivity for those species having their most significant population sources in the areas most protected. Under Scenario B, we used the regional Last of the Wild (Sanderson et al. 2002) representing areas of lowest human-caused landscape transformation (i.e., Human Footprint scores less than 10) as habitat nodes (N = 120). Under this scenario, we assume that we are evaluating connectivity for those species most sensitive to human-caused landscape transformation. We used the Human Footprint values as a cost surface for those species most sensitive to human-caused landscape transformation (Woolmer et al. 2008). Because the Human Footprint dataset is scaled from 0 (most wild, least influenced by human activity) to 100 (least wild, most impacted by human activity), we reclassified these values into the scale range accepted by the tool (0–1). Human Footprint values of 100 became 0 for the permeability value associated with the cost surface, and Human Footprint values of 0 became 1 (most permeable). We assigned no aggregation factor (i.e., value of '1') so that the original resolution (90 m) would be maintained. We accepted the default links Qn value (10) because we did not want to be either overly restrictive or liberal in our selection of potential linkages. The links Qn value represents the lowest percentile allocation values from which linkages are derived. Allocation zones were then grown from the source patches across the cost surface until they met, forming allocation boundaries. Based on patch-to-patch distance, cost-distance values for selected cells served as the midpoints for the first set of initial linkages. Within each allocation zone, cells exhibiting values less than the user-defined links Qn value were removed, which results in the final, proposed corridor.

Results The modeled Scenarios A and B include potential linkages and modeled corridors that, taken together, represent alternative plans for connectivity at the ecoregion scale. Of course, other plausible scenarios could arise from (a) inclusion of other patches (e.g., functional patches for specified focal species), (b) changes to model parameters, and (c) use of other software for modeling connectivity. Thus, the scenarios shown here provide a starting point for discussion rather than a finished plan for implementing connectivity. Each resulting landscape network is comprised of a number of subgraphs, which indicates that connectivity within the subregions is currently greater than among them or, by extension, throughout the entire ecoregion (Fig. 16.1). The landscape network resulting from connecting Gap status 1 protected lands (Scenario A) is comprised of four subgraphs – Adirondack Mountains, Green-White Mountains, Gaspé-New Brunswick, and Nova Scotia – indicating that within each of those four portions of the ecoregion, protected areas are more connected to one another than they are to protected areas elsewhere within the ecoregion. Not surprisingly, given the increased number of nodes, connecting the Last of the Wild (Scenario B) resulted in a more connected landscape network, with only three subgraphs; the Green-White Mountains and Gaspé-New Brunswick subregions from the results of Scenario A are combined (Fig. 16.2).

Discussion Two important conclusions about effective conservation strategies arise from these scenarios. First, it is clear that converting more of the Last of the Wild patches to GAP status 1 lands (e.g., through conservation fee purchases and/or easements) would increase structural connectivity at the ecoregional scale. A second strategy would be to improve functional connectivity among these patches using the linkages and potential corridors identified in this exercise (Fig. 16.2a) as a starting point for on-the-ground assessment of animal movements and for conservation action to protect corridors.

The amount of land required to achieve these connectivity goals may surprise conservation groups that work to acquire conservation rights or establish local conservation programs (e.g., The Nature Conservancy, the Wildlife Conservation Society) in linkage lands. Our results indicate that the most extensive corridors themselves could be as large as some of the largest existing protected areas (e.g., the Adirondack State Park). Allocation zones produced corridors ranging in area from 16 to nearly 1,000,000 ha, depending on the scenario on which they were based (Table 16.2). Similarly, some of the linkages are quite long (e.g., 300 km) and cross many jurisdictions (i.e., the Adirondack-Green Mountains linkage, Fig. 16.2b). Clearly, if connectivity is to be a priority for reserve network design, conservation organizations and governments will need to grasp the full extent of the land area

Fig. 16.2 Results of a connectivity modeling exercise in the Northern Appalachian/Acadian ecoregion showing multiple possible pathways (edges) and selected linkages within the modeled corridors. (**a**) Landscape network in which GAP status 1 protected areas are connected using the Human Footprint (90-m resolution; Woolmer et al. 2008) as a resistance surface. (**b**) Landscape network in which Last of the Wild patches are connected using the Human Footprint as a resistance surface

and transboundary cooperation that may be required. The areas identified in our analyses only serve as a guide; the actual areas required for connectivity may be lessened if existing GAP 2 and 3 protected lands (which have lower levels of conservation status) provided adequate habitat matrices and were thus included as suitable habitat nodes.

For Scenario A, the minimum spanning tree indicates that only 90 of the 392 modeled corridors (23%) are needed to minimally connect the 95 protected area

16 Modeling Ecoregional Connectivity

Table 16.2 Summary statistics for the two modeled landscape networks in the Northern Appalachian/Acadian ecoregion. Scenario A refers to the effort to build a landscape network for GAP status 1 protected areas; Scenario B to the effort to build a landscape network for Last of the Wild (HF < 10) patches

Metric		Scenario A: gap status 1 patches	Scenario B: last of the wild patches
Nodes		95	120
Edges		392	566
	Min. edge length (m)	3,460	2,958
	Max. edge length	312,118	241,001
	Average edge length	41,793	43,730
Linkages		392	566
	Min linkage length (m)	180	180
	Max linkage length	219,840	307,405
	Average linkage length	24,732	29,912
Corridors		392	566
	Min. corridor area (ha)	16	24
	Max. corridor area	878,510	786,063
	Average corridor area	39,766	33,108
Minimum spanning tree edges		90	116
	Min. edge length (m)	3,460	2,958
	Max. edge length	151,510	199,521
	Average edge length	34,615	34,470
	Total spanning tree length	3,115,313	3,998,564

nodes; for Scenario B, only 116 of the modeled 566 corridors (20.5%) are necessary to connect the 120 Last of the Wild nodes. These minimum spanning trees may be viewed as good news for conservation planners since they indicate that less than one-quarter of the identified linkages would have to be incorporated into some kind of conservation plan in order to achieve region-wide connectivity. On the other hand, a minimum spanning tree is just that: a minimum. Planning designed to achieve only the minimum does not take into account the conservation advantages of planning for redundancy or for future environmental change that alters the landscape (Theobald 2006). Planners must directly confront the philosophical tradeoff between efficiency and redundancy.

16.5 Lessons Learned

Over the last few decades, conservation biologists have been planning and implementing 'regional reserve networks' (Noss 1983; Soulé and Terborgh 1999). Along the way, much valuable research has been conducted to examine the assumptions that underlie the understanding of habitat connectivity and to develop systematic, repeatable methods to aid in its planning. Today, the dream of networked reserves has become closer to reality because of the powerful new computer-based tools available to plan for connectivity. Yet these tools rely upon spatial models with varying degrees of accessibility to conservation planners and varying data requirements. Even planners with relatively strong GIS skills may have neither the time nor ecological background to evaluate, choose, and implement the more complex of these tools. Furthermore, not every region will have a complete enough dataset to conduct meaningful connectivity analyses. The barrier that lies between connectivity that is merely modeled to connectivity that is an on-the-ground reality can be quite high. Although connectivity models are generally accessible to professionals with strong GIS backgrounds, technical expertise and availability of quality habitat data in GIS formats currently limits advanced connectivity modeling in many places around the world. In those places, understanding the concepts outlined in this chapter – and made explicit in other publications (cf., Crooks and Sanjayan 2006) – will help all regions to move towards a systematic basis for modeling functional connectivity. However, in many locations it is already possible to run the models, which will provide meaningful guidance for conservation planning.

Our experience indicates a number of key lessons that help simplify the process of planning for connectivity. First, in large and heterogenous ecoregions, the simplifying assumptions of using predefined 'patches' (i.e., existing and/or proposed protected areas or areas of low human influence) and broad measures of landscape transformation as indices of landscape permeability (i.e., the Human Footprint) provide a coarse-scale beginning to connectivity planning, which may provide insight into where on the landscape more detailed assessments should be focused.

Even when using coarse-scale assessments, ecoregional-scale conservation planners may want to employ focal species models for connectivity at more local scales. A problem with using generalized habitat suitability models at the ecoregional scale is that regional variability is seen not only in habitats but in animal behavior so that the suitability model may perform well in some areas of the ecoregion but not others. Using a coarse-scale method regionally combined with local, patch-to-patch connectivity exercises for focal species, as made possible by CorridorDesigner, might solve this problem. Focal species models will be more accurate locally, and such an approach will provide more flexibility in selecting focal species, particularly important or relevant for local-scale efforts. For example, lynx only occur in more northerly and mountainous areas of the Northern Appalachian/Acadian ecoregion and are very important for conservation in those areas but would not be suitable as a focal species for connectivity modeling throughout the ecoregion.

Connectivity modeling, despite its outward complexity and array of modeling choices (e.g., networks, circuits), is generally based on simple concepts of least-cost paths, graphs, and suitability. Conceptually, it is within the realm of any person with training in biological or environmental sciences to understand. Technically, a high level of expertise is required particularly when integrating ecological assumptions about a particular region with a chosen approach to modeling. Collaborations among academics and conservation practitioners are likely

16 Modeling Ecoregional Connectivity

to be the most effective means for developing ecoregional connectivity models. When modeling connectivity across an ecoregion, complexity and size of the dataset may face limits. For example, the number of patches that can be used as the basis for assessing connectivity has an upper limit. More research is needed on ecoregion-scale network modeling before it can be determined whether these limitations are due to the hardware, software, or resolution of aggregation for allocation zones.

Modeling exercises such as these should be taken only as guides to planning for connectivity and not as final answers. Designing and implementing a comprehensive conservation plan that includes consideration of connectivity certainly must build upon the results that emerge from such models, but ultimately experts and other stakeholders must be engaged to evaluate and revise the modeled results to achieve a plan that works on the ground.

References

Baldwin, R. F., & deMaynadier, P. G. (2009). Assessing threats to pool-breeding amphibian habitat in an urbanizing landscape. *Biological Conservation, 142*, 1628–1638.
Baldwin, R. F., Calhoun, A. J. K., & deMaynadier, P. G. (2006). Conservation planning for amphibian species with complex habitat requirements: A case study using movements and habitat selection of the wood frog *Rana sylvatica. Journal of Herpetology, 40*, 442–454.
Beier, P., & Noss, R. F. (1998). Do habitat corridors provide connectivity? *Conservation Biology, 12*, 1241–1252.
Beier, P., Majka, D. R., & Spencer, W. D. (2008). Forks in the road: Choices in procedures for designing wildland linkages. *Conservation Biology, 22*, 836–851.
Berger, J. (2004). The last mile: How to sustain long-distance migration in mammals. *Conservation Biology, 18*, 320–331.
Berven, K. A., & Grudzien, T. A. (1990). Dispersal in the wood frog (*Rana sylvatica*): Implications for genetic population structure. *Evolution, 44*, 2047–2056.
Burwell, W. B., III. (2009). Connectivity modeling for black bear (*Ursus americanus*) and bobcat (*Lynx rufus*) in the Champlain Valley, Vermont. Senior thesis, Middlebury College.
Carroll, C. (2006). Linking connectivity to viability: Insights from spatially explicit population models of large carnivores. In K. R. Crooks & M. Sanjayan (Eds.), *Connectivity conservation* (pp. 369–389). Cambridge, UK: Cambridge University Press.
Carroll, C. (2007). Interacting effects of climate change, landscape conversion, and harvest on carnivore populations at the range margin: Marten and lynx in the Northern Appalachians. *Conservation Biology, 21*, 1092–1104.
Chapin, T. G., Harrison, D. J., & Katnik, D. D. (1998). Influence of landscape pattern on habitat use by American marten in an industrial forest. *Conservation Biology, 12*, 1327–1337.
Circuitscape.org. (n.d.). *Welcome to the Circuitscape project!* Retrieved February 13, 2010, from Circuitscape.org Web site: http://www.circuitscape.org/
Clark, J. S. (1991). Disturbance and population structure on the shifting mosaic landscape. *Ecology, 72*, 1119–1137.
Compton, B. W., McGarigal, K., Cushman, S. A., & Gamble, L. R. (2007). A resistant-kernel model of connectivity for amphibians that breed in vernal pools. *Conservation Biology, 21*, 788–799.
CorridorDesign. (2009). *GIS tools and information for designing wildlife corridors.* Retrieved February 13, 2010, from CorridorDesign Web site: http://corridordesign.org/

Crooks, K. R., & Sanjayan, M. (Eds.). (2006). *Connectivity conservation*. Cambridge, UK: Cambridge University Press.

deMaynadier, P. G., & Houlahan, J. E. (2008). Conserving vernal pool amphibians in managed forests. In A. J. K. Calhoun & P. G. deMaynadier (Eds.), *Science and conservation of vernal pools in northeastern North America* (pp. 127–148). Boca Raton FL: CRC Press.

deMaynadier, P. G., & Hunter, M. L., Jr. (1998). Effects of silvicultural edges on the distribution and abundance of amphibians in Maine. *Conservation Biology, 12*, 340–352.

Eisenberg, J. F. (1981). *The mammalian radiations: An analysis of trends in evolution, adaptation and behavior*. Chicago, IL: The University of Chicago Press.

Forman, R. T. T., Sperling, D., Bissonette, J. A., Clevenger, A. P., Cutshall, C. D., Dale, V. H., et al. (2003). *Road ecology: Science and solutions*. Washington, DC: Island Press.

Gaines, M. S., & McClenaghan, L. R., Jr. (1980). Dispersal in small mammals. *Annual Review of Ecology and Systematics, 11*, 163–196.

Gobeil, J.-F., & Villard, M.-A. (2002). Permeability of three boreal forest landscape types to bird movements as determined from experimental translocations. *Oikos, 98*, 447–458.

Groves, C. R., Jensen, D. B., Valutis, L. L., Redford, K. H., Shaffer, M. L., Scott, J. M., et al. (2002). Planning for biodiversity conservation: Putting conservation science into practice. *BioScience, 52*, 499–512.

Haddad, N. M., & Tewksbury, J. J. (2006). Impacts of corridors on populations and communities. In K. R. Crooks & M. Sanjayan (Eds.), *Connectivity conservation* (pp. 390–415). Cambridge, UK: Cambridge University Press.

Harrison, D. J. (1992). Dispersal characteristics of juvenile coyotes in Maine. *Journal of Wildlife Management, 56*, 128–138.

Hitchings, S. P., & Beebee, T. J. C. (1998). Loss of genetic diversity and fitness in common toad (*Bufo bufo*) populations isolated by inimical habitat. *Journal of Evolutionary Biology, 11*, 269–283.

Hunter, M. L., Jr. (1991). Coping with ignorance: The coarse-filter strategy for maintaining biodiversity. In K. A. Kohm (Ed.), *Balancing on the brink of extinction: The Endangered Species Act and lessons for the future* (pp. 266–281). Washington, DC: Island Press.

Iriarte, J. A., Franklin, W. L., Johnson, W. E., & Redford, K. H. (1990). Biogeographic variation of food habits and body size of the American puma. *Oecologia, 85*, 185–190.

Kindlmann, P., & Burel, F. (2008). Connectivity measures: A review. *Landscape Ecology, 23*, 879–890.

Mace, R. D., Waller, J. S., Manley, T. L., Lyon, L. J., & Zuuring, H. (1996). Relationships among grizzly bears, roads and habitat in the Swan Mountains, Montana. *Journal of Applied Ecology, 33*, 1395–1404.

McRae, B. H., & Beier, P. (2007). Circuit theory predicts gene flow in plant and animal populations. *Proceedings of the National Academy of Sciences, 104*, 19885–19890.

McRae, B. H., Dickson, B. G., Keitt, T. H., & Shah, V. B. (2008). Using circuit theory to model connectivity in ecology, evolution, and conservation. *Ecology, 89*, 2712–2724.

Mech, L. D., Fritts, S. H., & Wagner, D. (1995). Minnesota wolf dispersal to Wisconsin and Michigan. *American Midland Naturalist, 133*, 368–370.

Merenlender, A. M., Huntsinger, L., Guthey, G., & Fairfax, S. K. (2004). Land trusts and conservation easements: Who is conserving what for whom? *Conservation Biology, 18*, 65–75.

Mladenoff, D. J., & Sickley, T. A. (1998). Assessing potential gray wolf restoration in the northeastern United States: A spatial prediction of favorable habitat and potential population levels. *Journal of Wildlife Management, 62*, 1–10.

Mladenoff, D. J., Sickley, T. A., & Wydeven, A. P. (1999). Predicting gray wolf landscape recolonization: Logistic regression models vs. new field data. *Ecological Applications, 9*, 37–44.

Morrison, M. L., Marcot, B. G., & Mannon, R. W. (2006). *Wildlife-habitat relationships: Concepts and applications*. Washington, DC: Island Press.

Newmark, W. D. (1987). A land-bridge island perspective on mammalian extinctions in western North American parks. *Nature, 325*, 430–432.

16 Modeling Ecoregional Connectivity

Newmark, W. D. (1995). Extinction of mammal populations in Western North American National Parks. *Conservation Biology, 9*, 512–526.

Noss, R. F. (1983). A regional landscape approach to maintain diversity. *BioScience, 33*, 700–706.

Powell, R. A., Zimmerman, J. W., & Seaman, D. E. (1997). *Ecology and behavior of North American black bears: Home ranges, habitat and social organization (Wildlife Ecology and Behavior Series, 4)*. London, UK: Chapman & Hall.

Pyke, C. R. (2005). Assessing suitability for conservation action: Prioritizing interpond linkages for the California tiger salamander. *Conservation Biology, 19*, 492–503.

Reh, W., & Seitz, A. (1990). The influence of land use on the genetic structure of populations of the common frog, *Rana temporaria*. *Biological Conservation, 54*, 239–249.

Ricketts, T. H. (2001). The matrix matters: Effective isolation in fragmented landscapes. *The American Naturalist, 158*, 87–99.

Rittenhouse, T. A. G., & Semlitsch, R. D. (2007). Postbreeding habitat use of wood frogs in a Missouri Oak-Hickory forest. *Journal of Herpetology, 41*, 645–653.

Rothermel, B. B., & Semlitsch, R. D. (2002). An experimental investigation of landscape resistance of forest versus old-field habitats to emigrating juvenile amphibians. *Conservation Biology, 16*, 1324–1332.

Sanderson, E. W., Jaiteh, M., Levy, M. A., Redford, K. H., Wannebo, A. V., & Woolmer, G. (2002). The Human Footprint and the last of the wild. *BioScience, 52*, 891–904.

Scott, J. M., Heglund, P. J., & Morrison, M. L. (Eds.). (2002). *Predicting species occurrences: Issues of accuracy and scale*. Washington, DC: Island Press.

Seidensticker, J., Christie, S., & Jackson, P. (Eds.). (1999). *Riding the tiger: Tiger conservation in human-dominated landscapes*. Cambridge, UK: Cambridge University Press.

Semlitsch, R. D. (2008). Differentiating migration and dispersal processes for pond-breeding amphibians. *Journal of Wildlife Management, 72*, 260–267.

Soulé, M. E., & Terborgh, J. (Eds.). (1999). *Continental conservation: Scientific foundations of regional reserve networks*. Washington, DC: Island Press.

Soulé, M. E., Estes, J. A., Miller, B., & Honnold, D. L. (2005). Strongly interacting species: Conservation policy, management, and ethics. *BioScience, 55*, 168–176.

Theobald, D. M. (2006). Exploring functional connectivity of landscapes using landscape networks. In K. R. Crooks & M. Sanjayan (Eds.), *Connectivity conservation* (pp. 416–443). Cambridge, UK: Cambridge University Press.

Theobald, D. M., Hobbs, N. T., Bearly, T., Zack, J. A., Shenk, T., & Riebsame, W. E. (2000). Incorporating biological information in local land-use decision making: Designing a system for conservation planning. *Landscape Ecology, 15*, 35–45.

Theobald, D. M., Norman, J. B., & Sherburne, M. R. (2006). *FunConn v1 user's manual: ArcGIS tools for functional connectivity modeling*. Fort Collins, CO: Colorado State University, Natural Resources Ecology Lab. Retrieved February 13, 2010, from: http://www.nrel.colostate.edu/projects/starmap/FUNCONN%20Users%20Manual_public.pdf

Urban, D., & Keitt, T. (2001). Landscape connectivity: A graph-theoretic perspective. *Ecology, 82*, 1205–1218.

Walther, G.-R., Post, E., Convey, P., Menzel, A., Parmesan, C., Beebee, T. J. C., et al. (2002). Ecological responses to recent climate change. *Nature, 416*, 389–395.

Williams, P., Hannah, L., Andelman, S., Midgley, G., Araújo, M. B., Hughes, G., et al. (2005). Planning for climate change: Identifying minimum-dispersal corridors for the Cape Proteaceae. *Conservation Biology, 19*, 1063–1074.

Wilson, E. O., & Peter, F. M. (Eds.). (1988). *Biodiversity*. Washington, DC: National Academy Press.

Woolmer, G., Trombulak, S. C., Ray, J. C., Doran, P. J., Anderson, M. G., Baldwin, R. F., et al. (2008). Rescaling the human footprint: A tool for conservation planning at an ecoregional scale. *Landscape and Urban Planning, 87*, 42–53.

Yellowstone to Yukon Conservation Initiative. (2010). *Making connections, naturally*. Retrieved February 13, 2010, from the Yellowstone to Yukon Conservation Initiative web site: http://www.y2y.net/

Chapter 17
A General Model for Site-Based Conservation in Human-Dominated Landscapes: The Landscape Species Approach

Michale J. Glennon and Karl A. Didier

Abstract It is widely recognized that parks and preserves cannot provide adequate habitat for the vast majority of wildlife species, and that alternative strategies are necessary for the long-term protection of biological diversity. Effective conservation planning often requires balancing a variety of competing interests with limited funding and creates inherent conflict if the needs of humans are not considered as part of the process. The Landscape Species Approach (LSA) of the Wildlife Conservation Society is an innovative approach to landscape-scale conservation planning which aims to create wildlife-based strategies for conserving large, wild ecosystems integrated in wider landscapes of human influence. This chapter describes the development and steps involved in the LSA approach, its application to the Adirondack Park in northern New York State, and advantages and disadvantages of the process.

Keywords Biological diversity • Conservation planning • Landscape Species Approach • Monitoring • Priority setting

17.1 Introduction

Effective conservation planning often involves making difficult decisions and balancing competing interests to achieve conservation goals, almost always in the context of limited funding. It is widely recognized that parks and preserves alone cannot effectively conserve all of the elements of biological diversity that should be

---

M.J. Glennon (✉)
Wildlife Conservation Society, 7 Brandy Brook Ave #204, Saranac Lake, NY 12983, USA
e-mail: mglennon@wcs.org

K.A. Didier
Wildlife Conservation Society, 907 NW 14th Avenue, Gainesville, FL 32601, USA
e-mail: kdidier@wcs.org

S.C. Trombulak and R.F. Baldwin (eds.), *Landscape-scale Conservation Planning*,
DOI 10.1007/978-90-481-9575-6_17, © Springer Science+Business Media B.V. 2010

conserved (Fischer et al. 2006). Protected areas are often not large enough, are seldom connected to other protected areas, and may be subject to negative human influences despite their protected status. Similarly, planning for conservation without taking the needs of humans into account creates inherent conflict, and biological diversity often loses in the long run.

In an effort to engage in effective conservation planning in the face of these constraints, The Wildlife Conservation Society (WCS) has developed an innovative approach to landscape-scale conservation planning that recognizes that animals do not acknowledge park boundaries and that aims to create wildlife-based strategies for conserving large, wild ecosystems that are integrated in wider landscapes of human influence. The Landscape Species Approach (LSA) is focused on addressing the ecological needs of and human threats to viable populations of a suite of species dubbed 'Landscape Species.'

The LSA was developed using 12 design and demonstration sites on four continents. Today, it has been applied, at least in a part, at a total of 28 land and seascapes across Africa, Asia, Latin America, and North America. Thus, this approach to landscape-scale conservation has broad geographic relevance. The LSA has no pre-defined scale for the area in which it will or should be applied. It has been applied by WCS in landscapes and seascapes as small as a few thousand (Glover's Reef Atoll, Belize) to more than 2 million (Coastal Patagonia, Argentina) square kilometers. In principle, the LSA could be adapted to even smaller spatial scales and used to enhance conservation efforts in places such as urban greenbelts. The only scalar limitation to applying the LSA is the availability of adequate information across the entire target area. Thus, for the purpose of applying the LSA, we define a 'landscape' as an area sufficient in size, composition, and configuration to support at least one ecologically functional population of all conservation features – species, communities, functions, and services – for the long term.

Our objectives in this chapter are to briefly review the steps involved in completing the LSA, discuss the gaps between the theory and on-the-ground reality as the LSA was applied in the Adirondack Park, and describe the advantages and disadvantages of the process as a whole.

17.2   The 10 Steps of the Landscape Species Approach

Each WCS project that uses the LSA proceeds through a series of 10 steps (Didier et al. 2009a; Table 17.1), similar to 'Systematic Conservation Planning' frameworks used by other authors and organizations (Groves et al. 2002). Several on-line technical manuals (www.wcslivinglandscapes.org) and published papers (Coppolillo et al. 2004; Didier et al. 2009a; Sanderson 2006; Treves et al. 2006) describe these steps and provide tools for completing them in detail.

17 A General Model for Site-Based Conservation in Human-Dominated Landscapes 371

Table 17.1 The 10 steps of conservation planning using the Landscape Species Approach (Didier et al. 2009a)

Step	References
1. Compile relevant information on the conservation context of the site.	Treves et al. (2006)
2. Use a conceptual model to set a broad goal and to describe threats and barriers to achieving it.	Wilkie and LLP (2004b)
3. Select a set of Landscape Species.	Coppolillo et al. (2004)
	Strindberg et al. (2006)
4. Set quantitative Population Target Levels for conserving Landscape Species.	Sanderson (2006)
5. Map Biological Landscapes for each Landscape Species.	Sanderson et al. (2002)
	Didier and the Living Landscapes Program (2006)
6. Map Human Landscapes for each important human activity.	Sanderson et al. (2002)
	Didier and the Living Landscapes Program (2006)
7. Map Conservation Landscapes for each Landscape Species.	Didier and the Living Landscapes Program (2008)
8. Assess the sufficiency of current and need for additional conservation areas.	In development.
9. Prioritize areas for action.	In development.
10. Develop a monitoring framework.	Wilkie and the Living Landscapes Program (2006)

17.2.1 Step 1: Assessments of Context, Stakeholders, and Threats

The initial step in the LSA or any conservation planning process usually involves a series of activities devoted to understanding the context for conservation in a landscape (Pressey and Bottrill 2008; Chap. 3).

One of the first decisions to be made is what geographic region and what flora and fauna are under consideration. For the LSA, planners should try to make a first approximation of the extent of the landscape, based on relevant social, political, and ecological boundaries. As a part of this decision-making process, planners should also discuss which elements of biological diversity they are interested in conserving. For example, in the planning process described in Section 17.3, we focus on the species and ecosystems that occur within the Adirondack Park of New York. However, as mentioned, a specific aim is of the LSA is to test and refine the relevance of these a priori boundaries for Landscape Species. For example, is the Adirondack Park sufficiently large to conserve the chosen Landscape Species?

After a first approximation of the landscape boundary is made, practitioners should then compile a set of basic contextual information for that landscape, including information on stakeholders, economic and social value of natural resources, governance and land-tenure systems, and biological diversity and threats to it (Pressey and Bottrill 2008).

Early in the process, it is important to identify a set of stakeholders who should be engaged for planning, not only because stakeholder participation is critical for acceptance of any planning products but also because stakeholders are often a critical source of information not otherwise available (e.g., where species or human activities occur) (Didier et al. 2009a; Chaps. 4 and 10). Until now, WCS had not developed its own or used a formal process for assessing stakeholder communities and identifying which ones to invite into the planning process, although recently it has begun pointing practitioners to formal processes developed by other organizations (e.g., Golder and Gawler 2005; Groves 2003; The Nature Conservancy 2000).

Particularly important in contextual analyses for new projects are threats assessments. Within the LSA, WCS has developed a method for identifying, ranking, and mapping threats (Treves et al. 2006; Wilkie and the Living Landscapes Program 2004a). A multi-stakeholder workshop is held for the purpose of generating a comprehensive list of human activities with the potential to negatively impact biological diversity in the region and to rank them in order of their perceived importance to those stakeholders participating.

The results of a successful threats assessment will indicate where within the landscape the most important human activities that threaten biological diversity occur, when they occur, whether they have changed in intensity over time, the relative severity of each threat, how long the system may require to recover if the threat were removed, and how urgent the need for management action may be (Wilkie and the Living Landscapes Program 2004a).

One of the requirements and key components of the threats assessment is its participatory nature. Bringing together a diverse set of stakeholders can help elucidate the relative roles of management capacity, stakeholder awareness, and policies or regulatory mechanisms in mitigating threats to biodiversity, and inviting a diverse set of stakeholders may serve to help reconcile conflicting interests. Likewise, one of the primary purposes for holding the workshop is to bring together the principal actors who may ultimately be required to work cooperatively to reduce threats and conserve biological diversity in the landscape or seascape of interest (Wilkie and the Living Landscapes Program 2004a). The complete steps of a threats assessment are detailed in Wilkie and the Living Landscapes Program (2004a) and include: (1) providing a step-by-step description of the task to be completed, (2) explaining what is meant by direct and indirect threats, (3) asking each participant to identify 3–7 threats to biological diversity in the landscape, (4) organizing human activities from all participants into groups, (5) voting to identify the highest priority threats for conservation to mitigate, (6) characterizing and mapping the highest priority threats, (7) reviewing and presenting threat maps, and (8) discussing results and additional steps that may be needed to complete the threat assessment.

17.2.2 Step 2: Development of a Conceptual Model

A conceptual model is a graphical representation of the goals, conservation features, causal network of threats to biological diversity, and priority conservation activities of any conservation project (Margoluis et al. 2008). Conceptual models are essentially a representation of what conservation managers think they know implicitly and, as such, they (1) explicitly define what needs to be influenced or changed as a result of project activities (i.e., the conservation features), (2) characterize and prioritize the factors that directly or indirectly threaten the species or landscapes that need to be conserved, (3) graphically represent how these threats, individually or in combination, cause the undesirable changes in the species or landscape, (4) demonstrate that the activities that are focused on reduce key threats and attain quantitative conservation targets, (5) provide a strategic framework for determining what to monitor to assess project effectiveness and to adapt project activities, and (6) offer a structure for reviewing and revising project assumptions and activities as conditions change over time (Wilkie and the Living Landscapes Program 2004b).

Conceptual models may be exceedingly simple or fairly complex but all are composed of four basic elements: goals, focal ecological features (with population targets levels), threats, and activities (Fig. 17.1). Table 17.2 provides definitions of these terms.

Wilkie and the Living Landscapes Program (2002) provide a brief overview of the process of creating conceptual models, and a full treatment of the methodology

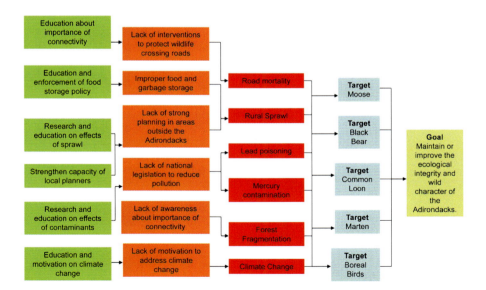

Fig. 17.1 Example of a partial and highly simplified conceptual model for the Adirondacks with goal (*yellow*), targets (*blue*), direct threats (*red*), indirect threats (*orange*), and interventions (*green*)

Table 17.2 Key terms and definitions we use in this paper. The exact words used differ from place to place and author to author, but the basic concepts are common to most conservation planning exercises

Term	Definition
Goal	A broad, visionary statement of what conservation wants to achieve at a particular place. Example: "Conserve the ecological integrity and wild character of the Adirondack Park."
Threat	A human or human-mediated activity which negatively impacts biodiversity or impedes our ability to reach our conservation goals and targets.
Direct threat	A threat which directly changes the abundance, quality, or extent of a conservation feature. Four major categories of direct threats, especially for species, include direct extraction (e.g., hunting), competition from exotic species, habitat/land-cover conversion, and pollution of habitat (Wilkie and the Living Landscapes Program 2004a).
Indirect threat	A social, economic, legal, or political factor that enables a direct threat to occur. Typical examples include "lack of alternative economic options", "lack of laws," "lack of enforcement," "lack of education/knowledge."
Landscape	An area sufficient in size, composition, and configuration to support at least one ecologically functional population of all conservation features for the long term.
Biodiversity feature	An element of biodiversity that a project aims to conserve, including species, ecosystems, habitats, subspecies, genes, ecological functions, ecosystem services, etc.
Focal biodiversity feature	A subset of conservation feature that a project will explicitly focus activities on. As it is typically impossible to focus activities on and collect information about all conservation features, projects typically have to select a "representative" and practical subset, the successful conservation of which will hopefully result in the conservation of most if not all conservation features. Landscape Species are focal conservation features.
Population target level	The state or condition of a biodiversity feature that a project wants to maintain or achieve. For Landscape Species, this is generally expressed in terms of a desired number of animals across the landscape (e.g., 4,000 elk), although PTLs can be far more detailed (e.g, a population of 3,000–5,000 elk, containing at least 10% reproductive females, at local densities no greater than $3/km^2$).
Conservation area	An area where conservation actions are taken (e.g., hunting enforcement) or actions are aimed to have an impact (e.g., new laws to outlaw hunting in particular places). Protected areas are considered one form of conservation areas.

is provided in Wilkie and the Living Landscapes Program (2004b). Usually, conceptual models are first built in draft form, but are refined and adjusted as other steps (e.g., selection of Landscape Species) are completed. As such, they serve as a repository for much of the planning information produced during the LSA.

When completed, conceptual models, although fluid and expected to change over time, provide a means for planning project priorities. Using a conceptual model, all members of a conservation project should be able to identify how and why any proposed intervention would have an impact (Wilkie and the Living Landscapes Program 2002). Conceptual models also provide a framework for developing a monitoring strategy that tracks changes in the model over time and allows for review and update of project priorities, which are key parts of measuring the effectiveness of conservation actions. (Monitoring frameworks are discussed further in Step 10, Section 17.2.8)

17.2.3 Step 3: Selection of Landscape Species

While most landscape-scale conservation projects have a broad goal or vision to conserve all or most of the biological diversity native to a place, it is impossible to dedicate sufficient resources to plan and act in such a way as to conserve all of it (Groves 2003). A process for selecting focal conservation features is commonly used, and many conservation NGOs have developed specific procedures for doing so, including The Nature Conservancy, World Wildlife Fund, and Conservation International (Bottrill et al. 2006). Within the LSA, WCS has developed a procedure for selecting a suite of focal conservation features called Landscape Species that should ensure that landscapes are large enough, sufficiently connected, and well configured to support functional populations of most other biological elements. In this sense, Landscape Species, as a group, are explicitly selected to serve as an 'umbrella' for conservation of all other features in the landscape (Lambeck 1997).

Landscape Species are defined as wildlife that typically require large, ecologically diverse areas to survive and often have significant impacts on the structure and function of natural ecosystems. Because of their habitat requirements and movement behavior, Landscape Species may be particularly threatened by human alteration and use of natural landscapes. Landscape Species are often cultural icons that can help generate a constituency for the conservation of biological diversity (Redford et al. 2000; Sanderson et al. 2002). WCS believes that planning conservation strategies to meet the needs of a suite of Landscape Species identifies the necessary area, condition, and configuration of habitats to meet the long-term ecological requirements for most species occurring in a wild landscape (Coppolillo and the Living Landscapes Program 2002). Thus, as noted above, no predefined rules are set for the extent of the landscape at which the LSA process might be applied. The boundaries of the potential site are determined by the needs of the wildlife species themselves.

The selection process is meant to identify an efficient set of species as focal features. To the degree that a selected set of Landscape Species appears insufficient to represent the broader set of conservation features or particular species are impractical for use (e.g., they are difficult to monitor), we recommend that planners consider adding other species to their set of focal conservation features, including broader and finer levels of biological organization (e.g., ecosystems, species assemblages, subspecies, or genotypes), special elements (e.g., threatened, endangered, or endemic species), and ecological processes (e.g., fire) (Groves 2003).

The process for selecting Landscape Species is described in detail in Coppolillo and the Living Landscapes Program (2002), Coppolillo et al. (2004), and Strindberg et al. (2006), and the process is facilitated by software, available at www.wcsliving-landscapes.org. Briefly, the selection process begins with identification of a set of candidate species. Although, in theory, any species can be considered a candidate, but it is practical to consider only those that will score highly on at least one or more of five selection criteria (Coppolillo and the Living Landscapes Program 2002). It is also important that the candidate pool be comprised of species that occupy the full range of habitat types in the target landscape.

Once the pool of candidate species has been selected, the next step is to score each, using data from local experts, field studies, and published literature, according to five selection criteria: (1) area requirements, (2) heterogeneity of habitat use, (3) vulnerability of the species to threats, (4) socio-economic significance, and (5) ecological functionality (Strindberg et al. 2006). The suite is then compiled by first selecting the candidate species with the highest composite score across the five criteria. Additional species are then added by iteratively selecting the candidate species that (1) is most complementary to the species already selected, in terms of habitats and threats they represent and (2) has a high composite score. As iterative selection proceeds, significant flexibility is given to planners in terms of choosing among candidate species that may have similar composite scores. Species are added to the suite until all threats and habitats have been represented by at least one Landscape Species.

No set number of Landscape Species is required to represent any particular landscape, as long as all of the important habitats and threats are represented by the final selected suite of species. Most of the landscapes on which WCS has applied this methodology have selected between three and eight Landscape Species. Landscape Species can come from any taxa. While in application at WCS landscapes, most selected species have been birds or mammals, other taxa including fish, invertebrates, amphibians, and reptiles have been selected. While to our knowledge, no plant species have been selected, they have been candidates. Species from these other taxa are not selected as often because they tend to score lower in terms of area requirements, heterogeneity of habitat use, or vulnerability to threats criteria (i.e., are not affected by multiple threats) or, commonly, not enough ecological information is available for them to complete the process.

17.2.4 Step 4: Establishing Population Target Levels for Landscape Species

Population target levels generally refer to the number of individuals needed to be saved across a landscape. Although many conservation biologists would prefer to leave it to policy makers to choose specific numbers, increasingly, policy makers look to scientists to objectively determine how many individuals are 'enough' (Sanderson 2006; Soulé et al. 2005; Tear et al. 2005). Although difficult, setting population target levels is often unavoidable so that choices with respect to natural resources can be justified and the success and cost of conservation efforts can be assessed (Groves 2003). Sanderson (2006) gives a detailed description of the many ways of setting population target levels for conservation. No single target level is correct for all times for any particular species, and setting population target levels is complicated by the fact that people's attitudes toward wildlife are highly variable and affect their feelings about what constitutes a 'desired population size.' For a variety of reasons, it may be desirable to conserve as many animals as possible or to maintain populations at current or historical baselines. Many different circumstances and desires can lead to different population target levels.

Sanderson (2006) provides a full discussion of the process and methodology of setting population target levels and highlights a number of potential criteria by which they may be determined, including demographic sustainability, ecological functionality, social dynamics, economic benefits, cultural benefits, and historical baselines. As a general rule, conservation should first ensure that the population is self-sustaining (demographic sustainability), then work to ensure that the population fully interacts with its environment (ecological functionality). Conservation efforts can then attempt to allow for human use above the levels necessary for ecological integrity and, finally, can work toward historical levels when humans had significantly lower impacts on ecological patterns and processes (Sanderson and the Living Landscapes Program 2006).

17.2.5 Steps 5–7: Mapping Biological, Human, and Conservation Landscapes

Once the key threats to wildlife within the focal landscape have been identified, a suite of Landscape Species with which to work chosen, and population targets for those species set, the next step is to undertake the mapping exercises needed to prioritize where conservation actions should be focused. This step consists of the construction of three important maps: (1) Biological Landscapes, (2) Human Landscapes (also referred to as Threat Landscapes), and (3) Conservation Landscapes.

A Biological Landscape is a map that represents the 'attainable' distribution of a Landscape Species, reflecting what habitats are important for the species and what its distribution would look like if conservation actions mitigated negative impacts of human activities (Didier et al. 2009a; Fig. 17.2). Biological Landscapes are typically expressed in abundance units (e.g., number of individuals, biomass) and represent 'habitat capacity' as opposed to actual abundances, or the capacity of the landscape to support a species throughout its life cycle.

Human (or Threat) Landscapes are maps of the distribution of human activities that affect Landscape Species (Fig. 17.3). Measures of vulnerability such as Human Landscapes are critical components of effective conservation planning (Wilson et al. 2005). As Biological Landscapes represent patterns in abundance, Human Landscapes are meant to represent patterns of how anthropogenic threats reduce species abundances. They typically are created first to reflect the distribution and relative intensity of human activities (e.g., relative number of hunters, concentration of pollutants) independent of a particular species, and then converted into maps of impact for particular species (i.e., reductions in abundance). In addition, they are often created in two versions: a 'Past' version that shows the spatial distribution of

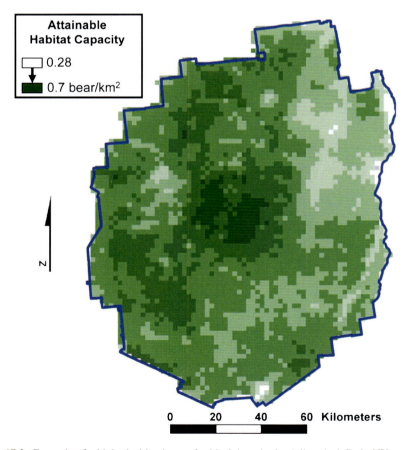

Fig. 17.2 Example of a biological landscape for black bear in the Adirondack Park, NY

17 A General Model for Site-Based Conservation in Human-Dominated Landscapes 379

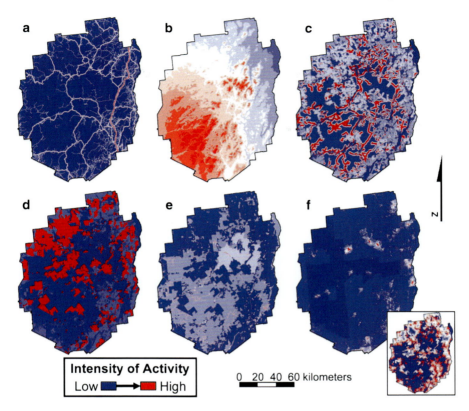

Fig. 17.3 Examples of human (Threats) landscapes for the Adirondack Park, NY, showing relative intensity of effects from (**a**) roads, (**b**) airborne contaminants, (**c**) hunting/poaching, (**d**) forest management, (**e**) recreation, and (**f**) development

human activities and impact up to the present, including recent impacts of ongoing activities, and a 'Future' version that forecasts human activities.

Most Biological and Human Landscapes are typically mechanistic models built from information in the literature or expert knowledge of the landscape. In a few cases, when sufficient field data have been available for the landscape and species in question, empirical and statistical modeling techniques (e.g., generalized additive models, Maximum Entropy models) have been used to generate such landscapes (Guisan and Zimmermann 2000; Phillips et al. 2006). Both empirical and mechanistic models, in fact, often take advantage of many sources of information, including field data, expert-opinions, and literature (Didier and the Living Landscapes Program 2006).

The combination of Biological and Human Landscapes can allow practitioners to produce additional maps, including the species' current (given the impact of human activities through present) and predicted future distributions (given the impacts of future human activities; Fig. 17.4). Conservation Landscapes are created by subtracting the three different distribution maps from one another (Fig. 17.5)

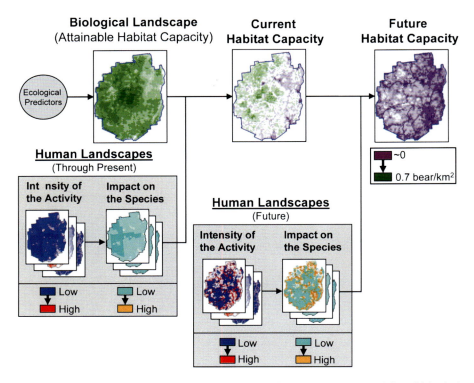

Fig. 17.4 Examples of current and future distributions of black bear constructed from biological and human landscapes in the Adirondack Park, NY

and depict the possible impacts of conservation actions across the study region. One version of the Conservation Landscape is created by subtracting the current distribution from the attainable (i.e., Biological Landscape) and represents the potential to increase populations by mitigating past threats (i.e., population recovery). A second version, created by subtracting the future from the current distribution, reflects the potential for preventing decreases by mitigating future threats (i.e., preventable loss; Didier et al. 2009a). Depending on the target species and the focal region that are the subject of the exercise, one or the other may be particularly useful and relevant.

For some target species, such as black bear (*Ursus americanus*) in the Adirondack Park, for example, the population is already at an ecologically functional level and not in danger of precipitous declines in the near future. In this case, prioritizing locations where actions should occur to prevent the decline of black bear populations in the future may be most useful. Other species, which may be rare and have already declined in the focal region – for example, the guanaco (*Lama guanicoe*) in the San Guillermo landscape of Argentina (Didier et al. 2009a) – may benefit as much or more from conservation actions aimed at areas where significant recovery or even recolonization is possible.

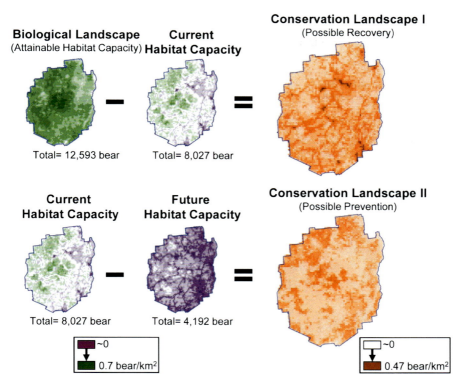

Fig. 17.5 Examples of Conservation Landscapes depicting potential benefits of interventions aimed at recovery (Conservation Landscape I) and prevention (Conservation Landscape II) in the Adirondack Park, NY

Didier et al. (2009b) provide a detailed description of the conceptual framework that underlies Biological, Human, and Conservation Landscapes. Didier and the Living Landscapes Program (2006, 2008) provide hands-on technical guidance in creating these maps, both available from the Living Landscapes Program website (www.wcslivinglandscapes.org).

17.2.6 Step 8: Estimating the Sufficiency of Existing Conservation Areas and Evaluating the Need for Additional Ones

When population target levels and species' distribution maps (attainable, current, and future) have been completed, these tools can be used to determine the sufficiency of existing conservation areas (do current abundances within protected areas meet population target levels?) and the need and possible impact of additional areas (what would happen if conservation actions were taken in this new area, and would overall targets be reached?). Maps of potential distributions can be compared with population target levels and used to estimate 'recovery' targets (e.g., how many individuals need to be added to the current population to reach the target?) and 'prevention targets' (e.g., what level of loss, measured in number of individuals, must be prevented to maintain the target?) Four outcomes of this process are possible (Didier et al. 2009a):

1. The current and future distribution maps for the species indicate that it is currently above the population target level, suggesting that additional conservation areas are not needed to reduce threats. In this case, practitioners might wish to review the target level or focus on monitoring and prevention of new threats.
2. The current distribution map indicates that although the species' is currently above the population target level, the future population is below it, suggesting that conservation efforts should focus on preventing future threats. Additional conservation areas may be needed or the effectiveness of activities occurring in existing ones improved.
3. The attainable population is above the population target level, but the current and future are below it, suggesting that new conservation areas may be needed and that conservation actions in existing or new areas need to both prevent future threats and mitigate impacts that have already occurred.
4. The attainable, current, and future populations are all below the population target level, suggesting that actions to mitigate both past and future threats are needed, but also that the current extent of the landscape needs to be expanded to reach target levels.

17.2.7 Step 9: Prioritize Areas for Action

Conservation Landscapes are critical tools for setting conservation priorities because they provide information on the possible impact of conservation activities in terms of adding animals to the current population or preventing future losses and, as such, can help practitioners decide where and when to invest resources. For example, the 'minimum' extent of the landscape needed to reach the target level for a particular Landscape Species can be determined by iteratively selecting those areas with the highest possible recovery or prevention impact (Didier et al. 2009a).

Though valuable, Conservation Landscapes do not incorporate all of the sources of information practitioners are likely to want to use in setting conservation priorities. For example, the costs of implementing conservation actions have not been included (Wilson et al. 2007). Similarly, practical constraints or particular opportunities that may make conservation easier or harder in any given location are not represented. Human judgment and expert opinion are, therefore, critical to setting conservation priorities (Carwardine et al. 2009; Didier et al. 2009a; Chap. 11).

Methods for setting site-specific priorities within the LSA have been drafted but have not yet been satisfactorily tested with field sites. It is likely that population targets and Conservation Landscapes will be used as inputs for decision support software such as Marxan or C-Plan (Ball and Possingham 2000; The University of Queensland n.d.), which can perform benefit-cost analyses to identify networks of conservation areas that efficiently meet quantitative targets for multiple biodiversity features such as Landscape Species. Although the methodology for this step has not yet been fully tested, the approach outlined here may allow for the inclusion of 'costs' in the priority setting process by incorporating them as land area, estimating monetary costs of implementing conservation actions, or estimating opportunity costs. Maps that identify both short- and long-term priority areas and that change as information improves can then be produced (Didier et al. 2009a).

17.2.8 Step 10: Monitoring Frameworks

The last step in the implementation of the LSA involves the critical step of monitoring the effectiveness of conservation actions and areas that are implemented. While difficult, monitoring is necessary because it permits (1) determination of whether or not the project is meeting its objectives and having a positive conservation impact, (2) identification of which actions lead to the success or failure of a particular conservation approach, (3) evaluation and revision of assumptions about why and where conservation efforts are needed, and (4) confidence that all participants in the project, from international NGO's to government staff to local residents, learn from the experience and use this knowledge to improve their implementation of future conservation programs (Ferraro and Pattanayak 2006; Margoluis et al. 2008; Stem et al. 2005; Wilkie and the Living Landscapes Program 2006).

Although costly, in order to demonstrate that LSA activities reduce threats and conserve wildlife and their habitat, monitoring at three key levels is needed: activities, threats, and conservation features (Fig. 17.6). Assessing how well actions are implemented is an example of performance monitoring, documenting changes in threats represents outcome monitoring, and tracking changes in the status of conservation features is an example of impact monitoring (Wilkie and the Living Landscapes Program 2006). Given that time, personnel, and funds are always limited, it is rare that monitoring can be implemented for every intervention, threat, and conservation target. A realistic approach to this challenge is to bring together a knowledgeable group of field staff and use a Delphi process to decide (1) which monitoring information is a priority and should, therefore, have resources allocated to it, (2) what level of precision is needed to feel confident in making a management decision based on the monitoring information, (3) what information would be highly useful but require additional funding to obtain, and (4) what information, while useful, would be unnecessary (Wilkie and the Living Landscapes Program 2006). The people involved in this discussion should address the tradeoffs associated with each choice, as well as the confidence associated with different qualitative

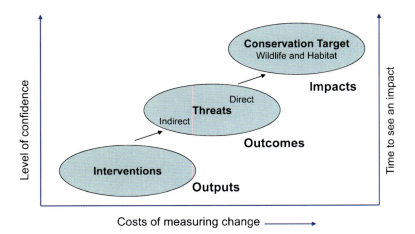

Fig. 17.6 Targets of monitoring efforts and the relative benefits and costs of monitoring at each level (Wilkie and the Living Landscapes Program 2006, used by permission)

and quantitative approaches to monitoring. Wilkie and the Living Landscapes Program (2006) provide additional details for creating monitoring frameworks.

17.3 Theory Versus Reality: an Adirondack Case Study

17.3.1 Challenges and Opportunities

Rarely does the conception of what a conservation planning process should entail match the reality of actually applying that process in a field-based situation, and the LSA is no exception. The LSA has been applied nearly in its entirety in the Adirondack Park, New York, and parts of it have been applied in other landscapes worldwide, including as of January 2009, 12 other terrestrial landscapes scattered across North America, Latin America, Africa, Asia, and two marine seascapes (Didier et al. 2009a). The Adirondack Park was one of the initial 'design and demonstration' landscapes, where the concepts and methods behind the LSA were developed and tested in situ. As such, the procedures used in the Adirondack Park are somewhat different than those more ideal steps described above. As one of the few sites that have completed most of the steps in the approach, the Adirondack experience can provide a valuable perspective on what worked well and what did not.

One of the challenges to conservation planning is that only in applying a method are all the limitations of the theory revealed. The approach, as envisioned in theory, often misses things that are important in practice. An example of this was in the process of selecting Landscape Species in the Adirondack Park, which was completed early in the design phase and prior to the development of the Landscape Species Selection software and associated technical manual (Strindberg et al. 2006).

17 A General Model for Site-Based Conservation in Human-Dominated Landscapes 385

Although the software is now available and has greatly simplified the process of selecting Landscape Species, its use for other landscapes suggest that human 'input' remains important in interpreting and enhancing computer-based output (Carwardine et al. 2009). Two specific problems arose in the Adirondack Park: habitat heterogeneity and 'monitorability.'

'Heterogeneity of Habitat Use' is one of the five criteria by which Landscape Species are selected. Candidate species that use more habitats receive a higher score for this criterion and are more likely to be selected. In theory, selecting Landscape Species with heterogeneous habitat requirements helps to identify the composition and configuration of habitat types necessary for successful conservation of diverse landscapes (Coppolillo and the Living Landscapes Program 2002). It also helps ensure that the suite of Landscape Species is smaller than it would otherwise be, as the suite as a whole must represent all important habitat stipulated by the planner.

It was found, however, that forcing Landscape Species to have heterogeneous habitat needs can bias selection toward wide-ranging generalists that use many habitats but, through their generalist nature, are not strongly affected by loss or degradation of any particular habitat. For example, in the initial run of the Landscape Species Selection, black bear was the only species needed to complete the suite. Individual black bears in the Adirondack Park to some degree use nearly all of the available habitats (e.g., deciduous forest, high- and low-elevation evergreen forest, wetlands) They are also affected, although not dramatically, by nearly all threats acting in the Park (e.g., hunting, poaching, unsustainable forest management, disturbance associated with recreation). They are, in many respects, what would appear to be a near perfect 'Landscape Species.'

However, in total, black bears are not particularly vulnerable in the Adirondack Park – they are fairly abundant throughout most of the park, their population is probably increasing, and they show no short-term sign of decreasing. They are also not particularly sensitive to changes in the extent or quality of any of the habitats they use – they can simply move elsewhere or rely on other resources.

Although at least technically only the black bear was needed to complete the landscape species suite, it was felt that to focus conservation on a single species would be misleading and ill-advised. Therefore, black bear was chosen as the initial Landscape Species, but the selection process was restarted without it, which eventually resulted in the inclusion of five additional species.

A second problem emerged with selection of Landscape Species. Some species that were selected were known to be difficult or impossible to monitor in the field. For example, the American Three-toed Woodpecker (*Picoides dorsalis*) was initially selected as a representative of low-elevation boreal habitat in the Adirondack Park. However, this species, although a good indicator of low-elevation boreal habitat when present, exhibits behavioral characteristics, such as periodic population irruptions in response to food sources created by recent fires, that make it particularly hard to find in some years. American Three-toed Woodpeckers will follow insect outbreaks and take advantage of recent fires and other disturbances that create newly dead trees and, as such, even under natural conditions are not always reliably present in the habitat with which they are typically associated. As such, it would be very hard to interpret the results of monitoring activities or to assess whether conservation activities were effective. Furthermore, such behavioral characteristics call into question whether the species is representative of others that have the same habitat requirements. The American Three-toed Woodpecker was kept as a Landscape Species, but several other low-elevation boreal birds were added, thus creating an assemblage of species to represent the habitat type.

These kinds of ad hoc modifications worked well for most steps in the LSA, and lessons learned have often been incorporated into the LSA for the benefit of other landscapes and planners, For example, subsequent to the application of the LSA in the Adirondack Park, the Landscape Species selection process and software was revised such that the 'heterogeneity of habitat use' criterion now favors species that *require* multiple habitats, rather than those that simply *use* multiple habitats as generalists do (Strindberg et al. 2006).

Monitorability, however, has not been incorporated as an explicit criterion in the software's algorithm, although guidance materials (Strindberg et al. 2006) now recommend that planners consider it. The software also now incorporates substantial flexibility and interactive processing so that planners can incorporate other criteria and opinions of stakeholders and experts into the selection process. As mentioned before, it has become clear that strict reliance on software and algorithms for selection of Landscape Species or any of the other steps in planning is ill-advised and that input of scientists and others is needed to ensure that acceptable and practical planning products emerge.

A second major challenge of applying the LSA, and certainly with other conservation planning methods, is that the theory envisions the use of better data than usually exists or is realistic to collect. This is exemplified by the procedures for setting population target levels and those for creating monitoring frameworks. Sanderson (2006) gives a thorough treatment of the possible methods for setting population target levels and the necessity of doing so in a transparent manner. Sanderson (2006) describes how although minimum viable population (MVP) estimates are a commonly used target, they are in many cases far below what should be considered desirable for many species, and other more ambitious targets, such as ecologically functional or historically representative levels should be articulated. Unfortunately, if estimates are available in the literature at all, they are usually MVP's. Although the notion of an ecologically functional or historically representative population level is appealing, finding actual numbers to support an estimate of those population levels can often be exceedingly difficult (Chap. 9). When practitioners are faced with an ambitious conservation planning methodology for which they cannot provide all of the necessary information, the result can be frustration, significant expenditure of time, and analytical results that are either incomplete or comprised of too much guesswork to be useful for real decision-making.

The last step in the LSA process, the construction and implementation of a monitoring framework, is also an example of where the data needed to complete conservation planning are unavailable or too expensive to collect. Although it is agreed that monitoring is a critical component of any conservation program, and although Wilkie and the Living Landscapes Program (2006) clearly articulate the reasons why monitoring at all three levels – target, threat, and intervention – is critical, funding for monitoring is often the most difficult to secure for the long time periods that are necessary to make monitoring data useful. Tracking populations of target species over the long term, in particular, does not have the 'sex appeal' of conservation projects that have immediate and demonstrable results such as purchasing land. Monitoring target species may be appealing if they are large 'charismatic megafauna.' These species, however, are often some of the most difficult to monitor in the field. Monitoring of threats and activities themselves is slightly more financially feasible in many cases but, in general, the reality of implementing a full monitoring program for targets, threats, and activities is probably only rarely met for all Landscape Species at a given site.

A third challenge of applying the LSA and conservation planning in general is that it simultaneously strives to incorporate all the complexity of real-world decision-making while making the process easily understandable (Hajkowicz et al. 2009). The mapping of Biological, Human, and Conservation Landscapes associated with the LSA in many ways exemplifies this challenge. Didier and the Living Landscapes Program (2006, 2008) provide details on how to map these landscapes, using GIS, expert knowledge, and spatial modeling techniques. However, as with population target levels, planners often balk at the apparent lack of data to create and validate the products (e.g., data on moose (*Alces alces*) sightings in the Adirondack Park are insufficient) and are often uncomfortable using 'educated guesswork' to complete the maps (e.g., moose are probably abundant in forests that have been disturbed because they contain abundant forage). In this sense, the theory of the mapping procedures can be too ambitious with respect to the available data. It is also often too complex to explain easily to stakeholders and, for this reason, risks stakeholders rejecting its results (Didier et al. 2009b; Hajkowicz et al. 2009). In many places, however, the theory is also too simple. For example, information on distribution of biological diversity (Biological Landscapes) and threats (Human Landscapes) are often less important for making decisions about where to work than is information about costs to implement conservation actions in different areas, opportunities for action, or political will (Naidoo and Ricketts 2006; Newburn et al. 2005). In total, the challenge for conservation planning and the LSA is that it is never possible to know exactly which criteria are truly the most important, relevant, or feasible to use in particular places, and a generalized framework that incorporates them all is complex and hard to communicate.

The process of developing the LSA in the Adirondack Park and elsewhere has resulted in important lessons for practitioners who seek to apply this or other conservation planning methodologies to achieving landscape-scale conservation planning. Based on the lessons learned applying the LSA in the Adirondack Park, it is clear that one goal should be to make the conservation planning framework of the LSA flexible: to have the complex, fully developed methodologies ready for those who want them, and simplified, resource-light tools for those who need them. It is also important to collect and make available information on the costs of doing conservation planning itself: how much time, money, and what kind of expertise are needed to complete various steps and tools (Didier et al. 2009a; Morrison et al. 2009). That way, those who seek to implement similar projects can better judge the costs versus the benefits and make decisions about which tools are most appropriate to use.

17.3.2 Implications and Conclusions

As noted in Didier et al. (2009b), the LSA has had several positive impacts, both in the Adirondack Park and elsewhere. One of the greatest strengths of the approach in the Adirondack Park has been the focus on Landscape Species themselves as a result of their selection. Because it was done before the selection software was developed, Landscape Species were selected in the Adirondack Park through a series of stakeholder meetings over a period of several years. Directly involving members of the scientific community as well as interested members of the general public resulted in a strong appreciation for WCS as an organization that involves local community members in conservation and is genuinely appreciative of their input. One of the most important outcomes of the participatory nature of the species selection process in the Adirondack Park is that the WCS Adirondack program became an integral player in all conservation issues involving these focal species and wildlife in general in the park. It has greatly served to distinguish the niche of WCS in the Adirondack Park from other environmental organizations as a distinctly science-based organization whose primary goal is to protect wildlife.

The participatory nature of the species selection process in the Adirondack Park has also spawned a number of important programs and efforts in which WCS is essentially participating in the co-management of wildlife species in the park. The early focus on black bears as a target species for the park has led to a suite of research and education activities that have resulted in policy changes for back-country food storage and dramatic declines in the number of negative human-bear conflicts reported in the High Peaks region of the park.

Similarly, a focus on the Common Loon (*Gavia immer*) as an important target in the park has led to long-term collaboration with the New York State Department of Environmental Conservation (NYSDEC) on research and education efforts and has ultimately resulted in increased protection for Common Loons from local and airborne contaminants, such as lead and mercury. Moose populations have been slowly increasing in the Adirondack Park since 1980 and their population trajectory has now reached the point of attracting the attention of NYSDEC as well as Adirondack residents. The early selection and focus on moose as Landscape Species has led to collaboration with NYSDEC on moose research to try to determine the current status of the population in the Adirondack Mountains, as well as its distribution and habitat affinities. Last, a focus on boreal birds as Landscape Species has also led to long-term research and monitoring efforts in collaboration with NYSDEC. Several boreal bird species are considered to be Species of Greatest Conservation Need (SGCN) under New York State's Comprehensive Wildlife Conservation Strategy through the State Wildlife Grants program of the U.S. Fish and Wildlife Service. Because of their selection as SGCN for the state and Landscape Species for the Adirondack Park, the WCS Adirondack program has been able to leverage funding to conduct long-term monitoring on a suite of species to inform and contribute toward the establishment of a long-term boreal wildlife conservation plan for the Adirondack region.

17 A General Model for Site-Based Conservation in Human-Dominated Landscapes 389

In addition to these three projects, several of the selected Landscape Species are also the focus of efforts to model and protect connectivity in the Black River Valley, which separates the Adirondack Park from the Tug Hill Plateau to the west. Black bear, American marten (*Martes americana*), and moose are part of a suite of species on which The Nature Conservancy, in collaboration with WCS, is focusing to inform long-term conservation of connectivity between these two biologically important regions of Northern New York State, thus exemplifying how the LSA can have cascading impacts at increasing spatial scales. As much the result of the species selected through this process as of any particular outcome, the direct focus on Landscape Species has raised the profile of this set of species in the Adirondack Park and provided a base from which to form long-term collaborations to work cooperatively to conserve them.

The specific outcomes of the LSA have also been used to inform other conservation initiatives in the Adirondack Park. The outcomes of early stage Conservation Landscapes were shared with the Adirondack Nature Conservancy for their potential use in ecoregional planning for the greater Northern Appalachian/Acadian ecoregion. Outcomes of Conservation Landscapes have also been used in several instances to provide information to the Adirondack Park Agency, the regional private land-use authority, for their use in project review. Common Loon, moose, and black bear, in particular, have been highlighted with respect to proposed residential developments and potential impacts to these species and their habitats in particular regions of the Park.

The LSA has also been used indirectly to support conservation in the Adirondack Park through presentations of the work for various audiences, including local college students, outdoor writers, and local government representatives. The above examples illustrate the important role that the LSA has played in applied conservation in the Adirondack Park. Through not only the explicit goal of the LSA – to provide a framework for setting conservation priorities – but also through serving as a springboard for collaborative management and protection efforts, the LSA has undoubtedly contributed significantly to long-term conservation of Adirondack wildlife and habitats. Although its application requires an investment of time and significant information, the LSA provides an extensive toolkit and methodology for applying site-based, spatially explicit conservation planning based directly on the needs of wildlife species. As such, it is useful for any conservation project that involves spatial planning and prioritization of goals and objectives.

17.4 Lessons Learned

Several lessons emerge from the development and application of the Landscape Species Approach. First, it is important to plan the planning (referred to as 'scoping' in Pressey and Bottrill 2008). It is best to start with a basic framework of conservation planning similar to the 10 steps outlined here or elsewhere (Groves 2003; Pressey and Bottrill 2008), and then decisions about what is most important to do first can be made in light of the time and money available for the project.

Second, when embarking on a conservation planning effort, it is critical to consider carefully the balance between complexity/realism of the models used and the need to communicate and explain those models with stakeholders (Hajkowicz et al. 2009). A good approach to any step in conservation planning is to first identify all the complex factors that may affect the decision or model (e.g., conservation value, threat, opportunities, costs), and then first to focus explicitly on the few that are most important, saving the others to address through longer-term efforts.

Third, it is clear that engaging in conservation planning has many benefits. One of the unheralded ones is that by simply facilitating a logical, participatory process of conservation and development planning, an organization can raise its profile and become an integral player not just in planning, but decision-making and implementation for conservation within a region.

Fourth, the value of expert opinion should not be discounted. Conservation planning tools by themselves cannot answer the ultimate conservation questions. Every tool is incomplete and is not a perfect fit for all situations, and thus, experts and stakeholders should be allowed to modify and manipulate outputs from the tools, fixing errors, and incorporating missing or additional decision-making criteria.

Fifth, participation of stakeholders is absolutely critical for successful conservation planning and especially for implementation of its results. With that said, decisions about who should participate and, more importantly, when in the process their participation should occur, need to be made strategically (McCulloch 2006). It is important to recognize that internal planning for an organization versus external planning for a set of stakeholders may need different levels of participation.

Finally, conservation planning always needs to be approached as a long-term, adaptive process that requires a significant investment of time and resources. Costs – both in terms of time and money – should be recorded both to better understand the current planning effort and design other efforts in the future. Outcomes – both in terms of successes and failures – should be noted. No planning effort will ever be complete or perfect the first time around, but with careful attention to learning how each individual effort could have been made better, the record of success for landscape-scale conservation planning efforts will steadily improve over time.

References

Ball, I., & Possingham, H. (2000). *MARXAN (V1.8.2): Marine reserve design using spatially explicit annealing, a manual.* Retrieved September 15, 2006, from University of Queensland, Marxan Web site: http://www.uq.edu.au/marxan/docs/marxan_manual_1_8_2.pdf

Bottrill, M., Didier, K., Baumgartner, J., Boyd, C., Loucks, C., Oglethorpe, J., et al. (2006). *Selecting conservation targets for landscape-scale priority setting: A comparative assessment of selection processes used by five conservation NGOs for a landscape in Samburu, Kenya.* Washington, DC: World Wildlife Fund.

Carwardine, J., Klein, C. J., Wilson, K. A., Pressey, R. L., & Possingham, H. P. (2009). Hitting the target and missing the point: Target-based conservation planning in context. *Conservation Letters, 2,* 3–10.

17 A General Model for Site-Based Conservation in Human-Dominated Landscapes 391

Coppolillo, P., & The Living Landscapes Program. (2002). *Selecting landscape species (Bulletin 4)*. Bronx, NY: Wildlife Conservation Society, Living Landscapes Program. Retrieved December 15, 2008, from http://www.wcslivinglandscapes.org/landscapes/bulletins.html

Coppolillo, P., Gomez, H., Maisels, F., & Wallace, R. (2004). Selection criteria for suites of landscape species as a basis for site-based conservation. *Biological Conservation, 115*, 419–430.

Didier, K., & The Living Landscapes Program. (2006). *Building biological and threats landscapes from ecological first principles, a step-by-step approach (Technical Manual 6)*. Bronx, NY: Wildlife Conservation Society, Living Landscapes Program. Retrieved February 13, 2010, from http://www.wcslivinglandscapes.org/landscapes/90119/bulletins/manuals.html

Didier, K., & The Living Landscapes Program. (2008). *Building conservation landscapes: Mapping the possible impact of your conservation actions (Technical Manual 7)*. Bronx, NY: Wildlife Conservation Society, Living Landscapes Program. Retrieved February 13, 2010, from http://www.wcslivinglandscapes.org/landscapes/90119/bulletins/manuals.html

Didier, K. A., Glennon, M. J., Novaro, A., Sanderson, E. W., Strindberg, S., Walker, S., et al. (2009a). The Landscape Species Approach: Spatially-explicit conservation planning applied in the Adirondacks, USA, and San Guillermo – Laguna Brava, Argentina, landscapes. *Oryx, 43*, 476–487.

Didier, K. A., Wilkie, D., Douglas-Hamilton, I., Frank, L., Georgiadis, N., Graham, M., et al. (2009b). Conservation planning on a budget: A possible "resource light" method for mapping priorities at a landscape scale? *Biodiversity and Conservation, 18*, 1979–2000.

Ferraro, P. J., & Pattanayak, S. K. (2006). Money for nothing? A call for empirical evaluation of biodiversity conservation investments. *PLoS Biology, 4*, 482–488.

Fischer, J., Lindenmayer, D. B., & Manning, A. D. (2006). Biodiversity, ecosystem function, and resilience: Ten guiding principles for commodity production landscapes. *Frontiers in Ecology and the Environment, 4*, 80–86.

Golder, B., & Gawler, M. (2005). Cross-cutting tool: Stakeholder analysis. Retrieved February 1, 2009, from World Wide Fund for Nature (WWF) Web site: http://assets.panda.org/downloads/1_1_stakeholder_analysis_11_01_05.pdf

Groves, C. (2003). *Drafting a conservation blueprint: A practitioner's guide to planning for biodiversity*. Washington, DC: Island Press.

Groves, C. R., Jensen, D. B., Valutis, L. L., Redford, K. H., Shaffer, M. L., Scott, J. M., et al. (2002). Planning for biodiversity conservation: Putting conservation science into practice. *BioScience, 52*, 499–512.

Guisan, A., & Zimmermann, N. E. (2000). Predictive habitat distribution models in ecology. *Ecological Modelling, 135*, 147–186.

Hajkowicz, S., Higgins, A., Miller, C., & Marinoni, O. (2009). Is getting a conservation model used more important than getting it accurate? *Biological Conservation, 142*, 699–700.

Lambeck, R. J. (1997). Focal species: A multi-species umbrella for nature conservation. *Conservation Biology, 11*, 849–856.

Margoluis, R., Stem, C., Salafsky, N., & Brown, M. (2008). Using conceptual models as a planning and evaluation tool in conservation. *Evaluation and Program Planning, 32*, 138–147.

McCulloch, C. S. (2006). Transparency: Aid or obstacle to effective defense of vulnerable environments from reservoir construction? Dam decisions and democracy in North East England. *Area, 38*(1), 24–33.

Morrison, J., Loucks, C., Long, B., & Wikramanayake, E. (2009). Landscape-scale spatial planning at WWF: A variety of approaches. *Oryx, 43*, 499–507.

Naidoo, R., & Ricketts, T. H. (2006). Mapping the economic costs and benefits of conservation. *PLoS Biology, 4*, 2153–2164.

Newburn, D., Reed, S., Berck, P., & Merenlender, A. (2005). Economics and land-use change in prioritizing private land conservation. *Conservation Biology, 19*, 1411–1420.

Phillips, S. J., Anderson, R. P., & Schapire, R. E. (2006). Maximum entropy modeling of species geographic distributions. *Ecological Modelling, 190*, 231–259.

Pressey, R. L., & Bottrill, M. C. (2008). Opportunism, threats, and the evolution of systematic conservation planning. *Conservation Biology, 22*, 1340–1345.

Redford, K., Sanderson, E., Robinson, J., & Vedder, A. (2000). *Landscape species and their conservation*. Bronx, NY: Paper Presented at Wildlife Conservation Society meeting.

Sanderson, E. W. (2006). How many animals do we want to save? The many ways of setting population target levels for conservation. *BioScience, 56*, 911–922.

Sanderson, E., & the Living Landscapes Program. (2006). *Setting population target levels for wildlife conservation: How many animals should we save? (Bulletin 8)*. Bronx, NY: Wildlife Conservation Society, Living Landscapes Program. Retrieved December 15, 2008, from http://www.wcslivinglandscapes.org/landscapes/bulletins.html

Sanderson, E., Redford, K., Vedder, A., Coppolillo, P., & Ward, S. (2002). A conceptual model for conservation planning based on landscape species requirements. *Landscape and Urban Planning, 58*, 41–56.

Soulé, M., Estes, J., Miller, B., & Honnold, D. (2005). Strongly interacting species: Conservation policy, management, and ethics. *BioScience, 55*, 168–176.

Stem, C., Margoluis, R., Salafsky, N., & Brown, M. (2005). Monitoring and evaluation in conservation: A review of trends and approaches. *Conservation Biology, 19*, 295–309.

Strindberg, S., Didier, K., & The Living Landscapes Program. (2006). *A quick reference guide to the Landscape Species Selection software, Version 2.1 (Technical Manual 5)*. Bronx, NY: Wildlife Conservation Society, Living Landscapes Program. Retrieved February 13, 2010, from http://www.wcslivinglandscapes.org/landscapes/90119/bulletins/manuals.html

Tear, T. H., Kareiva, P., Angermeier, P. L., Comer, P., Czech, B., Kautz, R., et al. (2005). How much is enough? The recurrent problem of setting measurable objectives in conservation. *BioScience, 55*, 835–849.

Treves, A., Andriamampianina, L., Didier, K., Gibson, J., Plumptre, A., Wilkie, D., et al. (2006). A simple, cost-effective method for involving stakeholders in spatial assessments of threats to biodiversity. *Human Dimensions of Wildlife, 11*, 43–54.

The Nature Conservancy. (2000). Stakeholder analysis exercise: A quick process for identifying stakeholders and developing community outreach strategies. Retrieved February 1, 2009, from http://conserveonline.org/workspaces/cbdgateway/era/standards/supportmaterials/std2sm/StakeholderAnalysisExercise.pdf/download

The University of Queensland (n.d.). The C-Plan conservation planning system. Retrieved February 1, 2010, from The Ecology Centre Web site: http://www.uq.edu.au/ecology/index.html?page=101951

Wilkie, D., & The Living Landscapes Program. (2002). *Using conceptual models to set conservation priorities (Bulletin 5)*. Bronx, NY: Wildlife Conservation Society, Living Landscapes Program. Retrieved December 15, 2008, from http://www.wcslivinglandscapes.org/landscapes/bulletins.html

Wilkie, D., & The Living Landscapes Program. (2004a). *Participatory spatial assessment of human activities – A tool for conservation planning (Technical Manual 1)*. Bronx, NY: Wildlife Conservation Society, Living Landscapes Program. Retrieved February 13, 2010, from http://www.wcslivinglandscapes.org/landscapes/90119/bulletins/manuals.html

Wilkie, D., & The Living Landscapes Program. (2004b). *Creating conceptual models – A tool for thinking strategically (Technical Manual 2)*. Bronx, NY: Wildlife Conservation Society, Living Landscapes Program. Retrieved February 13, 2010, from http://www.wcslivinglandscapes.org/landscapes/90119/bulletins/manuals.html

Wilkie, D., & The Living Landscapes Program. (2006). *Measuring our effectiveness – A framework for monitoring (Technical Manual 3)*. Bronx, NY: Wildlife Conservation Society, Living Landscapes Program. Retrieved February 13, 2010, from http://www.wcslivinglandscapes.org/landscapes/90119/bulletins/manuals.html

Wilson, K., Pressey, R. L., Newton, A., Burgman, M., Possingham, H., & Weston, C. (2005). Measuring and incorporating vulnerability into conservation planning. *Environmental Management, 35*, 527–543.

Wilson, K. A., Underwood, E. C., Morrison, S. A., Klausmeyer, K. R., Murdoch, W. W., Reyers, B., et al. (2007). Conserving biodiversity efficiently: What to do, where, and when. *PLoS Biology, 5*, 1850–1861.

Chapter 18
Integrating Ecoregional Planning at Greater Spatial Scales

Mark Anderson

Abstract Nature functions at many spatial and temporal scales, but it is at the larger spatial scales where traditional conservation practice has seen the least progress in terms of both conceptual understanding and implementation. This chapter reviews philosophical and practical perspectives for integrating ecoregional planning across larger landscapes. I draw upon recent work of The Nature Conservancy in developing a conservation plan that merges separate ecoregional work conducted in the Northern, Central, and Southern Appalachians, and thus creating a landscape for conservation planning that encompasses most of Eastern North America. I argue that addressing regional scale questions requires scientists to assemble regional datasets that have both the detail and credibility to create understanding of ecological processes and pathways. Further, the challenge of large-scale conservation is giving rise to an expanding repertoire of tools for protecting land and water. Conservationists are discovering ways to maintain a broad matrix of natural cover using a mosaic of permanent or temporary ownerships and easements, combined with best management practices and traditional reserves. Accordingly, implementing a vision of the dynamic conservation of nature at regional scales will require a cooperative of players that spans the geographic area, partnerships that will endure over time, and data that will support the measuring and monitoring of large scale dynamics.

Keywords Connectivity • Ecological datasets • Land protection • Large-scale processes • Regional conservation

---

M. Anderson (✉)
The Nature Conservancy, Eastern Regional Office, 11 Avenue de Lafayette,
5th floor, Boston, MA 02111-1736
e-mail: manderson@tnc.org

S.C. Trombulak and R.F. Baldwin (eds.), *Landscape-scale Conservation Planning*,
DOI 10.1007/978-90-481-9575-6_18, © Springer Science+Business Media B.V. 2010

18.1 Introduction

A significant ecological insight that emerged into the forefront of conservation planning in the 1990s was that nature functions at many spatial and temporal scales, and that the conservation of nature requires each scale – independently and together – be considered and addressed (Poiani et al. 2000). However, the challenges to acting on this insight are huge, as scientists are only beginning to understand how nature functions across a few scales, to say nothing of all possible scales at the same time (Adams 2006). Consider, for example, the rapid development of landscape ecology from a discipline that just a few decades back was rooted in studies of one-meter plots (Kareiva and Andersen 1988). Because of the multiple spatial scales across which critical ecological processes and even some individual species operate, conservation planners must become adept at shifting their focus from individual organisms to populations to natural communities to ecoregions to continents – and vice versa – in order to determine the proper scale for conservation action. Although these shifts in perspective can be confusing, they are necessary. Many conservationists have lost their belief that an exclusive focus on small-scale conservation actions will address the numerous and pressing issues of these times and are looking toward a landscape-scale vision that will offer a renewed sense of what is possible. Ultimately, the practice of conservation planning needs a vision – and a science – that inspires conservation action because it has a realistic chance to succeed.

For conservation, all scales matter, but it is at the greater spatial scales where traditional conservation practice has seen the least progress in terms of both conceptual understanding and implementation. This lack of progress is not surprising, as landscape-scale conservation planning requires planners to think about complex networks of conservation areas as well as all the land and water in between them. It demands that disturbances that occur periodically over long time spans be understood and planned for. It challenges conservation planners to develop extensive databases to support decision making at scales beyond the scope of one's local knowledge and confidence. Lastly, it demands that planners work together to forge relationships and cooperatives across distinct geographies simply because no single person can have for an entire project the intimate on-the-ground knowledge that is required to both design and implement an effective conservation strategy.

Regional conservation planning aims to match the scale of land use with the scale of the resources needed to be sustained. Only large natural regions can encompass critical resources like the full ranges of species, the entire distribution of an ecosystem type, the watershed of a great river, or an iconic landscape like the Appalachian Mountains. The frustrating paradox is that while regional resources are far beyond the power of most municipal governments to conserve on their own (Chap. 4), the development decisions made by local governments often represent the greatest threat to these resources (Chap. 2). Fortunately, access to information critical to conservation planning at regional scales is materializing fast, and new tools for cross-jurisdictional collaboration and ecological analyses are emerging rapidly that will help enable the challenge of regional scale thinking to be met.

18 Integrating Ecoregional Planning at Greater Spatial Scales

The perspectives offered by the authors in the preceding chapters are a testament to the development of a landscape-scale approach to conservation planning in recent times. Much of the work they describe addresses the issues I just raised: ecoregional-scale conservation planning is required to achieve comprehensive conservation goals. The work of The Nature Conservancy/Nature Conservancy Canada (TNC/NCC), detailed in Anderson et al. (2006) is a good example of many of the approaches they describe and advocate. Yet despite the large spatial extent of ecoregions and the challenges they pose for effective planning, they themselves are not the ultimate scale at which conservation planning needs to occur. The planning work that is done *within* ecoregions ultimately needs to transcend ecoregional boundaries so that plans seamlessly merge across larger landscapes.

In this chapter, I aim to review briefly a number of philosophical and practical perspectives for integrating ecoregional planning across larger landscapes. For this discussion, I draw upon recent work of TNC in developing a conservation plan that merges separate ecoregional work conducted in the Northern, Central, and Southern Appalachians and thus creating a landscape for conservation planning that encompasses most of Eastern North America. While the ecological realities that influence the outcome of this planning effort may be specific to this broader region, the perspectives my colleagues and I have developed in order to achieve our goal of integrating ecoregional planning at greater scales is not. The challenges we face are common to any conservation planning initiative that seeks to weave separate initiatives together across a larger landscape.

18.2 Merging Ecoregional Plans into Larger Landscapes

The last two decades saw a tremendous growth in ecoregional-scale assessments where conservation organizations, from the Appalachian Mountain Club to the World Wildlife Fund, progressed from a relatively ad hoc decision-making process for prioritizing lands to identifying portfolios of critical areas within ecological regions. The use of a homogeneous ecoregion as a framework for planning allowed conservationists to design portfolios of conservation areas that represented the characteristic species and ecosystems in a specified amount and configuration that best encompassed the total biological diversity of the region (Groves 2003). This body of work stands as a huge step forward for conservation and will rightly continue to be the backbone of conservation planning for decades to come.

When viewed collectively, the results of the many ecoregional plans revealed two important truths. First, the maps of key conservation areas displayed, in a tangible way, the full scope of the conservation challenges faced today – and it was larger than many had imagined. Second, when placed side-by-side, the maps triggered the realization that the ecoregions themselves, even as large as they are on their own, connect to form larger patterns, and conservation planners need to think about processes and species that cross ecoregions (Fig. 18.1). However, many of the lessons learned from working at the ecoregional scale, detailed in the preceding

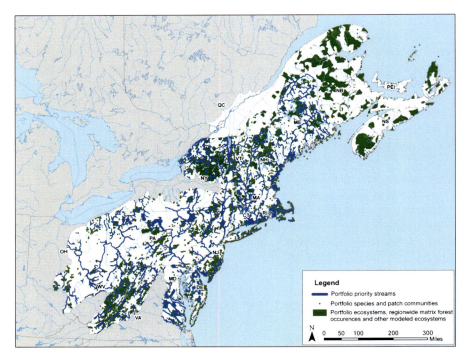

Fig. 18.1 An example of a large landscape vision. The Nature Conservancy and Nature Conservancy of Canada's portfolio of critical conservation areas for ecosystems, streams, and rare species for the Northeastern and Mid-Atlantic regions of the Appalachian Mountains. Note that the Canadian portion is incomplete, and does not yet include a stream assessment

chapters, remain true at greater scales and thus bear repeating. Working at greater spatial scales simply enlarges both their challenges and importance.

For better or (mostly) for worse, the spectre of global climate change has been the event that forced scientists to acknowledge the truly dynamic and spatially far-reaching nature of species distributions and community compositions (Chap. 15). The very ecoregions themselves only represent a particular moment in time, to which a look at maps of historic vegetation patterns will attest (Overpeck et al. 1992). Clusters of conservation areas can go a long way towards reversing the consequences of habitat fragmentation at finer spatial scales, but if ecological coherence is to be maintained at the landscape scale and the effective exchange of individuals and materials among sites for demographic, migratory, and ecological processes are to be restored, then the conceptualization and implementation of projects on even greater spatial and temporal scales is required. In short, the dynamic elements of nature must be provided for (Soulé and Terborgh 1999)

Conservation planning for a dynamic natural world may be thought of as a balance between areas with thriving breeding populations and functioning ecosystems (the core reserves; Chap. 14) and areas that connect those reserves and allow for movement and interchange (the connecting lands; Chap. 16). However, as several

18 Integrating Ecoregional Planning at Greater Spatial Scales

scientists have pointed out, connectivity is not just another goal of conservation; it is the natural state of things. Conservation scientists know far more about the deleterious effects of isolation and fragmentation than they do about how exchange happens in a wild landscape. Knowledge of what constitutes a barrier or corridor for various species is extremely limited, and models for dispersal of plants and animals through heterogeneous environments are not yet sufficient to be confident that implementation strategies based on them will be adequate for their task (Noss 1991). Regardless, it is strikingly clear that nature has always been a place of movement and interchange. The current distributions of organisms represent one point in a constantly changing mosaic. With the exception of certain islands, the majority of species present at any local site originally came from somewhere else; some may be a part of a large contiguous distribution, others disjunct remnants of a once continuous range, others refugees from past events, others recent colonists expanding their ranges, and a few may have evolved in situ. Thus biological diversity is a dynamic consequence of dispersal and isolation, combined with site conditions and local interactions (Cox and Moore 2000).

Consider the entire Appalachian Mountain chain, stretching through Eastern North America from Québec's high serpentine outcrops in the Northern Appalachians to the diverse cove forests of the Southern Appalachians in Tennessee and Georgia. The region comprises the core distribution of forest-dependent birds such as the Black-throated Blue Warbler (*Dendroica caerulescens*), dominant trees such as the sugar maple (*Acer sacharinum*), specialist mammals like the rock vole (*Microtus chrotorrhinus*), and declining amphibians such as the Jefferson salamander (*Ambystoma jeffersonianum*). Perhaps surprisingly, Virginia, in the Central Appalachians of the U.S., and New Brunswick, in the Northern Appalachians of Canada almost 1,500 km away, share 2,437 species, most of which are found all along the Appalachian Mountains (NatureServe Explorer 2008). Moreover, because temperate mixed forests are among the habitats projected to be most susceptible to climate change, future movement within this system is hypothesized to be dramatic and require high migration rates (Malcolm and Markham 2000).

Facilitating the migration of all species at all scales will require a strategy of optimizing the width and variety of natural habitats in linked landscapes to ensure that the full spectrum of native species can move throughout a landscape whose scope is measured in thousands of kilometers (Noss and Harris 1986). Essentially, this is a bet-hedging strategy, reflecting the paucity of our understanding of interchange within a fully connected landscape. Further, it is part of an emerging science, catalyzed by climate change concerns, that focuses on using models to make decisions under conditions of uncertainty, particularly with regards to landscape patterns, multi-species management, and the exact nature of the temporal changes being managed for (Burgman et al. 2005). In the Appalachian Mountains, the structure and location of potential landscape linkages are fairly clear (at least at the broad scale) and a number of new tools have been developed to quantity those patterns (Fig. 18.2; Chap. 16)

Most species, in addition to existing within a suitable temperature range, are constrained to the physical and chemical properties of the land or water itself.

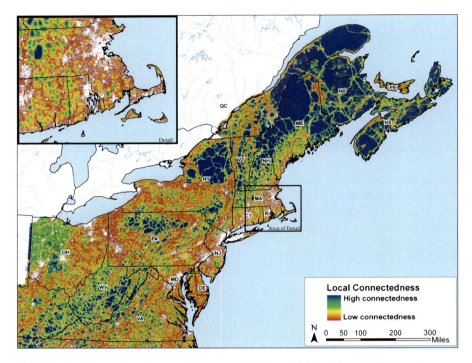

Fig. 18.2 The degree of local connectivity across the Eastern U.S. and Canada. High-scoring areas (*blue or green*) may offer the best places on which to focus initiatives for connectivity within and among ecoregions, but the map also highlights some substantial barriers along the Appalachian Mountain chain

Species that thrive in limestone environments, such as freshwater mussels or cave-dwelling species will ultimately need to find their way to other limestone settings, and the same will be true for coastal dunes, shale cliffs, high summits, and so on. Because of the close correlation between diversity and physical setting, many designs for conservation area networks already encompass the correct set of places for conservation action based on their physical setting, geology, remoteness, and intactness. The species and communities that currently inhabit these locations, however, will need to change over time to adapt to climatic changes. How will conservation planners facilitate such transitions?

Lastly, it must be acknowledged that the conditions that allow movement in a landscape may be quite different than those that allow and support breeding. Recognizing this, conservationists have vastly expanded their repertoire of tools for protecting land and water. Although strict reserves continue to form the substance of conservation area networks, it is now clearly understood that to succeed they need to be nested in a broader matrix of natural land cover. Maintaining this broader context requires a virtual quilt of permanent or temporary ownerships and easements, allowing for a range of uses but preventing conversion to development. Certified forestland with sustainable management practices, areas with use restrictions and

18 Integrating Ecoregional Planning at Greater Spatial Scales

enforced buffer zones created by policy revisions, and even commercially exploited lands where species-friendly management practices are enacted are all part of the configuration necessary to maintain diversity over the long term.

To account for this increased sophistication in land management and to understand how tracts of land with various levels of protection work together to create a larger conservation picture, older land classification systems have been revised to accommodate a wider range of purposes and activities. For example, expanding on the U.S. GAP analysis classification (Crist 2007), the Nature Conservancy's Conservation Management Status classification categorizes each tract of land that is secured against conversion to development by the (1) the intent of the owner, (2) the tenure of ownership, and (3) the potential for the owning entity to manage it effectively. Mapping this classification (Section 18.2.1) allows conservationists to distinguish land permanently reserved for the protection of biological diversity from land that is temporarily set aside for multiple purposes but that may facilitate species movement. In North America, the GAP and TNC classifications are roughly equivalent, but the latter system was designed for international use and it is helpful in understanding the extent of land that is theoretically intended for nature conservation but without the means to ensure its implementation (so-called 'paper parks').

18.2.1 Building the Science Foundation

Regional conservation planning has been called 'part vision and part science' (Soulé and Terborgh 1999) but to date it has been stronger on the vision and weaker on the science. This section focuses on the importance of developing a shared understanding of the region among a diverse set of state and international partners. Conflicting perspectives and partial knowledge can frustrate participants, as it becomes apparent in a conservation planning process that what is true in one's familiar portion of the region may not be true elsewhere. Thus, a first step, and a point of agreement in all landscape-scale conservation projects, is to establish a foundation of objective science information that all participants trust. Constructing a shared database of information for mapping ecological patterns and processes across large geographic areas literally establishes a common language that facilitates informed discussion. Moreover, it engages participants in a process of learning about the region that can be highly satisfying to the individuals.

It is useful to begin a large regional assessment using preexisting national or global datasets to map land use and land cover within the project region. This single macroscopic view of the world can establish the vital ecological and anthropogenic context for finer-scale studies of species, communities, and ecological processes (Scott et al. 1999). Eventually, however, if the planning process is to have credibility with local conservationists, the coordinators have to engage with the vast array of finer-scale datasets developed by state and provincial agencies, academic researchers, and NGOs. Scientists are notorious believers in their own data and confirmed skeptics of anyone else's information, and they are often particularly averse to large global datasets that lack the kind of detail that has long been the substance of ecology studied on a local scale. Therefore, the creation of an accurate and accepted regional database demands an investment to compile, understand, and incorporate data from high-quality, established local sources (Chap. 12). Furthermore, although joining and edge-matching local datasets, constructed in different ways and at different scales, is a time consuming process, ultimately this step seems to be a key to acceptance of the information and belief in the conclusions drawn from it.

Building Regional Datasets Detailed methods on constructing a regional database are beyond the scope of this chapter, but regardless of the purpose of the analyses or the sources of the data, the process must always begin with an understanding of the data content, the schema, the resolution, and the purpose for which each dataset was developed. Consider, for example, The Nature Conservancy's data layer of regional geology for the Appalachian Mountains, which was created from 17 separate state and provincial geologic agencies (Fig. 18.3). Collectively, these datasets originally recognized over 300 different geological categories. To create the map that combines these datasets, first the definition of each map category and the methods whereby each category was mapped were studied. Next, the hundreds of categories were collapsed into nine basic classes, and the state/province boundaries were carefully examined for areas of edge confusion, such as when a geologic formation disappears at a state border. Finally, all the maps were brought together with a single legend and

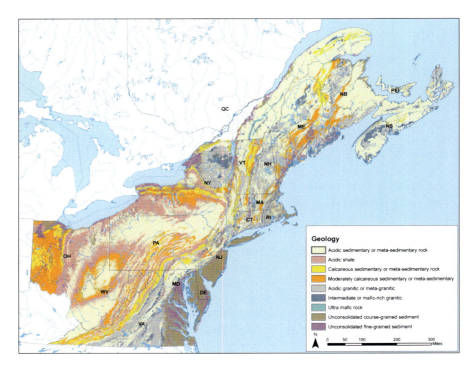

Fig. 18.3 Geology of the Eastern U.S. and Canada. This map derives from a searchable spatial database assembled by TNC/NCC from 17 state and provincial sources

18 Integrating Ecoregional Planning at Greater Spatial Scales 401

attribute table and dissolved into one GIS grid. The result was a database for the entire region that can be searched for specific geologic classes and formations and used to analyze any number of ecological patterns, especially the relationships between species distributions and underlying geology.

Maintaining Data Like the natural world, data about nature are also dynamic. Maintaining a regional dataset, especially one created from a variety of data sources, can be a complex process. Changes to the information need to be identified and tracked by the entities closest to the source (usually local area staff), entered into a spatial database with the necessary geometry and content attributes, and transferred to a central repository. There, the information is combined with inputs from other areas, processed, mapped, and checked for quality. Fortunately, the timing between revisions varies depending on the content of the dataset. For example, geologic data change slowly, and revisions to them happen approximately only every 25 years, primarily reflecting changes in sampling methodology. Land-cover maps have been updated about every 10 years as improvements to the technology develop or, as in the case of the Maritime Provinces of Canada, the province is remapped from orthophotos. The location of rare species are compiled and resurveyed much more frequently, as often as every year, requiring annual updates in order to keep them current and maximally useful for conservation planning. Quality checking complex multiple-source regional datasets is understandably tricky, and it is typically only done on datasets that are being actively used. An open, transparent process combined with periodic circulation and review by the data publishers can help to minimize errors, but mistakes are inevitable because it takes so many people to review the data thoroughly.

An example of a dataset that changes annually and the process developed to maintain it, is the TNC's regional 'Secured Lands' dataset. The dataset focuses on lands that are permanently secured from conversion to development (Fig. 18.4). In complete form, this large dataset contains over 150,000 individual tracts of conservation land and is maintained by TNC's Eastern Conservation Science office. However, its maintenance depends on annual inputs from 28 entities in 14 jurisdictions – two in each state and province. Annual revision to public lands are compiled by state or provincial agency staff, while changes to private lands, including activities by land trusts, universities, and conservation organizations, are compiled by state-based TNC staff or provincial-based NCC staff. New parcels of land, added to the dataset, need to be identified with respect to owner, interest types, management category, intent, tenure, management potential, GAP status, designation, and date of acquisition. Next, the state and provincial staff crosscheck the public and private revisions, and the final integrated datasets are compiled at the regional office and checked for errors or apparent differences in how a standard was applied across jurisdictions. The revised regional dataset is then circulated back to the states and provinces for quality checking before being accepted as the most recent Secured Lands dataset.

The final product allows users to assess objectively a variety of characteristics pertinent to the entire region: the extent of land conservation, the rate of new acquisition, what types of systems and species are being conserved, or what locations

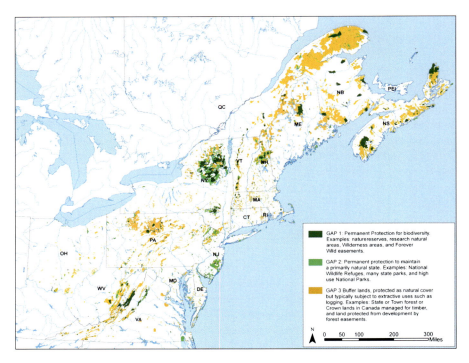

Fig. 18.4 Secured lands in Eastern North America. This map shows all lands that are permanently secured against conversion to development as of 2008. Green colors (GAP status 1 and 2) indicate that the primary intent of the securement is conservation of biological diversity. Yellow-brown (GAP status 3) indicates that the land is intended for multiple uses including forest management and recreation

should be prioritized for new conservation action. The excitement is in addressing regional-scale questions that go far beyond the capacity of individuals to know or comprehend without such a carefully maintained dataset. The dataset itself is posted annually for consumption by public agencies and other interested parties, and it is entered into a national collaborative (Protected Areas Database: PAD-US), which aims to create a matching, yet even greater-scale product for all of the U.S.

18.2.2 Fostering Partnerships

The design and management of regional networks of conservation areas by definition occurs at spatial and temporal scales that transcend the normal human life span and extend beyond traditional political boundaries. Accordingly, transforming such a vision into conservation action requires a cooperative team of partners that spans the geographic area and partnerships that will endure over time. The initial focus on developing the objective science information allows participants to maintain their own diverse organizational goals, and often the team that collaborates to create the information base is a good starting point for the formation of an enduring cooperative. More typically, however, the initial group expands to include others with the range of complementary skills needed to implement conservation action.

In reality, most conservation initiatives tend to revolve around specific projects with defined time lines and limited lifetimes. Long-term regional collaboration thus typically involves a sequential series of such projects, each building off of the work done before. In the case of the regional planning initiative for the Appalachian Mountain chain, TNC initially led a series of separate 2- to 3-year ecoregional assessments in the Northern Appalachians, Central Appalachians, Southern Appalachians, and High Alleghenies – each with its own independent team of scientists. Subsequent to the completion of those ecoregional assessments, we launched an effort to identify the most resilient examples of each ecosystem type within the combined region and the important areas for connectivity throughout the entire mountain chain. The latter project includes participants from Canada to Georgia, and this collaborative effort has proved enlightening as to the nuances involved in expanding conservation initiatives to encompass multiple ecoregions. For example, in the southern end of the region, the wood frog (*Rana sylvatica*), a species that reaches the southern limit of its range in the Southern Appalachians and exhibits breeding behavior there quite different than in the northern part of its range, is of conservation concern. A look at the species across the Appalachians is helping TNC/NCC put the concern into geographical context and better understand possible impacts from climate change.

A second insight that developed from a larger regional perspective was that although the intact Northern Appalachians and the diverse Southern Appalachian had national recognition and entire organizations devoted to their conservation, the critical Central Appalachian region was overshadowed and not distinct in the mind of the public. This remarkable ecoregion not only links the two ends of the Appalachian chain but is a global center of endemism itself. Although the 'hidden gem' approach may have kept the Central Appalachians intact in earlier times, the ecoregion has clearly been discovered by developers in the last decade, and amenities development is fast encroaching on almost every aspect of this remarkable ridge-and-valley landscape. The regional work has been effective in highlighting the Central Appalachians, which is now listed by the TNC's North America program as one of the most important landscapes for conservation on the continent.

18.2.3 Tracking Progress

One of the best ways for a group of collaborators to develop a shared vision for a region and a common goal is simply to track the progress of the group's initiatives – both separate and collective – in a manner that accounts for the work of all organizations and individuals. In Section 18.2.1, I briefly described the process TNC followed for creating and maintaining the secured lands database and noted

404 M. Anderson

the satisfaction conservation planners found in hearing about other projects similar to or complementary with their own within the larger region. The results of that collaborative allow TNC to map and measure the work of hundreds of organizations and to understand and clearly visualize how it adds up and generates progress towards achieving the regional vision (Fig. 18.5).

Periodic review or revisions of the datasets is a practical way to keep people and organizations involved in the process of conservation. Revisions to the secured lands dataset has proved especially fertile as a forum for individuals, organizations, and agencies to describe the on-the-ground conservation work they have accomplished in their portion of the region. The map derived from the dataset puts all the work into a spatial context, while the energy and commitment of the group tend to become recharged by hearing of projects outside of the day-to-day scope of any one participant but still within a shared region. Thus, the revisions provide a broader ecological and conservation relevance to all involved.

Combined with other regional spatial datasets on ecosystem types, species locations, landforms, and geology, we are beginning to reach the critical mass of information needed to measure and conceptualize how critical ecological processes flow across the ecoregions that comprise the Appalachian Mountain chain. Moreover, we can now start to rigorously identify important regional connections as they relate to known conservation features or secured lands and therefore allow us to focus future conservation action in the places that will make the most difference for biological diversity over time. While some of this work is being performed collaboratively by the Science Working Group of Two Countries, One Forest (Chap. 1), a number of independent researchers are exploring questions of their own using the datasets posted on the map service website (Chap. 12). This bodes well for enabling action and transferring knowledge to future generations of conservationists.

18.3 Lessons Learned

The natural world operates at many scales, and climate change is forcing scientists to pay attention to processes that happen across multiple ecoregions. The immense progress made over the last decade in designing conservation networks within entire ecoregions lays a strong foundation for planning at even greater scales. The challenge of maintaining connectivity among ecoregions has led to an increased sophistication in land and water conservation with a myriad of differing ownerships, easements, intents, and durations. Determining whether all this activity is having its intended result is not easy, but is possible. The largest challenges to conservation science at this scale are found in simply compiling and maintaining the information necessary to address key questions. Even within a large organization like TNC, the approach to building a store of data relevant across a large multi-ecoregional landscape has primarily been to aggregate datasets collected separately for subsets of the region, and will likely stay that way because of the value of this approach for encouraging long-term engagement by all of the people who need to

18 Integrating Ecoregional Planning at Greater Spatial Scales 405

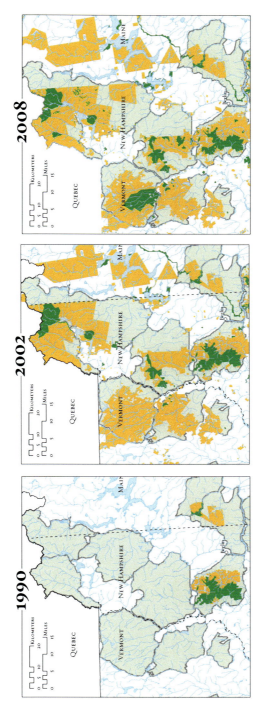

Fig. 18.5 Growth of the Secured Lands network in the Northeast United States. This representative area of Northern New England illustrates the substantial growth made in securing multiple-use lands (GAP status 3, light grey) intended for connectivity and buffers. Land secured primarily for conservation of biological diversity (GAP status 1 and 2, black) have also shown a moderate increase

be involved in a successful landscape-scale planning initiative. This highlights the critical importance of creating robust straightforward information systems that can be added to over time to fostering enduring partnerships.

Acknowledgments I would like to thank Rob Baldwin, Rodney Bartgis, Karen Beazley, Andy Finton, Graham Forbes, Louise Gratton, Nels Johnson, Bob Klein, Thomas Minney, Reed Noss, Conrad Reining, Steve Trombulak, and Barbara Vickery for insightful and ongoing arguments about connectivity; and Melissa Clark, Charles Ferree, Arlene Olivero, and Gillian Woolmer for sharing their understanding and mastery of regional datasets.

References

Adams, J. (2006). *The future of the wild: Radical conservation for a crowded world*. Boston, MA: Beacon Press.

Anderson, M. G., Vickery, B., Gorman, M., Gratton, L., Morrison, M., Maillet, J., et al. (2006). *The Northern Appalachian/Acadian ecoregion: Ecoregional assessment, conservation status and resource CD. The nature conservancy, eastern conservation science and the nature conservancy of canada: Atlantic and Quebec regions*. Retrieved January 31, 2010, from http://conserveonline.org/workspaces/ecs/napaj/nap

Burgman, M. A., Lindenmayer, D. B., & Elith, J. (2005). Managing landscapes for conservation under uncertainty. *Ecology, 86*, 2007–2017.

Crist, P. J. (2007). Mapping and categorizing land stewardship. In J. M. Scott (Ed.), *A handbook for GAP analysis* (pp. 119–133). Retrieved February 10, 2010, from ftp://ftp.gap.uidaho.edu/products/handbookpdf/CompleteHandbook.pdf

Cox, C. B., & Moore, P. D. (2000). *Biogeography: An ecological and evolutionary approach* (4th ed.). London: Blackwell.

Groves, C. (2003). *Drafting a conservation blueprint: A practitioner's guide to planning for biodiversity*. Washington, DC: Island Press.

Kareiva, P., & Andersen, M. (1988). Spatial aspects of species interactions: The wedding of models and experiments. In A. Hastings (Ed.), *Community ecology* (pp. 38–54). New York: Springer.

Malcolm, J. R., & Markham, A. (2000). *Global warming and terrestrial biodiversity decline*. Gland, Switzerland: World Wide Fund for Nature (WWF).

NatureServe Explorer. (2008). *NatureServe web services: Delivering biodiversity data to your desktop*. Retrieved February 19, 2010, from NatureServe Web site: http://services.natureserve.org/index.jsp

Noss, R. F. (1991). Landscape connectivity: Different functions at different scales. In W. E. Hudson (Ed.), *Landscape linkages and biodiversity* (pp. 27–39). Washington, DC: Island Press.

Noss, R. F., & Harris, L. D. (1986). Nodes, networks, and MUMs: Preserving diversity at all scales. *Environmental Management, 10*, 299–309.

Overpeck, J. T., Webb, R. S., & Webb, T., III. (1992). Mapping eastern North American vegetation change of the past 18 ka: No-analogs and the future. *Geology, 20*, 1071–1074.

Poiani, K. A., Richter, B. D., Anderson, M. G., & Richter, H. E. (2000). Biodiversity conservation at multiple scales: Functional sites, landscapes, and networks. *BioScience, 50*, 133–146.

Scott, J. M., Norse, E. A., Arita, H., Dobson, A., Estes, J. A., Foster, M., et al. (1999). The issue of scale in selecting and designing biological reserves. In M. E. Soulé & J. Terborgh (Eds.), *Continental conservation: Scientific foundations of regional reserve networks* (pp. 19–37). Washington, DC: Island Press.

Soulé, M. E., & Terborgh, J. (1999). The policy and science of regional conservation. In M. E. Soulé & J. Terborgh (Eds.), *Continental conservation: Scientific foundations of regional reserve networks* (pp. 1–17). Washington, DC: Island Press.

Index

A

Abies balsamea, 24, 332
Aboriginal people, 170
Acadia National Park, 74, 87
Accountability, 60, 62, 87, 207, 213
Acer rubrum, 26, 335
Acer saccharum, 338
Acidification, 103
Acid rain, 103, 287
Acipenser brevirostrum, 105
Acipenser oxyrhinchus, 105
Active listening, 60
Adaptive management, 56, 60–62, 251
Adelges tsugae, 24
Adelie Penguin, 145
Adirondack Mountains, 8, 68, 69, 101, 106,
    171–174, 181, 182, 243, 290, 361, 388
Adirondack park agency (APA), 389
Adirondack Park to Algonquin
    Provincial Park (A2A), 6
Adirondack State Park, 74, 360, 361
Aerial photograph, 258, 262
Africa, 2, 3, 283, 370, 384
Aggregate, 7, 17, 20–22, 27, 90, 144, 243,
    288, 293, 294, 360, 365, 404
Agricultural development, 21, 104, 109, 176,
    242, 260, 285
Agriculture, 3, 20, 21, 26, 100, 149, 174,
    177, 184–186, 196, 210, 212, 258,
    260, 271, 287
Airborne pollutant, 22, 287, 379, 388
Airborne pollution, 12
Air pollution, 12
Akaike information criterion (AIC), 356
Alaska, 130, 143, 149, 150, 154, 156, 327
Alasmidonta heterodon, 112
Alasmidonta varicosa, 112
Albatross, 142–145, 147, 149, 150, 152–154,
    157, 158

Alberta, 2, 3, 171, 183
Alca torda, 123
Alces alces, 168, 170–172, 333, 387
Alcids, 142–144
Aleutian Island chain, 147
Alewife, 106
Algal bloom, 328
Algonquin Provincial Park, 3, 6
Allagash River, 75
Allagash Valley, 41
Allagash Wilderness Waterway, 41, 74
Alosa aestivalis, 105
Alosa pseudoharengus, 106
Alosa sapidissima, 105
Altitudinal gradient, 25
Alytes, 24
Amazon basin, 20
Ambystoma, 217, 218, 221, 397
Ambystoma jeffersonianum, 397
Ambystoma laterale, 218, 221
Ambystoma maculatum, 221
Amenity development, 21, 22, 73, 91, 281,
    285, 289, 292–294, 296, 403
Amenity housing, 73, 282, 289, 293
American eel, 105
American marten, 194, 241, 307, 311, 389
American Procedures Act, 209 (Found as
    'Administrative Procedures Act')
American Revolution, 68
American shad, 105
American Three-toed Woodpecker, 385, 386
Ammodytes, 144
Amphibian, 19, 23, 218, 221–223, 228, 294,
    325, 330, 333–335, 338–341, 345, 353,
    359, 376, 397
Anadromous fish, 100, 105–107, 109
Anadromous Fish Restoration Act, 100
Anchovy, 144
Ancient Murrelet, 143

408   Index

Anguilla rostrata, 105
Angus King, 90
Annual precipitation, 328, 329
Anoxic condition, 328
Anthropogenic, 17, 20, 23, 28, 34, 108, 139, 141, 144, 148, 152, 155, 159, 191, 193, 223, 282–285, 288, 297, 378, 399
Anthropologist, 34
Anti-federalism, 41, 42
Antiquities Act, 74
Apiary damage, 177
Appalachian Mountain Club (AMC), 76, 79, 395
Appalachian Mountains, 6, 8, 76, 79, 124, 130, 243, 394–398, 400, 403, 404
Appalachian National Scenic Trail, 76
Aptenodytes patagonicus, 143
Aquatic, 20, 73, 92, 99–116, 287, 288, 310, 313, 352, 357
Aquila chrysaetos, 123
Aransas National Wildlife Refuge, 122
ArcGIS, 259, 360
Arctic Oscillation, 156
Arctic Tern, 140
Area requirement, 4, 6, 376
Area-weighted importance, 331, 335
Argentina, 370, 380
Army Corps of Engineers (ACOE), 219, 222, 224, 225
Ashmole's halo, 149
Asia, 147, 150, 156, 283, 370, 384
Assisted colonization, 343
Atlantic bluefin tuna, 140
Atlantic Canada, 140
Atlantic salmon, 103, 105, 106, 109
Atlantic sturgeon, 105, 106
Atlas, 148, 261, 264, 272–276, 334
Atmospheric deposition, 19, 20, 28
Attributes, 100–105, 143, 144, 216, 227, 263, 266, 268, 270, 276, 357, 401
Attrition, 20
Audobon's Shearwater, 148, 151
Audouin's Gulls, 153
Audubon Society, 132, 220–224, 226, 227
Australia, 2, 6
Autecology, 4
Average annual temperature, 327–329
Azores, 140

B
Bachman's Warbler, 121
Backpacking, 92
Bahamas, 151

Bald Eagle, 22
Balsam fir, 24, 332, 335, 337
Bangor Daily News, 39, 40
Bankable deal, 76
Barren Islands, 154
Base cation, 103
Base data, 260, 275
Batrachochytrium dendrobatidis, 24
Baxter State Park, 37–39, 41, 44, 74, 75, 81, 87, 360
Bay of Fundy, 149
Beaver, 105, 171
Bedrock, 111, 311
Behavioral shift, 176
Belize, 370
Berberis thunbergii, 23
Bering Sea, 147, 149
Bermuda Petrel, 142
Best development practices (BDP), 221, 224
Best-run, 241
Betula alleghaniensis, 335
Betula papyrifera, 332
Big A project, 45
Bigelow Range, 75
Bioaccumulate, 22, 287
Bioavailability, 153
Bioclimatic model, 330–333, 343, 344
Biofuels, 89, 90
Biogeochemical, 18, 28
Biogeochemistry, 287
Biogeographical province, 145
Biogeographic barrier, 23
Biogeography, 260
Biological community, 7
Biological diversity, 4, 17, 34, 53, 87, 107, 124, 145, 168, 206, 258, 282, 306, 326, 369, 395
Biological integrity, 4
Biological landscape, 377, 378, 380, 387
Biological organization, 4, 5, 376
Biological science, 9
Biomagnification, 152, 153
Biomagnify, 22
Biome, 7, 282, 284, 351
Bioproduct, 89
Biosentinel, 152
Bird, 23, 121, 139, 326, 353, 376
Bird conservation, 121–134
Bird conservation region (BCR), 125–128, 133
BirdLife International, 142, 147, 148, 158
Bison, 168
Bison bison, 168
Black basses, 105

Index 409

Black bear, 168, 170, 176–177, 181, 185, 191–193, 249, 352, 354, 355, 378, 380, 385, 388, 389
Black-capped Petrel, 142, 148, 152
Black cherry, 338
Black-footed Albatross, 153, 154
Black hickory, 332
Black-legged Kittiwake, 143, 145, 152
Black oak, 332
Black River Valley, 389
Black spruce, 332
Black-throated Blue Warbler, 123, 397
Black-throated Green Warbler, 338
Blanding's turtle, 218, 226
Blueback herring, 105
Blue-spotted salamander, 221, 226
Bobcat, 168, 182–185, 187, 192, 193, 355
Bobolink, 26
Body size, 191
Bogosl of Island, 147
Boise Cascade Mead, 88
Boobies, 142
Boreal, 12, 24, 25, 103, 122, 130, 172, 177, 183, 311, 385, 386, 388
Boreal forest, 103, 122, 130, 172, 183
Boreal toad, 24
Boundary, 3, 5–9, 13, 41, 55–59, 92, 108, 125, 127, 140, 141, 145–149, 152, 153, 158, 159, 189, 210, 225, 241–243, 246, 247, 258, 260, 264, 266–269, 288, 315, 316, 322, 327, 334, 344, 361, 370, 371, 375, 395, 400, 402
Boundary length modifier (BLM), 241, 315, 316
Bounty, 174, 176, 181, 184
Brachyramphus marmoratus, 143
Branch of Forestry, 41
Brazil, 20, 152, 154
Breeding Bird Survey (BBS), 332
Breeding range, 122, 148, 155
Brook floater, 112
Brookings institution, 73
Brook trout, 22, 114
Brown Pelican, 143, 145, 149
Buffer, 5, 111, 130, 228, 242, 288, 341, 399, 405
Buffer zone, 5, 228, 399
Bufo boreas, 24
Buldir Island, 147
Bullfrog, 23
Bureaucratic barrier, 61
Burnham Brook, 111
Burt's Bees, 81

C

CALFED Bay-Delta program, 207
California current system, 156, 157
Calonectris diomedea, 140
Campephilus imperialis, 121
Campephilus principalis, 121
Canada, 3, 20, 68, 108, 122, 140, 169, 238, 258, 284, 310, 332, 360, 395
Canada geographic information system (CGIS), 258
Canada lynx, 86, 182–183, 187, 189, 192, 194, 241, 311
Canada Warbler, 122, 123
Canadian land inventory (CLI), 258
Canadian Model Forest Network, 237
Canadian Wildlife Service, 132
Canary Islands, 140
Canis latrans, 168, 175–176, 357
Canis lupus, 68, 130, 168, 174–175, 241, 311, 352
Canis lycaon, 68, 130, 168, 174–175, 241, 311
Capacity, 58, 61, 81, 104, 111, 114, 115, 168, 172, 189, 212–214, 240, 247, 248, 259, 268–275, 321, 352, 372, 378, 402
Cape Breton, 170, 171, 182, 184
Carbon, 85, 90, 131, 133, 287, 297, 329, 354
Carbon dioxide (CO$_2$), 329, 330
Carbon sequestration, 85
Carbon storage, 131, 133
Caribbean, 6, 127, 133, 140, 148, 149, 159
Caribbean conservation corporation, 6
Caribou, 68, 130, 168–172, 175, 189, 190
Carnivore, 167–196, 288, 310–313, 354, 355
Carolina Parakeet, 121
Carya texana, 332
Case study, 6, 10, 35, 49, 59, 61, 74, 77, 81, 85, 108–115, 132–133, 217, 225, 326, 350, 384–389
Cassin's Auklet, 150, 156
Castor canadensis, 105, 171
Catskill state park, 2, 122
Cattails, 23
Cave-dwelling species, 398
Cellulosic ethanol, 90
Census, 3, 82, 260, 262, 266–268, 275, 292, 313, 322
Census block, 266, 267, 292
Center for International Earth Science Information Network (CIESIN), 284
Central America, 6, 8, 182
Central Appalachians, 397, 403
Central-place foraging, 155
Cephalopod, 144
Cepphus columba, 154

Cerorhinca monocerata, 154
Cerulean Warbler, 122, 131
Cervus elaphus, 168
2C1Forest Atlas, 273–275
Chainsaw, 37
Charadriiformes, 142
Charleston Bump, 151
Chatham Island Petrel, 142
Chemical stressor, 152
Chernobyl, 22
Chestnut blight, 24
Chile, 3
Chinese Crested Tern, 142
Chinstrap penguin, 145
Choristoneura fumiferana, 24
Christmas Island Frigatebird, 142
Chukchi Sea, 150
Chytridiomycosis, 24
Circuitscape, 359
Circulation model, 25, 328, 329, 334, 336, 339
Cistothorus platensis, 123
Citizen-science, 207, 221
City lights at night, 282
City of Bangor, 89
Civil unrest, 56
Clean Air Act, 210
Clean Water Act, 100, 109, 210
Clearcut, 36, 37
Clemmys guttata, 218
Clemmys insculpta, 218
Climate change, 5, 19, 54, 101, 130, 141, 178, 284, 326, 350, 396
Climate-change projection, 330, 333, 342
Climate-driven range shift, 326, 341
Climate-projection ensemble, 335
Climate refugia, 338, 342, 345
Climate space, 330, 332
Climatically suitable space, 335
Climatic condition, 186, 326, 330–332, 334, 342
Climatic gradient, 342
Climatic refugia, 332, 342
Climatic suitability, 326, 331, 332, 335, 343, 344
Climatic trend, 329
Clupeiformes, 144
Clutch, 143
CLUZ, 307
Coal-fired power plant, 22
Coalition, 40, 76, 92, 123–126, 130, 321
Coastal dune, 398
Coastal refugia, 25
Cohesion, 236, 252, 308, 309, 315
Collaboration, 13, 34, 59, 60, 114, 129, 158, 205–228, 275, 283, 364, 388, 389, 394, 403

Collaborative conservation planning, 206, 276
Collaborative organization, 59, 60
Collapse, 2, 74, 90, 91, 400
Colonization, 4, 105, 174, 176, 185, 287, 343, 350, 380
Colorado River, 104
Column mixing, 157
Co-management, 59, 388
Commander Islands, 147
Commercial fishery, 105, 144, 153, 157–159
Commercial-scale agriculture, 186
Commission for environmental cooperation (CEC), 11, 124, 125
Common Loon, 388, 389
Common Murre, 143, 156
Community-based forestry, 207
Compatibility, 258–260, 265–266, 272, 276
Complementarity, 306–308, 343
Complementary, 5, 304, 376, 403, 404
Complementary forms, 5
Complementary locations, 5
Complexity, 35, 38, 54–56, 58, 61, 62, 84–87, 126, 133, 159, 263, 288, 364, 365, 387, 390
Comprehensive Wildlife Conservation Strategy (CWCS), 388
Concept plan for the Moosehead Lake region, 78–84
Conceptual model, 305, 373–375
Conflict, 34, 36, 37, 39–40, 42, 45, 60, 68, 86, 87, 92, 107, 173, 174, 177, 182, 185, 187, 188, 206, 211, 212, 219, 249–252, 311, 370, 372, 388, 399
Conflict management, 60
Conflict resolution, 211, 251
Coniferous forest, 169, 183
Connecticut, 8, 100, 101, 105, 108–116, 181, 182
Connecticut River, 8, 101, 105, 108–116
Connectivity, 3, 19, 56, 103, 130, 192, 258, 285, 341, 349, 389, 397
Connectivity conservation, 5, 104, 130, 350, 389
Consensus, 60, 62, 113, 125, 206, 208, 209, 212–217, 220, 227, 228, 236, 247, 249–253
Consensus decision making, 206, 209, 217
Conservation action planning (CAP), 111–113
Conservation area, 5, 131, 188, 190, 191, 194, 307, 314, 317, 342, 381–383, 394–396, 398, 402
Conservation biology, 18, 92, 115, 275, 350
Conservation biology institute (CBI), 275
Conservation blueprint, 188

Index 411

Conservation easement, 36, 75–78, 82–86, 106, 265, 317
Conservation fund, 76, 78, 81, 132
Conservation goal, 3, 4, 7, 8, 13, 28, 35, 102, 131, 132, 169, 206, 225, 228, 239, 241, 243, 250, 252, 298, 304–309, 313–318, 320–322, 350–352, 369, 395
Conservation history, 34, 41
Conservation International, 375
Conservation landscape, 9, 25, 26, 28, 194, 195, 247, 283, 377–383, 387, 389
Conservation Management Status, 399
Conservation planning, 1, 18, 33, 54, 99, 123, 145, 168, 206, 235, 257, 281, 303, 325, 350, 369, 394
Conservation Planning Atlas, 272
Conservation Registry, 275
Conservation responsibility, 12
Conservation target, 9, 106, 112, 123, 188–190, 264, 305, 309, 311–312, 314, 373, 383
Conservation vision, 8, 48, 271
Contaminant, 142, 152–155, 159, 379, 388
Contaminant transport, 153, 159
Continental shelf, 144
Contopus cooperi, 122
Conuropsis carolinensis, 121
Convention on the Conservation of Antarctic Marine Living Resources (CCAMLR), 158
Converted landscape, 170
Co-occurring species, 188
Cook Inlet, 154
Coordinate system, 356
Core area, 5, 190–192, 194
Core conservation land, 196
Core-corridor-buffer, 5
Cormorant, 142, 144
Correlative bioclimatic model, 330, 331, 333, 334
Corridor, 5, 6, 12, 21, 37, 41, 75, 76, 78, 86, 92, 101, 192, 284, 285, 289, 304, 343, 350, 352, 354–357, 359, 361–364, 397
CorridorDesigner, 354, 364
Cory's Shearwater, 140
Cost, 48, 68, 76–78, 85, 87, 89–91, 105, 131, 133, 188, 206, 210, 211, 214, 224, 237, 258, 259, 261, 262, 273–275, 285, 306, 308, 312–315, 353, 356–357, 360, 361, 364, 377, 382–384, 387, 390
Cost surface, 285, 360, 361
Cougar, 168, 178–182, 185, 189, 191, 193, 352
Cougar Network, 181
County Rights Movement, 211

Coyote, 168, 170, 175–176, 183, 185, 187, 189, 357
C-Plan, 239, 307, 383
Crevice, 143
Critically imperiled, 311
Crop depredation, 177
Cross-country skiier, 172
Crosswalk, 314
Crude oil, 90
Cryphonectria parasitica, 24
Cuba, 151
Cultural history, 36
Cultural identity, 9
Cultural landscape, 2
Culturally defined boundary, 5
Cultural memory, 36, 43–45, 49, 50
Cultural reality, 57
Culture, 38, 43, 44, 48, 49, 61, 210
Current human footprint, 287–291, 295, 296
Current trends, 91, 293–294
Cycle, 62, 103, 106, 131, 140, 159, 287, 308, 378
Cygnus buccinator, 122

D

Dam, 41, 75, 114, 116
Database, 224, 238, 258, 260, 261, 264, 265, 271, 277, 284, 288, 394, 399–403
Data Basin, 275
DDT, 22, 23, 155
Deciduous forest, 23, 385
Decision making, 7–9, 48, 59, 60, 188, 206, 207, 209, 211–213, 217, 218, 226, 228, 229, 236–239, 241, 248–250, 371, 386, 387, 390, 394, 395
Decision-making process, 9, 59, 207, 209, 211, 217, 218, 226, 238, 250, 371, 395
Decision support, 236, 238–241, 253, 259, 309, 383, 394
Defenders of Wildlife, 275
Deforestation, 20, 102, 149, 173
Delphi process, 383
Demersal, 153
Demographic data, 189
Demographic sustainability, 189, 377
Dendroica caerulescens, 123, 397
Dendroica cerulea, 122
Dendroica kirtlandii, 122
Dendroica virens, 338
Development, 5, 21, 34, 54, 69, 101, 123, 148, 176, 206, 240, 258, 283, 310, 335, 373, 394
Development rights, 77

Diamond International Corporation, 70
Diamond Occidental, 45
Difference map, 294
Digital, 257, 259–263, 272
Digital elevation model (DEM), 262, 263
Diomedea empomophora, 143
Diomedea exulans, 143
Diomedea sanfordi, 143
Diomedeidae, 142, 157
Direct competition, 159, 176, 187
Disaggregate, 293
Disaggregated, 20–22, 27, 294
Disease, 12, 19, 24, 172, 341, 344
Dispersal, 5, 23, 149, 178, 193, 194, 330,
    335, 338, 340, 341, 343, 353,
    355–357, 360, 397
Dispositional ideals, 60
Dispute resolution, 206
Dissection, 20
Dissemination area (DA), 266, 267, 292
Dissolved oxygen, 109, 328
Distribution, 12, 25, 104, 130, 144, 168, 257,
    282, 304, 326, 377, 394
Disturbance regime, 102, 304, 344
DMTI Spatial, 268, 270
DNA analysis, 182
Dolichonyx oryzivorus, 26
Dominican Republic, 149, 152
Donor fatigue, 92
Downeast Coast, 74, 75
Downeast Lakes Forestry Partnership
    (DLFP), 78
Downeast Lakes Land Trust (DLLT), 78
Downeast Maine, 79, 242
Downscale, 329, 331, 333, 344
Ducks Unlimited, 84, 132
Dutch elm disease, 24
Dwarf wedgemussel, 112

E
Easement, 21, 36, 39, 48, 75–79, 81–87,
    91–93, 106, 131, 265, 304, 317, 321,
    393, 398, 404
Eastern Bering Sea, 147
Eastern brook trout, 114
Eastern Conservation Science Office, 401
Eastern gray wolf, 311
Eastern hemlock, 335–336
Eastern Meadowlark, 26
Eastern mole, 333–334
Eastern North America, 1, 6, 8, 11, 18, 24, 26,
    130, 167–196, 217, 218, 297, 314, 352,
    393, 395, 397, 402

Eastern spadefoot toad, 218
Eastern Tropical Atlantic, 140
Eastern white pine, 335–336
East Haddam, 111
Ecological character, 12, 310
Ecological connectivity, 3, 56, 349
Ecological envelope, 130
Ecological features, 276, 292, 304–312, 314,
    315, 318, 321, 351, 373
Ecological functionality, 189, 376, 377
Ecological functional role, 182, 195
Ecological Gifts Program, 84
Ecological integrity, 9, 100, 374
Ecological land unit, 241, 310
Ecological meaning, 3
Ecological pattern, 9, 377, 399, 401
Ecological process, 7, 17, 18, 20, 28, 67, 103,
    132, 168, 276, 354, 358, 376, 393, 394,
    396, 399, 404
Ecological reserve, 35, 238, 285, 307, 313,
    357, 360
Ecological stressor, 87
Ecological transition zone, 12
Ecological view, 4
Economic stressor, 87
Economic trend, 13, 56
Economy, 35, 37, 38, 40, 43, 71, 88, 90,
    186, 297
Ecoregion/ecoregional, 6–8, 10–12, 17,
    18, 20, 25–28, 33, 35, 43, 53–63, 68,
    69, 99–101, 103–104, 106, 121–134,
    145–149, 159, 167–170, 172–178, 181,
    183–196, 228, 235–253, 256–277, 281,
    284, 286–297, 303–305, 310–318, 321,
    326–329, 333–341, 344, 349–365, 389,
    393–406
Ecoregional assessment, 310, 334, 403
Ecoregional classification, 7
Ecoregional context, 62
Ecoregional health, 57
Ecoregional institution, 54, 57–63
Ecoregional resilience, 57
Ecoregional stakeholder, 55
Ecosystem, 2–4, 6, 7, 9, 10, 12, 17, 23–26, 28,
    34, 38, 42, 50, 54, 56, 68–70, 81, 85,
    87, 99–105, 108, 109, 112–114, 123,
    132, 139–141, 155, 157, 158, 173, 185,
    210, 218, 223, 260, 265, 287–289, 297,
    303, 305, 307, 310, 312–314, 326, 328,
    329, 342, 344, 351, 369–371, 374–376,
    394–396, 403, 404
Ecosystem health, 69, 81, 90, 173
Ecosystem management, 5, 53–56, 59, 206,
    211, 286

Index 413

Ecosystem services, 69, 85–86, 99, 100, 106, 107, 374
Ecozone, 7
Ectopistes migratorius, 121
Edge, 23, 109, 144, 179, 183, 267–268, 341, 357–358, 360, 362, 363, 400
Effective population size, 6, 350
Effluent, 287
Elanoides forficatus, 131
Electrical circuit, 358, 359
Electrical power infrastructure, 283
Electrical utility corridor, 284
Elevation, 130, 148, 170, 178, 182, 183, 260, 262, 263, 311, 331–333, 338, 342, 354, 385, 386
Elevational shift, 130
Elk, 4, 168, 175, 374
El Niño, 140, 155–157
Emergent, 23, 60, 62, 192, 271
Eminent domain, 42, 45
Emissions scenario, 25, 328, 331–336, 338–340
Empty habitat, 22
Emydoidea blandingii, 218
Endangered species, 124, 178, 207, 218, 221, 260, 265, 282, 303, 308, 310
Endangered Species Act (ESA), 4, 8, 23, 86, 100, 106, 172, 174, 206
Endangerment, 124, 126, 133, 186
Energy cost, 68, 89, 91
English language, 12, 236, 247
Environmental forcing, 142
Environmental gradient, 25, 345
Environmental history, 33–43, 50
Environmental Protection Agency (EPA), 7, 54, 219
Environmental Science Research Institute (ESRI), 259, 262, 263
Environmental threat, 12
Environmental variability, 155–157
Environment Canada, 84, 286
Erie Canal, 68
Erosion, 106, 109, 148, 287
Eschrichitius robustus, 140
Eskimo Curlew, 121–122
Essentialized images, 33, 36, 43, 45–47, 50
Estate taxes, 69
Eubranchipus, 218, 221
Euphagus carolinus, 122
Europe, 3, 130, 150, 168, 170, 172, 261, 265
European colonist, 184, 186
European colonization, 174, 287
European settlement, 100, 104, 105, 168, 169, 175, 288
European Union, 8

Even-aged management, 24
Evergreen, 356
Executive retreat, 73
Exocoetidae, 144
Exotic, 24, 104, 108, 159, 188, 284, 374
Exotic predator, 159
Expert
Expert judgment, 235–253
Expert opinion, 9, 247, 285, 289, 353, 379, 382, 390
Extinction, 4, 18, 24, 54, 122, 124, 186, 189, 287
Extinction-prone species, 188
Extinction rate, 4
Extirpation, 18, 23, 104–105, 173, 175, 182, 186, 193, 284, 287, 350
Extra-limital, 178
Exurban, 21, 22, 187, 193
Exurban sprawl, 195
Exxon Valdez, 154

F
Facilitated planning, 60
Fairy shrimp, 218, 221
Farm Bill, 76
Federal Advisory Committee Act (FACA), 209
Federal Employee Retirement Income Security Act, 71
Federal tax code, 71
Feeding frequency, 145
Feeds, 145
Fee simple acquisition, 76
Felid, 183–184
Fin and Feather Club, 47
Fire, 42, 68, 284, 287, 328, 329, 341, 344, 385
Firewood, 89
First Nation, 12, 248, 250
Fish, 42, 78, 100, 103–107, 109, 111, 113, 114, 132, 144, 153, 154, 156, 158, 220, 221, 328, 376, 388
Fisheries bycatch, 142, 144, 157, 158
Fishing, 44, 92, 153, 158
Flagship species, 355
Fledge, 143
Flood control, 103, 109
Floodplain, 101, 103, 107–109, 111–114, 310
Florida, 128, 132
Flow regime, 107, 116, 328
Flying fish, 144
Focal ecological features, 373
Focal species, 28, 168, 188, 190, 240–243, 260, 265, 311, 349–355, 359–361, 364, 388
Food web, 22, 102, 287
Foraging technique, 144

Forecast models, 291
Forest clearance, 185, 196
Forest cover, 109, 171, 176, 181, 186, 296, 352
Forest disease, 12
Forest industry, 70, 71, 84, 85, 89, 91, 92
Forest Inventory and Analysis (FIA), 331
Forestland, 22, 35, 39, 70, 77, 78, 81, 84, 85, 89, 92, 100, 131, 293, 398
Forest Legacy Program, 39, 76
Forest management, 49, 68, 71, 72, 77, 78, 85, 86, 88, 149, 209, 356, 379, 402
Forest ownership, 70–72, 77, 92
Forest Practices Act, 86
Forestry, 28, 37, 41, 42, 71, 76–79, 100, 207, 218, 221, 222, 224, 236, 258, 286, 293
Forestry Habitat Management Guidelines, 222, 224
Forest Society of Maine, 39, 76, 79
Forest Stewardship Council, 77, 224
Fourth Connecticut Lake, 108
Fragmentation, 19, 20, 53, 54, 56–57, 61, 63, 69, 84–87, 91, 92, 113, 186, 191, 193, 194, 210, 241, 242, 284, 294, 298, 308, 315, 316, 318, 353, 396, 397
Fratercula cirrhata, 154
Fraxinus americana, 336, 338
Freedom of Information Act, 209
Free-ranging hog, 174
Free-ranging sheep, 174
Fregata andrewsii, 142
Fregata magnificens, 143, 149
Fregata minor, 149
French language, 12, 236, 247
Freshwater, 99, 100, 106–108, 112, 113, 287
Freshwater mussel, 398
Frigatebird, 142–144, 149
Frontal eddy, 144
FunConn, 355, 360
Functional-action space, 58, 59
Functional connectivity, 192, 351–359, 361
Functional redundancy, 357
Funk Island, 151
Future condition, 9, 331
Future Human Footprint, 27, 288, 291–297

G
Gannets, 140, 142, 151
Gap Analysis Program, 258, 265, 286
GAP status, 289, 290, 297, 314–315, 318, 320–321, 334, 360–363, 401, 402, 405
Garbage raiding, 177
Gaspé Peninsula, 169–171, 182, 290
Gaspésie National Park, 170

Gau, 142
Gavia immer, 388
Gene flow, 87, 350–352, 358–359
General circulation model, 328, 329, 334, 336, 339
Genetic continuum, 175
Genetic impoverishment, 192
Gentoo Penguin, 145
GeoConnections, 262, 274
Geographic information system (GIS), 123, 236, 249, 257, 258, 262, 307
Geographic proxy, 292, 353, 356
Geolocator, 140
Geology, 260, 311, 343, 398, 400–401, 404
Geomorphology, 109, 114
GeoNova, 262
Geopolitical boundary, 55
Geoprocessing, 266
Georgia, 152, 397, 403
Georgia-Pacific, 78
Gifford Pinchot, 40
GIS clearinghouses, 261–263
GIS portal, 261–263
Glacially carved landscape, 12
Global average temperatures, 326–327
Global Human Footprint, 19, 27, 282–284, 288, 289
Global positioning system (GPS), 151
Global warming, 2, 311
Global weirding, 2
Glover's Reef Atoll, 370
Goal, 1–9, 47–49, 58, 60–62, 77, 102, 107, 108, 112–114, 123, 131–134, 147, 168, 169, 188, 194, 206, 208, 212, 228, 239, 241, 242, 255, 271, 303–309, 315–318, 320–322, 374
Golden Eagle, 123
Governance, 53–63, 68, 207–210, 217, 225, 259, 372
Governance fragmentation, 53, 54, 57
Governance structure, 58
Governmental policy, 56, 134
Governor's Task Force on Northern Forest Lands, 75
GRASS, 259
Grassland, 20, 175–176
Grassroots ecoregional conservation planning, 8
Gray whale, 140
Gray wolf, 311
Great Depression, 74, 91, 93
Greater Antilles, 148
Greater Northern Appalachians, 237, 240–242, 249, 389

Index 415

Greater Yellowstone Ecosystem, 4, 38
Great Frigatebird, 143, 149
Great Lakes, 128, 174, 178
Great Maine Forest Initiative, 93
Great Northern Paper Company, 75
Great Shearwater, 149
Great Smoky Mountains National Park, 40
Great War, 74
Green Mountains, 74, 294, 361
Grizzly bear, 4, 5, 130, 356
Ground burrow, 143
Group diplomacy, 60
Growth node, 22, 293
Grus americana, 122
Guanaco, 380
Guano, 155
Gulf of Alaska, 150, 154, 156
Gulls, 142–144, 153, 154
Gulo gulo, 168, 177–178
Gypsy moth, 24

H
Habitat, 17–23, 53–55, 86, 100, 102, 103,
    105–107, 122, 123, 126, 130–132,
    141–145, 147–149, 155–158, 171–186,
    189–195, 206–211, 217, 218, 221–224,
    226–228, 243, 249, 311, 329, 339, 350,
    352–356, 374–378, 385, 397
Habitat connectivity, 130, 350, 363
Habitat conservation plan, 8, 207
Habitat conversion, 194, 287
Habitat destruction, 19, 287
Habitat fragmentation, 210, 353, 396
Habitat generalist, 106, 190, 192, 356
Habitat linkages, 350
Habitat node, 357, 359, 360, 362
Habitat occupancy, 292, 355
Habitat patch, 7, 351–354, 358, 359
Habitat specialist, 192, 193
Habitat suitability, 285, 331–332, 339, 353,
    355–356, 360
Habitat suitability models, 249, 262, 264,
    356, 364
Habitat use, 182, 184, 352, 376, 385, 386
Habitat Workgroup of the New York-New
    Jersey Harbor Estuary Program, 216
Haiti, 149, 152
Haliaeetus leucocephalus, 22
Hancock Timber Resources Group, 71
Hardware, 259, 274–276, 365
Harvest control, 177, 186
Harvest regulation, 171
Hawaii, 148

Hawaiian Islands, 148, 153, 154
Headwater, 100, 103, 108, 111, 314
Heavy metals, 25, 287
Hedonic, 21, 292
Hemlock wooly adelgid, 24
Henry David Thoreau, 26
Herbicide, 37, 357
Herring, 105, 144
Heterogeneity, 376, 385, 386
Heterogeneous, 7, 25, 27, 190, 288, 385, 397
Heterogeneous mosaic, 7
Hexachlorobenzene, 155
Hexagon, 313
Hierarchical organization, 7
Hierarchical structure, 57, 208
High Alleghenies, 403
Highways, 72, 172, 175, 243, 270, 292, 356
Historical reality, 57
Historical trajectory, 9, 169
Historical trend, 36, 168–185
History, 2, 9, 13, 23, 33–50, 77, 78, 104, 108,
    111–113, 142–145, 153, 168, 173–175,
    185, 188, 195, 210, 211, 221, 250, 251,
    271, 288, 321, 352–354, 356, 357
Home heating oil, 89
Home range, 177, 191, 193, 194, 352, 353
Homestead Act, 68
Horizontal current, 155–156
Horizontal fragmentation, 57
Hostile environment, 195
Hotspot, 25, 124, 249, 353
Housing, 21, 73, 91, 187, 266, 282, 287, 292,
    295, 297, 298
Housing density, 73, 282, 284, 292
Human access, 281, 283, 289
Human attitudes, 167, 181, 195, 196
Human barrier, 175
Human-bear conflict, 388
Human conflict, 173, 177, 179, 180
Human dimension, 9
Human Footprint, 27, 185, 187, 189, 191, 193,
    281–298, 334, 349, 360, 362, 364
Human influence, 19, 22, 69, 73, 181,
    282–284, 288–290, 298, 364, 369, 370
Human Influence Index, 288
Human infrastructure, 20, 21, 186, 187, 193,
    295, 296, 356
Human intrusion, 177
Human Landscape, 179, 180, 191, 287, 371,
    377–380, 387
Human land use, 54, 260, 283, 285, 289, 290
Human-mediated stressor, 168
Human population, 3–4, 100, 195, 260, 266
Human population size, 281, 285, 292

416   Index

Human settlement, 20, 27, 173–177, 184, 186,
    190, 283, 288, 289, 292, 293, 296, 297
Human transformation, 4, 5, 12, 281, 283,
    284, 288, 290, 298
Hunt club, 173
Hunter, 172, 173, 378
Hunting, 2, 22, 26, 40, 41, 44, 92, 122, 132,
    170–173, 175–177, 181, 182, 186, 374,
    379, 385
Hutton's Shearwater, 148
Hybrid, 252
Hybridization, 183, 184
Hydrocarbons, 154, 287
Hydrological change, 101
Hydrology, 260, 298, 328, 341, 344, 354
Hydropower, 36, 89

I

Ice sheet, 104, 109, 329
Identity, 9, 34, 43, 44, 48–50, 182
Illinois, 131
Immigration, 12, 294
Imperial Woodpecker, 121
Imperiled, 311
Implementation, 1, 8, 9, 53, 54, 62, 99, 116,
    123, 125, 214–216, 222, 240, 248, 251,
    275, 309, 318, 383, 386, 394, 396, 397
Important Bird Area, 131, 134
Impoundment, 103
Indian midden, 176
Indian Township, 78
Indirect competition, 176
Industrial growth, 186
Informal partnership, 61, 62
Infrastructure, 20, 21, 109, 186, 187, 193, 195,
    281, 283, 295–298, 356
Inland Kaikoura Mountains, 148
Insect, 12, 24, 130, 330, 354, 385
Institutional amnesia, 62
Interdisciplinary, 9, 28, 50, 159
Intergovernmental institution, 58, 61
Intergovernmental Panel on Climate Change
    (IPCC), 331, 333
International Appalachian Trail, 304
International Organization of Standards, 263
International Union for Conservation of
    Nature (IUCN), 54, 265, 314
Interspecific relationship, 167, 196
Intervention, 188, 194, 195, 373, 375, 381,
    383, 386
Invasion, 18, 23, 28, 104–105, 147, 284
Invasive species, 23–24, 26, 73, 101,
    104, 105, 287

Invertebrate, 144, 154, 191, 376
Irreplaceability, 241, 243, 244, 250, 303–322
Irreplaceability analysis, 55, 243, 309, 321
Irreplaceable site, 12
IUCN protected area status, 265, 314
Ivory-billed Woodpecker, 121

J

Japan, 147, 149, 150
Japanese barberry, 23
Javelina, 354
Jefferson salamander, 397
John Baldacci, 93
John D. Rockefeller, Jr., 74
John McPhee, 21
John Muir, 40
Joint Ventures, 121, 124, 132–133
Jurisdictional boundary, 56, 58, 260
Jym St. Pierre, 38

K

Kamchatka Shelf, 147
Katahdin National Park, 41
Kaua'i, 148
Keystone species, 19, 105
Keystone threat, 17–28, 291
King George Island, 145
King Penguin, 143, 145
Kirtland's Warbler, 122
Krill, 144
Kudzu, 23
Kure Atoll, 154

L

Lake, 12, 21, 44, 73, 75, 78–86, 100, 103, 108,
    122, 128, 174, 178, 287, 293, 294, 355
Lakefront development, 78
Lake Hitchcock, 109
Lakeshore development, 285, 293
Lakeside camp, 73
Lama guanicoe, 380
Lampsilis cariosa, 112
Land and Water Conservation Fund, 76
Landbird, 125, 126, 132, 148
Land Conservation Movement, 74–84
Land cover, 17–24, 26–28, 102, 105, 123, 174,
    260, 263, 264, 268, 271, 281, 283, 284,
    286–289, 296, 314, 315, 331, 353, 356,
    360, 374, 398, 399, 401
Landform, 88, 102, 342, 404
Land for Maine's Future, 76, 83

Index 417

Land Information Ontario, 262
Landowner, 21, 22, 28, 35, 39–42, 77, 79, 84–87, 90, 92, 205, 206, 210–212, 222, 226, 227, 272, 294, 296, 298, 355
Landownership, 68, 69, 72, 77, 101, 238, 288, 293
Landsat, 263
Landscape, 1–10, 12–13, 20, 21, 25, 28, 34–43, 45, 49, 53, 67–70, 77, 78, 84–88, 92, 102, 103, 111, 123, 130, 167–170, 174, 176, 177, 179, 180, 184, 187, 190–196, 210, 218, 225, 236, 247, 249, 250, 260, 261, 265, 282–285, 287, 288, 290, 293, 294, 296–298, 307–309, 311, 343–344, 350–358, 377–387, 389, 394–398, 403, 404
Landscape alteration, 176
Landscape barrier, 351
Landscape boundaries, 7, 371
Landscape connectivity, 296, 343, 357
Landscape context, 7, 191
Landscape ecology, 9, 67, 87, 394
Landscape network, 353, 357–363
Landscape permeability, 356, 364
Landscape resistance, 349, 355–358, 360
Landscape-scale, 4–6, 50, 101, 102, 106, 116, 122, 142, 145, 205, 206, 236, 321, 322, 351, 370, 375, 394, 395, 399, 406
Landscape-scale conservation planning, 1–13, 27, 276, 308, 344, 369, 370, 387, 390, 394
Landscape-scale perspective, 4
Landscape Species, 370, 371, 374–378, 382–389
Landscape Species Approach, 369–390
Landscape transformation, 3, 281, 283, 285, 360
Land trust, 76, 78, 81, 82, 84, 111, 122, 132, 225, 276, 351, 401
Land Trust Alliance, 132
Land use, 8, 9, 12, 19–22, 24–26, 33–35, 40, 44, 54–56, 68, 74, 77, 81, 85, 92, 101–103, 111, 133, 186, 192, 205, 206, 211, 213, 214, 226, 236, 258, 260, 263, 283, 287–290, 355, 389, 394, 399, 405
Land-use change, 27, 28, 73, 167, 185, 193, 196, 258, 288, 293, 296, 297, 330, 351
Land-use designation, 195
Land-use economics, 9, 293
Land-use history, 9, 34, 37
Land use/land cover, 17, 19–24, 26–28, 266, 271, 286, 289, 296, 353, 356, 360
Land-use planning, 131–133, 188, 194, 196, 205–228, 273
Land Use Regulatory Commission (LURC), 37, 38, 79

Land-use transformation, 9, 285
Large-bodied mammal, 168, 190, 195
Large carnivore, 167–196, 354, 355
Large woody debris, 90, 102–103
Laridae, 142
Larus audouinii, 153
Larus michahellis, 154
Last of the Wild, 282, 327, 334, 360–363
Latent risk, 124
Latin America, 370, 384
Latitudinal gradient, 288, 343
Laysan Albatross, 153, 154
Lead, 388
Least-cost path, 353, 356–358
Least Tern, 143, 150
Legitimacy, 207, 213, 236, 239, 248, 251, 252
Leisure lot development, 91
Leopold, A., 3, 4, 25
Licensing, 257, 259, 263, 274
Life history, 121, 140, 142, 188, 195, 210, 221, 353, 354, 356, 357
Life history trait, 142–145, 174–175, 177
Life span, 177, 273, 275, 402
Limestone, 398
Linear sprawl, 85
Liquid biofuels, 90
Litter size, 177
Little Ice Age, 170, 173, 182, 184, 187
Little Tern, 150
Livestock depredation, 177
Living Landscapes Program (LLP), 371–377, 379, 381, 383–387
Local involvement, 61
Localized threat, 53–54
Local participation, 9
Locked in, 313, 318–320
Locked out, 313, 318, 319
Log drive, 37
Logging, 26, 39, 40, 44, 68, 79, 90, 110, 172, 176, 186
Logging road, 296, 356
Logistic regression, 356
Logit, 292
Long Island Sound, 108
Longleaf pine, 332
Longline fishing, 158
Louisiana, 131
Lower New England/Northern Piedmont, 100
Lymantria dispar, 24
Lynx, 180, 185, 249, 311, 364
Lynx canadensis, 86, 168, 182, 241, 311, 355
Lynx rufus, 168, 183–184, 355
Lythrum salicaria, 23

418 Index

M

Magnificent Frigatebird, 143, 149
Maine, 8, 20, 35, 68, 100, 132, 169, 208, 241, 259, 292, 304, 326, 352
Maine Audubon Society, 220–224, 226, 227
Maine Department of Environmental Protection, 219
Maine Department of Inland Fisheries and Wildlife (MDIFW), 218, 220, 221, 226–228
Maine Forest Biodiversity Project, 304
Maine Forest Service (MFS), 220, 224
Maine Guides, 38, 778
Maine Legislature, 41, 70
Maine Natural Areas Program (MNAP), 220
Maine's North Woods, 35, 37–39, 41, 42, 44, 49, 50, 69, 78
Maine State Planning Office, 219
Maine Times, 37
Maine Wildlife Society, 47
Maine Woods National Park and Preserve, 36, 38, 46
Mammal, 4, 19, 26, 130, 168, 169, 173, 175, 176, 179, 181, 184–196, 288, 307, 333–335, 338–341, 345, 351, 354, 376, 397
Mammalian conservation, 192
Mammalian species, 4
Man and Biosphere Program, 5
Mandatory Shoreland Zoning Act, 86
MapInfo, 259
Maple-beech-birch forest type, 332
MapServer, 274
Marbled Murrelet, 143, 149
Marine, 20, 100, 103, 107, 139, 141, 142, 144, 145, 147–149, 152, 153, 155, 158, 159, 384
Marine ecosystem, 108, 140, 156, 157, 287
Marine environment, 139, 142, 145, 149, 154, 155
Maritime provinces, 68, 90, 171, 401
Market hunting, 122, 173, 186
Martes americana, 194, 307, 355, 389
MARXAN, 237, 238, 241, 252, 303, 307–310, 312, 313, 315, 321, 383
Massachusetts, 68, 77, 100, 111, 114, 171–174, 181
Mathematical model, 28
Matrix forest blocks, 241, 242, 314
Matrix lands, 343
Maximum Entropy, 379
Mead Westvaco, 88
Mediation, 59

Mediterranean, 140
Megafauna, 310, 387
MegaFlyover, 283
Megatransect Project, 283
Menhaden, 144
Meningeal worm, 170–172
Mercury, 20, 22, 25, 152, 153, 155, 388
Mesic, 104, 183
Metadata, 263, 271, 274, 277
Mexico, 6, 11, 124, 125, 127, 133, 134, 145
Michael Kellett, 38
Microtus chrotorrhinus, 397
Midden, 173, 176
Midwestern U.S., 22, 131, 181
Midwife toad, 24
Migration, 87, 116, 122, 130, 140, 141, 149, 150, 156, 297, 353, 356, 357, 397
Mill, 70, 71, 89
Mill dam, 103
Millinocket, 40, 41, 47
Minimum spanning tree, 362, 363
Minimum viable population (MVP), 189, 191, 386
Mining, 20, 28, 260, 287
Mississippi, 131
Mississippi Alluvial Valley, 131
Mississippi River, 69
Model, 5, 9, 25, 49, 114, 123, 133, 209, 210, 238, 249, 262, 265, 272, 276, 283, 292–294, 296, 330, 331, 333, 334, 343, 355–357, 360, 361, 364, 369–390
Model averaging, 331
Model Forest of Newfoundland and Labrador, 284
Model selection, 356
Model specification, 356
Mole salamander, 217
Molokai, 148
Monetization, 85
Monitoring, 58, 61, 62, 87, 114, 192, 194, 195, 288, 306, 375, 382–384, 386–388
Montreal, 177, 294
Moose, 168, 170–172, 174, 175, 181, 185, 189, 191–193, 333, 334, 387–389
Moosehead Forest Conservation Project, 79, 80
Moosehead Lake, 21, 78–84
Mortality, 20, 122, 140, 147, 157–158, 172, 173, 175, 177, 182–184, 187, 191, 193, 356
Morus bassanus, 140
Mountain lion, 6, 178
Movement corridor, 6
Mt. Katahdin, 37, 75, 79

Index

Multiple jurisdictions, 8, 58, 140, 259, 261, 264
Multiple spatial scale, 8, 107, 140, 150, 394
Multi-scale approach, 54
Municipalities, 83, 86, 111, 259, 260

N

Naked islands, 154
Narrative, 2, 13, 34, 36, 50, 210, 212
National Conservation Easement Database, 265
National Environmental Policy Act (NEPA), 209
National Fish and Wildlife Foundation, 78
National Forest Management Act (NFMA), 209
National Forest System, 74, 100
National Geographic Society, 283, 284
National Land Cover Dataset (NLCD), 263, 268
National park, 2–4, 35–48, 74, 122, 170, 264, 288, 308, 350, 360
National Park Service, 4, 41, 46, 74, 288
National Wild and Scenic River, 74
National Wildlife Federation, 78
National Wildlife Refuge system, 122
National Zoological Park, 283
Natural Heritage Programs, 265
Natural history, 13, 142
Natural landscape, 2, 190, 282, 375
Natural Resources Council of Maine, 79
Natural Resources Protection Act (NRPA), 86, 218–221, 226
Nature Conservancy Canada (NCC), 11, 84, 241, 310, 311, 314, 395, 400, 401, 403
NatureServe, 265
Nature Trust of New Brunswick, 84
Neotropical, 125, 126
Net primary productivity, 282
New Brunswick, 10, 68, 84, 90, 169, 171, 173, 174, 177, 178, 181–183, 241, 242, 259, 290, 326, 397
Newell's Shearwater, 148
New England, 74–84, 88, 89, 93, 100, 101, 106, 108–111, 113, 114, 124, 171, 172, 174, 177, 221, 223, 224, 238, 405
New England Forestry Foundation (NEFF), 76–78
Newfoundland, 151, 156, 175, 284
New Hampshire, 10, 68, 69, 74, 100, 111, 169, 171, 178, 181, 182, 193, 224, 242, 259, 326, 332
New York, 2, 10, 68–70, 122, 169, 171–173, 175, 176, 181, 241, 259, 326, 328, 332, 355, 371, 384, 388, 389
New York State Department of Environmental Conservation (NYSDEC), 388
New Zealand, 2, 140, 141, 148

Nitrogen deposition, 284
No-analog assemblage, 326
Node, 21, 22, 282, 293, 296, 298, 357, 359–363
Non-dominating, 60
Non-game species, 129
Non-governmental institution
Non-government organization (NGO), 12, 54, 58, 59, 74, 76–84, 111, 129, 158, 237, 241, 245, 264, 285
Non-native, 23, 104
North America, 2, 3, 6, 8, 10, 11, 18, 22–26, 42, 68, 69, 89, 103, 105, 106, 122, 124, 125, 127, 130, 132, 133, 149, 153, 167–196, 217, 218, 258, 261, 264, 275, 287, 293, 297, 326, 334, 355, 370, 384, 399, 403
North American Bird Conservation Initiative (NABCI), 124, 127, 133
North American Environmental Atlas, 261, 264
North American Waterfowl Management Plan, 124, 133
North Atlantic Oscillation, 140, 156
North Central Lakes region, 294
Northeastern North America, 6, 168, 176, 181, 183–187, 190, 217, 218
Northeastern U.S., 10, 26, 42, 68, 90, 101–109, 114, 124–126, 178, 218, 332, 337
Northern Appalachian/Acadian ecoregion, 8, 18, 35, 68, 101, 123, 168, 236, 259, 284, 305, 326, 350, 389
Northern Appalachian Mountains, 6, 243
Northern Appalachians, 6, 20, 22, 24, 25, 140, 171, 172, 174, 178, 181, 183, 187, 237, 240–242, 249, 284, 397, 403
Northern boreal forest, 183
Northern California current, 156, 157
Northern Forest, 37, 39, 68–76, 84, 85, 87, 88, 91–93, 304
Northern Forest Alliance (NFA), 39, 304
Northern Forest Canoe Trail (NFCT), 304
Northern Forest Forum, 37
Northern Forest Lands Council (NFLC), 70, 75
Northern Forest Lands Study (NFLS), 75
Northern Gannet, 140, 151
Northern Hemisphere, 22, 105, 140, 141, 293, 343
Northern New England, 100, 124, 174, 290, 405
Northern Rocky Mountains, 6
Northern Royal Albatross, 143
Northern Vermont, 68, 169
North Maine Woods, 35, 38, 46
North-South biological corridor, 12
Northwestern Mexico, 6
Northwest Territory, 3

420 Index

North Woods of Maine, 35, 37–39, 41, 42, 44, 49, 50
Nova Forest Alliance (NFA), 237–240
Nova Scotia, 10, 84, 170–172, 174, 178, 182, 192, 237–242, 259, 262, 290, 326, 361
Nova Scotia Nature Trust, 84
Nuisance bear, 177
Nulhegan River, 111
Numenius borealis, 121
Nutrient, 25, 102, 113, 144, 155, 354
Nutrient cycle, 287

O

Oahu, 154
Oak-hickory forest, 307, 308
Occupancy analysis, 191
Occupancy modeling, 292
Oceanites oceanicus, 142
Ocean wanderer, 157, 159
Ohio Farmland Preservation Planning Program, 213
Oil spill, 141, 153, 154
Olfaction, 144
Oligotrophic, 144
Olive-sided Flycatcher, 122
Ontario, 2, 175, 178
Open-water barrier, 175
Ophiostoma ulmii, 24
Opportunistic decision, 9
Optimized decision, 9
Ordinal, 285, 356
Organochlorinc, 22, 153
Organizational affiliation, 61
Organizational capacity, 61
Orthophoto, 262, 401
Osmeridae, 144
Osprey, 22
Outcomes, 38, 39, 56, 60, 91, 112, 129, 188, 207–217, 222–225, 227, 237, 238, 248, 251, 310, 382, 383, 388–390, 395
Output, 207, 208, 214–217, 220–222, 228, 239–242, 266–268, 271–276, 285, 293, 322, 351, 356, 390
Outreach, 9, 220–222, 228, 259, 275–276
Overbrowsing, 173
Overexploit, 18
Overexploitation, 18, 20, 176, 178, 287, 288
Overhunting, 170, 171
Ownership fragmentation, 69, 84–87
Ownership pattern, 68, 72, 75, 238
Ozone, 19, 22, 287

P

Pacific Decadal Oscillation, 140, 156
Pacific Northwest, 6, 143, 149, 291, 293, 294, 297
Pacific Ocean, 140, 149, 156
Palustrine, 23
Pandion haliaetus, 22
Paper, 35, 37, 39, 42, 61, 68, 70, 71, 89, 247, 258, 272, 370, 374, 399
Paper birch, 332
Paper plantation, 37
Paralastrongylus tenius, 170
Parcelization, 72, 78, 297
Park boundary, 3, 41, 288, 370
Parks, 2, 34, 87, 122, 170, 264, 282, 304, 338, 350, 369, 399
Parks Canada, 264
Particulates, 287
Partnership, 11, 34, 50, 59, 61, 62, 78, 114, 132, 206–209, 211, 213–216, 223, 225, 227, 238, 274, 297, 402–403, 406
Partners in Flight (PIF), 123–125, 131–133
Paseo Pantera, 6
Passamaquoddy Tribe, 78
Passenger Pigeon, 121
Patagonia, 370
Pelagic, 140, 142, 143, 145, 149, 151–153
Pelecanus occidentalis, 143
Pelicans, 142–145, 149
Penalty, 312–314
Penguins, 142–145
Pennsylvania, 124, 171, 178, 182, 332
Penobscot River, 37, 39, 75
Per capita energy consumption, 4
Percival Baxter, 41, 74, 87
Perforation, 20
Permanently secured lands, 334
Permanent wetlands, 23
Petromyzon marinus, 106
pH, 109
Phaethon lepturus, 148
Phenological change, 344
Phenology, 329
Philopatric, 147
Phoebastria albatrus, 147
Phoebastria immutabilis, 153
Phoebastria nigripes, 153
Phragmites, 23
Phragmites australis, 23
Phytoplankton, 145
Picea mariana, 332
Picea rubens, 25, 337
Picoides dorsalis, 385
Pigeon Guillemot, 154

Index 421

Pilgrims, 68
Pinch point, 192, 359
Pine Barrens, 21
Pingree Associates, 77
Pingree, D., 77
Pingree Easement, 77–78, 81, 92
Pinus palustris, 332
Pinus strobus, 335
Piranga rubra, 338
Pittman-Robertson Wildlife
    Restoration Act, 173
Planning, 5, 8–9, 12, 19, 39, 49, 55, 59–61,
    113, 116, 123, 126, 127, 129, 168, 186,
    188–194, 207–209, 211, 212, 217, 236,
    241, 243, 246, 251, 253, 258, 259, 261,
    268, 274, 276, 304, 305, 312, 316, 364,
    365, 372, 375, 386, 389, 395, 399, 404
Planning decision, 8, 188, 210
Planning theory, 9
Planning unit, 125, 126, 241, 243, 244, 246,
    248–250, 252, 313–318
Plant, 5, 7, 10, 22–25, 89, 90, 104, 109, 113,
    123, 130, 132, 191, 221, 284, 287, 288,
    329, 330, 343, 351, 355, 359, 376, 397
Plasma biochemistry, 154
Pleistocene, 287
Plum Creek Timber Company, 78
Plymouth, 68
Policy, 9, 19, 23, 41–43, 75, 76, 122,
    132–134, 157, 159, 206, 207, 209,
    211, 213, 215–217, 220, 221, 224,
    226, 377, 388, 399
Policy outputs, 207
Pollutant, 22–23, 282, 287, 378
Polycentric ecoregional institution, 61
Polycentric institution, 58, 59, 61, 62
Polychlorinated biphenyl (PCB), 22
Polycyclic aromatic hydrocarbon, 154
Polygon, 264, 266, 313
Pool, 100, 110, 217, 218, 221, 222, 227
Pool-breeding amphibian, 19, 23, 218,
    221–223, 294, 359
Population decline, 172, 174, 175, 181, 184,
    192, 193
Population density, 285, 289, 292
Population viability, 19
Population viability analyses (PVA), 123
Populus tremuloides, 338
Post-progressive Era, 81
Potential habitat, 175, 332, 339
Precipitation, 22, 25, 111, 327–329
Predator, 4, 5, 22, 23, 103, 147, 148, 155, 159,
    170, 173–176, 187, 190, 195, 284, 287,
    344, 351

Predator control, 170
Preservation, 3, 40–42, 55, 216
Prestige, 154
Pribilof Islands, 147
Primary roads, 292
Prince Edward Island, 10, 177, 178, 182, 184,
    259, 326
Prince William Sound, 154
Priority, 93, 116, 123, 124, 126, 131–134, 240,
    298, 304–310, 313, 315, 316, 318, 321,
    354, 361, 372, 373, 383
Private land, 37, 40, 74, 81, 82, 84, 93, 122,
    134, 205–228, 236, 293, 298,
    307, 401
Private ownership, 68, 93
Private property, 206, 211, 228
Procellariiformes, 142, 151
Process-based, 107, 108, 116
Progressive Era, 74, 81
Property rights, 206, 211
Protected area, 2–6, 22, 24, 47, 86, 122, 123,
    133, 195, 237–240, 260, 261, 264, 265,
    274, 289, 297, 314, 315, 318, 320, 321,
    326, 338, 341, 343–345, 350, 360–362,
    364, 370, 381
Protected area networks, 6, 341, 344, 345
Protected Areas Database of the U.S. (PAD),
    261, 264, 402
Protocol, 87, 195, 252, 288
Provisioning rate, 145
Proxies, 215, 292, 296, 353, 356
Proximity, 70, 143, 149, 308, 315, 353
Prunus serotina, 338
Pterodroma cahow, 142
Pterodroma hasitata, 142
Pterodroma magentae, 142
Ptychoramphus aleuticus, 150
Public education, 9, 220
Public-private partnerships, 206
Public roads, 20, 69, 292, 294
Public sector, 75–76, 93
Pueraria spp., 23
Puffinus gravis, 149
Puffinus griseus, 140
Puffinus huttoni, 148
Puffinus lherminieri, 148
Puffinus newelli, 148
Pulp, 35, 37, 68, 71
Puma concolor, 168, 178–182, 352
Purple loosestrife, 23
Pursuit-diving seabird, 157
Pygoscelis adeliae, 145
Pygoscelis antarctica, 145
Pygoscelis papua, 145

Q

Quaking aspen, 338
Quantum GIS, 259
Québec, 10, 108, 169–172, 175, 178, 181,
182, 241, 242, 246, 247, 251, 259,
294, 326, 397
Quercus velutina, 332
Quincy Library Group, 212

R

Radiation, 19, 22
Radio collar, 172, 189
Rail, 79, 266, 289
Rain-fed wetland, 328
Ramsar Designation, 111
Rana catesbeiana, 23
Rana sylvatica, 217, 221, 352, 403
Random forest, 331–333
Range, 7, 18, 34, 53, 77, 100, 122, 142, 168,
212, 236, 273, 283, 309, 325, 376, 394
Range contraction, 288
Range expansion, 23, 176, 181, 332
Range shift, 12, 25, 187, 194, 325–345
Range-shift projection, 342–343
Rangifer tarandus, 68, 130, 168–170
Raptor, 22, 152
Raster, 260, 265, 267, 268, 276
Raver, A., 2
Razorbill, 123
Real Estate Investment Trust (REIT), 21, 71,
79, 85, 293
Reciprocity, 60, 212, 216
Recolonization, 105, 350, 380
Recovery, 74, 100, 102, 103, 105, 109, 168,
169, 171, 172, 174, 175, 181, 185, 186,
188, 189, 191, 192, 196, 261, 380–382
Recreation, 2, 23, 28, 36, 37, 44, 49, 69, 70,
77, 78, 258, 379, 385, 402
Recreational development, 72, 73
Red-legged Kittiwake, 147
Red List, 158
Red maple, 26, 335
Red Rock Lake, 122
Red spruce, 25, 337
Reduced snowpack, 328
Reference levels, 189
Reforestation, 26, 102
Refuges, 2, 122, 304, 343, 397
Refugia, 25, 103, 130, 131, 332, 338, 342, 345
Regionally relevant, 10
Regional planning, 59, 403
Regional reserve network, 351, 363
Registered Maine Guide, 78

Regulatory framework, 86
Reintroduction, 105, 174, 183, 189
Remote sensing, 243
Replaceability, 309, 320
Reproductive capacity, 168
Reptile, 376
Reserve network, 145, 325–345, 350, 351,
357, 361
Residential development, 101, 109, 292, 389
Residential growth, 73
Resilient, 23, 34, 48, 50, 56, 68, 92, 131, 212,
253, 342, 403
Resolution, 20, 27, 93, 206, 211–213, 243,
251, 260–261, 265, 276, 283, 284, 286,
288, 292, 310, 313, 331, 333, 344, 356,
360, 362, 365, 400
Resort, 21, 22, 79, 124, 295, 296
Resource availability, 61, 90
Resource management, 57, 206–208, 210,
211, 228, 251, 306
Restoration, 23, 104–105, 107, 108, 113, 116,
131, 173, 177, 189, 209
RESTORE: the North Woods, 36, 38, 45
Rhinoceros Auklet, 154
Rhode Island, 100, 124
Ridgetop, 297
Ringed boghaunter dragonfly, 2188
Riparian, 23, 101–107, 109, 111, 112, 304,
351, 352
Rissa brevirostris, 147
Rissa tridactyla, 143
River, 8, 37, 69, 100, 169, 243, 260, 287, 304,
356, 389, 394
River connectivity, 113
River morphology, 103
Road, 3, 20, 37, 68, 174, 260, 282, 343,
351, 379
Road density, 39, 174, 192, 193, 282, 283
Roads network, 289
Roaring Twenties, 74
Rock vole, 397
Rocky Mountain National Park, 24
Rocky Mountains, 6, 130
Roosevelt National Park, 41
Round Pond, 75
Roxanne Quimby, 44, 81
Royal National Park, 2
Royal Tern, 149
RTE species, 310–313
Runoff, 102, 104, 109, 111, 114, 115, 210, 217
Rural land, 21, 73, 91
Russia, 149, 156
Rusty Blackbird, 122, 123
Rynchops spp., 144

Index

S

Sagebrush Rebellion, 211
Salmonids, 105, 107
Salmo salar, 103
Salvelinus fontinalis, 114
Sand eel, 144
Sandwich tern, 147
San Guillermo, 380
Sardine, 144
Sarek National Park, 3
Satellite imagery, 258, 262, 263, 296
Satellite tag, 150, 151
Save the Tiger Fund, 283
Saybrook, 108
Scale, 1, 18, 34, 54, 69, 99, 121, 140, 181, 206, 236, 258, 283, 304, 338, 351, 370, 394
Scale of operation, 61
Scalopus aquaticus, 333
Scaphiopus holbrookii, 218
Scenario, 25, 27, 91, 178, 193, 194, 210, 235, 236, 239–242, 248, 251, 252, 261, 291–298, 311, 312, 317–319, 321, 328, 329, 331–340, 343, 349, 360–363
Scoping, 389
Seabird, 139–159
Seabird ecology, 139–159
Sea lamprey, 106
Sea level, 2, 328
Sea-level rise, 157, 329
Seamount, 144
Seaward Mountains, 148
Secondary road, 292
Second home, 72, 73, 289, 293
Secured Lands, 327, 401–405
Sedge Wren, 123
Self-identity, 48
Semipermanent, 23
Sensitivity analysis, 322
Sequoia National Park, 2, 122
Services-based, 107–108
Shale cliff, 398
Shifting mosaic, 350, 357
Shorebird, 132
Shorebird Conservation Plan, 123–125, 132, 133
Shorelines, 73, 77, 78, 154, 293, 297
Shortnose sturgeon, 105, 106
Short-tailed albatross, 147, 149, 150
Siberia, 108
Sierra Club, 6, 39
Sierra Nevada Mountains, 130
Significant vernal pool (SVP), 86, 218, 219, 221–224, 226, 227
Significant wildlife habitat, 218

Significant Wildlife Habitat Rules, 218, 221, 227
Silviculture, 26, 71
Simulated annealing, 308, 312, 315
Simulation, 308, 309, 315, 316, 318, 320, 328, 329, 336, 339, 344
Single-site focus, 3
Sink habitat, 6
Sir James Goldsmith, 70
Site selection, 123, 235–237, 241, 242
Ski area, 21, 285, 293
Skidder, 37
Skimmer, 142–144
Skua, 142, 144
Sky Islands, 6
Smelt, 144
Snow accumulation, 173
Snow depth, 184, 187
Snowfall, 183
Snowmobile, 40, 44
Snowmobiler, 172
Snowshoe hare, 183
Social capital, 59–61, 92, 212, 216, 223
Social carrying capacity, 172
Social geography, 25
Social landscape, 12
Social norm, 60
Social outcome, 207, 216, 222–225
Social reality, 49
Social science, 10, 34, 253
Social stressor, 87
Societal attitudes, 186
Socio-ecological perturbation, 54
Socio-ecological system, 54, 56, 58
Socio-economic condition, 67–93
Socio-political boundary, 58
Software, 28, 236, 237, 258, 259, 273–276, 307, 309, 321, 354, 355, 359, 361, 365, 376, 383–386, 388
Soil, 42, 68, 102, 108, 109, 111, 147, 258, 287, 308, 313, 331, 342, 343
Solution, 2, 34, 36, 38, 48, 54, 71, 122, 133, 158, 212, 213, 223, 227, 228, 236, 241, 248, 249, 252, 272, 274, 307–309, 311–321, 326
Songbird, 140, 141, 143, 152
Sooty shearwater, 140, 141, 154
Source habitat, 241, 311
South Africa, 65, 200, 324
South America, 3, 122, 127, 133, 140, 148, 181, 333
South Atlantic, 140, 149
South Carolina, 145, 147
Southeastern Canada, 10, 26

Southeast Ireland, 151
Southeast Scotland, 151
Southern Appalachians, 38, 40, 395, 397, 403
Southern Hemisphere, 68, 149
Southern Maine, 37, 183, 184
Southern Québec, 10, 172, 175, 181
Southern Royal Albatross, 143
Southern temperate forest, 12
Southwestern U.S., 6, 104
Space-demanding animal, 189
Spatial, 5, 19, 34, 54, 107, 124, 140, 208, 236,
        257, 282, 304, 331, 350, 370, 393
Spatial data, 257–262, 264–266, 268, 273,
        288, 400, 401, 404
Spatially explicit measures, 12
Spatially explicit population viability, 192
Spatial scale, 5, 8, 9, 13, 19, 21, 26, 34, 48, 54,
        107, 124, 140, 142, 144, 150, 152, 155,
        208, 225, 228, 249, 275, 283, 313, 331,
        332, 338, 353, 356, 357, 370, 389,
        393–406
Species-area relationship, 191
Species distribution, 26, 114, 262, 265, 326,
        329–333, 337, 342–344, 381, 396, 401
Species extirpation, 12, 284
Species interaction, 186
Species invasion, 18, 23, 28, 284
Species of Greatest Conservation Need
        (SGCN), 388
Species-specific habitat association, 185
Spednic Lake, 78
Sphenisciformes, 142
Spotted Owl, 6
Spotted salamander, 218, 221, 226
Spotted turtle, 218
Spring denning, 178
Spruce budworm, 24, 37, 69
Spruce-fir forest type, 111
Squid, 144
SRES A2, 328, 329, 333–335, 340
SRES A1B, 333
SRES A1FI, 331, 336, 339
SRES B1, 331, 333, 336, 339
Stakeholder, 8, 9, 13, 28, 55, 58, 59, 62, 87,
        113, 206–209, 211–214, 216–218, 220,
        222, 223, 225, 227, 228, 236–240, 244,
        245, 248, 249, 251, 253, 258, 259, 277,
        298, 305, 306, 311, 312, 318, 355, 365,
        371–372, 386–388, 390
Stakeholder fatigue, 62
State Programmatic General Permits
        (PGP's), 224
State Wildlife Grants program, 388
Statistics Canada, 3, 73, 266

Stepping stone, 343
Sterna bernsteini, 142
Sterna maxima, 147
Sterna paradisaea, 140
Sterna sandvicensis, 147
Sternidae, 142
Sternula albifrons, 150
Sternula antillarum, 143
Stewardship, 5, 71, 124, 220, 228
St. Lawrence Valley, 169, 172, 175
Stream, 42, 71, 75, 78, 101–103, 106, 107,
        110, 113, 114, 156, 328, 396
Strix occidentalis, 6
Structural connectivity, 192, 352, 361
Sturnella magna, 26
Subadult, 175, 177, 181
Subgraph, 361
Suburban-industrial corridor, 21
Suburbanization, 176, 186
Successional recovery, 186
Sugar maple, 338, 397
Summed-run, 241
Summed-summed-run, 241
Summer Tanager, 338
Summit, 307, 310, 398
Sunkhaze Meadows National
        Wildlife Refuge, 76
Surface water, 100, 102, 144, 158, 210, 217, 284
Surrogate, 188, 306, 310, 341, 352, 360
Sustainability, 9, 63, 189, 223, 377
Sustainable forest management, 77
Sustainable Forestry Initiative, 77, 224
Sustained-yield timber management, 68
Swallow-tailed Kite, 131
Sweden, 3
Sympatric species, 187
Synanthropic, 284
Synthliboramphus antiquus, 143
Systematic conservation planning, 9, 237, 243,
        303–322, 342, 370

T
Tactile, 144
Takings, 8, 9, 46, 206, 218, 227, 247, 292,
        343, 370
Tamarisk, 23
Tamarix, 23
Target, 9, 19, 34, 73, 102, 123, 168, 222, 238,
        257, 305, 370
Tauyskaya Bay, 156
Taxonomic confusion, 175
Tax Reform Act, 71
Tayassu tajacu, 354

Index 425

Technocratic approach, 55, 210
Temporal, 9, 27, 28, 145, 148, 236, 285, 350, 354, 397
Temporal scale, 13, 19, 34, 48, 54, 140, 144, 145, 154, 155, 351, 357, 394, 396, 402
Tennessee, 397
Tern, 140
Terrestrial, 19, 20, 92, 101–103, 140, 142, 144, 145, 147, 152, 155, 156, 158, 159, 176, 218, 222, 264, 287–289, 310, 313, 328, 357, 384
The Nature Conservancy (TNC), 6, 7, 11, 39, 54, 68, 76, 78, 79, 83, 101, 108–116, 123, 132, 241, 283, 286, 306, 307, 310, 318, 334, 361, 372, 375, 389, 395, 396, 399, 400
Theodore Roosevelt, 74
Thermal expansion, 329
Thick-billed Murre, 150
Threat, 3, 12, 13, 17–28, 36, 38, 41, 53–55, 58, 62, 63, 75, 100–102, 109, 111, 112, 116, 122, 126, 129, 133, 141, 145, 147, 148, 152, 168, 172, 174, 176, 180, 186, 188, 190, 193, 194, 218, 221, 241, 243, 257, 258, 265, 282–285, 287, 288, 291–298, 310, 311, 326, 351, 355, 370–373, 375–380, 382, 383, 385–387, 390, 394
Threat Landscape, 377–379
Threat vector, 28
Threshold, 114, 172, 190, 193–194, 196, 285, 357
Thunnus thynnus, 140
Tier 1, 241, 310, 314, 315, 318, 320, 321
Tier 2, 241
Tiger Conservation Landscapes, 283
TIGER/Line Files, 266, 268
Timber harvesting, 22, 37, 40, 87, 90, 103, 131, 149, 287, 296, 357
Timber Havesting Standards to Substantially Eliminate Liquidation Harvesting, 86
Timber industry, 6, 40
Timber investment management organization (TIMO), 78, 85
Timberland, 21, 41, 70–72, 77–79, 88
Timber sale, 72
Timber wolf, 130, 175
Tongoriro National Park, 2
Top-down approach, 55, 56
Top-down regulation, 207, 208, 311
Topography, 109, 168, 182, 192, 263, 311, 331, 342, 343, 353, 356
Torishima Island, 149
Tourism, 42, 46, 49, 79

Town planner, 19, 225
Township, 39, 70, 78, 175, 182, 313
Toxin, 23, 25, 26, 28, 122, 296
Transboundary, 8, 58, 139–159, 217, 260–262, 265–269, 271, 274, 276, 288, 292, 362
Transboundary threat, 53, 54
Transition probability, 292
Transplantation, 175
Transportation, 186, 287, 289, 356
Transportation network, 20, 23–25, 260
Trophic web, 152
Troposphere, 22
Trumpeter Swan, 122
Tsuga canadensis, 335
Tufted Puffin, 154, 157
Tug Hill Plateau, 389
Tundra, 177
Turbidity, 144, 328
Twenty-first century, 2, 49, 55, 68, 159, 326
Two Countries, One Forest (2C1Forest), 12, 26, 243, 244, 247, 249, 250, 258, 272–276, 288, 305, 310, 317, 321, 322, 334, 404
Typha, 23
Typhoon LLC, 78

U
Umbrella species, 106, 351, 354
UNESCO, 5
Ungulate, 167–196, 288
University of Maine, 220, 221, 223, 224
University of Massachusetts, 114
University of Vermont, 76
Unroaded area, 191
Upper Midwest, 291, 293, 294, 297
Upper St. Croix River, 78
Upwelling zone, 144
Urban area, 20, 22, 42, 91, 260, 289, 293, 294, 297, 313, 314, 318
Urbanization, 101, 109, 148, 149
Urban sprawl, 282
Uria aalge, 143
Uria lomvia, 150
Ursus americanus, 168, 176–177, 249, 352, 380
Ursus arctos, 4, 130, 356
U.S., 3–5, 8, 12, 20, 21, 23, 38, 40, 54, 72–74, 77–79, 81, 84, 91, 100, 103, 123–125, 127, 129–131, 134, 140, 156, 169, 173, 174, 207, 210, 211, 218, 240, 242, 259, 261, 262, 264–271, 284, 287, 291–293, 313, 314, 322, 326, 331, 332, 334, 335, 360, 397, 402
U.S. Census Bureau, 3, 262, 266
U.S. Federal Geographic Data Committee, 263

426 Index

U.S. Fish and Wildlife Service (USFWS), 78, 109, 113, 132, 219, 220, 388
U.S. Forest Service, 7, 39, 54, 73, 74, 76, 113, 114
U.S. Geological Survey, 113, 114
Utilitarian, 40, 43
UV-B radiation, 19

V

Vector, 28, 260, 265, 267, 268, 276
Vector Map Level 0, 261
Vegetation dynamics, 329
Vehicle collision, 172, 183
Vermivora bachmanii, 121
Vermont, 10, 69, 74, 100, 111, 169, 171, 181, 182, 224, 241, 242, 259, 294, 326, 355
Vernacular conservation, 36, 43, 47–48
Vernal pool, 86, 208, 217–228
Vernal Pool Working Group (VPWG), 208, 217–228
Vertical fragmentation, 56, 57
Vertical partnership, 61
Very Important Pool Program (VIP Program), 221–223
Viable effective population size, 6
View lots, 73
Vincente Perez Rosales National Park, 3
Virginia, 124, 397
Vulnerability, 12, 104, 147, 168, 186, 188, 243, 360, 376, 378
Vulnerable, 18, 19, 103, 111, 293, 296, 297, 311, 385

W

Wabanaki, 68
Wagner Forest Management, 78
Walk, 68, 266
Wall Street, 71
WalMart, 78
Wandering Albatross, 143, 152
Washington County, 78
Waterbird, 132, 134
Waterbird Conservation for the Americas, 123–126, 132, 133
Water flow, 103, 287
Waterfowl, 122, 132, 143, 258
Water pollution, 12, 109
Water Pollution Control Act Amendment, 4
Watershed, 8, 42, 73, 88, 101, 102, 104–106, 108–111, 113–116, 207, 247, 260, 289, 313, 394
Weeks Act, 41, 74

Western Hemisphere, 333
Western Indian Oceans, 140
Western Maine, 39, 74, 76, 77
Western Pacific garbage patch, 154
Western U.S., 174, 175, 284, 329
West Indian Breeding Seabird Atlas, 148
West Virginia, 124
Wetland, 23, 24, 78, 86, 100, 109, 111, 112, 122, 131, 132, 171, 207, 208, 210, 211, 214, 217–220, 222, 224, 298, 310, 328, 329, 351, 353, 356, 385
Wetland Conservation Plan, 219
White ash, 338
White Mountains, 74, 181, 361
White-red-jack pine forest type, 332
White-tailed Tropicbird, 148
Whooping Crane, 122
Wild and Scenic River System, 41
Wilderness, 3, 21, 36, 38, 41, 42, 69, 74, 76, 79, 108, 174, 178, 191, 193, 236, 294, 304, 307
Wilderness Act, 42
Wilderness area, 191, 304, 307
Wilderness Society, 36, 38, 76
Wildlands Network, 237, 240–242, 247, 249, 310, 311
Wildlands Project, 6, 310
Wildlands recreation, 23
Wildlife, 2, 4, 6, 20, 22–24, 26, 68, 74, 76, 77, 86, 90, 92, 122, 130–133, 140, 148, 150, 159, 173, 177, 184, 186–189, 193–195, 206, 207, 210, 211, 218, 221, 226, 227, 241, 258, 275, 285, 287, 293, 294, 297, 298, 307, 356, 370, 375, 377, 383, 388, 389
Wildlife Conservation Society (WCS), 6, 283–285, 334, 361, 370, 372, 375, 376, 388, 389
Wildlife ecology, 140, 159
William Cronon, 43
Williamsonia lintneri, 218
Wilsonia canadensis, 122
Wilson's Storm-petrel, 142
Wintering grounds, 122, 141
Wise Use Movement, 211
Wolf, 68, 174–176, 185, 187, 189, 193, 241, 311, 351, 352, 355
Wolf bounty, 174
Wolverine, 168, 177–178, 183, 185, 189
Wood, 36, 42, 68, 71, 78, 89, 102
Wood Buffalo National Park, 3
Wood energy, 89–91
Wood fiber, 36
Wood frog, 217, 218, 221, 352, 403

Index

Woodie Wheaton Land Trust, 78
Wood pellet, 90
Wood processors, 77
Wood products, 71, 89
Wood-to-energy, 90
Wood turtle, 218
Working forest, 76–78, 81, 85
Workshop, 112, 222, 227, 238–242,
    245–247, 250, 270,
    276, 372
World Climate Research Coupled
    Model Intercomparison
    Project, 333
World Database on Protected Areas
    (WDPA), 264
World War II, 41, 74

World Wildlife Fund, 7, 54, 283, 285, 286,
    375, 395

Y

Yellow birch, 335
Yellow lampmussel, 112
Yellow-legged Gull, 154
Yellowstone National Park, 3, 4
Yellowstone to Yukon (Y2Y), 6, 56, 351
Yoho National Park, 2
Yosemite National Park, 2, 122

Z

Zooplankton, 144, 156, 157